Selected Titles in This Series

215 **M. G. Nerurkar, D. P. Dokken, and D. B. Ellis, Editors,** Topological dynamics and applications, 1998

214 **Lewis A. Coburn and Marc A. Rieffel, Editors,** Perspectives on quantization, 1998

213 **Farhad Jafari, Barbara D. MacCluer, Carl C. Cowen, and A. Duane Porter, Editors,** Studies on composition operators, 1998

212 **E. Ramírez de Arellano, N. Salinas, M. V. Shapiro, and N. L. Vasilevski, Editors,** Operator theory for complex and hypercomplex analysis, 1998

211 **Józef Dodziuk and Linda Keen, Editors,** Lipa's legacy: Proceedings from the Bers Colloquium, 1997

210 **V. Kumar Murty and Michel Waldschmidt, Editors,** Number theory, 1998

209 **Steven Cox and Irena Lasiecka, Editors,** Optimization methods in partial differential equations, 1997

208 **Michel L. Lapidus, Lawrence H. Harper, and Adolfo J. Rumbos, Editors,** Harmonic analysis and nonlinear differential equations: A volume in honor of Victor L. Shapiro, 1997

207 **Yujiro Kawamata and Vyacheslav V. Shokurov, Editors,** Birational algebraic geometry: A conference on algebraic geometry in memory of Wei-Liang Chow (1911–1995), 1997

206 **Adam Korányi, Editor,** Harmonic functions on trees and buildings, 1997

205 **Paulo D. Cordaro and Howard Jacobowitz, Editors,** Multidimensional complex analysis and partial differential equations: A collection of papers in honor of François Treves, 1997

204 **Yair Censor and Simeon Reich, Editors,** Recent developments in optimization theory and nonlinear analysis, 1997

203 **Hanna Nencka and Jean-Pierre Bourguignon, Editors,** Geometry and nature: In memory of W. K. Clifford, 1997

202 **Jean-Louis Loday, James D. Stasheff, and Alexander A. Voronov, Editors,** Operads: Proceedings of Renaissance Conferences, 1997

201 **J. R. Quine and Peter Sarnak, Editors,** Extremal Riemann surfaces, 1997

200 **F. Dias, J.-M. Ghidaglia, and J.-C. Saut, Editors,** Mathematical problems in the theory of water waves, 1996

199 **G. Banaszak, W. Gajda, and P. Krasoń, Editors,** Algebraic K-theory, 1996

198 **Donald G. Saari and Zhihong Xia, Editors,** Hamiltonian dynamics and celestial mechanics, 1996

197 **J. E. Bonin, J. G. Oxley, and B. Servatius, Editors,** Matroid theory, 1996

196 **David Bao, Shiing-shen Chern, and Zhongmin Shen, Editors,** Finsler geometry, 1996

195 **Warren Dicks and Enric Ventura,** The group fixed by a family of injective endomorphisms of a free group, 1996

194 **Seok-Jin Kang, Myung-Hwan Kim, and Insok Lee, Editors,** Lie algebras and their representations, 1996

193 **Chongying Dong and Geoffrey Mason, Editors,** Moonshine, the Monster, and related topics, 1996

192 **Tomek Bartoszyński and Marion Scheepers, Editors,** Set theory, 1995

191 **Tuong Ton-That, Kenneth I. Gross, Donald St. P. Richards, and Paul J. Sally, Jr., Editors,** Representation theory and harmonic analysis, 1995

190 **Mourad E. H. Ismail, M. Zuhair Nashed, Ahmed I. Zayed, and Ahmed F. Ghaleb, Editors,** Mathematical analysis, wavelets, and signal processing, 1995

189 **S. A. M. Marcantognini, G. A. Mendoza, M. D. Morán, A. Octavio, and W. O. Urbina, Editors,** Harmonic analysis and operator theory, 1995

188 **Alejandro Adem, R. James Milgram, and Douglas C. Ravenel, Editors,** Homotopy theory and its applications, 1995

(*Continued in the back of this publication*)

Topological Dynamics and Applications

A Volume in Honor of
Robert Ellis

Robert Ellis

CONTEMPORARY MATHEMATICS

215

Topological Dynamics and Applications

A Volume in Honor of
Robert Ellis

Proceedings of a Conference
in Honor of the Retirement of Robert Ellis
April 5–6, 1995
University of Minnesota

M. G. Nerurkar
D. P. Dokken
D. B. Ellis
Editors

American Mathematical Society
Providence, Rhode Island

This volume is dedicated to Professor Robert Ellis for his pioneering work and influence in the development of the abstract theory of Topological Dynamics. It grew out of a conference in honor of his retirement, held on April 5 and 6, 1995 at the University of Minnesota.

1991 *Mathematics Subject Classification.* Primary 54H20, 28D05, 34C35, 28D15.

Library of Congress Cataloging-in-Publication Data

Topological dynamics and applications : a volume in honor of Robert Ellis : proceedings of a conference in honor of the retirement of Robert Ellis, April 5–6, 1995, University of Minnesota / M. G. Nerurkar, D. P. Dokken, D. B. Ellis, editors.

p. cm. — (Contemporary mathematics, ISSN 0271-4132 ; 215)

Includes bibliographical references.

ISBN 0-8218-0608-4

1. Topological dynamics—Congresses. I. Ellis, Robert, 1926– . II. Nerurkar, M. (Mahesh). III. Dokken, D. (Douglas P.). IV. Ellis, D. (David B.). V. Series: Contemporary mathematics (American Mathematical Society) ; v. 215.

QA611.5.T663 1998

514′.322—dc21

97-38715
CIP

∞ The paper used in this book is acid-free and falls within the guidelines
established to ensure permanence and durability.
Visit the AMS home page at URL: `http://www.ams.org/`

10 9 8 7 6 5 4 3 2 1   03 02 01 00 99 98

Contents

Part II. Applications and Other Dynamical Results

Part III. Applications to Differential Equations

Preface

The volume is dedicated to Professor Robert Ellis for his pioneering work and influence in the development of the abstract theory of Topological Dynamics. It grew out of a conference in honor of his retirement, held on April 5 and 6, 1995 at the University of Minnesota.

Prof. Robert Ellis in his life long work laid the foundation for an algebraic approach to topological dynamics. With the introduction of the notion of an enveloping semigroup R. Ellis began demonstrating how to capture the asymptotics of a dynamical system in terms of the algebraic features of a suitable compactification of the acting group. This work highlights the intrinsic interplay between algebra and analysis. Over the next twenty-five years Professor Ellis single handedly developed this theory, culminating in what is now known as the "Galois theory of minimal flows". During this period he proved deep structure theorems about dynamics of group actions which in turn gave a finer understanding of the equicontinuous and distal structure relations. This work provides ample evidence for his belief that topological dynamics is worthy of study as a discipline unto itself.

In addition to the abstract structure theory, R. Ellis's contributions to mathematics include papers encompassing other areas of mathematics like ergodic theory, theory of random walks and boundary theory of locally compact groups, actions of semi simple groups, combinatorial number theory and differential equations. This volume is divided into three parts in recognition of Ellis's influence on other mathematicians and other branches of mathematics.

Finally this preface would never be complete without mentioning his influence as a teacher on many of us. His unique style, enduring support and constant encouragement is experienced and cherished by many. We take this opportunity to express our respect, gratitude and love.

The Editors wish to thank Prof. Yingfei Yi for his editorial contribution, the excellent editorial staff at the A.M.S., in particular Ms. Donna Harmon and Ms. Christine M. Thivierge for their patience, advice and guidance.

Part I.
Topological Dynamics: Abstract Theory

Contemporary Mathematics
Volume **215**, 1998

Robert Ellis and the Algebra of Dynamical Systems

H. Furstenberg and E. Glasner

1. Introduction

From its inception dynamical systems theory was a branch of analysis, with the mathematical counterpart of physical dynamical systems taking the form of systems of ordinary differential equations, or vector fields on manifolds. In the hands of Poincaré and Birkhoff, the topological character of dynamical systems theory came to light and one development of this was the establishment of topological dynamics as a branch of dynamical systems theory. It is Robert Ellis who perceived the profound role that algebra was to play in the development of topological dynamics, and who shaped the instruments that proved to be powerful tools in this area.

To be sure, the setting was different from the classical one, with compactness being a crucial hypothesis, and the focus being more on "factor systems" than subsystems (e.g. periodic orbits, asymptotic cycles). The reason that the focus shifted to factors rather than subsystems is that a considerable part of the theory is devoted to studying minimal systems which do not have subsystems. While it is true that no "ergodic decomposition" exists to analyze a general (compact) system in terms of minimal ones, the minimal systems are in some sense still the basic building blocks, paralleling the role of simple groups in group theory.

Undoubtedly one of the inspirations for Ellis' ideas was Mackey's notion of "virtual" subgroup. Ergodicity is an approximation to transitivity of a group action, and while a strictly transitive action of a group T on a space can be represented by the action on a quotient T/A, A a subgroup of T, we can regard an ergodic action as coming from a "virtual subgroup" A of T. This idea has borne fruit in ergodic theory, but it has never had more than heuristic value, because in interesting situations virtual subgroups had no real counterparts. What Ellis showed was that in topological dynamics one could attach an actual group to a compact dynamical system and that a considerable part of the dynamics was reflected in the group and its relation to other groups. This idea would be appropriate for any point-transitive system, the analogue of an ergodic system, namely a system with a dense orbit (and in the metric case, a residual set of dense orbits). For such systems the dynamics "lives" on the acting group T. Namely if T acts on the compact space X, $(t, x) \to tx$, then X is determined by $C(X)$ which is faithfully represented on the dense orbit Tx_0 with $f \in C(X)$ corresponding to $f_0(t) = f(tx_0)$, the action of $t' \in T$ being determined by the algebra automorphism, $f_0(t) \mapsto f_0(tt')$. For simplicity we shall assume that T is discrete. We wish to associate to a T-invariant

1991 *Mathematics Subject Classifications*: Primary 54H20.

algebra of bounded functions on T a group that leaves it fixed (as functions of the form $f_0(t) = f(tA)$ are fixed by $t \mapsto ta, a \in A$). We cannot necessarily find this A in T but if we go to the Stone-Čech compactification βT we can locate the group there. To begin with we can extend the functions from T to βT, by definition of βT. Now βT has a natural semigroup structure extending the group structure of T, and there exists a minimal subset $M \subset \beta T$ invariant under T (or βT) on the left (actually there are 2^c of them). (M,T) is a *universal minimal system* for T and every point transitive system contains a factor of (M,T), which is identical to the system when the latter is minimal. We let $G = \mathrm{Aut}(M,T)$, i.e. the set of homeomorphisms of M which commute with the action of T. G can be identified with each of various subgroups inside M. Namely, let $J = \{e \in M : e^2 = e\}$. A basic result about compact semigroups, central to Ellis' work, affirms that $J \neq \emptyset$. Let $u \in J$. By the minimality of M, Tu is dense in M and each $\alpha \in G$ is determined by $\alpha(u)$. Since $\alpha(tu) = t\alpha(u)$, by continuity $\alpha(u^2) = u\alpha(u) \in uM$. It turns out that uM is a group with identity element u and we can check that $G \cong uM$. For any minimal (X,T) we have the homomorphism $\phi : (M,T) \to (X,T)$ and having chosen $u \in J$, set $x_0 = \phi(u)$. Then the functions $f_0(t) = f(tx_0)$ extend to functions on βT, or lift to functions on M with $f_0(tu) = f_0(t)$, or $f_0(pu) = f_0(p), p \in M$. We now set

$$\mathcal{G}(X) = \mathcal{G}(X, x_0) = \{\alpha \in G : \phi(\alpha u) = \phi(u) = x_0\}$$

which is also the subgroup of G which leaves fixed each element in the algebra of functions f_0 coming from the system (X,T), under the operation $\alpha f_0(p) = f_0(p\alpha)$. This is reminiscent of Galois' theory, and indeed the title of one of the chapters of Ellis' book [**E9**] is "The Galois theory of Distal Extensions".

Pointed minimal systems give rise to T-subalgebras of $C(M)$ and to each of these we associate the subgroup of $\mathrm{Aut}(M,T) \cong \mathrm{Aut}_T(C(M))$ leaving the subalgebra invariant. In the other direction, given a subgroup we may attach to it the algebra fixed by it. The correspondence is not 1-1, but there is a topology on G (weaker than the topology it inherits from $G = uM \subset \beta T$) such that the subgroups fixing the various algebras are closed, and conversely every closed subgroup is the subgroup fixing a T-subalgebra. This is called the τ-topology and it plays a fundamental role. A τ-closed subgroup of G doesn't determine a unique algebra—or a unique minimal system—but the correspondence is sufficiently well defined to be a useful tool, as we shall describe.

By way of the homomorphism $\phi : M \to X$, the semigroup βT acts on X and defines a compact semigroup of transformations, not necessarily continuous, of X. This is known as the *enveloping semigroup* $E(X,T)$ of the dynamical system (X,T). This notion was developed early in Ellis' work and became a useful tool independent of the Galois theory of dynamical systems. It is surprising that the usually non-metrizable enveloping semigroup of mostly discontinuous transformations can serve as a useful instrument in studying metric systems.

2. Distal Systems and the Origins of Ellis' Algebraic Theory

In a metric dynamical system (X,T) we say that a pair of points x_1, x_2 is *proximal* if $\inf_{t\in T} d(tx_1, tx_2) = 0$. A point $x_0 \in X$ is *distal* if it is not proximal to any point other than itself, and the system (X,T) is *distal* if every point is distal. The name distality originated with W. Gottschalk but the notion was introduced by Hilbert in an attempt to give a topological characterization of the concept of a rigid group of motions. The attempt to partially validate Hilbert's idea led to the conjecture that for compact metric spaces, minimal distal actions are equicontinuous, which would mean that they are isometries for an equivalent metric. Without the assumption of minimality it is easy to give counterexamples.

This problem attracted the attention of Robert Ellis and in a characteristic manner, for several years—believing the result to be true—he set himself the task of proving it. In fact when rumors had it that a counter example was found he jokingly retorted by asking not to be told about it. In fact Ellis had a good reason to be pleased with his persistent adherence to the positive approach. Some byproducts of his attempts by that time were:

(1) The proof of the fundamental joint continuity result (Ellis' joint continuity theorem), [**E1**].
(2) The introduction into topological dynamics of the basic apparatus of the enveloping semigroup $E(X,T)$ (the Ellis semigroup) and the derivation of its elementary properties, [**E3**].
(3) Two beautiful characterizations of distality: a minimal system (X,T) is distal iff $E(X,T)$ is a group iff $X \times X$ is the union of its minimal subsets, [**E2**].
(4) A proof that in fact distality implies equicontinuity when the phase space of the dynamical system is zero-dimensional and the group T belongs to a certain large class of acting groups, [**E2**].

Subsequently examples were provided which showed that even with the assumption of minimality, distality does not imply equicontinuity. One class of examples appears in [AGH] in the form of homogeneous space of a nilpotent group. Another closely related class of skew product transformations which defined distal $\mathbb{Z}$-actions which were not equicontinuous, was given by Furstenberg (it was also studied by Anzai and Kakutani), and became the model for the structure theorem for distal systems [**F1**]. Namely we observe that the toral transformation

$$\tau(x,y) = (x+\alpha, y+r(x)), \ x,y \in \mathbb{T}$$

defines a distal group of transformations since two points of the form $(x,y'),(x,y'')$ always remain exactly the same distance apart, and two points $(x',y'),(x'',y'')$ with $x' \neq x''$ never get closer together than the distance of x' to x''. This is the example of an *isometric extension* of a distal system (on $\mathbb{T}$). We can iterate this arbitrarily often, and the general structure theorem for minimal distal systems asserts that by this procedure, beginning with the trivial system, we can construct any such system.

The interest in this topic was stimulated by Ellis' work on distality and the tools he developed were crucial for later results.

3. Galois Correspondence and Structure Theory

The structure theorem proved in [**F1**] for minimal distal systems is one of a series of structure theorems in which Ellis' Galois correspondence is indispensable ([**E9**], [**E10**], [**E12**], [**EGS**], [**M**], [**V1**], [**V2**]). To formulate the most general of these we review several types of extensions of dynamical systems.

We say that a map $\pi : X \to Y$ defines a *homomorphism* of systems (X,T) and (Y,T) if $t\pi(x) = \pi(tx)$ for $x \in X, t \in T$. We then say (X,T) is an *extension* of

(Y,T). We say the extension is *proximal* if $\pi(x_1) = \pi(x_2)$ implies x_1 and x_2 are proximal, and we say that the extension is *isometric* if for some continuous $d(x',x'')$ defined for (x',x'') with $\pi(x') = \pi(x'')$ and satisfying the condition for a metric on each $\pi^{-1}(y) \subset X$, we have $d(tx',tx'') = d(x',x'')$. A special case of a proximal extension is an *almost 1-1 extension* of minimal systems for which there exists a $y \in Y$ with $\pi^{-1}(y)$ containing a single point of X.

Finally the extension is said to be *weakly mixing* if the subsystem $\tilde{X} = \{(x',x'') : \pi(x') = \pi(x'')\}$ with $t(x',x'') = (tx',tx'')$, has a dense orbit. Clearly we cannot interpolate an isometric extension between a system (Y,T) and a weakly mixing extension (X,T).

We now have the structure theorem for minimal systems:

Given a metric minimal system (X,T), there exists a countable ordinal η and a canonically defined commutative diagram (the canonical PI-Tower)

$$\begin{array}{ccccccccccc}
X = X_0 & \xleftarrow{\tilde{\theta}_1} & & X_1 & \cdots & X_\nu & \xleftarrow{\tilde{\theta}_{\nu+1}} & & X_{\nu+1} & \cdots & X_\eta = X_\infty \\
\pi_0\downarrow & \overset{\sigma_1}{\searrow} & & \pi_1\downarrow & & \pi_\nu\downarrow & \overset{\sigma_{\nu+1}}{\searrow} & & \pi_{\nu+1}\downarrow & & \pi_\infty\downarrow \\
pt = Y_0 & \xleftarrow[\rho_1]{} & Z_1 & \xleftarrow[\theta_1]{} Y_1 & \cdots & Y_\nu & \xleftarrow[\rho_{\nu+1}]{} & Z_{\nu+1} & \xleftarrow[\theta_{\nu+1}]{} Y_{\nu+1} & \cdots & Y_\eta = Y_\infty
\end{array}$$

where for each $\nu \leq \eta$, ρ_ν is isometric, $\theta_\nu, \tilde{\theta}_\nu$ are proximal and π_∞ is weakly mixing. For a limit ordinal ν, X_ν, Y_ν, π_ν etc. are the inverse limits (or joins) of $X_\iota, Y_\iota, \pi_\iota$ etc. for $\iota < \nu$.

We say that a minimal system (X,T) is a *strictly PI system* if there is a (countable) ordinal η and a family of systems $\{(W_\iota, w_\iota)\}_{\iota \leq \eta}$ such that (i) W_0 is the trivial system, (ii) for every $\iota < \eta$ there exists a homomorphism $\phi_\iota : W_{\iota+1} \to W_\iota$ which is either proximal or isometric, (iii) for a limit ordinal $\nu \leq \eta$ the system W_ν is the inverse limit of the systems $\{W_\iota\}_{\iota<\nu}$, and (iv) $W_\eta = X$. We say that (X,T) is a *PI-system* if there exists a strictly PI system $\tilde{X}$ and a proximal homomorphism $\theta : \tilde{X} \to X$.

If in the definition of PI-systems we replace proximal extensions by almost 1-1 extensions we get the notion of *AI-systems*. It is easy to see that in an AI system, a residual set of points are distal. The converse is also true and this is the content of the Ellis-Veech theorem:

If a metric minimal system has a single point which is distal, then it is an AI system ([**V1**],[**E10**]).

Finally, it is possible to show that if the system (X,T) is distal then one can dispense with the almost 1-1 extensions altogether so that the tower in the structure theorem consists only of isometric extensions. The towers of extensions occurring in all these theorems correspond to a single tower of groups under the Ellis Galois correspondence. Recall that the group $G = \mathrm{Aut}(M,T) \cong uM$ for the universal minimal system (M,T) was endowed with a topology called the τ-topology. In this topology G is compact but not Hausdorff. For any τ-closed subgroup $F \subset G$ we set

$$\mathcal{A}(F) = \{f \in C(M) : f(p\alpha) = f(p) \text{ for all } \alpha \in F\},$$

and for any subalgebra $\mathcal{F} \subset C(M)$ with $f(pu) = f(p)$ for all $f \in \mathcal{F}$ we define the Ellis group

$$\mathcal{G}(\mathcal{F}) = \{\alpha \in Mu : f(p\alpha) = f(p) \text{ for all } f \in \mathcal{F}\}.$$

If we begin with a system (X,T), and associate to it the algebra $\mathcal{F} \subset C(M)$ obtained from lifting $C(X)$ to $C(M)$ via a homomorphic map taking M to X (this depends on the choice of $x_0 \in X$ which is to be the image of the fixed $u \in J$). This defines for us the *Ellis group* attached to the system:

$$\mathcal{G}(X) = \mathcal{G}(X, x_0) = \mathcal{G}(\mathcal{F}).$$

The condition that F is τ-closed ensures that $\mathcal{G}(\mathcal{A}(F)) = F$. On the other hand $\mathcal{A}(\mathcal{F})$ contains $\mathcal{F}$ but need not be equal to it. Nonetheless the two algebras are close in the sense that they represent two systems one of which is a proximal extension of the other. Conversely two systems related in this way have the same Ellis group. We may speak of all the systems with $\mathcal{G}(X) = A$ as forming a *proximal class.*

We need one more notion, that of the *derived group.* For a τ-closed subgroup F of G the derived group F' is given by:

$$F' := \bigcap \{\tau\text{-closure}V : V\ \tau\text{-open neighborhood of } u \text{ in } F\}.$$

F' is a τ-closed normal (in fact characteristic) subgroup of F characterized as the smallest τ-closed subgroup H of F such that F/H is a compact Hausdorff topological group. Ellis showed that the quotient group F/F' is the key to the study and understanding of isometric extensions.

Suppose we have an extension $(X,T) \xrightarrow{\pi} (Y,T)$. Let $A = \mathcal{G}(X)$ and $B = \mathcal{G}(Y)$ so that $A \subset B$. We now interpolate between the two groups a third group AB'. B/B' is a compact Hausdorff topological group and B/AB' is a compact homogeneous space of this group. Corresponding to this group we can also interpolate an isometric extension, though not necessarily between X and Y but between two systems in the same proximality class:

$$X^* \to Z^* \to Y^*$$

with $\mathcal{G}(Z^*) = AB'$. Provided that $AB' \neq B$ we will have Z^* a genuine isometric extension where fibers can be identified with the homogeneous space B/AB'. Using this we can describe the groups occurring in the PI tower for any minimal system. Begin with G and form its derived group G', then its derived group $G^{(2)}$ and continue this transfinitely with $G^{\nu+1} = (G^\nu)'$ and for a limit ordinal λ, $G^\lambda = \bigcap_{\nu<\lambda} G^\nu$. The intersection of all the groups occurring in this way is G^∞. Now let $A = \mathcal{G}(X)$ for any minimal system X. We have

$$(*) \qquad G \supset AG^{(1)} \supset AG^{(2)} \supset \cdots \supset AG^{(\infty)} \supset A.$$

One can prove that $AG^{\nu+1} = A(AG^\nu)'$. By the foregoing this corresponds to interpolating an isometric extension between a system corresponding to AG^ν and one corresponding to $AG^{\nu+1}$. Thus modulo proximal extensions the tower (*) corresponds up to AG^∞ to a tower of isometric extensions. The last step $AG^\infty \supset A$ can be shown to correspond to a weakly mixing extension, thus giving—at least in outline—the result of the main structure theorem.

The ultimate step $AG^\infty \supset A$ is superfluous if $AG^\infty = A$, i.e., if $A \supset G^\infty$. This shows that a system is a PI system iff $\mathcal{G}(X) \supset G^\infty$. A (non-trivial) corollary of this is that a factor of a PI system is a PI system.

Another byproduct of this approach has to do with AI systems. Veech's original proof of his structure theorem required assuming that the point distal system has a dense G_δ subset of distal points. In Ellis' approach it is only necessary to assume a single distal point and the existence of a dense G_δ subset then follows as a consequence.

4. Further Applications of Ellis Galois Theory

Two minimal systems are said to be *disjoint* if their product is minimal. This implies that they have no non-trivial common factor. For an abelian group T it is easily seen that if X and Y are disjoint, $X \perp Y$, so are their maximal proximal extensions $X^* \perp Y^*$. Disjointness has a simple and natural formulation in terms of Ellis groups (for T abelian) : $X \perp Y$ iff $\mathcal{G}(X)\mathcal{G}(Y) = G$ ([**E9**]). In case one of the groups is normal this is equivalent to requiring that $\mathcal{G}(X)$ and $\mathcal{G}(Y)$ together generate G, and this in turn can be seen to be equivalent to the condition that X^* and Y^* have no common factor. A sufficient condition for normality of $\mathcal{G}(X)$ is that (X,T) be *regular*, i.e., that whenever $(x_1, x_2) \in X \times X$ has a minimal orbit closure in $(X \times X, T)$ there exists an automorphism of (X,T) taking x_1 to x_2.

Thus we have the following: *If T is abelian and (X,T) is regular, then $X \perp Y$ iff the systems X^* and Y^* have no common factor.*

A number of other conditions are given in [**EGS2**] under which the weaker condition of no common factor implies disjointness. For example this is true when one of the systems, say (X,T), is PI. A more general algebraic condition is that $\mathcal{G}(X)\mathcal{G}(Y) \supset G^\infty$, (still under the assumption that T is abelian). Under this hypothesis it is enough that the equicontinuous factors of X and Y have no common factors. Setting $A = \mathcal{G}(X), B = \mathcal{G}(Y)$, this latter assumption, which implies that these equicontinuous factors are disjoint, can be written $AG'BG' = G$ or $ABG' = G$. This is shown in [**EGS2**] to imply $ABG^\infty = G$. So if $AB \supset G^\infty$ then we have directly $AB = G$.

In [**EG1**] Ellis together with Glasner develop the notion of pure weak mixing, a strengthening of the condition of weak mixing. A system (X,T) is *pure weak mixing* if its join with any other system defines (up to proximal extension) a weak mixing extension of the latter system. This implies in particular the "multiplier property": that its join with any weak mixing system is weak mixing. That these conditions are equivalent (at least for $\mathbb{Z}$-systems) is non-trivial and the proof in [**EG1**] bases itself on Ellis' algebraic technique in general and on aspects of Ellis' work on cocycles in particular ([**E11**]). Here we just illustrate this by translating the two concepts to algebraic language. Let $A = \mathcal{G}(X)$ and $B = \mathcal{G}(Y)$. The condition that the join of X and Y be relatively weak mixing over Y is that $(A \cap B)B' = B$. This condition can be made simpler still: $(A \cap B)B' = B \Rightarrow AB = A(A \cap B)B' = AB'$ and $AB' = AB \Rightarrow B \subset AB' \Rightarrow B \subset (A \cap B)B' \Rightarrow B = (A \cap B)B'$. On the other hand, the condition that the join of X with a weak mixing Y is weak mixing, is that $BG' = B$ should imply $(A \cap B)G' = G$. Note that the condition $(A \cap B)B' = B$ implies $(A \cap B)G'$ contains BG'. But if $BG' = G$ then $(A \cap B)G' = G$.

This simple algebraic argument establishes the easier part of the equivalence assertion: pure weak mixing implies the multiplier property for weak mixing.

In general when dealing with multiplier property for a given property (e.g. weak mixing, minimal orbit closures) it is natural to form the maximal system with this property, i.e. those having no proper extension retaining it. Often the maximal systems are regular. A byproduct of the algebraic analysis in [**EG1**] is that the Ellis groups of the maximal weak mixing systems are not normal, implying that these systems are not regular.

An early example of the use of algebraic techniques appears in [**E3**] where Ellis characterizes the property that proximality be an equivalence relation in terms of the enveloping semigroup of the system: proximality is an equivalence relation in (X, T) iff $E(X, T)$ has a single minimal left ideal. This is a rare property. More common is the property that *regional proximality* is an equivalence relation, where x is regionally proximal to y (in the metric case) if for each $\epsilon > 0$ there exist $x', y' \in X, t \in T$ with $d(x, x') < \epsilon, d(y, y') < \epsilon$, and $d(tx', ty') < \epsilon$. When regional proximality is an equivalence relation for a system, it is precisely the relation that defines the maximal equicontinuous factor of the system. A sufficient condition for this is that T be abelian. A far reaching generalization of this has been given in [**AEE**] where it is shown that in group-theoretic terms a sufficient condition is $E \subset \mathcal{G}(X)G'$, where E is the Ellis group of the universal equicontinuous system (in particular this property depends only on the proximality class of X). In the abelian case $E = G'$ so this is valid. In general this condition is known to hold for distal systems (and therefore for proximal extensions of distal systems). Thus regional proximality is identical to the "equicontinuous structure relation " in these cases.

5. The Poisson System

The work of Ellis together with Dokken, [**D**], [**ED**], incorporates the philosophy that Ellis adopts in his approach to topological dynamics in developing a novel approach to the theory of bounded harmonic functions on a group. In [**F2**] the boundary theory of classical harmonic functions in symmetric domains was shown to be a special case of the boundary theory for functions on a group satisfying a group invariant mean-value property. The *Poisson space* on which the "boundary values" of such functions were defined was a compact space on which the group had a natural action, defining a dynamical system. Ellis and Dokken set themselves the problem of identifying this system, the Poisson system, corresponding to a probability measure μ on the group T defining the mean-value property of μ-harmonic functions.

Once again, the answer is to be found in terms of the Stone-Čech compactification βT when T is discrete, and more generally, in terms of the Gelfand space of the algebra of bounded left uniformly continuous functions on T. Here is a sketch of this idea. A μ-harmonic function on T is one that satisfies

$$f(t) = \int f(tt')d\mu(t')$$

for every $t \in T$. We write $f = \mu * f$. Note that this implies a series of mean-value properties, e.g. $f = \mu^{(n)} * f$, where $\mu^{(n)}$ is the n-fold convolution of μ, so that f is μ'-harmonic for $\mu' = \mu^{(n)}$ or any convex combination of these. These measures μ' sit in the non-compact space T, but they can also be regarded as measures on the compact space βT. In this space we can pass to the limit and we find that f is also ν-harmonic for idempotent measure ν on βT. It turns out that for an appropriate ν, the statement of this ν'-harmonicity is precisely the Poisson representation of f in terms of boundary values. The Poisson space will be a factor of the system defined on the support of ν.

Pursuing this idea Dokken and Ellis are able to give conditions for the triviality of the Poisson system, i.e. when constants are the only μ-harmonic functions. They also retrieve some of the results known for harmonic functions on solvable and semi-simple group T. Whether or not new results are obtained by this approach, it gives valuable insight into the theory of harmonic functions on groups.

6. The Enveloping Semigroup of a Dynamical System

Before developing his algebraic approach based on the existence of a universal minimal system, Ellis exploited the rich algebraic structure of compact semigroups in the form of the enveloping semigroup of a dynamical system. If T is a group of continuous transformations of a compact space X, T can be regarded as a subgroup of the semigroup X^X which is compact in the natural weak topology. The enveloping semigroup $E(X,T)$ is the closure of T in X^X. (See [**E3**] for the early use of the enveloping semigroup.) The significance of the enveloping semigroup is already made manifest by the fact mentioned earlier that the distality of a system is equivalent to the enveloping semigroup being a group. In general when $E(X,T)$ is not a group it possesses non-trivial left ideals ($I \subset E$ so that $EI = I$), and the minimal left ideals and the idempotents in these ideals of $E(X,T)$ play an important role for the dynamics of (X,T). For example, in a minimal system x and y are proximal iff $y = ux$ for some idempotent u in a minimal left ideal of $E(X,T)$.

For most systems $E(X,T)$ is a rather unwieldy object difficult to describe. One situation in which the enveloping semigroup is non-trivial but nonetheless manageable is that for T a subgroup of $GL(V)$ where V is a finite dimensional vector space, and $X = P(V)$, the corresponding projective space. This situation is analyzed in detail in Ellis' paper [**E15**], and it turns out that one can describe the general element in $E(X,T)$ and identify the minimal idempotents. One consequence of the analysis is a proof that, in this context, whenever T acts proximally it acts strongly proximally. Another consequence is a proof of the Borel density theorem, in which a key step is the analysis of measure preserving groups of projective transformations.

In the foregoing situation as well as in most situations the transformations of the enveloping semigroup are (mostly) discontinuous. There are exceptions, and when all elements of $E(X,T)$ are continuous we say that (X,T) is *weakly almost periodic*. This notion is the subject of the paper [**EN**] by Ellis and Nerurkar. One can prove that in this situation $E(X,T)$ has a unique minimal left ideal I which is two-sided and has a unique idempotent element so that I itself is a compact topological group. This idempotent element commutes with all elements of $E(X,T)$. If T has no equicontinuous actions then I must reduce to a single point. If we assume that (X,T) is point transitive then it possesses a unique minimal set which reduces to a singleton x_0. Ellis and Nerurkar prove furthermore that in case T is a semi-simple Lie group with finite center and no compact factors (in which case it has no equicontinuous factors), the orbit closure of any point consists of the orbit together with x_0.

Now these conditions all hold if we think of T as acting on the unit ball of a Hilbert space $\mathcal{H}$ via a unitary representation (we consider the ball as a compact set in the weak topology of the Hilbert space). The enveloping semigroup consists of linear transformations of norm ≤ 1, so these are all continuous and the system is weakly almost periodic. If we assume the representation is irreducible, the idempotent element in the enveloping semigroup which is necessarily an orthogonal projection in $\mathcal{H}$ must be 0. Thus 0 is in the orbit closure of x for any x and hence $x_0 = 0$. It follows now that $\overline{Tx} = Tx \cup \{0\}$ and from this one deduces a form of Moore's theorem: If $U : T \to \mathrm{Aut}(\mathcal{H})$ is a unitary representation of a semi-simple Lie group with finite center and no compact factors and $U(T)$ has no non-zero fixed vectors, then $< U(t)x, y > \to 0$ as $t \to \infty$.

The enveloping semigroup of a system has also been useful for aspects of combinatorics. For example, van der Warden's theorem on arithmetical progressions and various refinements as well as Hindman's theorem can be reformulated in terms of dynamical systems. Using the algebra of the enveloping semigroup one obtains new proofs of known theorems as well as results not obtained by other methods. This may be found in [**FW**], [**G2**], [**FK**] and [**BH**].

7. Concluding Remarks

The contribution of Robert Ellis to dynamical systems theory is not confined to ramifications of the algebraic methodology he introduced. For example a method of construction of minimal systems was described in [**E6**], where among other results a minimal system was constructed on the Klein bottle. This method of construction was later used by R. Peleg ([**P**]) to answer some questions about weakly mixing systems, and still later it inspired a general method of construction of minimal skew products ([**GW**], [**N**]). Linear differential systems were studied in [**EJ**] by means of cocycles in topological dynamics, using among other things constructions in [**GW**], thus completing a cycle. An elegant and powerful approach to questions of unique ergodicity on homogeneous systems was developed in [**EP**]. In [**E14**] an interesting attempt to treat ergodic theory in a topological way is made by passage to the, usually non-metrizable, Gelfand space of the Banach algebra $L^\infty(X,\mu)$.

The mathematical work of Ellis has had a profound influence on mathematics and it will undoubtedly continue to do so. We, his colleagues and friends, look forward to the continuing flow of mathematical ideas from Robert Ellis in years to come.

References

[A] J. Auslander, *Minimal systems and their extensions*, North-Holland, 1988.

[AEE] J. Auslander, R. Ellis, D.B. Ellis, *The regionally proximal relation*, Trans. Amer. Math. Soc. **347** (1995), 2139–2146.

[AGH] L. Auslander, L. Green and F. Hahn, *Systems on homogeneous spaces*, Princeton University Press, 1963.

[B] I. U. Bronstein, *Extensions of minimal transformation groups*, Sijthoff and Noordhoff, 1979.

[BH] V. Bergelson and N. Hindman, *Non-metrizable topological dynamics and Ramsey theory*, Trans. Amer. Math. Soc. **320** (1990), 293–320.

[D] D. Dokken, *μ-Harmonic functions on locally compact groups*, Journal d'Analyse Mathematique **52** (1989), 1–25.

[E1] R. Ellis, *Locally compact transformation groups*, Duke Math. Journal **24** (1957), 119–125.

[E2] R. Ellis, *Distal transformation groups*, Pacific J. of Math. **8** (1958), 401–405.

[E3] R. Ellis, *A semigroup associated with a transformation group*, Trans. Amer. Math. Soc. **94** (1960), 272–281.

[E4] R. Ellis, *Universal minimal sets*, Proc. Amer. Math. Soc. **11** (1960), 540–543.

[E5] R. Ellis, *Point trasitive transformation groups*, Trans. Amer. Math. Soc. **101** (1961), 261–287.

[E6] R. Ellis, *The construction of minimal discrete systems*, Amer. J. of Math. **87** (1965), 564–574.

[E7] R. Ellis, *Group-like extensions of minimal sets*, Trans. Amer. Math. Soc. **127** (1967), 125–135.

[E8] R. Ellis, *the structure of group-like extensions of minimal sets*, Trans. Amer. Math. Soc. **134** (1968), 261–287.

[E9] R. Ellis, *Lectures in topological dynamics*, Benjamin, New York, 1969.

[E10] R. Ellis, *The Veech structure theorem*, Trans. Amer. Math. Soc. **186** (1973), 203–218.

[E11] R. Ellis, *Cocycles in topological dynamics*, Topology **17** (1978), 111–130.

[E12] R. Ellis, *The Furstenberg structure theorem*, Pcific J. of Math. **76** (1978), 345–349.

[E13] R. Ellis, *Cohomology of groups and almost periodic extensions of minimal sets.*, Ergod. Th. and Dynam. Sys. **1** (1981), 49–64.

[E14] R. Ellis, *Topological dynamics and ergodic theory*, Ergod. Th. and Dynam. Sys. **7** (1987), 25–47.

[E15] R. Ellis, *The enveloping semigroup of projective systems*, Ergod. Th. and Dynam. Sys. **13** (1993), 635–660.

[ED] R. Ellis and D. Dokken, *The Poisson system associated with a measure*, Pacific J. of Math. **141** (1990), 79–103.

[EG1] R. Ellis and S. Glasner, *Pure weak mixing*, Trans. Amer. Math. Soc. **243** (1978), 135–146.

[EG2] R. Ellis and S. Glasner, *Iterated extensions*, Ergod. Th. and Dynam. Sys **2** (1982), 1–13.

[EGS1] R. Ellis, S. Glasner, L. Shapiro, *Proximal-Isometric systems*, Advances in Math. **17** (1975), 213–260.

[EGS2] R. Ellis, S. Glasner, L. Shapiro, *Algebraic equivalents of system disjointness*, Illinois J. Math. **20** (1976), 354–360.

[EG] R. Ellis and W.H. Gottschalk, *Homomorphisms of transformation groups*, Trans. Amer. Math. Soc. **94** (1960), 258–271.

[EJ] R. Ellis and R. Johnson, *Topological dynamics and linear differential systems*, Journal of Differential Equations **44** (1982), 21–39.

[EK1] R. Ellis and H. Keynes, *A characterization of the equicontinuous structure relation*, Trans. Amer. Math. Soc. **161** (1971), 171–183.

[EK2] R. Ellis and H. Keynes, *Bohr compactifications and a result of Fölner*, Israel J. of Math. **12** (1972), 314–330.

[EN] R. Ellis and M. Nerurkar, *Weakly almost periodic systems*, Trans. Amer. Math. Soc. **313** (1989), 103–119.

[EP] R. Ellis and W. Perrizo, *Unique ergodicity of systems on homogeneous spaces*, Israel J. of Math. **29** (1978), 276–284.

[F1] H. Furstenberg, *The structure of distal systems*, Amer. J. Math. **85** (1963), 477–515.

[F2] H. Furstenberg, *A Poisson formula for semi-simple Lie groups*, Annals of Math. **77** (1963), 335–383.

[FK] H. Furstenberg and Y. Katznelson, *Idempotents in compact semigroups and Ramsey theory*, Isreal J. of Math. **68** (1989), 257–270.

[FW] H. Furstenberg and B. Weiss, *Topological dynamics and combinatorial number theory*, J. Analyse Math. **34** (1978), 61–85.

[G1] S. Glasner, *Proximal systems*, Springer Verlag, Lecture Notes in Math. 517, 1976.

[G2] S. Glasner, *Divisible properties and the Stone-Čech compactification*, Can. J. Math. **32** (1980), 993–1007.

[GW] S. Glasner, B. Weiss, *On the construction of minimal skew products*, Israel J. Math. **34** (1979), 321–336.

[M] D. C. McMahon, *Weak mixing and a note on the structure theorem for minimal transformation groups*, Illinois J. of Math. **20** (1976), 186–197.

[N] M. Nerurkar, *Ergodic continuous skew product actions of amenable groups,*, Pacific J. Math **119 No.2** (1985), 343–363.

[P] R. Peleg, *Weak disjointness of transformation groups*, Israel J. of Math. **9** (1971), 330-336.

[V1] W. A. Veech, *Point-distal systems*, Amer. J. Math. **92** (1970), 205–242.

[V2] W. A. Veech, *Topological dynamics*, Bull. Amer. Math. Soc. **83** (1977), 775–830.

[Vr] J. de Vries, *Elements of topological dynamics*, Kluwer academic publishers., 1993.

INSTITUTE OF MATHEMATICS, HEBREW UNIVERSITY OF JERUSALEM
E-mail address: harry@math.huji.ac.il

DEPARTMENT OF MATHEMATICS , TEL AVIV UNIVERSITY
E-mail address: glasner@math.tau.ac.il

Contemporary Mathematics
Volume **215**, 1998

Weak Mixing and Pure Weak Mixing Minimal Flows

J. Auslander

ABSTRACT. We study weak mixing and pure weak mixing from the point of view of the Ellis groups of the flows. A dynamical characterization of regular pure weak mixing flows in terms of automorphisms is obtained.

We begin by briefly reviewing the algebraic theory of minimal flows, as developed by Ellis. For details, see [A] or [E1].

Let T be a discrete group, and let $\boldsymbol{\beta}T$ be the Stone-Čech compactification of T. Then T acts on $\boldsymbol{\beta}T$, and $(\boldsymbol{\beta}T, T)$ is a topologically transitive flow. If M is a minimal subset of $\boldsymbol{\beta}T$, then (M, T) is the universal minimal flow with acting group T (that is, every minimal flow with acting group T is a factor of (M, T).) Let u be an idempotent in M, and let $G = Mu$. Then G is a group with identity u; G may also be regarded as the group of automorphisms of M.

The group G can be provided with a compact T_1 (but in general not Hausdorff) topology, the τ-topology, with respect to which multiplication is unilaterally continuous, and inversion is continuous. We write $g_n \xrightarrow{\tau} g$ for convergence of a net $\{g_n\}$ in the τ-topology. It can be shown that $g_n \xrightarrow{\tau} g$ if and only if (in $\boldsymbol{\beta}T$) $g_n t_n \to g$ where $\{t_n\}$ is a net in T with $t_n \to u$.

If (X, T) is a minimal flow, and $x \in X$ with $xu = x$ then the Ellis group of (X, T) (with respect to the basepoint x) is defined by

$$\mathcal{G}(X, x) = \mathcal{G}(X) = [\alpha \in G | x\alpha = x].$$

Then $\mathcal{G}(X)$ is a τ-closed subgroup of G, and every τ-closed subgroup of G is the Ellis group of a minimal flow.

Let K be a subset of G which contains u. Let

$$K' = \cap[\tau \text{ -closure } (U \cap K) | U a \ \tau \text{ -neighborhood of the identity of } G].$$

Then $k \in K'$ if and only if there is a net $\{k_n\}$ in K with $k_n \xrightarrow{\tau} k$ and $k_n \xrightarrow{\tau} u$.

If B is a τ-closed subgroup of G, then B' is a normal subgroup of B, (B' is called $H(B)$ in [A] and [E2].)

The following lemma is proved in [E2] (and in the case AB is a group, in [G].) A (different) proof will be given at the end of this paper.

1991 *Mathematics Subject Classifications*: Primary 54H20.

LEMMA 1. *Let A and B be τ-closed subgroups of G. Then $A(AB)' = AB'$.*

Recall that a flow (X,T) is said to be weakly mixing if the product flow $(X \times X, T)$ is topologically transitive. If (X,T) is minimal, with T abelian, then (X,T) is weakly mixing if and only if (X,T) has no non-trivial equicontinuous factor, which in turn holds if and only if the regionally proximal relation $Q = X \times X$. (Recall that $(x,y) \in Q$ if and only if there are nets $\{x_n\}$ and $\{y_n\}$ in X and $\{t_n\}$ in T such that $x_n \to x$, $y_n \to y$ and $(x_n t_n, y_n t_n) \to \Delta_X$, the diagonal of X.) It is not difficult to see that if $g \in G'$ then $(x, xg) \in Q$.

In this case, weakly mixing can be characterized by the group of the flow, as was observed by Bob Ellis. Since this result is apparently not in the literature, we include a proof.

THEOREM 2. *Suppose (X,T) is a minimal flow, with T abelian. Let $x \in X$ with $xu = x$ and let $A = \mathcal{G}(X,x)$. Then (X,T) is weakly mixing if and only if $AG' = G$.*

PROOF. Suppose $AG' = G$. Then, if $g \in G$, $xg = xg'$ for some $g' \in G'$. Therefore $(x, xg) \in Q$ for every $g \in G$. Since T is abelian, xG is dense in X, and it follows easily that $Q = X \times X$. Therefore the maximal equicontinuous factor of (X,T) is trivial, so (X,T) is weakly mixing.

Conversely, if (X,T) is weakly mixing, then $Q = X \times X$. Let $g \in G$. Then $(x, xg) \in Q$. Let (M,T) be the universal minimal flow and let $\pi : M \to X$ be the homomorphism defined by $\pi(p) = xp$, for $p \in M$. Then, by Lemma 19, *(ii)* of [AG] there is an $h \in Mu = G$ with $\pi(h) = xg$ such that $(u,h) \in Q(M)$. By [AG], Theorem 11, *(ii)*, $h \in G'$. Therefore we have $xh = \pi(h) = xg$, so $gh^{-1} \in A$ and $g \in AG'$.

Ellis has proposed that this group theoretic characterization be taken as the definition of weak mixing in general– that is, a minimal flow (X,T) with Ellis group A will be called weak mixing (wm) if $G = AG'$. In this case, we will also say that the τ-closed subgroup A of G is wm.

In this spirit, we will say that a minimal flow (X,T) with Ellis group A is pure weak mixing (pwm) if $AB = AB'$ (equivalently $B \subset AB'$) for every τ-closed subgroup B of G. (As above we will say that a τ-closed subgroup A of G is pwm if it has this property.) Taking $B = G$, we see that if A is pwm, it is wm.

Pure weak mixing was defined dynamically by Ellis and Glasner ([EG]), who showed that their definition is equivalent to $(A \cap B)B' = B$ for every τ-closed subgroup B of G. It is easy to see that this is equivalent to our definition. (If $(A \cap B)B' = B$ then $AB = A(A \cap B)B' = AB'$. Now suppose $AB = AB'$. Then, if $b \in B$, $b = ab'$ for some $a \in A$ and $b' \in B'$. Then $a \in A \cap B$, and therefore $B \subset (A \cap B)B'$.)

THEOREM 3. *(i) Let A and C be τ-closed subgroups of G with $A \subset C$. If A is pwm, then C is pwm. (Therefore a factor of a pwm flow is pwm.)*

(ii) Let A_1 and A_2 be τ-closed subgroups of G.

(a) If A_1 is pwm and A_2 is wm, then $A_1 \cap A_2$ is wm.

(b) If A_1 and A_2 are both pwm then $A_1 \cap A_2$ is pwm.

(iii) Let A be a τ-closed normal subgroup of G. Then A is pwm if and only if $AC = AC'$ for all wm subgroups C of G.

PROOF. (*i*) Let B be a τ-closed subgroup of G. Then $B \subset AB' \subset CB'$ so $CB = CB'$.

(*iia*) Since A_1 is pwm, $A_2 \subset A_1A_2 = A_1A_2'$. Therefore $A_2 \subset (A_1 \cap A_2)A_2' \subset (A_1 \cap A_2)G'$. Then, since A_2 is wm, $G = A_2G' \subset (A_1 \cap A_2)G'G' = (A_1 \cap A_2)G'$.

(*iib*) Let B be a τ-closed subgroup of G. Since A_2 is pwm, $((A_1 \cap A_2) \cap B)(A_1 \cap B)' = ((A_1 \cap B) \cap A_2)(A_1 \cap B)' = A_1 \cap B$. Then $((A_1 \cap A_2) \cap B)B' \supset A_1 \cap B$. Also $((A_1 \cap A_2) \cap B)B' \supset B'$. Since B' is a normal subgroup of B, $B'((A_1 \cap A_2) \cap B)$ is a group, and we have $B'((A_1 \cap A_2) \cap B) \supset B'(A_1 \cap B) = B$ since A_1 is pwm.

(*iii*) Since G is wm, $G = AG = AG'$ so A is wm. Let B be a τ-closed subgroup of G. Then AB is a group, and since $A \subset AB$, AB is wm. Then $A(AB)' = A(AB) = AB$. On the other hand, $A(AB)' = AB'$, so $AB = AB'$. □

(We note that a proof of (*iib*) is contained in the proof of Proposition 2.2 in [EG].)

We conjecture that condition (*iia*) in Theorem 3 is actually a characterization of pwm groups. That is, if $A \cap B$ is wm, for every wm group B, then A is pwm. (For $T = \mathbb{Z}$, this is a consequence of Corollary 3.4 in [EG].)

THEOREM 4. *Let A be a normal τ-closed subgroup of G which is maximal in G. (That is, if C is a τ-closed subgroup of G with $A \subset C$ then $A = C$ or $C = G$.) Suppose A is wm. Then A is pwm.*

PROOF. Let B be a τ-closed subgroup of G. If $B \subset A$, then $AB = A = AB'$. Otherwise (since A is normal and maximal) $AB = G$. Moreover, $AB' = A(AB)' = AG' = G$. □

Examples of minimal flows whose Ellis groups are maximal are the graphic flows, studied in [AM], as well as the discrete times of certain horocycle flows.

If B is a τ-closed subgroup of G, we define, inductively $B_0 = B$, $B_1 = B'$, and for any ordinal number λ, $B_{\lambda+1} = B_\lambda'$. If λ is a limit ordinal, define $B_\lambda = \cap_{\nu<\lambda} B_\nu$.

In the proof of the following lemma, we make use of the "circle operation". If the group T acts on a space X, then there is a natural operation of T on 2^X, the space of closed subsets of X, provided with the Hausdorff topology. Hence there is an action of $\boldsymbol{\beta}T$ on 2^X. If $A \in 2^X$ and $p \in \boldsymbol{\beta}T$ this action is denoted by $A \circ p$. For further details, see [A].

LEMMA 5. *Suppose the τ-closed subgroup A is pwm. Let B be a τ-closed subgroup of G. Then for every ordinal number λ, $AB = AB_\lambda$. In particular, $AB = AB_\infty$.*

PROOF. Let $\lambda > 1$ be an ordinal number, and suppose $AB = AB_\nu$ for all $\nu < \lambda$. If λ is a successor ordinal, then clearly $AB = AB_\lambda$. Suppose λ is a limit ordinal. Let $\beta \in B$. Then $\beta \in Ab_\nu \subset A \circ b_\nu$ for some $b_\nu \in B_\nu$. Regarding $\{b_\nu\}$ as a net, there is a subnet (still called $\{b_\nu\}$) such that $b_\nu \to r \in M$. Then, if $b = ru$, we have $b_\nu \xrightarrow{\tau} b$. It follows easily that $b \in \cap[B_\nu | \nu < \lambda] = B_\lambda$. Also $\beta \in A \circ r$, so $\beta \in A \circ ru = A \circ b$. Then $\beta b^{-1} \in A \circ u \cap G = A$, so $\beta \in Ab \subset AB_\lambda$. Thus $B \subset AB_\lambda$ and it follows that $AB = AB_\lambda$. □

Another proof of Lemma 5 is in [EGS], Lemma 3.9.

We next consider the relation of pure weak mixing with the dynamics of automorphisms. If ϕ is an automorphism of (X,T), then the "prolongation" D_ϕ of ϕ

is the closure of the ϕ orbit relation. Hence if $x \in X$, then $y \in D_\phi(x)$ if there are nets $\{x_n\}$ in X and $\{k_n\}$ in $\mathbb{Z}$ such that $x_n \to x$ and $\phi^{k_n}(x_n) \to y$.

THEOREM 6. *Let (X, T) be minimal and pwm, and let $x \in X$ with $xu = x$. Let ϕ be an automorphism of (X, T). Then $(x, \phi(x)) \in D_{\phi\times\phi}(x, x)$.*

PROOF. Let $A = \mathcal{G}(X, x)$. Let $\beta \in G$ such that $x\beta = \phi(x)$, and let $B = [\beta]$, the τ-closed subgroup of G generated by β. Then, by Lemma 5, $xB = xAB = xAB_\infty = xB_\infty$, so $x\beta = xb$ for some $b \in B_\infty$. Now $B'_\infty = B_\infty$, so there is a net $\{b_n\}$ in B_∞ such that $b_n \xrightarrow{\tau} b$ and $b_n \xrightarrow{\tau} u$. Since $\{\beta^n\}_{n\in\mathbb{Z}}$ is τ-dense in B there is a net $\{k_i\}$ in $\mathbb{Z}$ such that $\beta^{k_i} \xrightarrow{\tau} b,u$. It follows that there are nets $\{t_i\}$ and $\{s_i\}$ in T such that $t_i \to u$, $s_i \to u$, $\beta^{k_i}t_i \to b$, and $\beta^{k_i}s_i \to u$ (convergence in $\boldsymbol{\beta}T$.) Then $(xs_i, xt_i) \to (xu, xu) = (x, x)$, $(\phi^{k_i} \times \phi^{k_i})(xs_i, xt_i) = (x\beta^{k_i}s_i, x\beta^{k_i}t_i) \to (x, xb) = (x, \phi(x))$. Therefore $(x, \phi(x)) \in D_{\phi\times\phi}(x, x)$. □

LEMMA 7. *Let (X, T) be minimal and let ϕ be an automorphism of (X, T). Suppose that $(x, \phi^k(x)) \in D_{\phi^k\times\phi^k}(x, x)$ for all $x \in X$ and $k \in \mathbb{Z}$. Then $D_\phi(x) \times D_\phi(y) = D_{\phi\times\phi}(x, y)$, for all $x, y \in X$.*

PROOF. Let $x \in X$ and $k \in \mathbb{Z}$. It is immediate that $(x, \phi^k(xt)) \in D_{\phi^k\times\phi^k}(x, xt)$ for all $t \in T$. Since D defines a closed relation, it follows that $(x, \phi^k(y)) \in D_{\phi^k\times\phi^k}(x, y) \subset D_{\phi\times\phi}(x, y)$, for all $x, y \in X$ and all $k \in \mathbb{Z}$. Since $D_{\phi\times\phi}$ is $\phi \times \phi$ invariant, we have $(\phi^m(x), \phi^n(y)) \in D_{\phi\times\phi}(x, y)$ for all integers m, n and all $x, y \in X$.

Now let $x, y \in X$, $x' \in D_\phi(x)$, $y' \in D_\phi(y)$. Then there are nets $\{x_j\}$, $\{y_j\}$ in X and $\{k_j\}$, $\{l_j\}$ in $\mathbb{Z}$ with $x_j \to x$, $y_j \to y$, $\phi^{k_j}(x_j) \to x'$, and $\phi^{l_j}(y_j) \to y'$. Then $(\phi^{k_j}(x_j), \phi^{l_j}(y_j)) \in D_{\phi\times\phi}(x_j, y_j)$ and it follows that $(x', y') \in D_{\phi\times\phi}(x, y)$. Hence $D_\phi(x) \times D_\phi(y) \subset D_{\phi\times\phi}(x, y)$. The reverse inclusion always holds, so the proof is completed. □

COROLLARY 8. *Let (X, T) be a pwm minimal flow, and let ϕ be an automorphism of (X, T). Then $D_\phi(x) \times D_\phi(y) = D_{\phi\times\phi}(x, y)$ for all $x, y \in X$.*

Recall that a minimal flow (X, T) is said to be regular if every almost periodic point in the product flow $(X \times X, T)$ is of the form $(x, \phi(x))$, where ϕ is an automorphism of (X, T).

THEOREM 9. *Let (X, T) be a regular minimal flow such that*

$$(x, \phi(x)) \in D_{\phi\times\phi}(x, x)$$

for all automorphisms ϕ and $x \in X$. Then (X, T) is pwm.

For the proof, we require a lemma.

LEMMA 10. *(i) Let v be an idempotent in M, let $\{b_i\}$ be a net in G, and let $\{\sigma_i\}$ be a net in $\boldsymbol{\beta}T$ such that $\sigma_i \to v$ and $b_i\sigma_i \to r \in M$. Then $b_i \xrightarrow{\tau} ru$.*

(ii) Let $\{\sigma_i\}$ and $\{b_i\}$ be nets in $\boldsymbol{\beta}T$ and G respectively such that $\sigma_i \to q \in M$ and $b_i\sigma_i \to r \in M$. Then $b_i \xrightarrow{\tau} r(qu)^{-1}$.

PROOF. (*i*) We may suppose $\sigma_i = t_i \in T$. If $s \in T$ we have $t_is \to vs$ and $b_it_is \to rs$. By choosing s in a neighborhood of u, we may assume that t_is is close to u and b_it_is is close to ru. Thus we can find a net $\{\tau_i\}$ in T such that $\tau_i \to u$ and $b_i\tau_i \to ru$. That is, $b_i \xrightarrow{\tau} ru$.

(*ii*) Let v be the idempotent in M such that $qv = q$. Then $q^{-1}u\sigma_i \to q^{-1}uq = v$ and $b_i quq^{-1}u\sigma_i = b_i u\sigma_i = b_i\sigma_i \to r$ so by (*i*) $b_i qu \xrightarrow{\tau} ru$ and $b_i \xrightarrow{\tau} r(qu)^{-1}$. □

PROOF OF THEOREM 9. Let $x \in X$ with $xu = x$ and let B be a τ-closed subgroup of G. It is sufficient to show that $xB \subset xB'$. Let $\beta \in B$. Since (X,T) is regular, $x\beta = \phi(x)$ for some automorphism ϕ. By hypothesis, $(x, \phi(x)) \in D_{\phi\times\phi}(x,x)$. Then there are nets $\{x_n\}$, $\{x'_n\}$ in X and $\{k_n\}$ in $\mathbb{Z}$ with $(x_n, x'_n) \to (x,x)$ and $(\phi^{k_n}(x_n), \phi^{k_n}(x'_n)) \to (x, \phi(x))$. Let q_n,$r_n \in M$ be such that $x_n = xq_n$ and $x'_n = xr_n$. Then we have $(xq_n, xr_n) \to (x,x)$ and $(x\beta^{k_n}q_n, x\beta^{k_n}r_n) \to (x, x\beta)$. Let (subnets of) $q_n \to q$, $r_n \to r$, $\beta^{k_n}q_n \to p$, and $\beta^{k_n}r_n \to m$, so $xq = xr = xp = x$ and $xm = x\beta$. Then (if $A = \mathcal{G}(X,x)$), pu,qu,$ru \in A$ and $mu \in A\beta$. Then, by Lemma 10, $\beta^{k_n} \xrightarrow{\tau} p(qu)^{-1} = a \in A \cap B$. Since $\beta^{k_n}r_n \to m$, also $\beta^{k_n} \xrightarrow{\tau} m(ru)^{-1} = \alpha\beta\alpha'$ with α, $\alpha' \in A$. Then $\beta^{k_n}a^{-1} \xrightarrow{\tau} u$ and $\beta^{k_n}a^{-1} \xrightarrow{\tau} \alpha\beta\alpha' a^{-1}$. Since $\beta^{k_n}a^{-1} \in B$, $\alpha\beta\alpha' a^{-1} = b' \in B'$. Also $xb' = \phi(x) = x\beta$. Since $\beta \in B$ is arbitrary, we have $xB \subset xB'$, as required.

COROLLARY 11. *Let (X,T) be a regular minimal flow. Then (X,T) is pwm if and only if $D_\phi(x) \times D_\phi(y) = D_{\phi\times\phi}(x,y)$ for all automorphisms ϕ and all $x, y \in X$.*

We conclude with a proof of Lemma 1.

PROOF OF LEMMA 1. It is sufficient to show that $(AB)' \subset AB'$. Let $\gamma \in (AB)'$. Then there are nets $\{\alpha_n\}$ and $\{\beta_n\}$ in A and B respectively with $\alpha_n\beta_n \xrightarrow{\tau} \gamma$ and $\alpha_n\beta_n \xrightarrow{\tau} u$. Hence there are nets $\{t_n\}$ and $\{s_n\}$ in T with $s_n \to u$, $t_n \to u$ and $\alpha_n\beta_n t_n \to \gamma$, $\alpha_n\beta_n s_n \to u$. Suppose (subnets of) $\beta_n t_n \to r \in M$ and $\beta_n s_n \to q \in M$. Then $\beta_n \xrightarrow{\tau} ru$ and also $\beta_n \xrightarrow{\tau} qu$. Thus ru, $qu \in B$. It follows that $(qu)^{-1}\beta_n \xrightarrow{\tau} u$ and $(qu)^{-1}\beta_n \xrightarrow{\tau} (qu)^{-1}(ru)$ so $(qu)^{-1}ru \in B'$. Now $\alpha_n\beta_n t_n \in A \circ \alpha_n\beta_n t_n = A \circ \beta_n t_n \to A \circ r$, so $\gamma \in A \circ r$. Therefore $\gamma \in A \circ ru$. Similarly $u \in A \circ qu$. Then $\gamma(ru)^{-1} \in A^\circ u \cap G = A$ and $qu \in A$ (so $qu \in A \cap B$.) Summarizing, we have $(qu)^{-1}ru \in B'$, so $ru \in quB' \subset AB'$, $\gamma(ru)^{-1} \in A$. Hence $\gamma \in A(ru) \subset A(AB') = AB'$.

I would like to thank Eli Glasner for discussions on the topic of this paper during my visit to Israel in March 1993. I would also like to thank the referee for a number of helpful comments and corrections.

References

[A] J. Auslander, *Minimal Flows and their Extensions*, Notas de Matematica, vol. 153, North-Holland Mathematical Studies, 1988.

[AG] J. Auslander and M. Guerin, *Regional proximality and the prolongation, to appear*, Forum Mathematicum.

[AM] J. Auslander and N. Markley, *Graphic flows and multiple disjointness*, Trans. Amer. Math. Soc. **292** (1985), 483-499.

[E1] R.Ellis, *Lectures on Topological Dynamics*, Benjamin, 1969.

[E2] R. Ellis, *Automorphism Groups of Flows*, Notes, University of Minnesota, 1995.

[EG] R. Ellis and E. Glasner, *Pure weak mixing*, Trans. Amer. Math. Soc. **243** (1978).

[EGS] R. Ellis, E. Glasner, and L. Shapiro, *Proximal Isometric (PI) flows*, Advances in Math. **17** (1975), 213-260.

[G] E. Glasner, *Proximal Flows*, Lecture Notes in Math., vol. 517, Springer-Verlag, 1976.

DEPARTMENT OF MATHEMATICS, UNIVERSITY OF MARYLAND, COLLEGE PARK, MD, 20742
E-mail address: jna@math.umd.edu

Contemporary Mathematics
Volume **215**, 1998

A Natural Family of Factors for Minimal Flows

E. Glasner, M. Mentzen and A. Siemaszko

ABSTRACT. Several definitions of a "natural family" of factors in topological dynamics are introduced and studied. Natural families for group extensions of rotations are described. Natural covering maps are shown to be regular, hence in particular normal, extensions. A kind of Furstenberg decomposition for normal RIC extensions is proved. The class of seminormal flows (analogous to the class of semisimple systems in ergodic theory) is defined and the decomposition theorem for normal RIC extensions is used to prove a structure theorem for this class.

Introduction

Given a dynamical system (either measure theoretical or topological), its family of factors can have a rich and complex structure. An interesting step towards a systematic classification of this family (in the measure theoretical case), was taken in [**JLM**]. It was shown there that for an ergodic system $(X, \mathcal{B}, \mu, T)$there always exists a unique minimal "natural" family of factors, $\mathcal{N}$, which includes all those factors arising from ergodic self-joinings and which has the following property: for every factor sub-algebra $\mathcal{A}$ of $\mathcal{B}$ there exists a "natural" cover $\hat{\mathcal{A}} \supset \mathcal{A}$ such that the corresponding factor map from the factor defined by $\hat{\mathcal{A}}$ to the one defined by $\mathcal{A}$, is a group extension. This natural subfamily of factors is strongly related to structure theory and can, in some cases, considerably simplify the study of the family of all factors. Two such cases are studied in [**JLM**]; the case of an ergodic group extension of a group rotation, and the case of what is called in [**JLM**], semisimple systems (these contain among others, simple as well as Gauss-Kronecker systems). In some cases the minimal family of natural factors coincides with the entire family of factors (see [**GL**] and [**Gl5**], where this is shown for Bernoulli systems). However, even in these cases, the mere fact that $\mathcal{N}$ consists of all factors is of great interest.

The purpose of the present paper is to study several analogues of the notion of natural family of factors in topological dynamics, or more precisely, in the theory of minimal dynamical systems (called here minimal flows). Our first approach (expounded in Section 2) is perhaps the most straightforward one. We define a self-joining of the minimal system (X, T) to be any minimal subset of $X \times X$. With this definition of joining we mimic the definition of natural family of factors given

1991 *Mathematics Subject Classification.* Primary 54H20.

The second and third authors were supported by KBN grant 2 P301 031 07 in [**JLM**]. As in the measure theoretical case we get in this way the existence of a unique smallest natural family of factors $\mathcal{N}$which includes all the factors arising from self-joinings. This family $\mathcal{N}$ has the property that for every factor Y of X there exists a unique natural cover $\tilde{Y} \in \mathcal{N}$ such that the map $\tilde{Y} \to Y$ is a regular (but not necessarily a group) extension. We characterize the least member of the smallest natural family of factors of the system (X, T) as the unique maximal regular factor of (X, T). At the end of Section 2 we consider an alternative approach. We call a subset W of $X \times X$ a B-set if it is closed, $T \times T$ invariant, topologically transitive and such that the union of the minimal subsets of W is dense in W. Now we enlarge the class of admissible self-joinings by allowing all B-sets to be joinings. The corresponding notion of a natural family of factors now has the property that the map $\pi : \tilde{Y} \to Y$ is again regular and in addition admits a decomposition $\pi = \omega \circ \kappa$ where κ is a group extension and ω a proximal one. We show that for this type of natural family the Kronecker factor is the least member of the smallest natural family. By a result of Bronstein, for a PI-flow X, and in particular for a distal flow, a B-set in $X \times X$ is necessarily minimal ([**B**], see also [**A**]) and the two notions of natural families coincide. Unlike the situation in ergodic theory, the largest zero-entropy factor of a minimal system need not be natural.

Section 3, motivated by the works [**JLM**], [**LM**], [**M**] and [**S**], deals with natural families of factors for a minimal group extension of a group rotation. We show by direct methods that for such a flow the family $\{X/F :\ F$ a closed normal subgroup of $G\}$, is a natural family of factors for the G-extension (X, T) of the group rotation $Z = X/G$ (Proposition 3.3).

In Section 4 we introduce our third version of natural families. The objects we work with here are Ellis groups. These groups classify minimal flows up to proximal extensions and therefore their study can lead to useful conclusions about the corresponding minimal flows. In fact, for minimal distal flows, these two categories are completely equivalent. For a fixed τ-closed Ellis group A, we define the smallest natural family $\mathcal{N}$ of ambient groups of A containing those groups that arise from the appropriate notion of a joining and closed under $\bigvee$ and show that this family has the property that each τ-closed group $B \supset A$ has a natural "core" $\tilde{B} \in \mathcal{N}$ with $\tilde{B} \lhd B$. The fact that $\tilde{B}$ is a normal subgroup of B focuses our attention on "normal extensions" of minimal flows. These are defined in Section 5 and a new structure theorem about such extensions (which generalizes results in [**Gl2**] and [**Gl3**]) is proved (Theorem 5.1).

Finally, in Section 6, we use the results obtained in the previous sections to introduce a large class of minimal flows (the class of seminormal flows) analogous to the class of semisimple ergodic systems introduced and studied in [**JLM**] and prove a structure theorem for these flows (Theorem 6.1).

We thank M. Lemańczyk for his interest and advice during the preparation of this work and the referee for several helpful suggestions.

1. General Backgrounds and Results

We briefly go over the theory and notation we shall be using. For more details see [**A**], [**E**], [**Gl1**], [**V1**].

Let T be a homeomorphism of a compact metric space (X, d). We will call the pair (X, T) a *flow*. When (X, T) is a flow and $A \subset X$, we write $o(A) = \{T^k x : x \in A,\ k \in Z\}$, and $\bar{o}(A) = \mathrm{cl}(o(A))$.

We say that a set $M \subset X$ is *T–minimal*, if M is closed, non-empty and invariant (i.e. $T^{-1}(M) = M$) and M has no proper subset with these properties. If X itself is T–minimal then (X,T) (or T) is called minimal. The set of all minimal subsets of $(X \times X, T \times T)$ is denoted by $\mathcal{M}_2(X,T)$. A point $x \in X$ is said to be *almost periodic* (a.p.) if $\bar{o}(x)$ is a minimal set.

We say that a closed, invariant set $N \subset X$ is a B–*set*, if it is topologically transitive (the only closed, invariant subset of N with non-empty interior is N itself, or equivalently it has a point with dense orbit) and it has a dense set of almost periodic points. The family of B–sets of $X \times X$ is denoted by $\mathcal{B}_2(X,T)$.

The *centralizer*, $C(X,T)$ of (X,T) is the set of all continuous $S : X \longrightarrow X$ which commute with T: $T \circ S = S \circ T$. Clearly $C(X,T)$ equipped with the topology of uniform convergence is a topological semigroup. The set of all invertible elements of $C(X,T)$ is denote by $\mathrm{Aut}(X,T)$. We endow $\mathrm{Aut}(X,T)$ with the topology induced by the metric given by

$$\bar{d}(S_1,S_2) = \sup_{x\in X} d(S_1x, S_2x) + \sup_{x\in X}(S_1^{-1}x, S_2^{-1}x),$$

which makes $\mathrm{Aut}(X,T)$ a topological group.

The simplest examples of minimal flows are minimal rotations on compact metrizable monothetic groups. If X is such a group and $\mathrm{cl}\{x_0^n : n \in Z\} = X$then the map $T(x) = x_0x$ is a minimal homeomorphism. In this case we can easily describe the centralizer of (X,T): $S \in C(X,T)$ iff there is an $a \in X$ such that $S(x) = ax$, $x \in X$. Actually, more is true:

$$\begin{aligned} &M \in \mathcal{M}_2(X,T) \text{ iff there exists } S \in C(X,T) \text{ such that} \\ &M = \mathrm{graph}(S) = \{(x,S(x)) : x \in X\}. \end{aligned} \tag{1}$$

A pair $(x,y) \in X \times X$ is called *distal* if there exists $\delta = \delta(x,y) > 0$ such that $d(T^nx, T^ny) > \delta$ for all $n \in Z$. If a pair is not distal, it is said to be *proximal*. A flow (X,T) is called *distal* if each pair $(x,y) \in X \times X$, $x \neq y$is distal. A minimal rotation is an example of a distal flow. Every distal flow can be decomposed into minimal pieces (see [**E**]). From this it is easy to deduce Ellis's result that if (X,T) is a minimal flow, then $(X \times X, T \times T)$ is decomposable into minimal pieces if and only if (X,T) is distal.

A minimal flow (X,T) is called *regular* if for every a.p. point $(x,y) \in X \times X$ there exists $S \in \mathrm{Aut}(X,T)$ such that $y = S(x)$.

Let (X,T) be any flow. We say that a flow (Y,S) is a *factor* of (X,T) if there is a continuous map $\pi : X \longrightarrow Y$ (called a *homomorphism*) such that $\pi(X) = Y$ and $\pi \circ T = S \circ \pi$. In such a case (X,T) (or sometimes π) is called an *extension* of (Y,S). It is easy to see that any factor of a minimal flow is also minimal. If the map π is a homeomorphism, we say that (X,T) and (Y,S) are *isomorphic* and call π an *isomorphism*. If $\pi : (X,T) \longrightarrow (Y,S)$ is a homomorphism, we can define a closed, $T \times T$–invariant equivalence relation $R_\pi \subset X \times X$ by $(x,y) \in R_\pi$ if $\pi(x) = \pi(y)$. Obviously, the quotient space (X_{R_π}, T_{R_π}) (here T_{R_π} denotes the quotient action of T on X_{R_π}) with the quotient topology is a metrizable compact space and is isomorphic to (Y,S). This allows us to picture factors of (X,T), as $T \times T$–invariant, closed equivalence relations (ICER's) on X also called *factor relations*. Conversely, given such a relation R we can define a homomorphism $\pi : (X,T) \longrightarrow (X_R, T_R)$ by $\pi(x) = [x]_R$ (here $[x]_R$ denotes an equivalence class of x). Note that if we have two factor relations R_i, $i = 1,2$, and a homomorphism $\pi : (X_{R_1}, T_{R_1}) \longrightarrow (X_{R_2}, T_{R_2})$

with $\pi_{R_2} = \pi \circ \pi_{R_1}$, then $R_1 \subset R_2$. If we have a family $\{R_i\}_{i \in I}$ of factor relations, by $\bigvee_{i \in I} R_i$ we denote the smallest factor relation containing all R_i's.

If $Y \subset X$ and R is a factor relation on $X \times X$, then by R_Y we denote the restriction of R to Y (i.e. $R_Y = R \cap (Y \times Y)$).

There is another possible way of talking about factors. Let us denote the algebra of all continuous functions $h : X \longrightarrow R$ by $\mathcal{C}(X)$. Given a factor relation R on X we can define a subalgebra $\mathcal{C}(X, R)$ by

$$\mathcal{C}(X, R) = \{p \circ h : \ h \in \mathcal{C}(X_R)\}.$$

Note that $\mathcal{C}(X, R)$ consists of all functions which are constant on the equivalence classes of R. Note also that $\mathcal{C}(X, R)$ is T–invariant (i.e. if $h \in \mathcal{C}(X, R)$, then $h \circ T \in \mathcal{C}(X, R)$), contains the constants and is uniformly closed. Conversely for such a subalgebra $\mathcal{A} \subset \mathcal{C}(X)$ set

$$R(\mathcal{A}) = \{(x, y) \in X \times X : h(x) = h(y) \text{ for all } h \in \mathcal{A}\}.$$

Clearly $R(\mathcal{A})$ is a factor relation. It is an easy corollary of the definitions that $R = R(\mathcal{C}(X, R))$ and $\mathcal{A} \subset \mathcal{C}(X, R(\mathcal{A}))$. Now $\mathcal{A}$ contains the constants and separates points of $X_{R(\mathcal{A})}$. Since it is closed, by Stone-Weierstrass theorem we get $\mathcal{A} = \mathcal{C}(X, R(\mathcal{A}))$. Therefore we conclude that there is a one-to-one correspondence between factor relations and closed T–invariant subalgebras of $\mathcal{C}(X)$ which contain the constants. Note that if $\{R_i\}_{i \in I}$ is a family of factor relations on X, then $R(\bigcap_{i \in I} \mathcal{C}(X, R_i)) = \bigvee_{i \in I} R_i$.

Next we describe a special class of factors. Let (X, T) be a minimal flow. Fix some compact subgroup G of $\mathrm{Aut}(X, T)$. We define a relation R_G in $X \times X$ setting $(x, y) \in R_G$ iff there exists a $g \in G$ such that $y = g(x)$. R_G is clearly $T \times T$–invariant and since G is compact, $R_G \subset X \times X$ is a closed set. To shorten notation we denote $[x]_G = [x]_{R_G}$, $X_G = X_{R_G}$ and $T_G = T_{R_G}$. We say that (X, T) is a *G–extension* of some (Y, S) if (Y, S) is isomorphic to (X_G, T_G). We will also call (X, T) a *group extension* of (Y, S) without specification of the group G. A group extension is called *free* if $g(x) = x$ for some $x \in X$ implies $g = e$ (e denotes an identity in G). In this paper we will consider only free group extensions.

The homomorphism $\pi : X \longrightarrow Y$ is called *isometric* if there exists a group extension $\rho : Z \longrightarrow Y$ and a homomorphism $\sigma : Z \longrightarrow X$ such that $\pi \circ \sigma = \rho$.

The notions of distality and regularity can be "relativized" (with respect to the factor). Let $\pi : (X, T) \longrightarrow (Y, S)$ be a homomorphism. Then π is said to be *distal* (*regular*) provided that the corresponding definitions hold for every $(x, y) \in R_\pi$. The homomorphism is called *proximal* if every pair $(x, y) \in R_\pi$ is a proximal pair; it is called a *weakly mixing* homomorphism if the flow $(R_\pi, T \times T)$ is topologically transitive.

The following theorem is due to W. H. Gottschalk ([**Go**]) and J. Auslander [**A**].

THEOREM 1.1. *Every regular distal flow is equicontinuous (is a group extension of a trivial flow.*

We will also use the relativized version of the above theorem ([**Gl2**]). This can be considered as a topological version of a theorem of Veech (see [**V2**], also [**JR**]; a proof of Veech's theorem is also contained in [**L**]).

THEOREM 1.2. *Let (X, T) be a minimal flow and let $\pi : X \longrightarrow Y$ be a regular distal homomorphism. Then π is a group extension.*

Now, we recall the basic notation and results of the universal theory of topological dynamics. As usuall βZ denotes the Stone-Čech compactification of Z, I a fixed arbitrary minimal left ideal of the semigroup βZ (replacing the customary M), J the set of idempotents in I, u a fixed arbitrary element of J, and $G = uI$ the maximal subgroup of I corresponding to u. The semigroup βZ acts on every flow (Z,T), and for $z \in Z$, $\bar{o}(z) = \{pz : p \in \beta Z\}$. A necessary and sufficient condition for z to be a.p. is that $vz = z$ for some $v \in J$. Thus JZ is the collection of all almost periodic points in Z. When z is a.p. then $\bar{o}(z) = \{pz : p \in I\}$.

Given a minimal flow (Z,T), we shall always choose a distinguished point z_0 in Z, such that $uz_0 = z_0$. Our convention is that under a homomorphism a distinguished point goes to a distinguished point. When (Z, z_0, T) is such a pointed minimal flow, its Ellis group $\mathcal{G}(Z, z_0)$ is defined by $\mathcal{G}(Z) = \mathcal{G}(Z, z_0) = \{\alpha \in G : \alpha z_0 = z_0\}$. The set G is equipped with a compact T_1 topology called the τ–topology with respect to which, all groups of the form $\mathcal{G}(Z, z_0)$ are closed. If we have a family $\{(Z_\sigma, z_\sigma)\}_{\sigma\in\Sigma}$ of pointed minimal flows we choose $x_0 \in \prod_{\sigma\in\Sigma} Z_\sigma$, $x_0(\sigma) = z_\sigma$, and set

$$\bigvee_{\sigma\in\Sigma} (Z_\sigma, z_\sigma) = (\bar{o}(x_0), x_0).$$

Observe that $\bigvee_{\sigma\in\Sigma}(Z_\sigma, z_\sigma)$ is minimal and

$$\mathcal{G}(\bigvee_{\sigma\in\Sigma} (Z_\sigma, z_\sigma)) = \bigcap_{\sigma\in\Sigma} \mathcal{G}(Z_\sigma, z_\sigma).$$

For every τ–closed subgroup F of G we let

(2) $$F' = \bigcap\{\tau\text{–closure}\, V\colon\ V \text{ is } \tau\text{–open neighbourhood of } u \text{ in } F\}.$$

Then F' is a τ–closed normal subgroup of F characterized as the smallest τ–closed subgroup H of F such that F/H is a compact Hausdorff topological group. One can iterate this operation to obtain the (possibly transfinite) sequence of "derived" groups $F'' = (F')', \ldots, F^{\alpha+1} = (F^\alpha)', \ldots$, where for a limit ordinal α, $F^\alpha = \bigcap\{F^\beta : \beta < \alpha\}$. For some ordinal η this process stabilizes (i.e $F^{\eta+1} = F^\eta$), and we denote $F^\infty = F^\eta$.

For a pointed minimal distal flow (X, x_0, T), $X = Gx_0$ and the family of pointed factors $(X, x_0, T) \xrightarrow{\pi} (Y, y_0, T)$ is in 1–1 correspondence with the family of τ–closed subgroups $F \supset A$, where $A = \mathcal{G}(X, x_0)$ and $F = \mathcal{G}(Y, y_0)$. (With caution, one can view the factor (Y, y_0, T) with the Ellis group $F = \mathcal{G}(Y, y_0)$ as the homogeneous space G/F).

In terms of Ellis groups, a minimal extension $(X,T) \xrightarrow{\pi} (Y,T)$ is proximal iff $\mathcal{G}(X, x_0) = \mathcal{G}(Y, y_0)$. It is isometric iff it is distal and $\mathcal{G}(Y, y_0)' \subset \mathcal{G}(X, x_0)$; if moreover $\mathcal{G}(X, x_0) \lhd \mathcal{G}(Y, y_0)$, then π is a group extension. The extension π is called *relatively incontractible* or RIC, if

$$\pi^{-1}(py_0) = p \circ u\pi^{-1}(y_0) = p \circ Fx_0 \text{ for every } p \in I,$$

where $F = \mathcal{G}(Y, y_0)$ (for the definition of the "circle" operation see for instance [**Gl1**]). RIC–extensions are open and have a dense set of almost periodic points in the relation R_π. Every distal extension is RIC. Every homomorphism from a minimal flow to the one-point flow is RIC.

If $\{A_i\}$ is any collection of τ–closed subgroups of G, we let $\bigvee A_i$ be the smallest τ–closed subgroup of G containing all the A_i's.

We next recall the construction and main results concerning PI-towers. If $\sigma : (X,T) \longrightarrow (Z,T)$ is a homomorphism of minimal flows, one constructs a canonically defined commutative diagram of homomorphisms of minimal flows (*the shadow diagram*),

$$\begin{array}{ccc} X & \xleftarrow{\ \theta^*\ } & X^* \\ {\scriptstyle\sigma}\downarrow & & \downarrow{\scriptstyle\sigma^*} \\ Z & \xleftarrow{\ \theta\ } & \sigma Z^* \end{array}$$

where σ^* is RIC and $\theta,\ \theta^*$ are proximal. The extensions θ and θ^* are isomorphisms if and only if σ is already RIC. In the above diagram $X^* = X \vee Z^*$ and θ^* (σ^*) is the projection onto the first (second) coordinate. It follows that if σ is regular then so is σ^*.

The main reason that RIC-extensions are so useful is the following theorem ([**EGS**], [**V1**]).

THEOREM 1.3. *Let $(X,T) \xrightarrow{\ \pi\ } (Y,T)$ be a RIC-extension of minimal flows. Then there exists a commutative diagram*

$$\begin{array}{ccc} X & \xrightarrow{\ \sigma\ } & Z \\ {\scriptstyle\pi}\downarrow & \swarrow{\scriptstyle\rho} & \\ Y & & \end{array}$$

where ρ is an isometric extension with $B = F'A$ $(B = \mathcal{G}(Z, z_0),\ F = \mathcal{G}(Y, y_0)$, and $A = \mathcal{G}(X, x_0))$. The flow Z is the largest isometric extension of Y within X, and ρ is an isomorphism iff π is a weakly mixing extension iff $B = F$.

Using the shadow construction and this theorem repeatedly one obtains the following structure theorem ([**EGS**], [**V1**]).

THEOREM 1.4. *Given a RIC homomorphism $\pi = \pi_0 : (X,T) \to (Y,T)$ of metric minimal flows, there exist a countable ordinal η and a canonically defined commutative diagram (the canonical PI-tower):*

$$\begin{array}{ccccccccccc} X = X_0 & \xleftarrow{\ \widetilde{\theta}_1\ } & & X_1 & \cdots & X_\nu & \xleftarrow{\ \widetilde{\theta}_{\nu+1}\ } & & X_{\nu+1} & \cdots & X_\eta = X_\infty \\ {\scriptstyle\pi_0}\downarrow & \searrow{\scriptstyle\sigma_1} & & {\scriptstyle\pi_1}\downarrow & & {\scriptstyle\pi_\nu}\downarrow & \searrow{\scriptstyle\sigma_{\nu+1}} & & \downarrow{\scriptstyle\pi_{\nu+1}} & & {\scriptstyle\pi_\infty}\downarrow \\ Y = Y_0 & \xleftarrow[\rho_1]{} & Z_1 & \xleftarrow[\theta_1]{} Y_1 & \cdots & Y_\nu & \xleftarrow[\rho_{\nu+1}]{} & Z_{\nu+1} & \xleftarrow[\theta_{\nu+1}]{} Y_{\nu+1} & \cdots & Y_\eta = Y_\infty \end{array}$$

where for each $\nu \le \eta$, π_ν is RIC, ρ_ν is isometric, θ_ν, $\widetilde{\theta}_\nu$ are proximal and π_∞ is RIC and weakly mixing. For a limit ordinal ν, X_ν, Y_ν, π_ν etc. are the inverse limits of X_ι, Y_ι, π_ι etc. for $\iota < \nu$. In terms of Ellis groups: $B = AF^\infty$ and $AB' = B$, where $A = \mathcal{G}(X)$, $B = \mathcal{G}(Y_\infty)$, and $F = \mathcal{G}(Y)$. The extension π is PI iff $X_\infty = Y_\infty$.

See [**EGS**] or [**Gl1**] for the definition of PI-flows.

2. A Natural Family of Factors of Minimal Flows

In this section we introduce two definitions of a natural family of factors (factor relations) for a minimal flow and prove some basic facts.

Given two minimal flows (X_i, T_i), and two factor relations R_i, $i = 1, 2$, on X_i respectively, the relation $\overline{R}_i$ in $X_1 \times X_2$ is defined by $((x_1, x_2), (y_1, y_2)) \in \overline{R}_i$ if $(x_i, y_i) \in R_i$. The equivalence classes of $\overline{R}_i$ are of the form $[x]_{R_1} \times X_2$ and $X_1 \times [x]_{R_2}$ respectively.

LEMMA 2.1. *Let $M \subset X_1 \times X_2$ be a minimal set and R_i, $i = 1, 2$, be two factor relations on X_i with natural homomorphisms π_i respectively. Then $(\overline{R}_1)_M = (\overline{R}_2)_M$ iff there exists an isomorphism $S : (X_1)_{R_1} \longrightarrow (X_2)_{R_2}$ such that*

$$(\pi_1 \times \pi_2)(M) = graph(S). \tag{3}$$

PROOF. ($\Rightarrow$) Because $(\overline{R}_1)_M = (\overline{R}_2)_M$ and the projections of M into both coordinates are onto, for each $x_1 \in X_1$ there exists an$x_2 \in X_2$ with

$$([x_1]_{R_1} \times X_2) \cap M = (X_1 \times [x_2]_{R_2}) \cap M.$$

Define S by setting $S([x_1]_{R_1}) = [x_2]_{R_2}$. Now, S is well defined, graph(S) is closed hence compact, so S is continuous. Since graph(S) $\subset M$ and M is minimal, (3) holds. Clearly, S is a bijection. Now take x_1, x_2, x_2', so that

$$([x_1]_{R_1} \times X_2) \cap M = (X_1 \times [x_2]_{R_2}) \cap M \tag{4}$$

and

$$([T_1(x_1)]_{R_1} \times X_2) \cap M = (X_1 \times [x_2']_{R_2}) \cap M. \tag{5}$$

We have $(T_2)_{R_2} \circ S([x_1]_{R_1}) = [T_2(x_2)]_{R_2}$ and $S \circ (T_1)_{R_1}([x_1]_{R_1}) = [x_2']_{R_2}$. If we act by $T_1 \times T_2$ on (4) and use (5), we get

$$[T_2(x_2)]_{R_2} = [x_2']_{R_2}.$$

Thus S is an isomorphism.

($\Leftarrow$) Suppose that $S : (X_1)_{R_1} \longrightarrow (X_2)_{R_2}$ is an isomorphism satisfying (3). Take $(x_1, x_2) \in ([x_1]_{R_1} \times X_2) \cap M$. Then $(\pi_1 \times \pi_2)(x_1, x_2) = ([x_1]_{R_1}, [x_2]_{R_2}) \in$ graph(S). We get $[x_2]_{R_2} = S([x_1]_{R_1})$, and therefore $(x_1, x_2) \in (X_1 \times S([x_1]_{R_1})) \cap M$. By symmetry of arguments $(\overline{R}_1)_M = (\overline{R}_2)_M$ and the proof is finished. □

When the equivalent conditions in Lemma 2.1 hold we say that M *induces* an isomorphism between $(X_1)_{R_1}$ and $(X_2)_{R_2}$. Notice that every isomorphism $S : Y_1 \to Y_2$ where $\pi_i : X_i \to Y_i (i = 1, 2)$, is induced by some minimal subset $M \subset X_1 \times X_2$; take M as any minimal subset of the closed invariant subset $\{(x_1, x_2) : S\pi_1(x_1) = x_2\}$. Conversely, for any minimal set $M \subset X_1 \times X_2$ we will show that there exist two smallest factor relations $P_i(M)$ on X_i respectively, such that

$$(\overline{P}_1(M))_M = (\overline{P}_2(M))_M. \tag{6}$$

Indeed, let (R^1_σ, R^2_σ), $\sigma \in \Sigma$, be the family of all pairs of factor relations satisfying

$$(\overline{R^1_\sigma})_M = (\overline{R^2_\sigma})_M.$$

Put

$$P_i(M) = \bigcap_{\sigma \in \Sigma} R^i_\sigma,$$

$i = 1, 2$. It is clear that $P_i(M)$ satisfy (6) since every equivalence class $[x]_{P_i(M)}$ is an intersection of all equivalence classes $[x]_{R^1_\sigma}$ and $[x]_{R^2_\sigma}$ respectively. There are other equivalent ways of defining $P_i(M)$. Note that $P_i(M)$ is the smallest factor relation containing $M \circ M^{-1}$ for $i = 1$ and $M^{-1} \circ M$ for $i = 2$ (here $M^{-1} = \{(x, y) \in X_2 \times X_1 : (y, x) \in M\}$. Finally, we can also define $P_i(M)$ using subalgebras of $\mathcal{C}(X_i)$. Consider two subalgebras $\overline{\mathcal{C}}_i \subset \mathcal{C}(M)$, $i = 1, 2$, defined by

$$\overline{\mathcal{C}}_i = \{h \in \mathcal{C}(M) : \ h \text{ depends only on the } i\text{-th coordinate}\}.$$

The intersection $\overline{\mathcal{C}}_1 \cap \overline{\mathcal{C}}_2$ can be considered as a subalgebra of $\mathcal{C}(X_1)$, in such a case we denote it by $\mathcal{C}_1(M)$, or as a subalgebra of $\mathcal{C}(X_2)$, then it is denoted by $\mathcal{C}_2(M)$. It is not difficult to see that $P_i(M) = R(\mathcal{C}_i(M))$, $i = 1, 2$.

Now, let (X, T) be a minimal flow. Assume that a family $\mathcal{N}$ of factor relations stisfies the following conditions.

(a) Δ_X(the diagonal) $\in \mathcal{N}$;

(b) $\{R_\lambda\}_{\lambda \in \Lambda} \subset \mathcal{N} \Longrightarrow \bigvee_{\lambda \in \Lambda} R_\lambda \in \mathcal{N}$.

DEFINITION 1. A family $\mathcal{N}$ of factor relations satisfying (a) and (b) is said to be *natural* if it in addition satisfies the following conditions:

(c) $P_i(M) \in \mathcal{N}, i = 1, 2$, for all $M \in \mathcal{M}_2(X, T)$;

(d) $(S \times S)(R) \in \mathcal{N}$ for all R, R_1, $R_2 \in \mathcal{N}, R_1 \subseteq R$ and for all isomorphisms $S : X_{R_1} \to X_{R_2}$, where $(S \times S)(R)$ is defined by: $(x_1', x_2') \in (S \times S)(R)$ iff $[x_i']_{R_2} = S[x_i]_{R_1}, (i = 1, 2)$ for some $(x_1, x_2) \in R$.

REMARK 1. For any family of factor relations $\mathcal{N}$ satisfying (a) and (b) and for each factor relation R there exists the biggest factor relation $\widetilde{R} \in \mathcal{N}$ with $\widetilde{R} \subset R$. We will call $\widetilde{R}$ the $\mathcal{N}$-*core* of R. The corresponding factor map $\pi : X_{\widetilde{R}} \longrightarrow X_R$ is the $\mathcal{N}$-*cover* of X_R .

REMARK 2. Given a natural family of factor relations $\mathcal{N}$ for X, $\bigvee \mathcal{N}$ is the largest element of $\mathcal{N}$ (or the least factor in the corresponding family of factors).

REMARK 3. If we take the intersection of all natural families of factor relations for X, we get the smallest natural family of factor relations.

It is possible to characterize natural family in an alternative way. Let $\mathcal{N}$ be a family of factor relations satisfying (a) and (b).

PROPOSITION 2.1. *The family $\mathcal{N}$ is natural iff for every factor relations R_1, R_2 and each $M \in \mathcal{M}_2(X, T)$ satisfying $(\pi_{R_1} \times \pi_{R_2})(M) = graph(S)$ for an isomorphism $S : X_{R_1} \longrightarrow X_{R_2}$ we have $(\pi_{\widetilde{R}_1} \times \pi_{\widetilde{R}_2})(M) = graph(\widetilde{S})$ for an isomorphism $\widetilde{S} : X_{\widetilde{R}_1} \longrightarrow X_{\widetilde{R}_2}$. In other words iff whenever a minimal M induces an isomorphism between R_1 and R_2 it also induces an isomorphism betwen their $\mathcal{N}$-cores.*

PROOF. ($\Rightarrow$) Denote $\pi_i = \pi_{R_i}$ and $\widetilde{\pi}_i = \pi_{\widetilde{R}_i}$ for $i = 1, 2$. Assume that $M \in \mathcal{M}_2(X, T)$ and $(\pi_1 \times \pi_2)(M) = \mathrm{graph}(S)$ for some isomorphism $S : X_{R_1} \longrightarrow X_{R_2}$. By (c), $R_i \supset \widetilde{R}_i \supset P_i(M) \in \mathcal{N}$. Let $\bar{S} : X_{P_1(M)} \longrightarrow X_{P_2(M)}$ be the isomorphism given by Lemma 2.1. Let $\widetilde{S}$ be the restriction of $\overline{S}$ to $X_{\widetilde{R}_1}$. It remains to show that $\widetilde{S}$ is an isomorphism between $X_{\widetilde{R}_1}$ and $X_{\widetilde{R}_2}$. By (d), $(\widetilde{S} \times \widetilde{S})(\widetilde{R}_1) \in \mathcal{N}$ hence $(\widetilde{S} \times \widetilde{S})(\widetilde{R}_1) \subset \widetilde{R}_2$. Thus $\widetilde{R}_1 \subset (\widetilde{S} \times \widetilde{S})^{-1}(\widetilde{R}_2)$. But, again by (d), $(\widetilde{S} \times \widetilde{S})^{-1}(\widetilde{R}_2) \in \mathcal{N}$ and therefore $\widetilde{R}_1 \supset (\widetilde{S} \times \widetilde{S})^{-1}(\widetilde{R}_2)$. We conclude that $(\widetilde{S} \times \widetilde{S})(\widetilde{R}_1) = \widetilde{R}_2$.
($\Leftarrow$) Take $M \in \mathcal{M}_2(X, T)$. By Lemma 2.1, there exists an isomorphism S :

$X_{P_1(M)} \longrightarrow X_{P_2(M)}$ with $(\pi_{P_1(M)} \times \pi_{P_2(M)})(M) = \mathrm{graph}(S)$. Then for some isomorphism $\widetilde{S} : X_{P_1(M)^\sim} \longrightarrow X_{P_2(M)^\sim}$ satisfying $\widetilde{p}_2 \circ \widetilde{S} = S \circ \widetilde{p}_1$, where $\widetilde{p}_i$, $i = 1, 2$, denote the corresponding homomorphisms from $X_{P_i(M)^\sim}$ onto $X_{P_i(M)}$, we have $(\pi_{P_1(M)^\sim} \times \pi_{P_2(M)^\sim})(M) = \mathrm{graph}(\widetilde{S})$. But $P_i(M)$ are the smallest with such a property, therefore

$$P_i(M) = P_i(M)^\sim \in \mathcal{N}.$$

Now take $R_i \in \mathcal{N}$, $i = 1, 2$, and an isomorphism $S : X_{R_1} \longrightarrow X_{R_2}$. Take $M \in \mathcal{M}_2(X, T)$ satisfying (3) (any minimal set in $\{(x_1, x_2) \in X \times X : \pi_{R_2}(x_2) = S\pi_{R_1}(x_1)\}$). Assume that $R_1 \subset R \in \mathcal{N}$. By $S' : X_R \longrightarrow X_{(S\times S)(R)}$ denote the isomorphism S considered on the equivalence classes of R. Then $(\pi_R \times \pi_{(S\times S)(R)})(M) = \mathrm{graph}(S')$. Therefore $(\pi_{\widetilde{R}} \times \pi_{((S\times S)(R))^\sim})(M) = \mathrm{graph}(\widetilde{S'})$, where $\widetilde{S'}$ is an isomorphism between $X_{\widetilde{R}}$ and $X_{((S\times S)(R))^\sim}$. Since $\widetilde{R} = R$, we get $\widetilde{S'} = S'$, but this forces $(S \times S)(R) = ((S \times S)(R))^\sim \in \mathcal{N}$. □

PROPOSITION 2.2. *For a family $\mathcal{N}$ of factor relations satisfying (a),(b) and (c), condition (d) is equivalent to*

(e) *If $S : X_R \to X_{R'}$ is an isomorphism and $R \in \mathcal{N}$ then $R' \in \mathcal{N}$.*

PROOF. The implication (e) ⇒ (d) is clear. For the converse implication suppose M induces an isomorphism from X_R to $X_{R'}$ for $R \in \mathcal{N}$. Let $S : P_1(M) \to P_2(M)$ be the isomorphism induced by M, then $(S \times S)(R) = R' \in \mathcal{N}$ by (d). □

COROLLARY 2.1. *Let us suppose that (X, T) is a minimal flow with the property that for each two factor relations R_i, $i = 1, 2$, if (X_{R_1}, T_{R_1}) is isomorphic to (X_{R_2}, T_{R_2}), then $R_1 = R_2$. Then*

$$\mathcal{N} = \left\{ \bigvee_{i \in I} P_1(M_i) : \{M_i\}_{i\in I} \subset \mathcal{M}_2(X, T) \right\}$$

is a natural family of factor relations for (X, T).

PROOF. Clearly $\mathcal{N}$ satisfies conditions (a) and (b). If $M \in \mathcal{M}_2(X, T)$ then by assumption $P_1(M) \in \mathcal{N}$ and since M induces an isomorphism between $P_1(M)$ and $P_2(M)$ we have $P_1(M) = P_2(M) \in \mathcal{N}$. Thus (c) is also satisfied. Now (e) follows trivially from our assumption. □

Also as a corollary of Proposition 2.1, we obtain the following result, which is an analogue of the one in the measure-theoretic context (see [**JLM**]).

PROPOSITION 2.3. *Let $\mathcal{N}$ be a natural family of factor relations for a minimal flow (X, T). For each factor relation R on X the homomorphism $\pi : X_{\widetilde{R}} \longrightarrow X_R$ is regular. Furthermore if π is distal, it is a group extension.*

PROOF. Take $\widetilde{x}_i \in X_{\widetilde{R}}$, $i = 1, 2$ with $\pi(\widetilde{x}_i) = x \in X_R$, where $(\widetilde{x}_1, \widetilde{x}_2)$ is almost periodic. Let $M = \overline{o}(\widetilde{x}_1, \widetilde{x}_2) \in \mathcal{M}_2(X_{\widetilde{R}}, T_{\widetilde{R}})$. Clearly $(\pi \times \pi)(M) = \Delta_{X_R} = \mathrm{graph}(\mathrm{Id}_{X_R})$. Proposition 2.1 yields $\widetilde{S} \in \mathrm{Aut}(X_{\widetilde{R}}, T_{\widetilde{R}})$ with $M = \mathrm{graph}(\widetilde{S})$. Therefore $\widetilde{S}(\widetilde{x}_1) = \widetilde{x}_2$. Thus π is regular. The second part of the proof follows immediately from Theorem 1.2. □

Since any factor of a distal flow is distal, we have the following.

COROLLARY 2.2. *Let $\mathcal{N}$ be a natural family of factor relations for a minimal distal flow (X,T). Then for each factor relation R on X the homomorphism $\pi : X_{\widetilde{R}} \longrightarrow X_R$ is a group extension.*

LEMMA 2.2. *Let (X,T) be a minimal flow. Let R_1, R_2, K be factor relations on X. Assume that $M \in \mathcal{B}_2(X,T)$. If M induces isomorphisms $S : X_{R_1} \longrightarrow X_{R_2}$ and $\overline{S} : X_K \longrightarrow X_K$, then M induces an isomorphism between $X_{R_1 \cap K}$ and $\longrightarrow X_{R_2 \cap K}$.*

PROOF. Take $x \in X$ and define $\widetilde{S}([x]_{R_1 \cap K}) = S([x]_{R_1}) \cap \overline{S}([x]_K)$. All we need to show is that $\widetilde{S}$ is indeed a map from $X_{R_1 \cap K}$ to $X_{R_2 \cap K}$. Assume that $S([x]_{R_1}) = [y]_{R_2}$ and $\overline{S}([x]_K) = [\overline{y}]_K$. Since $M \in \mathcal{B}_2(X,T)$, there exists $\tilde{y} \in X$ such that $(x,\tilde{y}) \in M$. Since M induces S and $\overline{S}$, we have $[y]_{R_2} = [\tilde{y}]_{R_2}$ and $[\overline{y}]_K = [\tilde{y}]_K$. Therefore $\widetilde{S}([x]_{R_1 \cap K}) = [\tilde{y}]_{R_2} \cap [\tilde{y}]_K = [\tilde{y}]_{R_2 \cap K}$. □

LEMMA 2.3. *Let K be a factor relation of a minimal flow (X,T). If X_K is regular then*

$$\mathcal{N} = \{R : \ R \text{ is a factor relation on } X \text{ and } R \subset K\}$$

is a natural family of factor relations.

PROOF. Obviously, $\mathcal{N}$ satisfies (a) and (b) of Definition 1. Now, if R is a factor relation on X, then the $\mathcal{N}$-core of R is equal to $R \cap K$. To finish the proof take two factor relations R_1, R_2 on X and $M \in \mathcal{M}_2(X,T)$ which induces an isomorphism $S : X_{R_1} \longrightarrow X_{R_2}$. Since M must induce an isomorphism of X_K with itself, by Lemma 2.2, M induces an isomorphism $S' : X_{R_1 \cap K} \longrightarrow X_{R_2 \cap K}$. Clearly $S \circ p_1 = p_2 \circ S'$, where p_i denotes the homomorphism $X_{R_i \cap K} \longrightarrow X_{R_i}$, for $i = 1,2$. In view of Proposition 2.1, $\mathcal{N}$ is natural.

□

PROPOSITION 2.4. *Let (X,T) be a minimal flow.*
1. *There exists a unique maximal regular factor Y of X.*
2. *The factor Y is the least member of the smallest natural family of factors of X.*

PROOF. (1) Let X_1 and X_2 be two regular factors of X, corresponding to the factor relations R_1 and R_2. Let $Y = X_1 \vee X_2$ be the factor corresponding to the factor relation $R_1 \cap R_2$. Then Y is isomorphic to the subflow $\bar{o}(x_1,x_2)$ of the product flow $X_1 \times X_2$, where (x_1,x_2) is some almost periodic point in $X_1 \times X_2$. To show that Y is regular we examine an almost periodic point $((z_1,z_2),(w_1,w_2)) \in Y \times Y$. Now (z_i,w_i) is a.p. point of $X_i \times X_i$, $i = 1,2$, so by regularity of X_i, there exist $\phi_i \in \mathrm{Aut}(X_i), i = 1,2$ with $\phi_i(z_i) = w_i$. Let $\phi : Y \longrightarrow X_1 \times X_2$ be defined by $\phi(u,v) = (\phi_1(u),\phi_2(v))$. Clearly $\phi(Y)$ is a minimal subset of the flow $X_1 \times X_2$ and since $\phi(z_1,z_2) = (w_1,w_2) \in Y$, we conclude that $\phi : Y \longrightarrow Y$ is an automorphism of Y, and Y is a regular flow. The same argument works for a joining of any family of regular factors. Therefore $Y = \bigvee X_\nu$ is a regular factor of X, where $\{X_\nu\}$ is the collection of all regular factors of X.

(2) Let K be the factor relation corresponding to Y. Since the family $\mathcal{N}$ defined in Lemma 2.3 only factor relations contained in K, it remains to show that $K \in \mathcal{N}$ for every natural family $\mathcal{N}$. Let $\pi : X_{\tilde{K}} \longrightarrow X_K$ be the $\mathcal{N}$-cover. By regularity

of Y, for an a.p. point $(\tilde{y}_1, \tilde{y}_2) \in \tilde{Y} \times \tilde{Y}$ ($\tilde{Y} = X_{\tilde{K}}$), there exists $\phi \in \mathrm{Aut}(Y)$ with $y_2 = \phi(y_1)$, where $y_i = \pi(\tilde{y}_i)$. In view of Proposition 2.1, there exists an automorphism $\tilde{\phi} \in \mathrm{Aut}(\tilde{Y})$ satisfying $\pi \circ \tilde{\phi} = \phi \circ \pi$. Put $\tilde{y} = \tilde{\phi}(\tilde{y}_1)$, then

$$\pi(\tilde{y}) = \pi(\tilde{\phi}(\tilde{y}_1)) = \phi(\pi(\tilde{y}_1)) = \phi(y_1) = y_2 = \pi(\tilde{y}_2).$$

Thus $(\pi \times \pi)(\tilde{y}_1, \tilde{y}_2) = (\pi \times \pi)(\tilde{y}_1, \tilde{y}) = (y_1, y_2)$ and $(\tilde{y}_1, \tilde{y}_2)$, $(\tilde{y}_1, \tilde{y})$ are almost periodic. Assume that $v, w \in J$ (J is the set of all indempotents in the fixed minimal ideal I, see page 5) with $v(\tilde{y}_1, \tilde{y}) = (\tilde{y}_1, \tilde{y})$, $w(\tilde{y}_1, \tilde{y}_2) = (\tilde{y}_1, \tilde{y}_2)$ and put $\tilde{y}_0 = v\tilde{y}_2$. Now, since

$$\pi(\tilde{y}) = \pi(v\tilde{y}) = v\pi(\tilde{y}) = v\pi(\tilde{y}_2) = \pi(v\tilde{y}_2) = \pi(\tilde{y}_0)$$

and $(\tilde{y}, \tilde{y}_0)$ is an almost periodic point ($v(\tilde{y}, \tilde{y}_0) = (\tilde{y}, \tilde{y}_0)$), by Proposition 2.3 we obtain that $\tilde{y}_0 = \tilde{\psi}(\tilde{y})$ for some $\tilde{\psi} \in \mathrm{Aut}(\tilde{Y})$. Finally

$$\tilde{y}_2 = w\tilde{y}_2 = wv\tilde{y}_2 = w\tilde{y}_0 = w\tilde{\psi}(\tilde{y}) = w\tilde{\psi} \circ \tilde{\phi}(\tilde{y}_1) = \tilde{\psi} \circ \tilde{\phi}(w\tilde{y}_1) = \tilde{\psi} \circ \tilde{\phi}(\tilde{y}_1),$$

so $X_{\tilde{K}}$ is regular and the proof is complete. □

Now we modify our definition of natural family to obtain a stronger result about natural covers. Using this we will be able to identify the least element in this new natural family as the Kronecker factor. Call a family of factors of a minimal flow (X, T) B-*natural* if in (c) of the original definition of natural family one replaces the minimal joinings $M \in \mathcal{M}_2(X, T)$ by joinings $N \in \mathcal{B}_2(X, T)$. Notice that if a family is B-natural, it is natural, therefore all assertions we made for natural families remain valid for B-natural families and that for PI flows natural = B-natural.

LEMMA 2.4. *Let $\pi : X \longrightarrow Y$ be a regular homomorphism of minimal flows. Let Γ_π be the group of automorphisms S of X satisfying $\pi \circ S = \pi$. If Γ_π is compact then the homomorphism $\omega : X/\Gamma_\pi \longrightarrow Y$ is proximal.*

PROOF. Put $Z = X/\Gamma_\pi$. Denote the corresponding homomorphism from X to Z by κ, (hence for every $x \in X$, $\kappa(x) = \Gamma_\pi x$ and $\omega(\kappa(x)) = \pi(x)$). Take $x \in X$ with $ux = x$ and put $z = \kappa(x)$, $y = \pi(x)$. Denote $B = \mathcal{G}(Y, y)$ and $F = \mathcal{G}(Z, z)$. All we have to show is that $B \subset F$.

Take $\beta \in B$, hence $\pi(\beta x) = \beta\pi(x) = \pi(x)$. Since $(x, \beta x)$ is a.p. ($u(x, \beta x) = (x, \beta x)$), by regularity of π, there exists $S \in \mathrm{Aut}(X)$ such that $S(x) = \beta x$. We have $\pi(x) = \pi(\beta x) = \pi(S(x))$, hence, by minimality of X, $\pi \circ S = \pi$ and $S \in \Gamma_\pi$. Therefore we have $\beta z = \beta\kappa(x) = \kappa(\beta x) = \kappa(S(x)) = \kappa(x)$ and $\beta \in F$. □

THEOREM 2.1. *Let (X, T) be a minimal flow, $\mathcal{N}$ a B-natural family of factors for (X, T). Then for any factor relation R the homomorphism $\pi : X_{\tilde{R}} \longrightarrow X_R$ decomposes as $\pi = \omega \circ \kappa$, $X_{\tilde{R}} \xrightarrow{\kappa} Z \xrightarrow{\omega} X_R$; where ω is a proximal extension and κ is a group extension.*

PROOF. Let us denote $Y = X_R$, $\widetilde{Y} = X_{\widetilde{R}}$. Let

$$\begin{array}{ccc} X & & \\ \Big\downarrow{\scriptstyle \widetilde{\pi}} & & \\ \widetilde{Y} & \xleftarrow{\theta^*} & (\widetilde{Y})^* \\ \Big\downarrow{\scriptstyle \pi} & & \Big\downarrow{\scriptstyle \pi^*} \\ Y & \xleftarrow{\theta} & Y^* \end{array}$$

be the shadow diagram of the homomorphism π. By Proposition 2.3 the homomorphism π is regular, thus it easily follows (as in **[Gl3]**, Lemma 4), that also π^* is regular. By construction π^* is RIC. Now use Theorem X.2.1 in **[Gl1]** to construct the diagram

$$\begin{array}{ccc} (\widetilde{Y})^* & \xrightarrow{\eta^*} & Z^* \\ \Big\downarrow{\scriptstyle \pi^*} & \swarrow{\scriptstyle \rho^*} & \\ Y^* & & \end{array}$$

where ρ^* is the maximal isometric extension of Y^* under $(\widetilde{Y})^*$. By Theorem 3 of **[Gl3]**, ρ^* is a group extension and η^* is RIC and weakly mixing. In particular then $R_{\eta^*} \in \mathcal{B}_2((\widetilde{Y})^*)$. Since the image of a B–set under a homomorphism is again a B–set, we conclude that $\widetilde{N} = (\theta^* \times \theta^*)(R_{\eta^*}) \in \mathcal{B}_2(\widetilde{Y})$.

Let

$$\mathcal{L} = \{L \subset X \times X \ : \ L \text{ is closed invariant and } (\widetilde{\pi} \times \widetilde{\pi})(L) = \widetilde{N}\}.$$

Let N be a minimal element in $\mathcal{L}$; then clearly $N \in \mathcal{B}_2(X)$. By the commutative nature of our diagrams it follows that $(\pi \circ \widetilde{\pi} \times \pi \circ \widetilde{\pi})(N) = \Delta_Y$ and we conclude that $\widetilde{N} = (\widetilde{\pi} \times \widetilde{\pi})(N) = \text{graph}(\widetilde{S})$ for some $\widetilde{S} \in \text{Aut}(\widetilde{Y}, T)$ and in particular $\widetilde{N}$ is minimal. However as $\Delta_{\widetilde{Y}} \subset \widetilde{N}$, we have $\widetilde{N} = \Delta_{\widetilde{Y}}$. Thus $\widetilde{N} = (\theta^* \times \theta^*)(R_{\eta^*}) = \Delta_{\widetilde{Y}}$. Since θ^* is a proximal extension, we conclude that $\Delta_{(\widetilde{Y})^*}$ is the only minimal subset in R_{η^*}. As η^* is RIC – hence R_{η^*} is a B-set – this implies that $R_{\eta^*} = \Delta_{(\widetilde{Y})^*}$, i.e. η^* is an isomorphism and it follows that π^* is a group extension. Now by Lemma 3 in **[Gl3]**, the group Γ_π of automorphisms σ of $\widetilde{Y}$ satisfying $\pi \circ \sigma = \pi$ is topologically isomorphic with the group $\Gamma_{\theta \circ \pi^*}$ of automorphisms of $(\widetilde{Y})^*$; the latter is a compact group and so we put $Z = \widetilde{Y}/\Gamma_\pi$ and let

$$\begin{array}{ccc} \widetilde{Y} & \xrightarrow{\kappa} & Z \\ \Big\downarrow{\scriptstyle \pi} & \swarrow{\scriptstyle \omega} & \\ Y & & \end{array}$$

be the diagram with the corresponding quotient maps. By Lemma 2.4, ω is proximal. This completes the proof of the theorem. □

We call the maximal equicontinuous factor of a flow the *Kronecker factor*.

PROPOSITION 2.5.

(1) *The Kronecker factor Y, of a minimal flow (X, T) is B-natural, for every B-natural family of factors.*

(2) *The Kronecker factor is the least member of the smallest B-natural family of factors of X.*

PROOF. (1) By Theorem 2.1 we have

$$\widetilde{Y}\xrightarrow{\kappa}Z\xrightarrow{\omega}Y,$$

where κ is a K–extension ($K \subset \mathrm{Aut}(\widetilde{Y})$ is compact) and ω is proximal. Put $\pi = \omega \circ \kappa$.

Let $S : Y \longrightarrow Y$ be an automorphism of Y and let $\tilde{S} : \tilde{Y} \longrightarrow \tilde{Y}$ be a lift of S to an automorphism of $\tilde{Y}$. Then for $\tilde{y} \in \tilde{Y}$ and $k \in K$, $(\tilde{S}\tilde{y}, \tilde{S}(\tilde{y}k))$ is an almost periodic point in $\tilde{Y} \times \tilde{Y}$, so that also $(k \times k)(\tilde{S}\tilde{y}, \tilde{S}(\tilde{y}k))$ is an almost periodic point of $Z \times Z$. On the other hand,

$$\omega(\kappa(\tilde{S}(\tilde{y}k))) = \pi(\tilde{S}(\tilde{y}k)) = S\pi(\tilde{y}k) = S\pi(\tilde{y}) = \pi(\tilde{S}\tilde{y}) = \omega(\kappa(\tilde{S}\tilde{y})),$$

and it follows that $\kappa(\tilde{S}(\tilde{y}k))$ and $\kappa(\tilde{S}\tilde{y})$ are also proximal points in Z. Therefore $\kappa(\tilde{S}(\tilde{y}k)) = \kappa(\tilde{S}\tilde{y})$ and we can define $\hat{S} : Z \longrightarrow Z$ unambiguously by $\hat{S}(\kappa\tilde{y}) = \kappa(\tilde{S}\tilde{y})$. Since ω is a proximal extension, it follows that over every minimal set in $Y \times Y$ there exists a unique minimal set in $Z \times Z$, so that our observation above yields an isomorphism of $\mathrm{Aut}(Y)$ onto $\mathrm{Aut}(Z)$. By Lemma 3 of [**Gl3**], this is a topological isomorphism and since $\mathrm{Aut}(Y) = Y$ is a compact group, so is $\mathrm{Aut}(Z)$. We find that $Z/\mathrm{Aut}(Z)$ is a factor of Z which is necessarily a proximal flow. Finally, since a proximal minimal flow is a one point flow, we have $\mathrm{Aut}(Z)$ acting transitively on Z and conclude that ω is 1-1.

Therefore π is a K–extension and we can finish the proof as follows. Take $\widetilde{y}_1, \widetilde{y}_2 \in \widetilde{Y}$ and put $\pi(\widetilde{y}_i) = y_i$. We know that $Sy_1 = y_2$ for some $S \in \mathrm{Aut}(Y)$. Let $\widetilde{S} \in \mathrm{Aut}(\widetilde{Y})$ with $\pi \circ \widetilde{S} = S \circ \pi$. Then $\pi(\widetilde{y}_2) = Sy_1 = \pi(\widetilde{S}\widetilde{y}_1)$ and there exists $k \in K$ such that $(\widetilde{S}\widetilde{y}_1)k = \widetilde{y}_2$. Therefore the group $\mathrm{Aut}(\widetilde{Y})$ acts transitively on $\widetilde{Y}$ and by Theorem 1.1, $\widetilde{Y}$ is an equicontinuous factor. Thus $Y = \widetilde{Y}$, which finishes the proof of (1).

(2) Since a B-set in a Kronecker system is necessarily minimal, we see that the proof of Lemma 2.3 implies that the family of factor relations $\{R : R \subset K\}$ is B-natural. Now conclude as in the proof of Proposition 2.4 (2). □

EXAMPLE 1. Take (X, T) to be the minimal Chacon transformation. Then (X, T) is a regular flow and also a prime flow (i.e. has no nontrivial factors), and it follows that the smallest natural family of factors for (X, T) is $\mathcal{N} = \{X\}$. In contrast, the smallest B-natural family of factors (as well as the smallest measure theoretical one) is $\mathcal{N}_B = \{X, \mathrm{point}\}$. If we take $Y = X \times Z$, where Z is an irrational rotation, then we still have $\mathcal{N} = \{Y\}$, but $\mathcal{N}_B = \{Y, Z\}$. In general, the Kronecker factor is the least member of the smallest natural family of factors for (X, T) iff it is also the largest regular factor of X. Comparing Propositions 2.4 and and recalling that for PI-flows natural = B-natural, we conclude that for PI-flows, and in particular for distal flows, the Kronecker factor and the largest regular factor coincide.

EXAMPLE 2. As established in [**JLM**], in the measure theoretical case the map $\widetilde{Y} \longrightarrow Y$ is always a group extension. A direct corollary of this is that the Pinsker factor (i.e. the largest zero-entropy factor) is always natural. It turns out that the minimal flow constructed in [**BGK**] serves also as an example of a regular, weakly mixing flow (X,T) of positive entropy with natural and B-natural families $\{X\}$ and $\{X,\ \text{point}\}$ respectively, with the property that its maximal zero-entropy factor $X \longrightarrow Z$ is neither a natural nor a B-natural factor. We omit the details.

PROBLEM 1. Find a nontrivial example of a minimal flow (X,T) all of whose factors belong to the smallest family of natural factors. Also a B-flow all of whose factors belong to the smallest family of B-natural factors. In particular is this the case for the shift on $\{0,1\}^Z$?

3. Group Extensions of Minimal Rotations

In this section we describe an example of a natural family for a group extension of minimal rotation. The proof we present here involves, in some sense, only the basic methods. In section 4 we will obtain the same result as a corollary of the more abstract theory of Ellis groups.

First we show how to reduce an arbitrary group extension of minimal rotation to a minimal flow (Section 3.1). Then we describe minimal subsets in the Cartesian square of such a flow (Section 3.2) and using this we construct a natural family of factors (Section 3.3).

3.1. Minimal Subsets of Group Extensions of Minimal Rotations.

In what follows we let (X,T) be a minimal rotation of a compact monothetic group. We will study properties of a group extension $(\overline{X},\overline{T})$ (not necessarily minimal) of (X,T).

Suppose that $(\overline{X},\overline{T})$ is a G–extension of (X,T). Recall that G is a compact (in the topology of uniform convergence) subgroup of $\mathrm{Aut}(\overline{X},\overline{T})$. Let $M \subset \overline{X}$ be $\overline{T}$–minimal. Denote by $\pi : \overline{X} \longrightarrow X$ the factor homomorphism given by $\pi(\overline{x}) = [\overline{x}]_G$. Because T is minimal, $\pi(M) = X$. Put

$$H = \{g \in G:\ g(M) = M\}.$$

Clearly H is a closed subgroup of G. Observe that if $g \in G$, then $g(M)$ is minimal, hence either $g(M) = M$ or $g(M) \cap M = \emptyset$. Therefore

$$g \in H \text{ iff there exists } \overline{x} \in M \text{ such that } g(\overline{x}) \in M. \tag{7}$$

For $\overline{x} \in \overline{X}$ put

$$M_{\overline{x}} = \{g \in G:\ g^{-1}(\overline{x}) \in M\}.$$

$M_{\overline{x}}$ is not empty because $\pi(M) = X$, i.e. each $[\overline{x}]_G$ contains at least one element of M ($GM = \overline{X}$ and each minimal subset of $\overline{X}$ is of the form $g(M)$). We also have

LEMMA 3.1. *For each $\overline{x} \in \overline{X}$ there exists a $g = g_{\overline{x}} \in G$ such that*

$$M_{\overline{x}} = gH.$$

PROOF. Fix $\overline{x} \in \overline{X}$. Take $g \in M_{\overline{x}}$. We will show that $M_{\overline{x}} = gH$. Assume that $g' \in M_{\overline{x}}$. Then ${g'}^{-1}(\overline{x}) \in M$ and $g^{-1}(\overline{x}) \in M$. Thus $g^{-1}g'({g'}^{-1}(\overline{x})) = g^{-1}(\overline{x}) \in M$. This and (7) imply that $g^{-1}g' \in H$ and therefore $g' \in gH$.

On the other hand if $h \in H$ then it is clear that $(gh)^{-1}(\overline{x}) = h^{-1}g^{-1}(\overline{x}) \in M$. Thus $gh \in M_{\overline{x}}$. □

By Lemma 3.1, there exists a map $\beta : \overline{X} \longrightarrow G/H$ given by

$$\beta(\overline{x}) = M_{\overline{x}} = g_{\overline{x}}H. \tag{8}$$

Here G/H is understood as the homogeneous space of right cosets since H is not necessarily normal. In what follows we will consider the quotient topology on G/H. The map β has the following obvious properties:

(9) $$\beta(g(\overline{x})) = g\beta(\overline{x}), \quad g \in G,$$

(10) $$\beta^{-1}(H) = M,$$

(11) $$\beta^{-1}(gH) = g(\beta^{-1}(H)), g \in G,$$

(12) β is constant on each minimal subset of $\overline{X}$.

Let $\mathcal{M}(\overline{X})$ denote the set of all minimal subsets of $\overline{X}$. Let $2^{\overline{X}}$ be the space of non-empty, closed subset of $\overline{X}$ with the topology induced by the Hausdorff metric. (Recall that if (Z,d) is a metric space then we define the Hausdorff metric d' in 2^Z by the formula $d'(A,B) = \max\left(\sup_{x\in B}\{d(A,x)\}, \sup_{y\in A}\{d(B,y)\}\right)$. Recall also that if (Z,d) is compact then so is $(2^Z, d')$.)

LEMMA 3.2. *If $g_n \longrightarrow g$, then $g_n M \longrightarrow gM$ (in $2^{\overline{X}}$).*

PROOF. Suppose that $g_n \longrightarrow g$. Take a sequence $\{m_n\}$ of elements from M. Choosing a subsequence we may assume that $m_n \longrightarrow m$, for some $m \in M$. Then $g_n m_n \longrightarrow gm$, which implies that $g_n M \longrightarrow gM$. □

From this lemma it follows that $\mathcal{M}(\overline{X})$ is closed in $2^{\overline{X}}$ and on $\mathcal{M}(\overline{X})$ we will consider the topology induced from $2^{\overline{X}}$.

LEMMA 3.3. *The map $\beta : \mathcal{M}(\overline{X}) \longrightarrow G/H$ is a continuous bijection.*

PROOF. By (12), β is well defined and by (12) and (9), it is 1–1. It follows from (11) and (10)that β is onto.

Now suppose that $g_n M \longrightarrow gM$ and $\beta(g_n M) = g_n H \longrightarrow g'H$. Choosing a subsequence we can assume that $g_n \longrightarrow g'h$ for some $h \in H$. By Lemma 3.2 we have that $g_n M \longrightarrow g'hM = g'M$, then $gM = g'M$. Therefore $\beta(g_n M) \longrightarrow g'H = \beta(g'M) = \beta(gM)$. □

Define a metric $\widetilde{d}$ on $\mathcal{M}(\overline{X})$ setting

$$\widetilde{d}(gM, g'M) = d(\beta(gM), \beta(g'M)) = d(gH, g'H),$$

where d is the quotient metric on G/H induced by an invariant metric on G. As a corollary of the above consideration we have the following.

PROPOSITION 3.1. *The metric $\widetilde{d}$ is equivalent to the Hausdorff metric on $\mathcal{M}(\overline{X})$ and $\beta : (\mathcal{M}(\overline{X}), \widetilde{d}) \longrightarrow (G/H, d)$ is an isometry.*

THEOREM 3.1. *Set $\widehat{T} = \overline{T}\mid_M$. Note that $H \subseteq \mathrm{Aut}(\widehat{T})$. Then $\widehat{T}$ is an H-extension of T.*

PROOF. It is enough to show that $\overline{T}_G$ is isomorphic to $\widehat{T}_H$. Let $J : M_H \longrightarrow \overline{X}_G$ be given by the formula $J([\overline{x}]_H) = [\overline{x}]_G$. First we will show that J is one-to-one. If $\overline{x}, \overline{y} \in M$ and $[\overline{x}]_G = [\overline{y}]_G$, then $\overline{y} = g(\overline{x})$ for some $g \in G$. This implies $g \in H$ (use (7)) and consequently $[\overline{x}]_H = [\overline{y}]_H$.

Now take $[\overline{y}]_G \in \overline{X}_G$. Because $\pi(M) = X$, there is an $\overline{x} \in M$ such that $[\overline{x}]_G = [\overline{y}]_G$. This implies that J is a bijection.

The continuity of J is obvious because $[\overline{x}]_H \subset [\overline{x}]_G$, $\overline{x} \in M$. Since M_H is compact, J is a homeomorphism.

Next,

$$J \circ \widehat{T}_H([\overline{x}]_H) = J([T\overline{x}]_H) = [T\overline{x}]_G,$$
$$\overline{T}_G \circ J([\overline{x}]_H) = \overline{T}_G([\overline{x}]_G) = [T\overline{x}]_G.$$

Thus $\widehat{T}_H$ is isomorphic to $\overline{T}_G$ and the proof is complete. □

3.2. Minimal Subsets of Cartesian Squares of Minimal Group Extensions of Rotations.

Assume that $(\overline{X}, \overline{T})$ is a minimal G–extension of a rotation (X,T). Consider $(\overline{X} \times \overline{X}, \overline{T} \times \overline{T})$. Take $M \in \mathcal{M}_2(\overline{X}, \overline{T})$. Clearly the projection of M onto the first and onto the second coordinate is just $\overline{X}$. Besides $(\pi \times \pi)(M) = \text{graph}(S)$ for some $S \in \text{Aut}(X)$. Obviously $(\text{graph}(S), T \times T)$ is isomorphic to (X,T). Observe that if $\overline{W} = (\pi \times \pi)^{-1}(\text{graph}(S))$ and if $\tau : \overline{W} \longrightarrow X$ is given by $\tau(\overline{x}, \overline{y}) = \pi(\overline{x})$, then τ is continuous, $\tau \circ (\overline{T} \times \overline{T}) = T \circ \tau$ and $\tau^{-1}(x)$ is a $G \times G$–orbit of an $(\overline{x}, \overline{y}) \in \tau^{-1}(x)$. Thus $(\overline{W}, \overline{T} \times \overline{T})$ is a free $G \times G$–extension of (X,T) and we can apply results of Section 3.1.

Put $H = \{(g,g') \in G \times G : (g,g')(M) = M\}$. By results of Section 3.1, $\mathcal{M}(\overline{W})$ with the topology induced by the Hausdorff metric is homeomorphic to $(G \times G)/H$. Now, we will examine the structure of the group H.

Denote by $\pi_i : G \times G \longrightarrow G$ the projection $\pi_i(g_1, g_2) = g_i$, $i = 1,2$. Then

$$\pi_i(H) = G, \quad i = 1,2. \tag{13}$$

Indeed, fix $g \in G$ and $(\overline{x}, \overline{x}') \in M$. Put $x = \pi(\overline{x})$, $x' = \pi(\overline{x}')$. There exists $\overline{x}'' \in \overline{X}$ such that $(g\overline{x}, \overline{x}'') \in M$. Since $\pi(g\overline{x}) = \pi(\overline{x}) = x$, $\pi(\overline{x}'') = x'$. Thus $\overline{x}'' = g'\overline{x}'$ for some $g' \in G$ and we have $(g\overline{x}, g'\overline{x}') = (g,g')(\overline{x}, \overline{x}') \in M$. This implies $(g,g') \in H$, hence $\pi_1(H) = G$. Similarly, $\pi_2(H) = G$.

Put

$$H_1 = \{g \in G : (g,e) \in H\}, \qquad H_2 = \{g \in G : (e,g) \in H\}.$$

Fix $h_1 \in H_1$. Given $g \in G$, by (13), we can find $j \in G$ so that $(g,j) \in H$, hence $(gh_1, j) \in H$. Since $(g^{-1}, j^{-1}) \in H$, $(gh_1g^{-1}, e) \in H$. Consequently,

$$H_1 \text{ and } H_2 \text{ are normal closed subgroups of } G. \tag{14}$$

Besides we have

$$\text{if } (g_1,g_2), (g_1',g_2) \in H \text{ then } g_1'g_1^{-1} \in H_1 \tag{15}$$

and

$$\text{if } (g_1,g_2), (g_1,g_2') \in H \text{ then } g_2'g_2^{-1} \in H_2; \tag{16}$$

besides

(17) $$(g_1, g_2) \in H \text{ iff } g_1H_1 \times g_2H_2 \subset H.$$

Define ξ by

$$\xi(gH_1) = \pi_2((gH_1 \times G) \cap H).$$

Now, (14)–(17) imply directly that

LEMMA 3.4. *The map ξ establishes an isomorphism of topological groups between G/H_1 and G/H_2 and moreover*

$$H = \bigcup_{g \in G} gH_1 \times \xi(gH_1).$$

Let us define a map $\overline{S} : \overline{X}_{H_1} \longrightarrow 2^{\overline{X}}$ by

$$\overline{S}([\overline{x}]_{H_1}) = \Pi_2(([\overline{x}]_{H_1} \times \overline{X}) \cap M),$$

where $\Pi_2 : \overline{X} \times \overline{X} \longrightarrow \overline{X}$ denotes the projection onto the second coordinate.

PROPOSITION 3.2.

$$M = \bigcup_{\overline{x} \in \overline{X}} [\overline{x}]_{H_1} \times \overline{S}([\overline{x}]_{H_1}),$$

moreover $\overline{S}$ establishes an isomorphism between $\overline{X}_{H_1}$ and $\overline{X}_{H_2}$.

PROOF. First, we show that $\overline{S}$ is a map between $\overline{X}_{H_1}$ and $\overline{X}_{H_2}$. Take $(\overline{x}, \overline{y}) \in ([\overline{x}]_{H_1} \times \overline{X}) \cap M$. We show that $\overline{S}([\overline{x}]_{H_1}) = [\overline{y}]_{H_2}$. To this end take another $\overline{y}'$ with $(\overline{x}, \overline{y}') \in ([\overline{x}]_{H_1} \times \overline{X}) \cap M$. Obviously

$$(\pi \times \pi)(\overline{x}, \overline{y}) = (\pi \times \pi)(\overline{x}, \overline{y}') = [\overline{x}]_G \times S([\overline{x}]_G),$$

where $\text{graph}(S) = (\pi \times \pi)(M)$. From the proof of Theorem 3.1 we deduce that there exists $(g_1, g_2) \in H$ such that $(g_1, g_2)(\overline{x}, \overline{y}) = (\overline{x}, \overline{y}')$. Thus $g_1 = e$ and $g_2 \in H_2$. This forces $\overline{S}([\overline{x}]_{H_1}) \subset [\overline{y}]_{H_2}$. The opposite inclusion is obvious. Then $\overline{S}$ is obviously a bijection which commutes with $\overline{T}$ (considered on $\overline{X}_{H_1}$ and $\overline{X}_{H_2}$ respectively). Since M is compact and $\text{graph}(\overline{S})$ is equal to the natural projection of M to $\overline{X}_{H_1} \times \overline{X}_{H_2}$, $\overline{S}$ is continuous, hence it is an isomorphism of flows. □

3.3. Natural Family of Factors for Minimal Group Extensions of Rotations.

Now we are in a position to construct a natural family of factors for $(\overline{X}, \overline{T})$. Put

$$\mathcal{N} = \{R_F \subset \overline{X} \times \overline{X} : F \text{ is closed normal subgroup of } G\}.$$

PROPOSITION 3.3. *The family $\mathcal{N}$ is a natural family of factor relations for $(\overline{X}, \overline{T})$.*

REMARK 4. To prove this proposition we use the results of Section 3.1 and 3.2 and apply Corollary 2.2. The only property of a minimal rotation that we use in Section 3.1 and 3.2 is regularity. However, if we would like to apply Corollary 2.2 ((X, T) is distal) the assumption that (X, T) is minimal forces (X, T), by Theorem 1.1 to be a minimal rotation.

PROOF. First, $\Delta_{\overline{X}} = R_{\{e\}} \in \mathcal{N}$. Then notice that if $R_{F_i} \in \mathcal{N}$, $i \in I$, then

$$\bigvee_{i \in I} R_{F_i} = R_F,$$

where F is the closed subgroup generated by the union of the groups $F_i, i \in I$. Since F is normal, the condition (b) of the definition of natural family is satisfied.

Now take $M \in \mathcal{M}_2(\overline{X}, \overline{T})$. From Proposition 3.2 we get that $P_i(M) = R_{H_i}$, where H_i are defined in Section 3.2 for M. Thus $P_i(M) \in \mathcal{N}$.

Now take $R_{F_i} \in \mathcal{N}$, $i = 1, 2$, an isomorphism $\overline{S} : \overline{X}_{R_{F_1}} \longrightarrow \overline{X}_{R_{F_2}}$ and $R_{F_1} \subset R_F \in \mathcal{N}$. Assume that $M \in \mathcal{M}_2(\overline{X}, \overline{T})$, $M = \text{graph}(S)$. Lemma 3.4 yields a topological group isomorphism $\xi : G/F_1 \longrightarrow G/F_2$. Now consider $\overline{S}$ acting on the equivalence classes of R_F. By the form of S, $(S \times S)(R_F) = R_{\xi(F/F_1)} \in \mathcal{N}$, and since $\xi(F/F_1)$ is a normal subgroup of G/F_2, the condition (d) is also satisfied. □

COROLLARY 3.1. *For any factor $(\overline{X}_R, \overline{T}_R)$ of $(\overline{X}, \overline{T})$ there exists a largest closed normal subgroup F of G such that*

$$\pi : (\overline{X}_F, \overline{T}_F) \longrightarrow (\overline{X}_R, \overline{T}_R),$$

(where π denotes the corresponding homomorphism) is a group extension.

PROOF. The group F is defined by taking the $\mathcal{N}$-core R_F of R. By Corollary 2.2, π is a group extension. □

It is not difficult to see that π is a Γ_π-extension, where $\Gamma_\pi = \{S \in \text{Aut}(X) : \pi \circ S = \pi\}$.

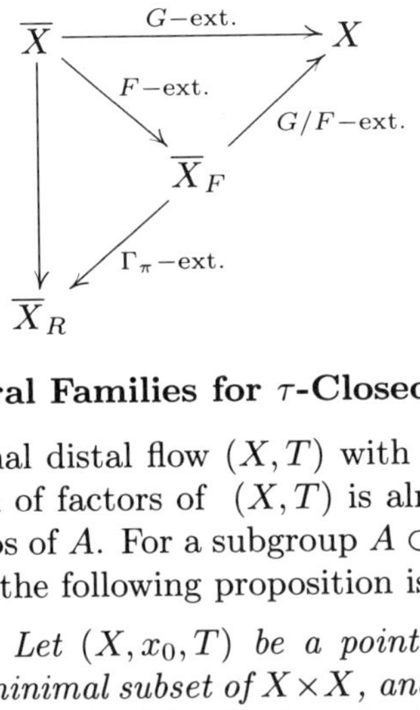

4. Natural Families for τ-Closed Groups

Recall that for a minimal distal flow (X, T) with Ellis group A, all the information about the collection of factors of (X, T) is already present at the level of Ellis τ-closed ambient groups of A. For a subgroup $A \subset G$ and $\gamma \in G$ let us denote $A^\gamma = \gamma A \gamma^{-1}$. The proof of the following proposition is an easy exercise.

PROPOSITION 4.1. *Let (X, x_0, T) be a pointed minimal distal flow with Ellis group A. Let M be a minimal subset of $X \times X$, and let $(x_0, \gamma x_0) \in M$ for some $\gamma \in G$. Then the factors Y_i defined by the invariant closed equivalence relations $R_i = P_i(M), i = 1, 2$, are the factors corresponding to the τ-closed subgroups $B_1 = A \vee A^\gamma$ and $B_2 = A \vee A^{\gamma^{-1}}$ respectively, and the map $Y_1 \ni [gx_0]_{R_1} \mapsto S([gx_0]_{R_1}) = [g\gamma x_0]_{R_2} \in Y_2$ is the isomorphism induced by M. Moreover, if $\pi_i : (X, x_0) \to (Y_i, y_i)$, $i = 1, 2$, are two pointed factors with $\mathcal{G}(Y_i, y_i) = B_i$, then M induces an isomorphism $S : Y_1 \to Y_2$ iff $B_1 = (B_2)^\gamma$.*

Motivated by this proposition we now define a natural family for a τ-closed group A.

DEFINITION 2. Let A be a τ-closed subgroup of G. A collection $\mathcal{N}$ of τ-closed subgroups $F \subset G$ will be called a *family* for A if it satisfies the following conditions:

(a$'$) $A \subseteq F$, for all $F \in \mathcal{N}$;

(b$'$) $\mathcal{N}$ contains A and is closed under $\bigvee$.

It is called a *natural family* for A if in addition it satisfies

(c$'$) $\forall_{\gamma \in G}\ A_{\uparrow\gamma} = A \vee A^{\gamma} \in \mathcal{N}$;

(d$'$) $\forall_{\gamma \in G} \forall_{F \in \mathcal{N}}\ F^{\gamma} \supset A \Longrightarrow F^{\gamma} \in \mathcal{N}$.

Given $\mathcal{N}$, a family for A and a τ-closed group $B \supset A$ put

$$\widetilde{B} = \bigvee\{F \in \mathcal{N} :\ F \subset B\},$$

the *$\mathcal{N}$-core* of B.

PROPOSITION 4.2. *A family $\mathcal{N}$ of τ-closed subgroups for A is a natural family for A if and only if:*

(e$'$) *for any τ-closed group $B \supset A$ and $\gamma \in G$, $B^{\gamma} \supset A$ implies $A \subset (\widetilde{B})^{\gamma} = \widetilde{B^{\gamma}}$.*

PROOF. First we show that a natural family $\mathcal{N}$ satisfies (e$'$). Assume B and B^{γ} contain A. Then $B^{\gamma} \supset A \vee A^{\gamma}$. By property (c$'$) of natural families, $A \vee A^{\gamma} \in \mathcal{N}$ and it follows that $\widetilde{B^{\gamma}} \supset A \vee A^{\gamma}$. Hence $(\widetilde{B^{\gamma}})^{\gamma^{-1}} \supset A$ and by property (d$'$), $(\widetilde{B^{\gamma}})^{\gamma^{-1}} \in \mathcal{N}$. Now $(\widetilde{B^{\gamma}})^{\gamma^{-1}} \subset (B^{\gamma})^{\gamma^{-1}} = B$, whence $(\widetilde{B^{\gamma}})^{\gamma^{-1}} \subset \widetilde{B}$ and $\widetilde{B^{\gamma}} \subset (\widetilde{B})^{\gamma}$. In particular, $A \subset (\widetilde{B})^{\gamma}$ and again by (d$'$), $(\widetilde{B})^{\gamma} \in \mathcal{N}$. Finally $B^{\gamma} \supset (\widetilde{B})^{\gamma}$ and $(\widetilde{B})^{\gamma} \in \mathcal{N}$ imply $\widetilde{B^{\gamma}} \supset (\widetilde{B})^{\gamma}$, so that $\widetilde{B^{\gamma}} = (\widetilde{B})^{\gamma}$.

For the other direction, to show (c$'$), put $B = A \vee A^{\gamma}$, then $B^{\gamma^{-1}} \supset A$ and (e$'$) implies $(A^{\gamma^{-1}} \vee A)^{\sim} = (B^{\gamma^{-1}})^{\sim} = (\widetilde{B})^{\gamma^{-1}}$. In particular $(\widetilde{B})^{\gamma^{-1}} \supset A$ hence $\widetilde{B} \supset B$ and it follows that $\widetilde{B} = B$, whence $B = A \vee A^{\gamma} \in \mathcal{N}$. Now for (d$'$), given $F \in \mathcal{N}$ and $\gamma \in G$ with $F^{\gamma} \supset A$, we have by (e$'$), $(F^{\gamma})^{\sim} = (\widetilde{F})^{\gamma} = F^{\gamma}$ hence, $F^{\gamma} \in \mathcal{N}$. □

PROPOSITION 4.3. *Let $\mathcal{N}$ be a natural family for the τ-closed group A. Then for $B \supset A$ a τ-closed group, the $\mathcal{N}$-core $\widetilde{B}$ of B is a normal subgroup of B: $\widetilde{B} \lhd B$.*

PROOF. Let $\gamma \in B$. Then $A \subset B^{\gamma} = B$ and this implies $A \subset (\widetilde{B})^{\gamma} = \widetilde{B^{\gamma}} = \widetilde{B}$. □

Since clearly the intersection of any collection of natural families for A is again a natural family, it follows that there exists a unique smallest natural family for A. It is not hard to see that in fact this smallest family is the following one:

$$\mathcal{N} = \{\bigvee_{\gamma \in \Gamma} A^{\gamma} :\ u \in \Gamma \subset G\}.$$

EXAMPLE 3. Let $(\overline{X}, \overline{T})$ be a minimal metric distal flow with the Ellis group A. Assume that $A \lhd F$, where $F = AG'$, so that F is the Ellis group of the Kronecker factor X of $\overline{X}$. Morever, assume that $F' \subset A$, so that $\overline{X}$ is a group extension of X with compact fiber group $K = F/A$. Given $\gamma \in G$ (this defines a minimal set $M = \bar{o}(\overline{x}_0, \gamma\overline{x}_0) \subset \overline{X} \times \overline{X}$), we have that $B = A \vee A^{\gamma^{-1}} \subset F$, is an element of $\mathcal{N}$ and $B \lhd F$, $F' \subset B$. Since $A \lhd F$ and $B \subset F$, $A \lhd B$.

Moreover $B' \subset A$. Thus the natural factor Y corresponding to B is a group extension of X and is given as the quotient of $\overline{X}$ by the compact normal subgroup $L = B/A \lhd F/A = K$. Since likewise for every $B = \bigvee_\Gamma A^\gamma$, $A \lhd B \lhd F$, we see that the smallest natural family of factors for $\overline{X}$ consists of group extensions $Y = \overline{X}/L \to X$ with B as above and $L = B/A \lhd F/A = K$. We thus obtained an alternative proof of Proposition 3.3.

5. Normal Extensions

DEFINITION 3. An extension $(X,T)\xrightarrow{\pi}(Y,T)$ of minimal flows will be called *normal*, if for some $x_0 \in X$, $A = \mathcal{G}(X, x_0) \lhd \mathcal{G}(Y, \pi(x_0)) = F$. The flow (X,T) is *normal* if $A \lhd G$.

DEFINITION 4. Let $\pi : (X,T) \longrightarrow (Y,T)$ be an extension of minimal flows. We say that π is *a virtually weakly mixing extension* if for some $x_0 \in X$ with $ux_0 = x_0$, $F'A = F$, where $A = \mathcal{G}(X, x_0)$ and $F = \mathcal{G}(Y, \pi(x_0))$.

The two above definitions do not depend on the choice of x_0. Let

$$\begin{array}{ccc} X & \xleftarrow{\theta^*} & X^* \\ {\scriptstyle\pi}\downarrow & & \downarrow{\scriptstyle\pi^*} \\ Y & \xleftarrow{\theta} & Y^* \end{array}$$

be the associated RIC shadow diagram, then in terms of this diagram, π is a virtually weakly mixing extension iff π^* is a weakly mixing extension in the usual sense, iff no nontrivial intermediate isometric extension exists between X^* and Y^*.

THEOREM 5.1. *Let $(X,T)\xrightarrow{\pi}(Y,T)$ be a normal RIC extension of metric minimal flows; $A = \mathcal{G}(X) \lhd \mathcal{G}(Y) = F$. Then there exists a commutative diagram:*

$$\begin{array}{ccc} X & \xrightarrow{\sigma} & Z \\ {\scriptstyle\pi}\downarrow & \swarrow{\scriptstyle\rho} & \\ Y & & \end{array}$$

where ρ is a group extension and σ a virtually weakly mixing extension. In terms of Ellis groups;

$$B = \mathcal{G}(Z) = AF' = AF^\infty, \quad AB' = B \lhd F.$$

In particular a minimal flow (X,T) whose Ellis group is normal is a virtually weakly mixing extension of its maximal equicontinuous factor.

PROOF. Let

$$Y = Y_0 \xleftarrow{\rho_1} Z_1 \xleftarrow{\theta_1} Y_1 \xleftarrow{\rho_2} Z_2$$

be the first three stages of the bottom row of the PI-tower associated with π. Thus

$$\mathcal{G}(Z_1) = AF' = \mathcal{G}(Y_1), \quad \mathcal{G}(Z_2) = AF''.$$

Denote by ϕ the map from $Z_2 \longrightarrow Y$, and by ψ the map from $Y_1 \longrightarrow Y$. As in [**Gl4**], we can assume, by passing to some larger proximal extensions, that ϕ is RIC; in particular $\overline{JR_\phi} = R_\phi$. By Lemma 2.4 in [**Gl2**], for every non-empty relatively open $V \subset \overline{JR_\phi} = R_\phi$ with $(\rho_2 \times \rho_2)(V) = R_\psi$ we have $\overline{o}(V) = R_\phi$. By Lemma 2.1 and Lemma 2.2 in [**Gl2**], the collection

$$\mathcal{V} = \{V \subset R_\phi : V \text{open and } (\rho_2 \times \rho_2)(V) = R_\psi\}$$

has a countable subbasis $\{V_i\}_{i=1}^\infty$. Let $S = \bigcap_{i=1}^\infty o(V_i)$, a dense G_δ subset of R_ϕ. Now choose some $(x, x') \in S$, put $L = \overline{o}(x, x')$ and $V = R_\phi \backslash L$. If ρ_1 is trivial, then π is a weakly mixing extension and we are done. Otherwise ρ_1 is nontrivial, and it is then easy to see that $V \neq \emptyset$. Suppose $V \in \mathcal{V}$, then there exists an i with $V_i \subset V$ and since $(x, x') \in S$, there exists $n \in Z$ such that $(T \times T)^n(x, x') \in V_i \subset V$. This contradicts the definition of V and we conclude that $V \notin \mathcal{V}$. Thus there exists $(y_1, y_2) \in R_\psi$ with $(y_1, y_2) \notin (\rho_2 \times \rho_2)(V)$; i.e. $\rho_2^{-1}(y_1) \times \rho_2^{-1}(y_2) \subset L$.

Let β be an arbitrary element of B, and write $(y_1, y_2) = (ay_0, by_0)$, $(x, x') = (pz_0, qz_0)$, for some $a, b, p, q \in I$, where y_0 is the base point in Y_1 and z_0 the base point in Z_2. Since

$$\rho_2(az_0) = a\rho_2(z_0) = ay_0 = y_1, \ \rho_2(b\beta z_0) = b\beta\rho_2(z_0) = b\beta y_0 = by_0 = y_2,$$

there exist $r, s \in I$ with

$$r(pz_0, qz_0) = (az_0, b\beta z_0), \ s(pz_0, qz_0) = (az_0, bz_0).$$

Multiplying on the left with u throughout these equations, we can assume that p, q, r, s are elements of G and denoting $a^{-1}b = \gamma$, $a^{-1}rs^{-1}a = \delta$, we have

$$\delta(z_0, \gamma z_0) = (z_0, \gamma\beta z_0).$$

Thus $\delta z_0 = z_0$ hence $\delta \in AF'' = \mathcal{G}(Z_2)$, and $\gamma\beta z_0 = \delta\gamma z_0$ hence $\gamma^{-1}\delta^{-1}\gamma\beta \in AF''$. However by assumption $A \lhd F$ and as $(az_0, bz_0) \in R_\phi$, $\gamma = a^{-1}b \in F$, and we conclude that also $\beta \in AF''$. Since β was an arbitrary element of $B = AF'$, we now have $AF' = AF''$. This implies that ρ_2 is an isomorphism and moreover $AF' = AF^\infty$ so that the PI-tower for π collapses to the diagram:

$$\begin{array}{ccccc}
X = X_0 & & \longleftarrow & & X_1 = X_\infty \\
\downarrow{\scriptstyle \pi_0} & \overset{\sigma_1}{\searrow} & & & \downarrow{\scriptstyle \pi_1} \\
Y = Y_0 & \xleftarrow{\rho_1} & Z_1 & \xleftarrow{\theta_1} & Y_1 = Y_\infty
\end{array}$$

Here π_1 is the RIC shadow of the map $\sigma : X \to Z_1$ and is weakly mixing, i.e. σ is virtually weakly mixing. The proof is completed. $\square$

6. Seminormal Flows

DEFINITION 5. A τ-closed group A will be called *seminormal*, if for every $\gamma \in G$

$$A_{\downarrow\gamma} A' = A.$$

Here $A_{\downarrow\gamma} = A \cap A^\gamma$. Clearly every normal subgroup of G is seminormal.

PROPOSITION 6.1. *For a seminormal τ-closed group A, the collection of τ-closed groups*

$$\mathcal{N} = \{F \supset A : F'A = F\}$$

is a natural family for A.

PROOF. We verify that conditions (b') and (c') and (d') of Definition 2 are satisfied.

(b') Let $\{F_i\}_{i\in I} \subset \mathcal{N}$, then for $F = \bigvee_{i\in I} F_i$,

$$F'A \supset F_i'A = F_i \Rightarrow F'A \supset F.$$

Thus $F'A = F$ and $F \in \mathcal{N}$.

(c') For $\gamma \in G$ we have by the seminormal property:

$$(A_{\uparrow\gamma})'A_{\downarrow\gamma} \supset A'A_{\downarrow\gamma} = A,$$
$$(A_{\uparrow\gamma})'A_{\downarrow\gamma} \supset (A^\gamma)'A_{\downarrow\gamma} = (A'A_{\downarrow\gamma^{-1}})^\gamma = A^\gamma.$$

Hence $(A_{\uparrow\gamma})'A_{\downarrow\gamma} = A_{\uparrow\gamma}$ and a fortiori $(A_{\uparrow\gamma})'A = A_{\uparrow\gamma}$.

(d') Let $F \in \mathcal{N}$ satisfy $F^\gamma \supset A$. We have to show that F^γ is in $\mathcal{N}$, i.e. $(F^\gamma)'A = F^\gamma$. Now by the seminormal property of A, $A = A'A_{\downarrow\gamma}$, hence

$$\begin{aligned}(F^\gamma)'A &= (F^\gamma)'A'A_{\downarrow\gamma} = (F^\gamma)'A_{\downarrow\gamma} = [F'(A^{\gamma^{-1}} \cap A)]^\gamma \\ &= [F'A_{\downarrow\gamma^{-1}}]^\gamma \supset [F^\infty A_{\downarrow\gamma^{-1}}]^\gamma = [F^\infty A^\infty A_{\downarrow\gamma^{-1}}]^\gamma = [F^\infty A]^\gamma = F^\gamma,\end{aligned}$$

where, in the last line, we use Lemma X.4.1 of [**Gl1**] to get

$$A = A'A_{\downarrow\gamma^{-1}} = A^\infty A_{\downarrow\gamma^{-1}}$$

and

$$F = F'A = F^\infty A.$$

□

PROPOSITION 6.2. *Let A be a seminormal group and let B be any τ-closed group containing A. Then $\widetilde{B} = B^\infty A$.*

PROOF. Since $B^\infty \lhd B$, from the proof of Lemma X.4.1 in [**Gl1**] it follows that $(B^\infty A)'A = (B^\infty)'A = B^\infty A$, so $B^\infty A \in \mathcal{N}$, whence $B^\infty A \subset \widetilde{B}$.

On the other hand we have $(\widetilde{B})'A = \widetilde{B}$, hence $\widetilde{B} = (\widetilde{B})^\infty A$. Therefore $\widetilde{B} \subset B^\infty A$. □

DEFINITION 6. Call a minimal flow (X, T) *seminormal*, if for some $x_0 \in X$ with $ux_0 = x_0$, (hence for every such x_0), $A = \mathcal{G}(X, x_0)$ is a seminormal group.

Equivalently, (X, T) is seminormal iff for every minimal self-joining $M \subset X\times X$ the projection maps $\pi_i : M \longrightarrow X$, $i = 1, 2$, are virtually weakly mixing extensions.

PROPOSITION 6.3.

(1) A minimal flow with normal Ellis group is seminormal. In particular every regular flow is seminormal.

(2) Every purely weakly mixing flow (PWM flows, see [**EG**] *for the definition and examples) is seminormal.*

PROOF. (1) Immediate from the definition of seminormal.

(2) By definition (X, T) is PWM if for every τ-closed group B one has $B'(A \cap B) = B$, where A is the Ellis group of (X, T) (or in terms of flows: for every minimal (Y, T) and a minimal joining $M \subset X\times Y$, the projection map $M \longrightarrow Y$ is virtually weakly mixing). Now for $\gamma \in G$ take $B = A^\gamma$; then by PWM, $(A^\gamma)'(A \cap A^\gamma) = A^\gamma$. Conjugating with γ^{-1} we get $A'(A^{\gamma^{-1}} \cap A) = A$, which is the definition of seminormal. □

Thus among the seminormal flows are all minimal group rotations, all graphic flows (the analogue in topological dynamics of minimal-self-joining), all group extensions of graphic flows, all regular flows and every flow which is proximally equivalent to one of those flows, i.e. have the same Ellis group as one of them.

THEOREM 6.1. *Let (X,T) be a metric seminormal flow and $\pi : X \longrightarrow Y$ a factor map. Then there exists a diagram*

$$\begin{array}{ccccc} X & \xleftarrow{\theta^*} & X^* & \xrightarrow{\eta} & Z \\ {\scriptstyle\pi}\downarrow & & {\scriptstyle\pi^*}\downarrow & \swarrow{\scriptstyle\rho} & \\ Y & \xleftarrow{\theta} & Y^* & & \end{array}$$

where θ and θ^* are proximal extensions, π^* is RIC, ρ is a group extension and η is a virtually weakly mixing extension.

PROOF. Let

$$\begin{array}{ccc} X & \xleftarrow{\theta^*} & X^* \\ \downarrow{\scriptstyle\pi} & & \downarrow{\scriptstyle\pi^*} \\ Y & \xleftarrow{\theta} & Y^* \end{array}$$

be the RIC shadow diagram associated with π, and let $\mathcal{G}(X) = \mathcal{G}(X^*) = A$, $\mathcal{G}(Y^*) = B$. By Theorem 1.3 there exists a diagram

$$\begin{array}{ccc} X^* & \xrightarrow{\eta} & Z \\ {\scriptstyle\pi^*}\downarrow & \swarrow{\scriptstyle\rho} & \\ Y^* & & \end{array}$$

where ρ is the largest isometric extension of Y^* within X^*; thus $\mathcal{G}(Z) = B'A$.

There exists a minimal flow $\overline{Y}$ which is an extension of Y and such that $\mathcal{G}(\overline{Y}) = \widetilde{B}$. Since $\widetilde{B} \lhd B$, the shadow homomorphism of $\overline{Y} \longrightarrow Y$ is normal and RIC. Now, by Theorem 5.1, applied to the shadow homomorphism of $\overline{Y} \longrightarrow Y$, we have

$$B'\widetilde{B} = B''\widetilde{B} = B^{\infty}\widetilde{B}$$

and

$$B'\widetilde{B} \lhd B.$$

By Proposition 6.1 (as A is seminormal), $(\widetilde{B})'A = \widetilde{B}$, hence $B'A \supset \widetilde{B}$. Thus $B'\widetilde{B} \supset B'A = B'A\widetilde{B} = B'\widetilde{B}$ and $B'\widetilde{B} = B'A$. Hence

$$(B'A)'A = (B'\widetilde{B})'A \subset B''\widetilde{B} = B'\widetilde{B} = B'A.$$

Since $\mathcal{G}(Z) = B'A$, this means that η is a virtually weakly mixing extension. Finally, since $B'A = B'\widetilde{B} \lhd B$, we conclude that ρ is a group extension.

References

[A] J. Auslander, *Minimal Flows and their Extensions*, North-Holland Mathematics Studies **153**; Notas de Matemática **122 North-Holland** (1988).

[B] I. Bronstein, *A characteristic property of PD-extensions*, Bull. Akad. Stiince RSS Moldaven **3** (1977), 11–15.

[BGK] F. Blanchard, E. Glasner and J. Kwiatkowski, *Minimal self-joining and positive topological entropy*, Monatsh. fur math. **120** (1995), 205–222.

[E] R. Ellis, *Topological Dynamics*, Benjamin, New York (1969).

[EG] R. Ellis and S. Glasner, *Pure weak mixing*, Trans. Amer. Math. Soc. **243** (1978), 135–146.

[EGS] R. Ellis, S. Glasner and L. Shapiro, *Proximal-Isometric Flows*, Advances in Math **17** (1975), 213–260.

[Gl1] S. Glasner, *Proximal flows*, Springer Verlag Lecture Notes in Math. **517** (1976).

[Gl2] ______, *A topological version of a theorem of Veech and almost simple flows*, Ergod. Th. & Dynam. Syst. **10** (1990), no. 3, 463–482.

[Gl3] ______, *Regular PI metric flows are equicontinuous*, Proc. Amer. Math. Soc **114** (1992), 269–277.

[Gl4] ______, *Topological ergodic decompositions and applications to products of powers of a minimal transformation*, J. Analyse Math. **64** (1994), 241–262.

[Gl5] ______, *For Bernoulli transformations the smallest natural family of factors consists of all factors*, Proceedings of the 1993 Alexandria conference, Cambridge University Press, 1995, pp. 291–297.

[Go] W. Gottschalk, *Transitivity and equicontinuity*, Bull. Amer. Math. Soc. **54** (1984), 982–984.

[GL] P. Gabriel and M. Lemańczyk, *On Z^2-cocycle extensions of measure-preserving actions, and applications*, preprint..

[JLM] A. del Junco and M. Lemańczyk, M. K. Mentzen, *Semisimplicity, Joinings and Group Extensions*, Studia Mathematica **112 (2)** (1995), 141–164.

[JR] A. del Junco D. Rudolph, *On ergodic action whose self-joinings are graphs*, Ergod. Th. & Dynam. Sys. **7** (1987), 229-247.

[L] M. Lemańczyk, *Ergodic compact abelian group extensions of rotations*, Toruń (1990).

[LM] M. Lemańczyk and M. Mentzen, *Compact subgroups in the centralizer of natural factors an ergodic group extension of a rotation determine all factors*, Ergod. Th. & Dynam. Sys **10** (1990), 763–776.

[M] M. K. Mentzen, *Ergodic properties of group extensions of dynamical systems with discrete spectra*, Studia Mathematica **101 (1)** (1991), 19–31.

[S] A. Siemaszko, *On group extensions of 2-fold simple ergodic actions*, Studia Mathematica **110 (1)** (1994), 53–64.

[V1] W. Veech, *Topological Dynamics*, Bull. Amer. Math. Soc **83** (1977), 775–242.

[V2] ______, *A criterion for a process to be prime*, Monathshefte Math **94** (1982), 335–341.

[Z] R. Zimmer,, *Ergodic Theory and Semisimple Groups*, Birkhauser (1984).

School of Mathematical Sciences, Tel Aviv University, 69978 Tel Aviv, Israel
E-mail address: `glasner@math.tau.ac.il`

Faculty of Mathematics and Computer Science, Nicholas Copernicus University, ul. Chopina 12/18, 87-100 Toruń, Poland
E-mail address: `mentzen@mat.uni.torun.pl`

Department of Applied Mathematics, Olsztyn Agricultural and Technical University, ul. Oczapowskiego 1, 10-572 Olsztyn – Kortowo, Poland
E-mail address: `artur@moskit.art.olsztyn.pl`

Contemporary Mathematics
Volume **215**, 1998

Topological Ergodic Decomposition and Homogeneous Flows

E. Akin and E. Glasner

ABSTRACT. A topological analogue for the ergodic decomposition in ergodic theory is described. Next this is applied to the study of homogeneous flows. A flow (X, Φ) is called homogeneous if there exists a commuting action (X, T) which is minimal. Using Robert Ellis' machinery some basic features of the theory of homogeneous flows are developed.

0. Introduction

A main problem in topological dynamics is to extend the rich theory of minimal flows to other classes of flows. For example, a basic particular case is to determine for which flows the lack of non-trivial equicontinuous factors is equivalent to weak mixing.

In this paper, following a suggestion of Hillel Furstenberg, we study the class of homogeneous flows. Let Φ be a group; we call an action of Φ on a compact Hausdorff space X *homogeneous* if there exists a second action of a countable group T on X such that $\phi t = t\phi$ for every $t \in T$ and $\phi \in \Phi$, and such that the flow (X, T) is minimal. (This notion is introduced in [FW]).

Some important examples of homogeneous flows arise as follows.

1. Let (X, T) be a minimal flow $\Phi \subset T$ a central subgroup; then (X, Φ, T) is a homogeneous flow.
2. Start with a minimal cascade (Y, σ) (i.e. $\sigma : Y \to Y$ is a homeomorphism), and form the n-fold product $X = Y \times Y \cdots \times Y$. Fix a vector $(k_1, k_2, \dots, k_n) \in \mathbb{Z}^n$, let $\phi = \sigma^{k_1} \times \sigma^{k_2} \cdots \times \sigma^{k_n}$ and put $\Phi = \{\phi^n : n \in \mathbb{Z}\}$. T will be the group
$$\{\sigma^{j_1} \times \sigma^{j_2} \cdots \times \sigma^{j_n} : (j_1, j_2, \dots, j_n) \in \mathbb{Z}^n\}.$$
Clearly then (X, Φ, T) is a homogeneous flow.
3. In the latter example take $\phi = \sigma \times \sigma^2 \cdots \times \sigma^n$ and let $\mathcal{T} \subset T$ be the subgroup generated by $\sigma \times \sigma \cdots \times \sigma$ and ϕ. Then as was shown in [G1], the flow $(Z, \Phi, \mathcal{T})$—where $Z \subset X$ is the smallest Φ-invariant closed subset of X containing the diagonal—is a homogeneous flow.

1991 *Mathematics Subject Classification.* 54H20.

4. Finally if (X,T) is a minimal flow and ϕ an element of the group $Aut(X,T)$ of continuous automorphisms of the flow (X,T), then for $\Phi = \{\phi^n : n \in \mathbb{Z}\}$ the flow (X, Φ, T) is a homogeneous flow.

In section 1 we extend results in [G1] to introduce a topological "ergodic decomposition" of a general metrizable flow (X, Φ). We show that J. Auslander's "prolongation" relation D_Φ coincides on most of the space with the Φ orbit closure relation and that consequently on an almost 1-1 extension X^* of X, the relation D_Φ is an equivalence relation. An interesting corollary of these results is that for a metric regular minimal flow (X,T) with Ellis group A, to each τ-closed subgroup $F \supset A$ corresponds—up to almost 1-1 extension of (X,T)—a factor (Y,T) with Ellis group F. In section 2 we identify the maximal equicontinuous factor of a metrizable homogeneous flow (X, Φ) and show in particular that when (X, Φ) is topologically transitive the maximal equicontinuous factors of (X,T) and (X, Φ) coincide. When T is commutative we also show that (X, Φ) is weakly mixing when and only when this maximal equicontinuous factor is trivial and that when in addition (X, Φ) is topologically transitive, the regionally proximal relation Q_Φ is an equivalence relation.

For a treatment of the Ellis group associated with a group of automorphisms of a minimal system we refer to Auslander and Markley's paper [AM]. We thank Daniel Penazzi for helpful comments.

1. Ergodic Decomposition for Metrizable Flows

In this section we are dealing with a general metrizable flow (X, Φ); i.e. X is a metrizable compact space and Φ is a group of homeomorphisms of X. Recall that

$$D_\Phi = \{(x, x') \in X \times X : \exists x_i \in X, \phi_i \in \Phi, \text{ with } (x_i, \phi_i x_i) \to (x, x')\}.$$

Also it is easy to check that

$$D_\Phi[x] = \bigcap \{\bar{\mathcal{O}}_\Phi(U) : U \text{ an open neighborhood of } x\},$$

and if we let

$$\mathcal{O}_\Phi = \{(x, \phi x) : \phi \in \Phi\}$$

then

$$D_\Phi = \bar{\mathcal{O}}_\Phi.$$

However, for convenience, we will by abuse of notation use in the sequel $\bar{\mathcal{O}}_\Phi[x]$, rather than $\overline{\mathcal{O}_\Phi[x]}$, to denote the orbit closure of x.

Let X_c be the set of continuity points of the map $x \mapsto D_\Phi[x]$ from X into the space 2^X of closed subsets of X with the Hausdorff topology.

THEOREM 1. *Let (X, Φ) be a metrizable flow. Then there exists a dense G_δ Φ-invariant subset $X_g \subset X_c$ such that:*

1. $$D_\Phi[x] = \bar{\mathcal{O}}_\Phi[x], \quad \textit{for all } x \in X_g.$$
2. *The set D_Φ defines an equivalence relation on X_g; i.e. for $x, x' \in X_g$ either $D_\Phi[x] \cap X_g = D_\Phi[x'] \cap X_g$ or they are disjoint.*
3. *There exists an almost 1-1 extension $(X^*, \Phi) \xrightarrow{\theta} (X, \Phi)$ with $|\theta^{-1}(x)| = 1$ for every $x \in X$,*
4. *The set D_Φ defines a closed Φ-invariant equivalence relation on X^*.*

5. *The map* $(X^*, \Phi) \xrightarrow{\pi} (X^*/D_\Phi, \mathrm{id})$, *where* id *is the identity map on* X^*/D_Φ, *is an open homomorphism. Thus* $X^*_c = X^*$ *and therefore* $X^{**} = X^*$,
6. (X^*/D_Φ) *is isomorphic to the quasifactor* $(\mathcal{D}, \Phi)$ *of* (X, Φ), *where*

$$\mathcal{D} = \text{closure}\{D_\Phi[x] : x \in X_0\} \subset 2^X$$

and each point of $\mathcal{D}$ *is a fixed point of* Φ.

7. *If the map* $(X^*, \Phi) \xrightarrow{\theta} (X, \Phi)$ *is an isomorphism i.e. when* $x \mapsto D_\Phi[x]$ *is continuos, then* D_Φ *is an equivalence relation on* X,
8. X_g *is invariant under every automorphism of* (X, Φ),
9. $X_g = X$ *iff* X *is a union of* Φ*-minimal sets and the partition into* Φ*-minimal sets is a closed* Φ*-invariant equivalence relation.*

PROOF. Some parts of the proof were already developed in [G1] for related purposes. For uniformity of notations and completeness we repeat some of it here. Although it will not be used in the proof, we include a discussion of the relation D^*_Φ defined by

$$D^*_\Phi[x] = \bigcap\{\mathcal{O}_\Phi[U] : U \text{ an open neighborhood of } x\}.$$

Notice that D_Φ is a symmetric closed Φ invariant subset of $X \times X$ and for every $x \in X$, $D^*_\Phi[x] \subset D_\Phi[x]$ and $\bar{\mathcal{O}}_\Phi[x] \subset D_\Phi[x]$.

LEMMA 1.

1. *The function* $x \mapsto D_\Phi[x] = \{x' \in X : (x, x') \in D_\Phi\}$ *from* X *into* 2^X, *is upper-semi-continuous; i.e. if* $x_i \to x$ *then* $\limsup D_\Phi[x_i] \subset D[x]$. *There exists a* Φ *invariant dense* G_δ *subset* X_c *of* X *of continuity points for this map.*
2. *For* $x \in X_c$ *we have:*

$$x' \in D_\Phi[x] \Longrightarrow D_\Phi[x] \subset D_\Phi[x'].$$

Thus if $D_\Phi[x] \supset D_\Phi[x']$ *then* $D_\Phi[x] = D_\Phi[x']$ *and* $D_\Phi[x]$ *is minimal.*

PROOF. Upper-semi-continuity is easy to check and the existence of X_c follows from the general properties of such maps. For part (2), observe that $x' = \lim \phi_i x_i$ for some sequences $x_i \to x$ and $\phi_i \in \Phi$. Hence

$$D_\Phi[x] = \lim D_\Phi[x_i] = \lim D_\Phi[\phi_i x_i] \subset D_\Phi[x'].$$

□

LEMMA 2.

1. *The map* $x \mapsto \bar{\mathcal{O}}_\Phi[x]$ *is lower-semi-continuous (i.e.* $\lim x_i = x$ *implies* $\limsup \bar{\mathcal{O}}_\Phi[x_i] \supset \bar{\mathcal{O}}_\Phi[x]$*). There exists therefore a* Φ *invariant dense* G_δ *subset* X_{or} *of* X *of continuity points for this map;*
2. $$D^*_\Phi[x] = \{x' \in X : \bar{\mathcal{O}}_\Phi[x'] \supset \bar{\mathcal{O}}_\Phi[x]\};$$
   i.e. $D^*_\Phi = \bar{\mathcal{O}}_\Phi^{-1} = \{(x, x') : (x', x) \in \mathcal{O}_\Phi\}$.
3. *If* $x_0 \in X_{or}$ *then* $\bar{\mathcal{O}}_\Phi[x] \supset \bar{\mathcal{O}}_\Phi[x_0]$ *implies* $\bar{\mathcal{O}}_\Phi[x] = \bar{\mathcal{O}}_\Phi[x_0]$*; i.e.* $\bar{\mathcal{O}}_\Phi[x_0]$ *is maximal. In particular* $\bar{\mathcal{O}}_\Phi[x_0] \supset D^*_\Phi[x_0]$.

PROOF. That our map is lower-semi-continuous is clear and again the existence of X_{or} follows. Part (2) is easy to verify. If $x_0 \in X_{or}$ and $\bar{\mathcal{O}}_\Phi[x] \supset \bar{\mathcal{O}}_\Phi[x_0]$ then there exists a sequence $\phi_i \in \Phi$ with $\phi_i x \to x_0$. By continuity at x_0 we get: $\bar{\mathcal{O}}_\Phi[\phi_i x] = \bar{\mathcal{O}}_\Phi[x] \to \bar{\mathcal{O}}_\Phi[x_0]$, hence $\bar{\mathcal{O}}_\Phi[x] = \bar{\mathcal{O}}_\Phi[x_0]$. The last assertion now follows from (2). □

Set
$$X_g = \{x \in X : \bar{\mathcal{O}}_\Phi[x] = D_\Phi[x]\}.$$

LEMMA 3. *With the above notation*

1. $X_g = X_{or} \subset X_c$.
2. *Restricted to* X_g, D_Φ *is an equivalence relation; i.e. for* $x, x' \in X_\Phi$ *either* $D_\Phi[x] \cap X_g = D_\Phi[x'] \cap X_g$ *or they are disjoint.*

PROOF. (1) If $x = \lim x_n$ then
$$\bar{\mathcal{O}}_\Phi[x] \subset \limsup \bar{\mathcal{O}}_\Phi[x_n] \subset \limsup D_\Phi[x_n] \subset D_\Phi[x].$$
Similar inclusions hold for the operation lim inf and thus for a point $x \in X_g$ all of these sets coincide. We conclude that $X_g \subset X_{or}$ and $X_g \subset X_c$. Finally if $x \in X_{or}$ and $x' \in D_\Phi[x]$ then there exists sequences $x_n \to x$ and $\phi_n \in \Phi$ such that $\phi_n x_n \to x'$. Hence
$$x' = \lim \phi_n x_n \in \lim \bar{\mathcal{O}}_\Phi[x_n] = \bar{\mathcal{O}}_\Phi[x].$$
Thus for $x \in X_{or}$, $D_\Phi[x] \subset \bar{\mathcal{O}}_\Phi[x]$. Since the converse inclusion holds as well we get an equality; i.e. $x \in X_g$.

(2) For $x \in X_g$ and $y \in D_\Phi[x] \cap X_g$, $\bar{\mathcal{O}}_\Phi[y] \subset D_\Phi[x] = \bar{\mathcal{O}}_\Phi[x]$. By maximality of $\bar{\mathcal{O}}_\Phi[y]$, $D_\Phi[x] = \bar{\mathcal{O}}_\Phi[x] = \bar{\mathcal{O}}_\Phi[y]$. Now if $y \in D_\Phi[x] \cap D_\Phi[x'] \cap X_g$ for $x, x' \in X_g$, then
$$D_\Phi[x] = \bar{\mathcal{O}}_\Phi[x] = \bar{\mathcal{O}}_\Phi[y] = \bar{\mathcal{O}}_\Phi[x'] = D_\Phi[x'].$$
□

In analogy with ergodic theory, the points of X_g will be called the *ergodic points* of the flow (X, Φ).

In the spaces 2^X and $X \times 2^X$ respectively let
$$\mathcal{D} = \text{closure}\{D_\Phi[x] : x \in X_g\},$$
$$\mathcal{X} = \text{closure}\{(x, D_\Phi[x]) : x \in X_g\}.$$
The proof of the next Lemma is straightforward.

LEMMA 4. *The map* $\theta : (x, K) \mapsto x$, $(\mathcal{X}, \Phi) \xrightarrow{\theta} (X, \Phi)$, *is an almost 1-1 extension:* $x \in X_c$ *implies* $\theta^{-1}(x) = \{(x, D_\Phi[x])\}$. *The map* $\pi : (x, K) \mapsto K$ *is a homomorphism* $(\mathcal{X}, \Phi) \xrightarrow{\pi} (\mathcal{D}, \mathrm{id})$.

LEMMA 5. *Let*
$$X^* = \{(x, K) : x \in K \in \mathcal{D}\},$$
then $X^* = \mathcal{X}$ *and* π *is an open map.*

PROOF. Clearly $\mathcal{X} \subset X^*$. Let (x, K) be an arbitrary point in X^*, i.e. $x \in K \in \mathcal{D}$; then there exists a sequence $x_i \in X_g$ with $D_\Phi[x_i] \to K$. Since x is in K there exist points $x_i' \in D_\Phi[x_i]$ such that $x_i' \to x$. Now $D_\Phi[x_i] = \bar{\mathcal{O}}_\Phi[x_i]$ and we can find $\phi_i \in \Phi$ such that $\phi_i x_i \to x$. Thus

$$\lim(\phi_i x_i, D_\Phi[\phi_i x_i]) = \lim(\phi_i x_i, D_\Phi[x_i]) = (x, K),$$

so that (x, K) is in $\mathcal{X}$. This proves the first assertion of the lemma and the second follows easily. □

This proves parts (1)-(3) of Theorem 1.

To prove (4) consider $(x', K') \in D_\Phi[(x, K)]$. There exist sequences $\phi_i \in \Phi$ and $(x_i, K_i) \in X^*$ such that $(x_i, K_i) \to (x, K)$ and $\phi_i(x_i, K_i) \to (x', K')$. Now $\phi_i K_i = K_i \to K'$ so that $K' = K$, $x' \in K$ and it follows that $D_\Phi[(x, K)] \subset \{(y, K) : y \in K\} = K \times \{K\}$. On the other hand $D_\Phi[x_i] \to K$ for some sequence $x_i \in X_g$ and since for every i, $D_\Phi[x_i] = \bar{\mathcal{O}}_\Phi[x_i]$, we can find for every $y \in K$ a sequence $\phi_i \in \Phi$ with $\phi_i(x_i, D_\Phi[x_i]) \to (y, K)$. Thus also $D_\Phi[(x, K)] \supset \{(y, K) : y \in K\} = K \times \{K\}$ and we have

$$D_\Phi[(x, K)] = \{(y, K) : y \in K\} = K \times \{K\}.$$

Thus

$$D_\Phi = R_\pi := \{(x^*, x_1^*) \in X^* \times X^* : \pi(x^*) = \pi(x_1^*)\},$$

and in particular we see that D_Φ is an equivalence relation on X^*. Clearly now $(X^*/D_\Phi, \Phi)$ is isomorphic to $(\mathcal{D}, \Phi)$ and this completes the proof of parts (5) and (6); parts (7),(8) and (9) of the theorem are clear. This completes the proof of Theorem 1. □

In the proof of the next theorem and throughout the rest of the paper we apply R. Ellis' universal theory of minimal flows to the flow (X, T) and other related T-flows. Thus our notations refer to the usual objects in the Stone-Čech compactification βT of the group T : M is a left minimal ideal in βT, J the set of idempotents in M, u a fixed idempotent in M, and G the maximal subgroup uM of M. In a minimal flow (X, T) our choice of a base point $x_0 \in X$ is always of one for which $ux_0 = x_0$ (see e.g. [G3]). The Ellis group of the pointed flow (X, x_0, T) is defined by:

$$\mathcal{G}(X, x_0, T) := \{\alpha \in G : \alpha x_0 = x_0\}.$$

Given a homogeneous flow (X, Φ, T) with distinguished point $x_0 = ux_0 \in X$, following [AM], we define the τ-closed subgroup

$$F = \mathcal{G}(\Phi) := \tau\text{-}\operatorname{closure}\{\beta \in G : \exists \phi \in \Phi \text{ with } \phi x_0 = \beta x_0\}.$$

The distinguished point x_0 is an arbitrary point in X_g satisfying $ux_0 = x_0$. (For this we have first to choose $x_0 \in X_g$ and only then choose $u \in J$ with $ux_0 = x_0$). Clearly the Ellis group of (X, x_0, T), $A = \mathcal{G}(X, x_0, T)$ is a subgroup of F. By Theorem 2 of [AM] (which can certainly be applied to the more general situation considered here)

$$D_\Phi[x_0] = \bigcup \{p \circ F x_0 : p \in M, px_0 = x_0\}.$$

Applying Theorem 1 to the given homogeneous flow (X, Φ, T) we get the following.

THEOREM 1$'$. *Let (X, Φ, T) be a metrizable homogeneous flow. Then there exists a T and Φ invariant dense G_δ subset $X_g \subset X_c$ such that:*

1. $$D_\Phi[x] = \bar{\mathcal{O}}_\Phi[x], \quad \textit{for all } x \in X_g,$$
2. *D_Φ defines equivalence relation on X_g; i.e. for $x, x' \in X_g$ either $D_\Phi[x] \cap X_g = D_\Phi[x'] \cap X_g$ or they are disjoint.*
3. *There exists a T-minimal almost 1-1 extension $(X^*, \Phi, T) \xrightarrow{\theta} (X, \Phi, T)$ with $|\theta^{-1}(x)| = 1$ for every $x \in X_c$.*
4. *The set D_Φ defines a closed T and Φ invariant equivalence relation on X^*.*
5. *The map $(X^*, \Phi, T) \xrightarrow{\pi} (X^*/D_\Phi, \mathrm{id}, T)$, where* id *is the identity map on X^*/D_Φ, is an open homomorphism.*
6. *The flow $(X^*/D_\Phi, T)$ is isomorphic to the quasifactor $(\mathcal{D}, T)$ of (X, T), where*
$$\mathcal{D} = \text{closure}\{D_\Phi[x] : x \in X_0\} \subset 2^X.$$
7. *If the map $(X^*, \Phi) \xrightarrow{\theta} (X, \Phi)$ is an isomorphism i.e. when $x \mapsto D_\Phi[x]$ is continuous, then D_Φ is an equivalence relation on X.*
8. *Let $Y = \mathcal{D}$ and $y_0 = D_\Phi[x_0]$ then*
$$\mathcal{G}(Y, y_0, T) = \mathcal{G}(\Phi).$$

PROOF. The only parts we need to show are the assertions about the T action. Now by part (8) of Theorem 1 the subset X_g is T-invariant and it is not hard to see that the T-actions induced on $\mathcal{D}$ and $\mathcal{X}$ are minimal. To show (8) we observe that since x_0 is in X_c we have $u \circ D_\Phi[x_0] = D_\Phi[x_0]$. Also, since $\phi D_\Phi[x_0] = D_\Phi[x_0]$ for every $\phi \in \Phi$, it follows easily that $\beta \circ D_\Phi[x_0] = D_\Phi[x_0]$ for every $\beta \in F = \mathcal{G}(\Phi)$. Thus $F = \mathcal{G}(\Phi) \subset \mathcal{G}(Y, y_0, T)$.

Conversely if for some $\beta \in G$, $\beta y_0 = y_0$, then $\beta \circ D_\Phi[x_0] = D_\Phi[x_0]$, hence $\beta x_0 \in D_\Phi[x_0]$. Again by Theorem 2 of [AM],
$$D_\Phi[x_0] = \bigcup \{v \circ F x_0 : v \in J, v x_0 = x_0\}.$$
Thus for some $v \in J, v x_0 = x_0$, $\beta x_0 \in v \circ F x_0$. Hence $\beta x_0 \in u \circ F x_0 \cap G x_0 = F x_0$, where the latter equality holds since F is τ-closed. This implies $\beta \in F$ and the proof is complete. □

PROBLEM 1. For homogeneous flows is the converse of part (7) true; i.e. is $x \mapsto D_\Phi[x]$ continuous when D_Φ is an equivalence relation on X?

One case in which $X = X^*$ is the case when (X, T) is distal. In this case, by [AM], D_Φ is an equivalence relation and moreover the homomorphism $\pi : (X, T) \to (Y, T) = (X/D_\Phi, T)$ being distal, is open. Thus in this case $X = X^*$ and the map $x \mapsto D_\Phi[x]$ is continuous.

Another case is when (X, Φ) is topologically transitive. In this case $D_\Phi = X \times X$.

A simple example where $X \neq X^*$ is obtained as follows. Let $(\mathbb{T}, R_\alpha)$ be an irrational rotation of the circle and $(Z, \sigma) \to (\mathbb{T}, R_\alpha)$ be the extension obtained by splitting the points of a single orbit into two. Take $X = Z \times Z$, $\phi = \sigma \times \sigma$, Φ the group generated by ϕ and T the group generated by $\sigma \times \mathrm{id}$ and $\mathrm{id} \times \sigma$. It is easy to check that in that case D_Φ is not an equivalence relation.

The following corollary of Theorem $1'$ partially answers a question raised during the Ellis conference, April 95 in Minneapolis. We recall that the minimal flow (X,T) is *regular* if every minimal subset N of the product flow $(X\times X, T\times T)$ is a graph of an automorphism ϕ of (X,T): $N=\{(x,\phi(x)) : x\in X\}$. For a regular flow (X,T), the Ellis group $A=\mathcal{G}(X,T)$ does not depend on the choice of the base point x_0.

PROPOSITION 1. *Let (X,T) be a metric, minimal, regular flow. Let $A=\mathcal{G}(X,T)$ be the Ellis group of (X,T). Then for every τ-closed group $A\subset F\subset G$, there exists a minimal metric almost 1-1 extension $(X^*,T)\xrightarrow{\theta}(X,T)$ and a factor map $(X^*,T)\xrightarrow{\pi}(Y,T)$ such that for some point $y_0\in Y$, with $uy_0=y_0$, $\mathcal{G}(Y,y_0)=F$.*

PROOF. Since (X,T) is regular, A is a normal subgroup of F and the group $\Phi=F/A$ acts as a group of automorphisms of the flow (X,T); i.e. (X,Φ,T) is a homogeneous flow.

Theorem $1'$ now applies and we obtain the corresponding maps

$$(X^*,T)\xrightarrow{\theta}(X,T)\quad\text{and}\quad(X^*,T)\xrightarrow{\pi}(\mathcal{D},T).$$

We put $Y=\mathcal{D}$ and let $y_0=D_\Phi[x_0]\in Y$, where x_0 is some point in X_g satisfying $ux_0=x_0$ (again for this we have to first choose $x_0\in X_g$ and only then choose an idempotent $u\in J$ with $ux_0=x_0$). Since x_0 is in X_c we have $uy_0=y_0$.

It remains to show that $\mathcal{G}(Y,y_0)=F$. However, since clearly $F=\mathcal{G}(\Phi)$, this follows directly from part (8) of Theorem $1'$. □

REMARK.

1. We do not claim (and it is probably not true in general) that the flow (X^*,T) is regular.
2. For the general minimal flow (X,T) (not necessarily regular or metrizable) and a τ-closed group $A\subset F\subset G$ $(A=\mathcal{G}(X,T))$, there always exists a minimal *proximal* extension $(X^*,T)\xrightarrow{\theta}(X,T)$ and a factor map $(X^*,T)\xrightarrow{\pi}(Y,T)$, with $\mathcal{G}(Y,T)=F$.

2. The Maximal Equicontinuous Factor of a Homogeneous Flow

THEOREM 2. *Let (X,Φ,T) be a homogeneous flow in which D_Φ is an equivalence relation so that $(X,\Phi,T)\xrightarrow{\pi}(Y,\mathrm{id},T)$, where $Y=X/D_\Phi$, is a homomorphism of homogeneous flows.*

(1) *The extension $(X,T)\xrightarrow{\pi}(Y,T)$ is full (see [EGS] Section 4); i.e. there exists a commutative diagram*

$$\begin{array}{ccc} X & & \\ {\scriptstyle\pi}\downarrow & \searrow{\scriptstyle\sigma} & \\ Y & \underset{\rho}{\longleftarrow} & Z \end{array}$$

where ρ is an isometric extension with $B=F'A=\mathcal{G}(Z,z_0)$, $F=\mathcal{G}(Y,y_0)$, and $A=\mathcal{G}(X,x_0)$. The flow (Z,T) is the largest isometric extension of (Y,T) within (X,T).

(2) *The relation*

$$R_\sigma=\{(x,x')\in X\times X : \sigma(x)=\sigma(x')\},$$

is Φ-invariant. Thus the Φ-action on X induces a Φ-action on Z such that (Z, Φ, T) is a homogeneous flow and $(X, \Phi, T) \xrightarrow{\sigma} (Z, \Phi, T)$ is a homomorphism.

(3) *The flow (Z, Φ) is the largest equicontinuous factor of (X, Φ) and $(Z, T) \xrightarrow{\rho} (Y, T)$ is a group extension where the compact acting group K is the closure of Φ in $Aut(Z, T)$.*

PROOF. (1) Since every point of the form $(x, \phi x) \in X \times X$ is a T-almost periodic point it follows that in the relation

$$R_\pi = \{(x, x') \in X \times X : \pi(x) = \pi(x')\} = D_\Phi = \text{closure}\{(x, \phi x) : x \in X, \phi \in \Phi\},$$

the T-almost periodic points are dense. Now ([E], Propositions 5.4, 6.3), (see also [EGS],Theorem 4.3) implies that π is full.

(2) Since the relation R_σ is the smallest closed equivalence relation containing the T relative regionally proximal relation $Q_T(\pi)$, it is clear that R_σ is Φ-invariant. Thus Φ acts on Z; clearly $(X, \Phi, T) \xrightarrow{\sigma} (Z, \Phi, T)$ is a homomorphism and part (2) is proved.

(3) We see that symmetrically T acts on the largest equicontinuous factor say (W, Φ) of (X, Φ), so that (W, T) is a minimal flow and Φ acts as a precompact group of automorphisms of (W, T). If we denote by K the closure of Φ in the group $Aut(W, T)$, then clearly $(W/K, T)$ is isomorphic to (Y, T). Thus by definition, $(W, T) \to (Y, T)$ is a group extension and the maximality of (Z, T) as an isometric extension of (Y, T) within (X, T) shows that (Z, T) factors onto (W, T). Note that the minimality of (W, T) implies that the action of K on W is free.

For the other direction consider the T-minimal group extension $(\tilde{Z}, T) \xrightarrow{\tilde{\rho}} (Y, T)$ (say with compact group K) that corresponds to the isometric extension $(Z, T) \xrightarrow{\rho} (Y, T)$ and let λ be the map $(\tilde{Z}, T) \xrightarrow{\lambda} (Z, T)$. Consider the set

$$K_\Phi = \{k \in K : \exists \phi \in \Phi, \ \lambda \circ k = \phi \circ \lambda\}.$$

One can easily check that K_Φ is a subgroup of K and that on Z

$$\Phi = \{\lambda \circ k \circ \lambda^{-1} : k \in K_\Phi\}.$$

Since the action of K_Φ on $\tilde{Z}$ is equicontinuous, it follows that Φ acts equicontinuously on Z. Hence (W, Φ) factors onto (Z, Φ). Thus $Z = W$ and the proof of Theorem 2 is complete. □

The reader should be warned here that in general for individual "Φ ergodic components" $X_y = \pi^{-1}(y)$, $y \in Y$, the largest equicontinuous factor of the flow (X_y, Φ) may be larger than $Z_y = \rho^{-1}(y)$. To illustrate this consider the following example. Let $(\mathbb{T}, \psi)$ be the cascade defined on the torus $\mathbb{T}^2 = \mathbb{R}^2/\mathbb{Z}^2$ by $\psi(x, y) = (x + \alpha, y + x)$, where $\alpha \in \mathbb{R}$ is irrational. We take $X = \mathbb{T}^4$, $\phi = \psi \times \psi$, $\Phi = \{\phi^n : n \in \mathbb{Z}\}$ and let T be the group $\{\psi^m \times \psi^n : m, n \in \mathbb{Z}\}$. It is easy to check that for the homogeneous flow (X, Φ, T), with notations as in Theorem 2, $Y = \mathbb{T}$, $\pi : \mathbb{T}^4 \to \mathbb{T}$ is given by $\pi(x, y, x', y') = x - x'$ and $Z = \mathbb{T}^2$ with $\sigma(x, y, x', y') = (x, x')$ (so that $\rho(x, x') = x - x'$). However for a fixed $\beta \in \mathbb{T} = Y$, the Φ-flow on $\pi^{-1}(\beta) = \{(x + \beta, y, x, y') : x, y, y' \in \mathbb{T}\}$ has $(\mathbb{T}^2, \alpha \times \beta)$ as its largest equicontinuous factor with factor map $\zeta(x + \beta, y, x, y') = (x, y - y')$ whereas $\rho^{-1}(\beta) = \{(x + \beta, x) : x \in \mathbb{T}\}$ is isomorphic to $(\mathbb{T}, \alpha)$.

PROBLEM 2. In the context of Theorem 2, is it true that the Φ regionally proximal relation Q_Φ is an equivalence relation; i.e. is $Q_\Phi = Q_T(\pi) = R_\sigma$? (See Proposition 4 below).

Next we examine the case when (X, Φ) is topologically transitive. First a very general and (surprisingly easy to prove) observation.

PROPOSITION 2. *Let X be a compact Hausdorff space, T and Φ discrete groups and T be commutative. Let (X,T) and (X, Φ) be two commuting actions. If (X,T) is weakly mixing and (X,Φ) is topologically transitive then (X,Φ) is also weakly mixing. Thus if both groups are commutative and both actions are topologically transitive, then one is weakly mixing iff so is the other.*

PROOF. Let U, U', V, V' be non-empty open sets in X. By the assumption, there exists $t \in T$ such that

$$tU \cap U' \neq \emptyset \quad , \quad tV \cap V' \neq \emptyset.$$

Set $U_0 = U \cap t^{-1}U'$ and $V_0 = V \cap t^{-1}V'$. Since weak mixing implies topological transitivity, there exists $s \in T$ with $W := U_0 \cap sV_0 \neq \emptyset$. Use the topological transitivity of (X, Φ) to find an element $\phi \in \Phi$ with $O := \phi W \cap tW \neq \emptyset$. Now

$$\phi U \supset \phi W \supset O \subset tW \subset tU_0 \subset U'.$$

Similarly,

$$\phi V \supset \phi s^{-1} W = s^{-1}\phi W \supset s^{-1}O \subset s^{-1}tW = ts^{-1}W \subset tV_0 \subset V'.$$

Thus,

$$\phi U \cap U' \neq \emptyset \quad , \quad \phi V \cap V' \neq \emptyset,$$

and we are done. □

PROPOSITION 3. *Let (X, Φ, T) be a metric homogeneous flow. Suppose (X, Φ) is topologically transitive. Then the largest equicontinuous factor (Z,T) of the minimal flow (X,T) serves also as the largest equicontinuous factor (Z, Φ) of (X, Φ). In particular this factor is trivial for (X,T) iff it is trivial for (X, Φ). When T is commutative then (X, Φ) is weakly mixing if and only if (X, Φ) has no non-trivial equicontinuous factors.*

PROOF. Since (X, Φ) is topologically transitive $D_\Phi = X \times X$. In particular it is an equivalence relation and Theorem 2 applies. In the notations of Theorem 2, $Y = X/D_\Phi$ is the trivial one point flow and (Z,T) is the largest equicontinuous factor of (X,T). Now by Theorem 2, (Z, Φ) is also the largest equicontinuous factor of (X, Φ).

When (X, Φ) is weakly mixing, clearly (Z, Φ) is trivial. Conversely, if (Z, Φ) is trivial then (Z,T) is trivial and (X,T) is weakly mixing. When T is commutative, Proposition 2 applies and we conclude that (X, Φ) is also weakly mixing. □

PROPOSITION 4. *Let X be a compact Hausdorff space, T and Φ discrete groups. Let (X,T) and (X, Φ) be two commuting actions both topologically transitive.*

(1) *If T is commutative then $Q_T \subset Q_\Phi$. Thus if also Φ is commutative then $Q_T = Q_\Phi$.*

(2) *If T is commutative and acts minimally on X which is metrizable, then $Q_T = Q_\Phi$ and in particular Q_Φ is an equivalence relation.*

PROOF. (1) Recall the definition of the T-regionally proximal relation:

$$Q_T = \bigcap \{\mathrm{closure}\, T\alpha : \alpha \text{ a neighborhood of } \Delta \text{ in } X \times X\},$$

where Δ is the diagonal in $X \times X$. Let U, V be open subsets in X with $U \times V \subset t^{-1}\alpha$ for some $t \in T$ and choose $s \in T$ such that $W =: sV \cap U \neq \emptyset$. Next choose $\phi \in \Phi$ with $O =: \phi W \cap tW \neq \emptyset$. Now

$$\phi U \supset \phi W \supset O \subset tW \subset tU$$

and

$$\phi V \supset \phi s^{-1} W = s^{-1}\phi W \supset s^{-1} O \subset s^{-1} t W = t s^{-1} W \subset tV.$$

Since $t(U \times V) \subset \alpha$, we have

$$\emptyset \neq \phi(U \times V) \cap t(U \times V) \subset \phi(U \times V) \cap \alpha$$

so that $(U \times V) \cap \phi^{-1}\alpha \neq \emptyset$. This implies closure $\Phi\alpha \supset$ closure $T\alpha$ whence $Q_T \subset Q_\Phi$.

(2) By Proposition 3 the largest equicontinuous factor $(X, T) \xrightarrow{\pi} (Z, T)$ of the minimal flow (X, T) coincides with the largest equicontinuous factor (Z, Φ) of (X, Φ). Since for the commutative group T and the metric minimal flow (X, T), the equicontinuous structure relation E_T satisfies $E_T = Q_T = R_\pi$, we conclude that $Q_\Phi \subset E_\Phi = R_\pi = Q_T$. Thus $E_\Phi = Q_\Phi = Q_T = R_\pi$ and Q_Φ is an equivalence relation. □

PROBLEM 3. Is the assumption that T is commutative in Propositions 2-4 necessary?

References

[AM] J. Auslander and N. Markley, *Dynamical properties of automorphisms of minimal flows*, preprint (1995).

[E] R. Ellis, *The Veech structure theorem*, Trans. Amer. Math. Soc.. **186** (1973), 203-218.

[EGS] R. Ellis, S. Glasner and L. Shapiro, *Proximal-Isometric Flows*, Advances in Math. **17** (1975), 213-260.

[FW] H. Furstenberg and B. Weiss, *Topological Dynamics and Combinatorial Number Theory*, Journal d'Analyse Math. **34** (1978), 61–85.

[G1] E. Glasner, *Topological ergodic decompositions and applications to products of powers of a minimal transformation*, Journal d'Analyse Math. **64** (1994), 241-262.

[G2] E. Glasner, *A topological version of a theorem of Veech and almost simple flows*, Ergod. Th. and Dynam, Sys. **10** (1990), 463-482.

[G3] S. Glasner, *Proximal Flows*, Springer Verlag, Lecture Notes in Math. 517, 1976.

Department of Mathematics , CUNY, New York, NY, USA
E-mail address: `etkcc@cunyvm.cuny.edu`

Department of Mathematics , Tel Aviv University, Tel Aviv, Israel
E-mail address: `glasner@math.tau.ac.il`

Contemporary Mathematics
Volume **215**, 1998

On the Proximal and Regionally Proximal Relation of an Extension between Minimal Flows

D. Penazzi

Dedicated to Robert Ellis for his constant encouragement and support.

ABSTRACT. Algebraic conditions for the relativized proximal and regionally proximal relations to be equivalence relations are provided. In the case of the proximal relation, the condition is also sufficient.

1. Introduction

The proximal and regionally proximal relations are in general not equivalence relations. Recently, results have been obtained ([**AEE**],[**E4**]) that provide conditions for these relations of a flow to be equivalence relations. We generalize those results to relations on the extension of a given flow. We stress the algebraic point of view and the theory of groups of Topological Dynamics, surely two of the three greatest contributions of Robert Ellis to the area. (The other being the enveloping semigroup).

In section 1 we introduce the definitions and notations we will use. They are almost all standard in the field. In this paper, at several places we shall deal with full extensions, so in section 2 we prove that the concept of full extension is independent of the base points. We also introduce a new algebra $\mathcal{F}'$ that corresponds to the group F', and which satisfies $\mathcal{F}^{\#} = \mathcal{F}^{*} \cap \mathcal{F}'$. This allow us to give a new proof of the result $F^{\#} = F^{*}F'$, which we are going to use frequently. In section 3, we give a necessary and sufficient condition for the proximal relation of an extension between minimal flows to be an equivalence relation, and using it, give another proof of the famous Auslander's result that if the relation is closed then it is an equivalence relation. In section 4, we give suficient conditions under which the regionally proximal relation is an equivalence relati on. The theorems in this section generalize to the relative case theorems in [**EK**] and [**AEE**].

I would like to thank, in addition to Robert Ellis for his selfless dedication and insights; David Ellis and Joseph Auslander, for many helpful conversations, specially in regard to section 4.

1991 *Mathematics Subject Classification.* 54H20.

The author was supported in part by grants from CONICET and SeCYTUNC

The Algebraic Theory of Flows: All the following definitions and notations are standard in the subject except for the notation v_p (see Notation 1.1) and $\mathfrak{E}_p(\mathcal{F})$ and $\mathfrak{J}_p(\mathcal{F})$ (Definition 1.3)

All topological groups are assumed to be discrete. If X is a space, $\Delta(X)$ or simply Δ, will denote the diagonal in $X \times X$. For the proofs of statements below, see [**E2**]. By βT we shall denote the Stone-Cech compactification of the discrete group T. The representation of T as a set of ultrafilters is the one that we are going to use: A *filter* on T is a set p of subsets of T with the property that: i) $\emptyset \notin p$; ii) $T \in p$; iii) $A \subseteq B$ and $A \in p \Rightarrow B \in p$ and iv) $A, B \in p \Rightarrow A \cap B \in p$. An ultrafilter is a maximal filter. It then follows ([**E2**], Chapter 8) that βT is the set of ultrafilters on T. We shall use the following notation throughout the paper. Set,

(a) $h(t) = \{V \subseteq T : t \in V\} \in \beta T$;

(b) if $U \subseteq T$, define $h(U) = \{p \in \beta T : U \in p\}$.

Note that $h(\{t\}) = \{h(t)\}$. A basis of the topology of βT is given by subsets of βT of the form $h(U)$, $U \subseteq T$. The sets $h(U)$ are not only open but also closed ("clopen"). The set βT can be given a structure of a semigroup, in which the left multiplication L_p is continuous for all $p \in \beta T$, but in general the right multiplication R_q is continuous only when $q \in T$. The semigroup structure in the representation of βT as ultrafilters is given by the following: Define for $V \subseteq T$ and $p \in T$, the subset Vp of T by $Vp = \{t \in T \text{ with } Vt^{-1} \in p\}$. Then it turns out that $pq = \{V : Vp \in q\}$. Since $p \in \beta T$ is a collection of subsets of T, it makes sense to form the products $pt = \{Vt \text{ with } V \in p\}$ and $tp = \{tV \text{ with } V \in p\}$. Then $ph(t) = pt$ and $h(t)p = tp$, hence we will simply drop the h. However, a little caution must be used, since if $V \subseteq T$, $Vh(t)$ is NOT equal to Vt, but $Vh(t) = t^{-1}V$. Note then that $V \in pq \iff Vp \in q$ and $t \in Vp \iff Vt^{-1} \in p$. Let $\mathcal{C} = \mathcal{C}(\beta T)$ denote the set of continuous real-valued functions on βT. We will use the algebraic point of view (see [**E2**]) and denote $al(X, x_0)$ by $al(X)$ if x_0 is understood. If $\mathcal{A}$ is a T-subalgebra of $\mathcal{C}$, the space of $\mathcal{A}$ will be denoted $|\mathcal{A}| = sp(\mathcal{A})$. Any element of βT induces a function on $\mathcal{C}$ by $p(f) = f(p)$. If $\mathcal{A} \subseteq \mathcal{C}$ is a subalgebra, we can talk of the restriction of the function p to $\mathcal{A}$, and denote it by $p|_{\mathcal{A}}$. It can be shown that $p|_{\mathcal{A}}$ may be identified with x_0p. We will frequently use $pq|_{\mathcal{A}} = p|_{\mathcal{A}}q$, which follows from the identification $p|_{\mathcal{A}} \leftrightarrow x_0p$, or simply because pq is the composition $q \circ p$. We will frequently use the notation $q \underset{\mathcal{A}}{=} p$ to indicate $q|_{\mathcal{A}} = p|_{\mathcal{A}}$. Another important property is that there exists a T-homomorphism $\varphi : X \mapsto Y$ with $\varphi(x_0) = y_0$ if and only if $al(Y, y_0) \subset al(X, y_0)$, ([**E2**], 10.7).

Let K be a minimal right ideal of βT. An element $v \in K$ is called an *idempotent* if $vv = v$. Let M denote the universal minimal flow. We will fix an idempotent u of M, and form the set $G = Mu$, which is a group. The set of idempotents of M will be denoted J.

NOTATION 1.1. If $p \in M$, denote by v_p the unique idempotent in M with $pv_p = p$ and by p^{-1} the unique element q of M with $qp = pq = v_p$.

We refer to [**E2**], 3.5, for the existence and uniqueness of v_p and p^{-1}. Define also, for any subset K of M, the set: $\mathfrak{A}(K) = \{f \in \mathcal{C} : fk = f \text{ for all } k \in K\}$ and let $\mathfrak{A}(u) = \mathfrak{A}(\{u\})$. Recall that (X, T) is minimal if and only if there exists an idempotent v in M and $x_0 \in X$ with $al(X, x_0) \subset \mathfrak{A}(v)$ ([**E2**], 10.2). The idempotent v will satisfy $x_0v = x_0$. Since we have leeway in choosing the x_0, (because X is minimal, so every point is transitive) and we can always take one with $x_0u = x_0$ (by picking any x and taking $x_0 = xu$), we shall assume that all the algebra associated to a given minimal flow is a subalgebra of $\mathfrak{A}(u)$. If $\mathcal{A}$ is a T-subalgebra of $\mathfrak{A}(u)$ we will denote the group of $\mathcal{A}$ by $A = \mathfrak{G}(\mathcal{A}) = \{\alpha \in G : f\alpha = f \ \forall f \in \mathcal{A}\}$. Note that $\mathcal{A} \subset \mathcal{B}$ implies $B \subset A$ and that $\mathfrak{G}(\mathfrak{A}(A)) = A$. In [**E1**], Ellis introduced a topology to G, called the τ-topology. This topology is compact, makes *both* right and left multiplication by any element of G continuous and the inversion map is also continuous. However, it doesn't make G a topological transformation group, because this topology is not T_2. (it is T_1, though). In this topology, F is a τ-closed subgroup of G if and only if there exists a T-subalgebra $\mathcal{F} \subseteq \mathfrak{A}(u)$ with $F = \mathfrak{G}(\mathcal{F})$ ([**E2**], 11.12). A characterization of the τ-topology is that it is the topology generated by declaring that the sets $(A, U) := \{\beta \in G : A\beta \cap U \neq \emptyset\}$, where $A, U \subseteq T$, $A \in \alpha$ and $U \in u$, are the basic neighborhoods of α.

DEFINITION 1.2. Given a τ-closed subgroup F of G, define :

$$F' = \cap_V cls_\tau(V \cap F)$$

where the intersection is over all τ-open neighborhoods V of u.

Notice that, F' is normal in F and it has the property that $\alpha \in F'$ iff there exists $\{\alpha_i\}_i \subset F$ with $\alpha_i \underset{i}{\to}_{\tau} \alpha$ and $\alpha_i \underset{i}{\to}_{\tau} u$. The following definition is also not standard:

DEFINITION 1.3. If $\mathcal{F}$ is a T-subalgebra of $\mathfrak{A}(u)$ and $p \in M$, set

$$\mathfrak{E}_p(\mathcal{F}) := \{q \in M : q \underset{\mathcal{F}}{=} p\}$$

and

$$\mathfrak{J}_p(\mathcal{F}) := \{v \in J : pv \underset{\mathcal{F}}{=} p\}$$

Given a T-subalgebra $\mathcal{F}$ of $\mathfrak{A}(u)$, one can define two new algebras:

DEFINITION 1.4. Set,

$$\mathcal{F}^* := \cap\{\mathcal{F} : \mathcal{F} \subseteq \mathfrak{A}(p, v)\}$$

where $\mathfrak{A}(p, v) = \{f \in \mathfrak{A}(u) : fpv = fp\}$. Observe that $\mathcal{F} \subseteq \mathfrak{A}(p, v)$ iff $v \in \mathfrak{J}_p(\mathcal{F})$ and

$$\mathcal{F}^{\#} = \{f \in \mathcal{F}^* : pf|_F \ : F \to I\!R \text{ is } \tau\text{-continuous for all } p \in M\}.$$

The fundamental property of these algebras is that the extension $\mathcal{F} \subseteq \mathcal{A}$ is distal if and only if $\mathcal{A} \subseteq \mathcal{F}^*$ (see [**E2**], 12.5.2) and the extension $\mathcal{F} \subseteq \mathcal{A}$ is equicontinuous iff $\mathcal{A} \subseteq \mathcal{F}^{\#}$ (see [**E2**], 14.21).

REMARK 1.5. We will denote $\mathfrak{G}(\mathcal{F}^{\#})$ by $F^{\#}$, and $\mathfrak{G}(\mathcal{F}^*)$ by F^*. However, note that the notation is misleading, since F^* and $F^{\#}$ depend on the algebra $\mathcal{F}$ rather than on the group F.

2. PI-Flows and Full Extensions

A natural question is, what kind of algebra do we obtain if in the definition of $\mathcal{F}^{\#}$ we allow f to be an *arbitrary* element of $\mathfrak{A}(u)$ such that $pf|_{\mathcal{F}}$ is τ-continuous for all $p \in M$. In [**E2**], it is proved that $\mathcal{F}^{\#} = \mathfrak{A}(F') \cap \mathcal{F}^*$, so one might think that this new algebra is indeed $\mathfrak{A}(F')$. Although we can't prove this, we prove that the group of this new algebra is F'.

DEFINITION 2.1. Given a T-subalgebra $\mathcal{F}$ of $\mathfrak{A}(u)$, define

$$\mathcal{F}^{\tau} = \{f \in \mathfrak{A}(u) : f|_F \ : F \to I\!R \ \text{ is} \, \tau - \text{continuous } \},$$

where $F = \mathfrak{G}(\mathcal{F})$ and

$$\mathcal{F}' = \{f \in \mathfrak{A}(u) : vf|_F : F \to I\!R \text{ is } \tau\text{-continuous} \ \text{ for } \ \text{all } v \in J\}.$$

REMARK 2.2. We use the notations $\mathcal{F}'$ and $\mathcal{F}^{\tau}$, to emphazise the fact that these are algebras. However, it is clear from the definitions that $\mathcal{F}'$ and $\mathcal{F}^{\tau}$ depend on F and not on $\mathcal{F}$.

LEMMA 2.3. *With the above notation, we have*

$$\begin{aligned} \mathcal{F}' &= \{f \in \mathfrak{A}(u) : pf|_F \ : F \to I\!R \text{ is } \tau\text{-continuous } \forall \ p \in M\} \\ &= \{f \in \mathfrak{A}(u) : pf|_F \ : F \to I\!R \text{ is } \tau\text{-continuous } \ \forall \ p \in \beta T\}. \end{aligned}$$

PROOF. Clearly $\mathcal{F}'$ contains the second set above, which contains in turn the third set. So it is enough to prove that $\mathcal{F}'$ is contained in the third set.

Let $p \in \beta T$. Then, $up \in M$, hence there exists a $v \in J$ with $upv = up$. Let $(\alpha_i) \subseteq F$ with $\alpha_i \underset{\tau}{\to} \alpha$. Then, since right multiplication is τ-continuous, we have that $\alpha_i(upu) \underset{\tau}{\to} \alpha(upu)$, hence $\alpha_i(upu)v \underset{\tau_v}{\to} \alpha(upu)v$, i.e. $\alpha_i p \underset{\tau_v}{\to} \alpha p$, since $\alpha_i(upu)v = \alpha_i(upv) = \alpha_i(up) = \alpha_i p$, and the same for α.

But $(\alpha_i p) \subseteq Fv$ and $f|_{Fv}$ is τ_v-continuous, hence $f(\alpha_i p) \to f(\alpha p)$, in other words, $pf(\alpha_i) \to pf(\alpha)$. □

COROLLARY 2.4. *With the above notation*

$$\mathcal{F}^{\#} = \mathcal{F}^* \cap \mathcal{F}'.$$

PROPOSITION 2.5. *We have the following inclusions:*

$$\mathfrak{A}(F) \subseteq \mathcal{F}' \subseteq \mathfrak{A}(F') \subseteq \mathcal{F}^{\tau}.$$

PROOF. For the first inclusion, note that if $f \in \mathfrak{A}(F), \alpha \in F$ and v is an idempotent, then $(vf)(\alpha) = f\alpha(v) = f(v)$, hence vf is a constant on F, hence τ-continuous, and f is then in $\mathcal{F}'$.

For the second inclusion, let $f \in \mathcal{F}'$, $\alpha \in F'$ and $p \in \beta T$. Then pf is τ-continuous. Since $\alpha \in F'$, there exists $\{\alpha_i\}_i \subset F$ with $\alpha_i \underset{\tau}{\to} \alpha$ and $\alpha_i \underset{\tau}{\to} u$. Hence, since pf is τ-continuous, we have both $(pf)(\alpha_i) \to (pf)(\alpha)$ and $(pf)(\alpha_i) \to (pf)(u)$. Thus, $pf(\alpha) = pf(u)$. So we have proved that for all, $f \in F', pf(\alpha) = pf(u)$. Hence,

$$f\alpha(p) = f(\alpha p) = (pf)(\alpha) = (pf)(u) = f(up) = fu(p),$$

i.e., $f\alpha = fu = f$ for all $\alpha \in F'$, and this implies that $f \in \mathfrak{A}(F')$.

For the third inclusion, let $f \in \mathfrak{A}(F')$. Denote the canonical projection from F onto F/F' by π. Since $f \in \mathfrak{A}(F')$, then $f\alpha = f$ for all $\alpha \in F'$. Hence, $f|_F$ projects to the quotient, i.e., there exists $g : F/F' \to I\!R$ such that $f = g \circ \pi$. But $f \in \mathcal{C}(\beta T)$, thus g is also continuous, in the usual topology. Since F/F' is compact Hausdorff, the usual and the τ-topologies are equal. So, g is continuous in the τ-topology, hence so is $f|_F$. □

DEFINITION 2.6 ([**EGS**]). An extension $\pi : (X, x_0) \to (Y, y_0)$ of minimal flows is called *full* if

$$\mathfrak{G}(Y, y_0)^{\#} \subseteq \mathfrak{G}(X, x_0)\mathfrak{G}(Y, y_0)'$$

The definition seem to depend on the base points chosen. We will prove shortly that this is not so. We need to use the fact that $F^{\#} = F^*F'$, a proof of which can be found combining 14.14 of [**E2**] and 3.10 of [**EGS**]. However, we give a simpler proof now.

COROLLARY 2.7. *With have the above notation*

$$\mathfrak{G}(\mathcal{F}^{\tau}) \subseteq F' = \mathfrak{G}(\mathcal{F}').$$

PROOF. $\mathfrak{G}(\mathcal{F}^{\tau}) \subseteq F' \subseteq \mathfrak{G}(\mathcal{F}')$ is obvious by 2.5. Now, apply 2.4 to $\mathfrak{A}(F)$. We then have that $\mathfrak{A}(F)^{\#} \subseteq \mathfrak{A}(F)'$. But $\mathfrak{A}(F)' = \mathcal{F}'$ by 2.2. So $\mathfrak{G}(\mathcal{F}') \subseteq \mathfrak{G}(\mathfrak{A}(F)^{\#})$. However, by ([**W**] III.1.13 a) and [**EGS**] 5.12), $\mathfrak{A}(F) \subseteq \mathfrak{A}(u)$ is full, i.e. $\mathfrak{G}(\mathfrak{A}(F)^{\#}) \subseteq \mathfrak{G}(\mathfrak{A}(u))\mathfrak{G}(\mathfrak{A}(F))' = \{u\}F' = F'$, so we have $\mathfrak{G}(\mathcal{F}') \subseteq F'$. □

COROLLARY 2.8. *We have the following equality*

$$F^{\#} = F^*F'.$$

PROOF. By 2.7, 2.4 and the fact ([**E2**] 13.11) that $\mathfrak{G}(\mathcal{A} \cap \mathcal{F}^*) = \mathfrak{G}(\mathcal{A})\mathfrak{G}(\mathcal{F}^*)$ for any T-subalgebra $\mathcal{A}$ of $\mathfrak{A}(u)$. (For another proof see [**E2**],14.14). □

COROLLARY 2.9. *With the above notation*

$$\mathfrak{G}(\mathfrak{A}(F)^{\#}) = F'.$$

PROOF. As in the proof of 2.7, we have the inclusion $\mathfrak{G}(\mathfrak{A}(F)^{\#}) \subseteq F'$. On the other hand, by 2.8, $\mathfrak{G}(\mathfrak{A}(F)^{\#}) = F'\mathfrak{G}(\mathfrak{A}(F)^*) \supseteq F'$. □

Now we prove that the notion of "full extension" is independent of base point.

LEMMA 2.10. *The following equivalence holds:*

$$F^{\#} \subseteq AF' \iff (q^{-1}Fq)^{\#} \subseteq (r^{-1}Ar)(q^{-1}Fq)' \;\forall\, r, q \in M : r \in \mathfrak{E}_q(\mathcal{F}).$$

PROOF. ($\Leftarrow$) Take $q = r = u$.

($\Rightarrow$) Let $v = v_q$, $w \in J$, $p \in M$ such that $\mathcal{F}q \subseteq \mathfrak{A}_v(p, w)$ and $f \in \mathcal{F}$. Then $fqp \in \mathfrak{A}(w)$ and so $fqpw = fqp \;\forall\, f \in \mathcal{F}$, which proves that $\mathcal{F} \subseteq \mathfrak{A}(qp, w)$, and then of course $\mathcal{F}^* \subseteq \mathfrak{A}(qp, w)$ also. So if $h \in \mathcal{F}^*$, we have $hqpw = hqp$, i.e., $hq \in \mathfrak{A}_v(p, w)$. So, we have shown that

$$\mathcal{F}^*q \subseteq (\mathcal{F}q)^*. \tag{1}$$

Applying (1) with $\mathcal{F}$ replaced by $\mathcal{F}q$ and q replaced by $q^{-1}u$, we obtain $(\mathcal{F}q)^*q^{-1}u \subseteq ((\mathcal{F}q)q^{-1}u)^* = (\mathcal{F}u)^* = \mathcal{F}^*$. Hence $(\mathcal{F}q)^* \subseteq \mathcal{F}^*q$. This with (1) gives us

$$\mathcal{F}^*q = (\mathcal{F}q)^*. \tag{2}$$

Now,

$$q^{-1}F^*q = q^{-1}\mathfrak{G}(\mathcal{F}^*)q = \mathfrak{G}(\mathcal{F}^*q) \overset{(2)}{=} \mathfrak{G}((\mathcal{F}q)^*) = (\mathfrak{G}(\mathcal{F}q))^* = (q^{-1}Fq)^*. \tag{3}$$

Also, let $f \in \mathcal{F}'$, $w \in J$ and let $\alpha_i \underset{\tau}{\to} \alpha$ with $\alpha_i \in \mathfrak{G}(\mathcal{F}q) = q^{-1}Fq$. Then, let $\beta_i \in F$ such that $\alpha_i = q^{-1}\beta_i q$. Because both left and right multiplication are τ-continuous, we get that $\beta_i \underset{\tau}{\to} q\alpha q^{-1}u$. Then,

$$\begin{aligned} wfq(\alpha_i) &= wfq(q^{-1}\beta_i q) = wf(\beta_i q) = \\ &= qwf(\beta_i) \to qwf(q\alpha q^{-1}) = wfq(\alpha v) = wfq(\alpha). \end{aligned}$$

Since $fq \in \mathfrak{A}(v)$, we get that

$$fq \in \{g \in \mathfrak{A}(v) : wf|q^{-1}Fq \text{ is } \tau - \text{continuous}\} = (\mathcal{F}q)'.$$

Thus, $\mathcal{F}'q \subseteq (\mathcal{F}q)'$, and hence, as in (2), we obtain $\mathcal{F}'q = (\mathcal{F}q)'$. Consequently, we obtain

$$q^{-1}F'q = (q^{-1}Fq)'. \tag{4}$$

Now, by hypothesis $r \in \mathfrak{E}_q(\mathcal{F})$, thus $rq^{-1} \in \mathfrak{E}_v(\mathcal{F})$. So if $\alpha \in F$ we obtain

$$rq^{-1}\alpha|_{\mathcal{F}} = rq^{-1}|_{\mathcal{F}}\alpha = v|_{\mathcal{F}}\alpha = v\alpha|_{\mathcal{F}} = \alpha|_{\mathcal{F}} = u|_{\mathcal{F}}.$$

Thus, $rq^{-1}\alpha \in F$, i.e., $rq^{-1}F \subseteq F$, or $rq^{-1}F = F$ since F is a group. Thus

$$q^{-1}F = r^{-1}F \tag{5}$$

and, taking inverses and multiplying by v, $Fq = Frv$. Using this and (5), we get: $q^{-1}Fq = r^{-1}Frv = (rv)^{-1}F(rv)$, and so by (4):

$$q^{-1}F'q = \left((rv)^{-1}F(rv)\right)' = (rv)^{-1}F'rv = r^{-1}vF'rv = r^{-1}F'rv.$$

In the same way, by (3), $q^{-1}F^*q = r^{-1}F^*rv$. Thus

$$(q^{-1}Fq)^{\#} \overset{2.8}{=} (q^{-1}Fq)^*(q^{-1}Fq)' = r^{-1}F^*rvr^{-1}F'rv = r^{-1}F^*F'rv = r^{-1}F^{\#}rv.$$

Now, by hypothesis, $F^{\#} \subseteq AF'$, thus

$$(q^{-1}Fq)^{\#} \subseteq r^{-1}AF'rv = r^{-1}Arr^{-1}F'rv = (r^{-1}Ar)(q^{-1}Fq)'.$$

□

COROLLARY 2.11. *The notion of a "full extension" does not depend on the chosen base point.*

PROOF. Let $\pi : (X, x_0) \to (Y, y_0)$ be full. Take $y_i \in Y$ and $x_i \in X$ with $\pi(x_i) = y_i$, $i = 1, 2$ and $p, q \in M$ with $x_1 = x_0 p$, $y_1 = y_0 q$. Then

$$\mathfrak{G}(Y, y_1) = q^{-1}\mathfrak{G}(Y, y_0)q, \mathfrak{G}(X, x_1) = p^{-1}\mathfrak{G}(X, x_0)p,$$

and

$$p|_{al(Y,y_0)} = y_0 p = \pi(x_0 p) = \pi(x_1) = y_1 = y_0 q = q|_{al(Y,y_0)},$$

so that $p \in \mathfrak{E}_q(al(Y, y_0))$.

Since $\pi : (X, x_0) \to (Y, y_0)$ is full we have $\mathfrak{G}(Y, y_0)^{\#} \subseteq \mathfrak{G}(X, x_0)\mathfrak{G}(Y, y_0)'$, so by 2.10 $(q^{-1}\mathfrak{G}(Y, y_0)q)^{\#} \subseteq (p^{-1}\mathfrak{G}(X, x_0)p)(q^{-1}\mathfrak{G}(Y, y_0)q)'$, i.e. $\mathfrak{G}(Y, y_1)^{\#} \subseteq \mathfrak{G}(X, x_1)\mathfrak{G}(Y, y_1)'$, which says that $\pi : (X, x_1) \to (Y, y_1)$ is full. □

DEFINITION 2.12. (A) An extension $\mathcal{F} \subseteq \mathcal{A}$ is EP (equicontinuous extension of a proximal extension) iff there exists a T-subalgebra $\mathcal{F}_1$ of $\mathfrak{A}(u)$ with $\mathcal{F} \subseteq \mathcal{F}_1 \subseteq \mathcal{A}$ and :

i) $\mathcal{F} \subseteq \mathcal{F}_1$ is proximal;

ii) $\mathcal{F}_1 \subseteq \mathcal{A}$ is equicontinuous;

(B) An extension $\mathcal{F} \subseteq \mathcal{A}$ is PEP (proximally equivalent to an EP) iff there exists a T-subalgebra $\mathcal{F}_2$ of $\mathfrak{A}(u)$ with $\mathcal{F} \subseteq \mathcal{F}_2$ EP and:

iii) $\mathcal{F}_2$ and $\mathcal{A}$ are proximally equivalent.

The following result is known, and a proof for the absolute case can be found in [**G**].

PROPOSITION 2.13. *The extension $\mathcal{F} \subseteq \mathcal{A}$ is PEP iff $F' \subseteq A$.*

PROOF. ($\Rightarrow$) Let $\mathcal{F}_1$ and $\mathcal{F}_2$ be as in the definition of PEP. We then have that $F = F_1$, by (i). Hence, $F' = F_1'$ and $F' = F_1' \subseteq F_1'F_1^* = F_1^{\#} \subseteq F_2 = A$, (by (ii) and (iii)).

($\Leftarrow$) Consider $\mathcal{F} \subseteq \mathfrak{A}(F) \subseteq \mathfrak{A}(F)^{\#} \cap \mathcal{A}$. Clearly $\mathcal{F} \subseteq \mathfrak{A}(F)$ is proximal and $\mathfrak{A}(F) \subseteq \mathfrak{A}(F)^{\#} \cap \mathfrak{A}(A)$ is equicontinuous. In [**EGS**], section 7, it's proved that the extension $\mathfrak{A}(F)^{\#} \cap \mathfrak{A}(AF') \subseteq \mathfrak{A}(AF')$ is proximal. In our case, since $F' \subset A$, we have that $AF' = A$. So, we obtain that the extension $\mathfrak{A}(F)^{\#} \cap \mathfrak{A}(A) \subseteq \mathfrak{A}(A)$ is proximal. Since the extension $\mathcal{A} \subseteq \mathfrak{A}(A)$ is proximal as well, we have that $\mathcal{A}$ and $\mathfrak{A}(F)^{\#} \cap \mathfrak{A}(A)$ are proximally equivalent. □

PROPERTY 2.14. *The extension $\mathcal{F} \subseteq \mathcal{A}$ is EP implies $\mathcal{A} \subseteq \mathcal{F}'$ and $A \subseteq F'$ implies $\mathcal{F} \subseteq A$ is PEP.*

PROOF. We only need to prove the first implication, since the second is true by 2.13 and 2.7. If $\mathcal{F} \subseteq \mathcal{A}$ is EP, then take $\mathcal{F}_1$ as in the definition of EP. Then by (ii) and 2.4, $\mathcal{A} \subseteq \mathcal{F}_1^{\#} \subseteq \mathcal{F}_1' = \mathcal{F}'$, the last equality by condition i) and 2.2. □

Recall from [**EGS**] that an extension is called a strict-PI extension if it is a tower of alternate proximal and equicontinuous extensions, and it is PI if it is proximally equivalent to a strict-PI extension. It is known (see [**EGS**], Section 7)) that an extension $\mathcal{A}$ of $\mathcal{F}$ is PI iff $F^{\infty} \subseteq A$.

PROPOSITION 2.15. *With above notation*

(a) $\mathcal{F}^{\infty} = \mathfrak{A}(F^{\infty})$,

(b) $\mathcal{A} \subseteq \mathcal{F}^{\infty} \iff \mathcal{F} \subseteq \mathcal{A}$ *is PI.*

PROOF. If we apply the first and second inclusions of 2.5 to $\mathcal{F}^{\infty}$, we have that $\mathfrak{A}(F^{\infty}) \subseteq (\mathcal{F}^{\infty})' = \mathcal{F}^{\infty} \subseteq \mathfrak{A}((F^{\infty})') = \mathfrak{A}(F^{\infty})$. Now (b) is clear from (a). □

3. The Proximal Relation

DEFINITION 3.1. Given an extension $\pi : X \mapsto Y$, define

$$R(\pi) = R(X, Y) = \{(x, z) : \pi(x) = \pi(z)\}.$$

Two points $x, z \in X$ are called (π-)*proximal* if $\pi(x) = \pi(z)$ and there exists $\{t_i\}_i \subset T$ with $\lim xt_i = \lim zt_i$. The *proximal relation* $P(\pi) = P(X, Y) = P(\mathcal{A} :$

$\mathcal{F}$) is $\{(x,z) \in X \times X : x$ and z are π-proximal $\}$. When $Z = \star - a$ one point space, we set $P(\pi) = P(X)$. Note that $P(\pi) = P(X) \cap R(\pi)$ and π is *proximal* if $R(\pi) = P(\pi)$.

REMARK 3.2. It follows that $P(X,Y)$ is reflexive and symmetric, but in general it is not necessarily transitive. Moreover, it need not be closed. However, Auslander, in [**A1**], proved that if $P(X,Y)$ is closed, then it is an equivalence relation. An example of a (non-minimal) flow X in which $P(X)$ is an equivalence relation but not closed can be found in [A1], while [**S**] has an example of a minimal flow where P is a non closed equivalence relation. We are mainly interested in minimal flows, so from now on we shall assume that the flows are minimal. We will use the algebraic point of view and denote $P(X,Y)$ as $P(\mathcal{A} : \mathcal{F})$, where $\mathcal{A}$, $\mathcal{F}$ are the algebras of X and Y respectively. In [**E2**] a necessary and sufficient condition in terms of the groups of $\mathcal{A}$ and $\mathcal{F}$ is given for $P(\mathcal{A} : \mathcal{F})$ to be a closed equivalence relation. The hypothesis of being a closed relation could not be removed until recently. Finally, in [E4] a necessary and sufficient condition for $P(\mathcal{A})$ to be an equivalence relation was obtained in terms of the group of $\mathcal{A}$. We give here a necessary and sufficient condition in terms of the Ellis group of $\mathcal{A}$ and a group induced by $\mathcal{F}$ for $P(\mathcal{A} : \mathcal{F})$ to be an equivalence relation, and prove that this condition is the same as the one in [**E4**] when $\mathcal{F} = I\!R$. Also, we use this characterization to give another proof of the fact that if $P(\mathcal{A} : \mathcal{F})$ is closed, then it is an equivalence relation. (Note however that Auslander's result is valid for non-minimal flows as well).

DEFINITION 3.3. Given another minimal right ideal K of βT, and $v \in J(M)$, let's denote by v^K the unique idempotent in K which is equivalent to v. That is, $vv^K = v$ and $v^K v = v^K$, (for the existence and uniqueness of v^K, see again [**E2**]).

LEMMA 3.4. *If $v \in \mathfrak{J}_p(\mathcal{F})_v$ then $pv^K \underset{\mathcal{F}}{=} p$.*

PROOF. Observe that

$$p \underset{\mathcal{F}}{=} pv \Rightarrow pv^K \underset{\mathcal{F}}{=} pvv^K = pv \underset{\mathcal{F}}{=} p.$$

□

DEFINITION 3.5. Let $\mathcal{N}(\mathcal{F})$ be the τ-closed subgroup of F generated by the set of elements of the form $pv^K p^{-1}u$, where $p \in M$, $v \in \mathfrak{J}_p(\mathcal{F})$ and K is a minimal right ideal of βT.

REMARK 3.6. Note that $v \in \mathfrak{J}_p(\mathcal{F}) \Rightarrow pv^K p^{-1}u \in F$ since it is clearly in G, and by 3.4 if $v \in \mathfrak{J}_p(\mathcal{F})$, then $pv^K \underset{\mathcal{F}}{=} p$. Hence $pv^K p^{-1}u \underset{\mathcal{F}}{=} pp^{-1}u = u$.

DEFINITION3.7. Let,

$$P^* = P^*(\mathcal{A} : \mathcal{F}) = \{(p,q) \in M \times M : (p|_{\mathcal{A}}, q|_{\mathcal{A}}) \in P(\mathcal{A} : \mathcal{F})\}.$$

Thus, $(p,q) \in P^* \iff p \underset{\mathcal{F}}{=} q$ and there exists $r \in \beta T$ such that $pr \underset{\mathcal{A}}{=} qr$. Let us recall, however, that a necessary and sufficient condition for (p,q) to be in $P^*(\mathcal{A} : \mathcal{F})$ is the existence of a minimal right ideal K of βT, and of an idempotent $\eta \in K$, with $p\eta \underset{\mathcal{A}}{=} q\eta$ and $p \underset{\mathcal{F}}{=} q$. (see [**E2**]). Clearly $P^*(\mathcal{A} : \mathcal{F})$ is an equivalence relation iff $P(\mathcal{A} : \mathcal{F})$ is so.

Also, let's recall (see [**E2**]) that the only almost periodic points in $P(\mathcal{A} : \mathcal{F})$ are the elements of the diagonal. In particular:

LEMMA 3.8. *Let $(p,q) \in P^*(\mathcal{A}:\mathcal{F})$. If (p,q) is an almost periodic point in $M \times M$, then $p \underset{\mathcal{A}}{=} q$.*

LEMMA 3.9. *If $(p, pv^K v_p) \in P^*(\mathcal{A}:\mathcal{F})$, for all $p \in M$, $v \in \mathfrak{J}_p(\mathcal{F})$ and K a minimal right ideal in βT, then $\mathcal{N}(\mathcal{F}) \subseteq A$.*

PROOF. Let $p \in M$, $v \in \mathfrak{J}_p(\mathcal{F})$ and K a minimal right ideal in βT. Then, $(p, pv^K v_p) \in P^*(\mathcal{A}:\mathcal{F})$, by the hypothesis. Since $(p, pv^K v_p) = (p, pv^K v_p)v_p$, we have that $(p, pv^K v_p)$ is an almost periodic point. Thus, by 3.8, we have that $p \underset{\mathcal{A}}{=} pv^K v_p$, which implies $u \underset{\mathcal{A}}{=} pv^K p^{-1} u$. Hence, $pv^K p^{-1} u \in A$ and $\mathcal{N}(\mathcal{F}) \subseteq A$. □

LEMMA 3.10. *If $\mathcal{N}(\mathcal{F}) \subseteq A$ and $(p,q) \in P^*(\mathcal{A}:\mathcal{F})$, then $p \underset{\mathcal{A}}{=} qv_p$.*

PROOF. Let $(p,q) \in P^*(\mathcal{A}:\mathcal{F})$. Then there exists a minimal right ideal K and an idempotent $\eta \in K$ with $p\eta \underset{\mathcal{A}}{=} q\eta$.

Let $v = v_p$. Since $v^K, \eta \in K$ and η is an idempotent, we have that $\eta v^K = v^K$. On the other hand, $p = pv^K$ (3.4 with "$\mathcal{F}$"$= \mathfrak{A}(u)$). Hence

$$p = p(\eta v^K) = (p\eta)v^K \underset{\mathcal{A}}{=} (q\eta)v^K = q(\eta v^K) = qv^K. \tag{6}$$

Since $qv \underset{\mathcal{F}}{=} pv = p \underset{\mathcal{F}}{=} q$, we have that $v \in \mathfrak{J}_q(\mathcal{F})$, so $qv^K q^{-1} u \in \mathcal{N}(\mathcal{F})$. By hypothesis, $\mathcal{N}(\mathcal{F}) \subseteq A$, thus $qv^K q^{-1} u \underset{\mathcal{A}}{=} u$ and $qv^K v \underset{\mathcal{A}}{=} qv$. So, by (6), $p \underset{\mathcal{A}}{=} qv$. □

THEOREM 3.11. *The set $P(\mathcal{A}:\mathcal{F})$ defines an equivalence relation if and only if $\mathcal{N}(\mathcal{F}) \subseteq A$.*

PROOF. ($\Rightarrow$) Let $p \in M$, $v \in \mathfrak{J}_p(\mathcal{F})$ and K be a minimal right ideal in βT. Then $p \underset{\mathcal{F}}{=} pv^K$ by 3.4 and since $pv^K = (pv^K)v^K$ we have

$$(p, pv^K) \in P^*(\mathcal{A}:\mathcal{F}). \tag{7}$$

On the other hand, $(pv^K)v_p = (pv^K v_p)v_p$ and $pv^K v_p \underset{\mathcal{F}}{=} pv_p = p \underset{\mathcal{F}}{=} pv^K$. Hence

$$(pv^K, pv^K v_p) \in P^*(\mathcal{A}:\mathcal{F}). \tag{8}$$

Now (7), (8) and the hypothesis imply that $(p, pv^K v_p) \in P^*(\mathcal{A}:\mathcal{F})$. So $\mathcal{N}(\mathcal{F}) \subseteq A$ by 3.9.

($\Leftarrow$) Let $(p,q),(q,r) \in P^*(\mathcal{A}:\mathcal{F})$. Then, by 3.10, $q \underset{\mathcal{A}}{=} rv_q$ and $p \underset{\mathcal{A}}{=} qv_p$ thus $p \underset{\mathcal{A}}{=} (rv_q)v_p = rv_p$. We also have that $p \underset{\mathcal{F}}{=} q$, $q \underset{\mathcal{F}}{=} r$, so $p \underset{\mathcal{F}}{=} r$. Hence, $(p,r) \in P^*(\mathcal{A}:\mathcal{F})$. □

COROLLARY 3.12. *The condition "$P(\mathcal{A}:\mathcal{F})$ is an equivalence relation" is a group condition on $\mathcal{A}$, i.e., if $A = B$, and $P(\mathcal{A}:\mathcal{F})$ is an equivalence relation, then so is $P(\mathcal{B}:\mathcal{F})$.*

QUESTION 3.13. Is the above condition a group condition on $\mathcal{F}$?

COROLLARY 3.14. *Let $\mathcal{F} \subseteq \mathcal{H} \subseteq \mathcal{B} \subseteq \mathcal{A}$. Then, if $P(\mathcal{A}:\mathcal{F})$ is an equivalence relation, so is $P(\mathcal{B}:\mathcal{H})$.*

PROOF. Notice that $\mathfrak{J}_p(\mathcal{H}) \subseteq \mathfrak{J}_p(\mathcal{F})$ implies that $\mathcal{N}(\mathcal{H}) \subseteq \mathcal{N}(\mathcal{F})$. Since $A \subseteq B$, we have that $\mathcal{N}(\mathcal{F}) \subseteq A \Rightarrow \mathcal{N}(\mathcal{H}) \subseteq B$. □

The following corollary was proved in [**A1**], for the absolute case, although that proof can be easily modified to take care of the general case. This proof is shorter but valid only for minimal flows, while in [**A1**] it is valid for any general flow.

COROLLARY 3.15. *If $P(\mathcal{A} : \mathcal{F})$ is closed, then it is an equivalence relation.*

PROOF. Let $p \in M$, $v \in \mathfrak{J}_p(\mathcal{F})$ and K a minimal right ideal in βT. Thus $p \underset{\mathcal{F}}{=} pv^K$ by 3.4. Since $pv^K = (pv^K)v^K$, we have that $(p, pv^K) \in P^*(\mathcal{A} : \mathcal{F})$. However, $P^*(\mathcal{A} : \mathcal{F})$ is βT-invariant (since it is T-invariant and, by the hypthesis, closed), so $(p, pv^K)v_p \in P^*(\mathcal{A} : \mathcal{F})$. Thus, we have that $(p, pv^K v_p) \in P^*(\mathcal{A} : \mathcal{F})$ and hence $\mathcal{N}(\mathcal{F}) \subseteq A$ by 3.9. □

REMARK 3.16. By 13.13 of [**E2**], $P(\mathcal{A} : \mathcal{F})$ is closed iff $F^* \subseteq A$. Since $N(\mathcal{F}) \subseteq F^*$ (directly from the definition, or else, because $P(\mathcal{F}^* : \mathcal{F}) = \Delta$ is an equivalence relation), we can get yet another proof of the previous result.

PROPOSITION 3.17. *The subgroup $\mathcal{N}(\mathcal{F})$ is normal in F.*

PROOF. Let $\alpha \in F$, $p \in M$, $v \in \mathfrak{J}_p(\mathcal{F})$ and K a minimal right ideal of βT. Then $\alpha \underset{\mathcal{F}}{=} u$ and thus we have both $p \underset{\mathcal{F}}{=} \alpha p$ and $\alpha pv \underset{\mathcal{F}}{=} pv \underset{\mathcal{F}}{=} p$ so that $\alpha pv \underset{\mathcal{F}}{=} \alpha p$, i.e. $v \in \mathfrak{J}_{\alpha p}(\mathcal{F})$, which implies that $\alpha(pv^K p^{-1}u)\alpha^{-1} \in \mathcal{N}(\mathcal{F})$. Since $\alpha(pv^K p^{-1}u)\alpha^{-1} = (\alpha p)v^K(\alpha p)^{-1}$, we are done. □

DEFINITION 3.18. Let $\hat{P}$ be the τ-closed normal subgroup of G generated by the set of elements vu^K, where $v \in J(M)$ and k is a minimal rigth ideal of βT.

REMARK 3.19. In [**E4**], it is proved that $P(\mathcal{A})$ is an equivalence relation if and only if $\hat{P} \subseteq A$. In fact we have the following:

PROPOSITION 3.20.

$$\hat{P} = \mathcal{N}(I\!R).$$

PROOF. Since

$$(vu^K)(uv^K u) = v(u^K u)v^K u = vu^K v^K u = v(u^K v^K)u = vv^K u = vu = u,$$

and

$$(uv^K u)(vu^K) = uv^K(uv)u^K = uv^K vu^K = u(v^K v)u^K = uv^K u^K = uu^K = u,$$

we have that $(vu^K)^{-1} = uv^K u$. Thus $vu^K \in \mathcal{N}(\mathcal{F})$ and $\hat{P} \subseteq \mathcal{N}(I\!R)$ by 3.17. On the other hand, again using $(vu^K)^{-1} = uv^K u$ we have that $uv^K u \in \hat{P}$. Now let $p \in M$. Then, since $\hat{P}$ is normal, we conclude $pv^K p^{-1}u = (pu)(uv^K u)(pu)^{-1} \in \hat{P}$. Thus $\mathcal{N}(I\!R) \subseteq \hat{P}$. □

4. The Regionally Proximal Relation

DEFINITION 4.1. Given $\pi : X \mapsto Y$, two points $x, z \in X$ are called (π)-*regionally proximal* if there exists $\{t_i\}_i \subset T$ and $\{x_i\}_i; \{z_i\}_i \subset X$ with $\lim x_i t_i = \lim z_i t_i$, $\pi(x_i) = \pi(z_i)$ and $\lim x_i = x$, $\lim z_i = z$. The *regionally proximal* relation $Q(\pi) = Q(X, Y) = Q(\mathcal{A} : \mathcal{F})$ is simply $\{(x, z) \in X \times X : x$ and z are regionally proximal.$\}$.

REMARK4.2. Note that $P(\pi) \subseteq Q(\pi) \subseteq R(\pi)$. As before, if we set $Q(X) = Q(X, \star)$, then we have that $Q(\pi) \subseteq Q(X) \cap R(\pi)$. However, in general the equality does not hold.

DEFINITION 4.3. If $\mathcal{A} \subseteq \mathfrak{A}(u)$ is a T-subalgebra, then since $\mathcal{A} \subseteq \mathcal{C}(M)$, $\mathcal{A}$ induces a new topology on M. We will denote this new topology by $\mathcal{P}_{\mathcal{A}}$.

REMARK 4.4. Notice that $p_i \xrightarrow{\mathcal{P}_{\mathcal{A}}} p \iff p_i|_{\mathcal{A}} \to p|_{\mathcal{A}}$. Since $(p_i t)|_{\mathcal{A}} = p_i|_{\mathcal{A}} t$, it is clear that right multiplication by an element of T is $\mathcal{P}_{\mathcal{A}}$-continuous.

LEMMA 4.5. *If $\mathcal{F} \subseteq \mathcal{A} \subseteq \mathcal{B}$, and $\mathcal{A} \subseteq \mathcal{B}$ is regionally proximal (in particular, if it is proximal) and $Q(\mathcal{B} : \mathcal{F})$ is an equivalence relation, then so is $Q(\mathcal{A} : \mathcal{F})$.*

PROOF. Let $(p_1|_{\mathcal{A}}, p_2|_{\mathcal{A}})$ and $(p_2|_{\mathcal{A}}, p_3|_{\mathcal{A}}) \in Q(\mathcal{A} : \mathcal{F})$. Then, there exists $(q_1|_{\mathcal{B}}, q_2|_{\mathcal{B}})$ and $(\hat{q}_2|_{\mathcal{B}}, q_3|_{\mathcal{B}}) \in Q(\mathcal{B} : \mathcal{F})$ with $q_i|_{\mathcal{A}} = p_i|_{\mathcal{A}}$, $i = 1, 2, 3$ and $\hat{q}_2|_{\mathcal{A}} = p_2|_{\mathcal{A}}$. ([**E2**], pg 133). Thus, $q_2|_{\mathcal{A}} = \hat{q}_2|_{\mathcal{A}}$ and since $\mathcal{A} \subseteq \mathcal{B}$ is regionally proximal we have that $(q_2|_{\mathcal{B}}, \hat{q}_2|_{\mathcal{B}}) \in Q(\mathcal{B} : \mathcal{A})$. Since $Q(\mathcal{B} : \mathcal{A}) \subseteq Q(\mathcal{B} : \mathcal{F})$, we have that $(q_2|_{\mathcal{B}}, \hat{q}_2|_{\mathcal{B}}) \in Q(\mathcal{B} : \mathcal{F})$. Since we also have $(q_1|_{\mathcal{B}}, q_2|_{\mathcal{B}})$, $(\hat{q}_2|_{\mathcal{B}}, q_3|_{\mathcal{B}}) \in Q(\mathcal{B} : \mathcal{F})$ and $Q(\mathcal{B} : \mathcal{F})$ is an equivalence relation, we obtain that $(q_1|_{\mathcal{B}}, q_3|_{\mathcal{B}}) \in Q(\mathcal{B} : \mathcal{F})$. Hence $(p_1|_{\mathcal{A}}, p_3|_{\mathcal{A}}) = (q_1|_{\mathcal{A}}, q_3|_{\mathcal{A}}) \in Q(X, Y)$. □

REMARK 4.6. By the previous result, we can always assume that the algebra of X is of the form $\mathfrak{A}(A)$. This implies that $M \mapsto X$ is open. Thus,

$$\{h(U) : h(U) \text{ is } \mathcal{P}_{\mathcal{A}} - \textit{open}\}$$

is a basis for the $\mathcal{P}_{\mathcal{A}}$-topology.

DEFINITION 4.7. Define for $U, V \subseteq T$, and $p, r \in M$:

$$\begin{aligned} [U, V] =& \{m \in M : Um \cap V \neq \emptyset\} \\ [U, V]_r =& [U, V] \cap \mathfrak{E}_r(\mathcal{F}) \\ [p]_r^{\mathcal{A}} =& \cap\, cls_{\mathcal{P}_{\mathcal{A}}}[U, V]_r \end{aligned}$$

where the intersection is over all $V \in v_p$ and $U \in p$ is such that $h(U)$ is $\mathcal{P}_{\mathcal{A}}$-open.

The sets $[U, V]$ and $[p] = \cap cls[U, V]$ were first introduced in [**EK**]. We will denote $[p]_r^{\mathfrak{A}(u)}$ by $[p]_r$.

We need the following characterization:

PROPOSITION 4.8. *With this notation, we have*

$$q \in [p]_r^{\mathcal{A}} \iff \exists\ \{t_i\}_i \subset T, \{q_i\}_i \subset \mathfrak{E}_r(\mathcal{F}) \textit{such that} \begin{cases} q_i \xrightarrow{\mathcal{P}_{\mathcal{A}}} q \\ q_i t_i \xrightarrow{\mathcal{P}_{\mathcal{A}}} p \\ t_i \longrightarrow v_p \end{cases}$$

PROOF. ($\Rightarrow$) Let $v = v_p$ and $q \in [p]_r^{\mathcal{A}}$. Take

$$I = \{(U, V, W) \in p \times v \times q : h(U), h(W) \text{ are } \mathcal{P}_{\mathcal{A}} \text{ open }\}$$

and $i = (U, V, W) \in I$.

We then have that $W \cap [U, V]_r \neq \emptyset$ and hence there exists $q_i \in h(W)$ with $Uq_i \cap V \neq \emptyset$ and $q_i \in \mathfrak{E}_r(\mathcal{F})$. Thus, there exists $t_i \in V : q_i t_i \in U$, i.e., putting all together, we have $q_i t_i \xrightarrow{\mathcal{P}_{\mathcal{A}}} p$; $t_i \longrightarrow v_p$ and $q_i \xrightarrow{\mathcal{P}_{\mathcal{A}}} q$.

($\Leftarrow$) Let $q \in M$ satisfy the conditions on the right hand side. Let $(U, V, W) \in \{p \times v \times q : h(U), h(W) \text{ are } \mathcal{P}_{\mathcal{A}} \text{ open}\}$. Then, there exists i_1, i_2, i_3 such that $i \geq i_1 \Rightarrow t_i \in V$; $i \geq i_2 \Rightarrow q_i \in h(W)$ and $i \geq i_3 \Rightarrow q_i t_i \in h(U)$. Let $i \geq i_1, i_2, i_3$. Then, $t_i \in U q_i \cap V$; $U q_i \cap V \neq \emptyset$ and $q_i \in [U, V]$. Since $\{q_i\}_i \subset \mathfrak{E}_r(\mathcal{F})$ we then have $q_i \in [U, V]_r$, so $W \cap [U, V]_r \neq \emptyset$ and $q \in [p]_r^{\mathcal{A}}$ □

LEMMA 4.9. *Let $p \in M$. Then: $[U, V]_p = [Ut, Vt]_p$ for all $U, V \subseteq T$ and $t \in T$.*

PROOF. By Lemma 2.5 of [**EK**], $[U, V] = [Ut, Vt]$. □

LEMMA 4.10. *For all $p, r \in M$ and $w \in J$, we have*

$$[p]_r^{\mathcal{A}} = [pw]_r^{\mathcal{A}}.$$

PROOF. Let $x \in [p]_r^{\mathcal{A}}$ and take $U \in pw$, $V \in w$ with $h(U)$ being $\mathcal{P}_{\mathcal{A}}$-open. We want to show that $x \in cls_{\mathcal{A}}[U, V]_r$. So let $v = v_p$ and $\{t_i\}_i \subset T : t_i \to w$. Hence $vt_i \to vw = w$ and $pt_i \to pw$. Bt $h(U)$ and $h(V)$ are open in the usual topology of M, so there exists an i such that $pt_i \in h(U)$ and $vt_i \in h(V)$. Let $t = t_i$. Then $p \in h(Ut^{-1})$ and $v \in h(Vt^{-1})$. This says that $Vt^{-1} \in v$ and since $h(Ut^{-1})$ is a $\mathcal{P}_{\mathcal{A}}$-neighborhood of p (because right multiplication is continuous), we obtain $x \in cls_{\mathcal{A}}[Ut^{-1}, Vt^{-1}]_r \overset{4.9}{=} cls_{\mathcal{A}}[U, V]_r$ i.e. $x \in [pw]_r^{\mathcal{A}}$. So, we have proved that $[p]_r^{\mathcal{A}} \subseteq [pw]_r^{\mathcal{A}}$ and by symmetry we are done. □

DEFINITION 4.11. Let S be a subset of $M \times M$. Then:

$$\mathfrak{E}_{\mathcal{A}}(S) := \{(p, q) \in M \times M : \text{there exists } (r, \hat{r}) \in S : (p|_{\mathcal{A}}, q|_{\mathcal{A}}) = (r|_{\mathcal{A}}, \hat{r}|_{\mathcal{A}})\}.$$

DEFINITION 4.12. Let $Q^*(\mathcal{A} : \mathcal{F})$ ($= Q^*$ if there is no possibility of confusion) denote the relative regionally proximal relation of (M, T) with respect to $\mathcal{F}$ when M is provided with the $\mathcal{P}_{\mathcal{A}}$-topology.

That is,

$$(p, q) \in Q^* \iff \exists \{p_i\}_i, \{q_i\}_i \subset M, \{t_i\}_i \subset T : \begin{cases} p_i|_{\mathcal{F}} = q_i|_{\mathcal{F}} \\ p_i \xrightarrow{\mathcal{P}_{\mathcal{A}}} p \\ q_i \xrightarrow{\mathcal{P}_{\mathcal{A}}} q \\ \lim_{\mathcal{A}} p_i t_i = \lim_{\mathcal{A}} q_i t_i. \end{cases}$$

Also, we will simply denote by Q or $Q(M, \mathcal{F})$ the relative regionally proximal relation of (M, T) with respect to $\mathcal{F}$ when M is provided with the usual topology. That is $Q = Q(\mathfrak{A}(u), \mathcal{F})$. We will use the standard notation pQ^* to denote $\{q : (p, q) \in Q^*\}$.

LEMMA 4.13. *With the introduced notation,*

i) $Q^*(\mathcal{A} : \mathcal{F}) = \{(p, q) \in M \times M : (p|_{\mathcal{A}}, q|_{\mathcal{A}}) \in Q(\mathcal{A} : \mathcal{F})\}$,

ii) Q^* is T (hence βT)-right invariant;

iii) $Q^* = \mathfrak{E}_{\mathcal{A}}(Q)$;

iv) $A(pQ^*) = pQ^*$.

PROOF. The statement i) is obvious, ii) follows from i) and iv) follows from iii). Now, if $\mathcal{F} \subseteq \mathcal{A} \subseteq \mathcal{B}$, then $Q(\mathcal{B} : \mathcal{F})|_{\mathcal{A}} = Q(\mathcal{A} : \mathcal{F})$ by ([**E2**], pg 133). Taking $\mathcal{B} = \mathfrak{A}(u)$, we have iii). $\square$

PROPOSITION 4.14. *Furthermore,*

$$[pw]_p^{\mathcal{A}} \subseteq pQ^* \textit{for} \quad \textit{all } p \in M, w \in J.$$

PROOF. By 4.10we need only to prove that $[p]_p^{\mathcal{A}} \subseteq pQ^*$. Let $q \in [p]_p^{\mathcal{A}}$. Then there exists $\{t_i\}_i \subset T, \{q_i\}_i \subset \mathfrak{E}_p(\mathcal{F})$ with: $q_i \xrightarrow{\mathcal{P}_{\mathcal{A}}} q$; $q_i t_i \xrightarrow{\mathcal{P}_{\mathcal{A}}} p$ and $t_i \longrightarrow v_p$. From this it follows that $pt_i \to pv_p = p$, so $\lim_{\mathcal{A}} pt_i = p|_{\mathcal{A}} = \lim_{\mathcal{A}} q_i t_i$. Since $p \xrightarrow{\mathcal{P}_{\mathcal{A}}} p$ and $p|_{\mathcal{F}} = q_i|_{\mathcal{F}}$, we have that $(p, q) \in Q^*$. $\square$

DEFINITION 4.15. Let $\mathcal{F}$ be a T-subalgebra of $\mathfrak{A}(u)$. Set,

$$H = H_{\mathcal{F}} = \{\alpha \in F : (\alpha, u) \in Q(M, \mathcal{F})\}.$$

REMARK 4.16. By 14.18 of [**E2**], we have $F' \subseteq H \subseteq F^{\#}$.

LEMMA 4.17. *If $H \subseteq AF'$, then $AH = AF'$.*

PROOF. Notice that $F' \subseteq H \Rightarrow AF' \subseteq AH \subseteq A(AF') = AF'$. $\square$

LEMMA 4.18. *With H as defined above, we have*

$$i)\ H^{-1} = H$$

$$ii)\ \alpha H = H\alpha \textit{ for } \quad \textit{all } \alpha \in F.$$

PROOF. i) Let $\alpha \in H$. Then, $(\alpha, u) \in Q^*$, so by 4.13ii), $(u, \alpha^{-1}) = (\alpha\alpha^{-1}, \alpha^{-1}) \in Q^*$ and $\alpha^{-1} \in H$.

ii) Let $\gamma = \beta\alpha$, with $\beta \in H$. Then $(\beta, u) \in Q$ and so $(\beta\alpha, \alpha) \in Q$ and $(\alpha^{-1}\gamma, u) = (\alpha^{-1}\beta\alpha, \alpha^{-1}\alpha) \in Q$ i.e., $\alpha^{-1}\gamma \in H$. So, $\gamma = (\alpha)(\alpha^{-1}\gamma) \in \alpha H$ and $H\alpha \subseteq \alpha H$ for all $\alpha \in F$. Then, using i), we have $\alpha^{-1}H \subseteq H\alpha^{-1}$ for all $\alpha \in F$, or, since F is a group, $\alpha H \subseteq H\alpha$ for all $\alpha \in F$. That is, $\alpha H = H\alpha$ for all $\alpha \in F$. $\square$

LEMMA 4.19. *With above notation,*

$$AH = \{\gamma \in F : (\gamma, u) \in Q^*(\mathcal{A}, \mathcal{F})\}.$$

PROOF. Let $\gamma = \alpha\beta$ with $\alpha \in A$ and $\beta \in H$. Then, $(\gamma|_{\mathcal{A}}, u|_{\mathcal{A}}) = (\beta|_{\mathcal{A}}, u|_{\mathcal{A}})$. Since $(\beta, u) \in Q$, we get that $(\gamma, u) \in \mathfrak{E}_{\mathcal{A}}(Q) = Q^*$ (by 4.13iii)).

Conversely, let $(\gamma, u) \in Q^*$. Then, by 4.13iii), there exists $(p, q) \in Q : p|_{\mathcal{A}} = \gamma|_{\mathcal{A}}$ and $q|_{\mathcal{A}} = u|_{\mathcal{A}}$. Since $(p, q) \in Q$, then $(pu, qu) \in Q$ and $pu(qu)^{-1} \in H$. On the other hand, $pu|_{\mathcal{A}} = \gamma|_{\mathcal{A}}u = \gamma|_{\mathcal{A}}$, or, in other words, $\gamma(pu)^{-1} \in A$. Since also $qu \in A$, we obtain that $\gamma = \gamma(pu)^{-1}pu(qu)^{-1}qu \in AHA = AH$, the last equality follows from 4.18ii). $\square$

COROLLARY 4.20. *With the above notation*

$$H \subseteq AF' \textit{ if and only if } F^{\#} \subseteq AF'.$$

PROOF.
($\Leftarrow$) This is obvious.
($\Rightarrow$) This follows from the previous lemma, 4.17 and 5.1 of [**E3**]. $\square$

PROPOSITION 4.21. *With the notation as above*

$$AHp = pQ^* \cap Mv_p = pQ^*v_p, \text{ for all } p \in \mathfrak{E}_u(\mathcal{F}).$$

PROOF. Let $q \in AHp$. We then have that $v_q = v_p$ and that there exists $\alpha \in AH$ such that $q = \alpha p$. So, by 4.19, $(\alpha, u) \in Q^*$, and $(q,p) = (\alpha p, p) \in Q^*$, i.e., $q \in pQ^*$. Since $v_q = v_p$, $q \in pQ^* \cap Mv_p$ and we have proved that $AHp \subseteq pQ^* \cap Mv_p$.

The inclusion $pQ^* \cap Mv_p \subseteq pQ^*v_p$ is obvious and to prove $pQ^*v_p \subseteq AHp$, take $q \in pQ^*v_p$. Then, there exists $r \in pQ^*$ scuh that $q = rv_p$. Thus, $(r,p) \in Q^*$ and $(q,p) = (rv_p, pv_p) \in Q^*$. Then, $(qp^{-1}u, u) \in Q^*$, i.e., $qp^{-1}u \in AH$, so $q = qp^{-1}up \in AHp$. $\square$

LEMMA 4.22. *If $F^\# \subseteq AF'$, then $\alpha Q^* u = AF'\alpha$, for all $\alpha \in G$.*

PROOF. Note that $\alpha Q^*u \overset{4.21}{=} AH\alpha \overset{4.17}{=} AF'\alpha$. $\square$

The following is a generalization of Theorem 2.10 of [**EK**].

THEOREM 4.23. *Suppose that :*
I) $F^\# \subseteq AF'$ and
II) $\alpha Q^* \subseteq [\alpha]^{\mathcal{A}}_\alpha$ for all $\alpha \in F$.
Then:
(1) *$(pu)Q^* \subseteq pQ^*$ for all $p \in \mathfrak{E}_u(\mathcal{F})$;*
(2) *$(\alpha, \beta) \in Q^* \cap F \times F$ implies $\alpha Q^* = \beta Q^*$;*
(3) *$pQ^* = (pu)Q^*$ for all $p \in \mathfrak{E}_u(\mathcal{F})$;*
(4) *$Q(\mathcal{A} : \mathcal{F})$ is an equivalence relation on $\mathfrak{E}_u(\mathcal{F})$.*

PROOF. (1) Since $pu|_\mathcal{F} = p|_\mathcal{F}u = u|_\mathcal{F}u = u|_\mathcal{F} = p|_\mathcal{F}$, we have that $[x]^{\mathcal{A}}_p = [x]^{\mathcal{A}}_{pu}$ for any x. Now, since $pu \in F$, we have by I) that $(pu)Q^* \subseteq [pu]^{\mathcal{A}}_{pu} = [pu]^{\mathcal{A}}_p \subseteq pQ^*$, the last inclusion by 4.14.

(2) Assume $(\alpha, \beta) \in Q^* \cap F \times F$. Then $\alpha \in \beta Q^* u$, so by 4.22, we have $\alpha \in AF'\beta$. Now let $x \in \alpha Q^*$. Thus, $xu \in (\alpha Q^*)u$, so again by 4.22,

$$xu \in AF'\alpha \subseteq AF'AF'\beta = AF'\beta = \beta Q^* u.$$

Thus $\beta \in (xu)Q^*$. However, since $x|_\mathcal{F} = \alpha|_\mathcal{F} = u|_\mathcal{F}$ implies by (1) that $(xu)Q^* \subseteq xQ^*$, we have that $\beta \in xQ^*$, and hence $x \in \beta Q^*$, that is, $\alpha Q^* \subseteq \beta Q^*$. By symmetry, we are finished.

(3) By (1), we only need to show that $pQ^* \subseteq (pu)Q^*$, so let $x \in pQ^*$. Then $(xu, pu) \in Q^* \cap F \times F$, so by (2), $(xu)Q^* = (pu)Q^*$. Since $x \in (xu)Q^*$ (because $(x|_\mathcal{A}, xu|_\mathcal{A}) = (x|_\mathcal{A}, x|_\mathcal{A}u) \in P(\mathcal{A} : \mathcal{F}) \subseteq Q(\mathcal{A} : \mathcal{F})$), we have $x \in (pu)Q^*$.

(4) Let $(p,q) \in Q^*$ with p, $q \in \mathfrak{E}_u(\mathcal{F})$. Then $(pu, qu) \in Q^* \cap F \times F$, so

$$pQ^* \overset{(3)}{=} (pu)Q^* \overset{(2)}{=} (qu)Q^* \overset{(3)}{=} qQ^*.$$

Let now $(q,r) \in Q^*$ with $q|_\mathcal{F} = r|_\mathcal{F} = u|_\mathcal{F}$. As above, we also have $qQ^* = rQ^*$, thus $pQ^* = rQ^*$ and $(p,r) \in Q^*$. $\square$

The hypothesis $\alpha Q^* \subseteq [\alpha]_\alpha^{\mathcal{A}}$ involves Q^* too much. We would like to eliminate it, or at least, replace it with a hypothesis not involving Q, preferably an hypothesis about $\mathcal{F}$ alone.

LEMMA 4.24. *With the notation as bove*

$$A[p]_r \subseteq [p]_r^{\mathcal{A}}.$$

PROOF. Let $\alpha \in A$ and $q \in [p]_r$. Then there exists $\{q_i\}_i \subset \mathfrak{E}_r(\mathcal{F})$, $\{t_i\}_i \subset T$ with $q_i \to q$, $q_i t_i \to p$ and $t_i \to v_p$. Then, $\alpha q_i \to \alpha q$ and $\alpha q_i t_i \to \alpha p$, which implies that $q_i|_{\mathcal{A}} = \alpha q_i|_{\mathcal{A}} \to \alpha q|_{\mathcal{A}} = q|_{\mathcal{A}}$, and similarly, $q_i|_{\mathcal{A}} t_i \to p|_{\mathcal{A}}$. And $q_i \in \mathfrak{E}_r(\mathcal{F})$, so $q \in [p]_r^{\mathcal{A}}$ □

THEOREM 4.25. *If $\mathcal{F} \subseteq \mathcal{A}$ is full and $AF'p \subseteq A[p]_p$ for all $p \in \mathfrak{E}_u(\mathcal{F})$, then Q^* is an equivalence relation on $\mathfrak{E}_u(\mathcal{F})$.*

PROOF. Let $\alpha \in F$. If $p \in \alpha Q^*$ (hence $p \in \mathfrak{E}_u(\mathcal{F})$, because $\alpha \in F$), then $(p, \alpha) \in Q^*$. Thus $(pu, \alpha) = (p, \alpha)u \in Q^*$, i.e. $pu \in \alpha Q^* u = AF'\alpha$ by 4.22. So $p \in AF'\alpha v_p \subseteq A[\alpha v_p]_{\alpha v_p}$ (by the second hypothesis). But $[\alpha v_p]_{\alpha v_p} = [\alpha]_{\alpha v_p}$ (by 4.10) and $\alpha v_p|_{\mathcal{F}} = v_p|_{\mathcal{F}}$ because $\alpha \in F$. However, since $p \in \mathfrak{E}_u(\mathcal{F})$, we have that $v_p|_{\mathcal{F}} = u|_{\mathcal{F}}$ and so $\alpha v|_{\mathcal{F}} = \alpha|_{\mathcal{F}}$. Thus, $p \in A[\alpha]_\alpha$.

Hence, we have shown that $\alpha Q^* \subseteq A[\alpha]_\alpha \subseteq [\alpha]_\alpha^{\mathcal{A}}$. Since the extension is full, we are done by 4.23. □

COROLLARY 4.26. *If $\mathcal{F} \subseteq \mathcal{A}$ is full, and $\mathcal{F}$ satisfies the property $F'p \subseteq [p]_p$ for all $p \in \mathfrak{E}_u(\mathcal{F})$, then $Q(\mathcal{A} : \mathcal{F})$ is an equivalence relation.*

REMARK 4.27. We used the hypothesis of being "full" twice in the proof: first to prove that $\alpha Q^* \subseteq A[\alpha]_\alpha$, and then to complete the proof.

Can we give conditions under which $F'p \subseteq [p]_p$ is satisfied? In general we cannot. However, if we add the hypothesis of being "full", (which we are going to need anyway to prove the hypothesis $AF'p \subseteq A[p]_p$). We need to introduce a new object:

DEFINITION 4.28. Giving F the discrete topology makes the pair (F, M) (and also $(F, \mathfrak{E}_u(\mathcal{F}))$ into a (left) transformation group. The set of almost periodic points of this transformation group will be denoted $ap(F, M)$. We will denote the operation of taking τ-interior of the τ-closure by ic_τ, and the τ-closure of the τ-interior of the τ-closure by cic_τ. Also, for $p \in M$, $V \subseteq T$ and a T-subalgebra, $\mathcal{F} \subseteq \mathfrak{A}(u)$, define

$$h_p(V) = h(V) \cap \mathfrak{E}_p(\mathcal{F}).$$

Let $p \in \mathfrak{E}_u(\mathcal{F})$. Set,

$$L_{\mathcal{F}}(p) = L(p) = \cap_{V \in p} cic_\tau h_u(V) u.$$

The set $L(p) = L_{I\!R}(p)$ was introduced in [**AEE**].

REMARK 4.29. Notice that $L(p) \subseteq F$.

The next property points out the importance of L.

LEMMA 4.30. *We have,*

$$F' L(p) = L(p).$$

PROOF. $\supseteq$ The inclusion $F'L(p) \supseteq L(p)$ is obvious. To prove the other inclusion, observe that $F'cls_\tau U = cls_\tau U$ for all τ-open set U, so if $V \in p$ we have $F'cic_\tau V = cic_\tau V$. Thus, by the definition of $L(p)$, $F'L(p) \subseteq F'cic_\tau V \subseteq cic_\tau V \subseteq L(p)$. □

LEMMA 4.31. *If $q \in L(p)$ then $p \in [q]_q$.*

PROOF. Let $q \in L(p)$. By 4.29, $q \in F$ and so $u = v_q$. Let $V \in p$, $W \in q$, $U \in u$. Thus (W, Uu) is a τ-neighborhood of q. Since $q \in L(p)$ we obtain $(W, Uu) \cap h_u(V)u \neq \emptyset$. This means that there exists $r \in h_u(V)$ with $Wru \cap Uu \neq \emptyset$. Since $Wru \cap Uu = (Wr \cap U)u$, we obtain $Wr \cap U \neq \emptyset$, hence $r \in [W, U]_u$. Also since $r \in h(V)$, we have $h(V) \cap [W, U]_u \neq \emptyset$ and so $p \in [q]_u$. On the other hand, $q \in F$, hence $q \in \mathfrak{E}_u(\mathcal{F})$, i.e. $[q]_u = [q]_q$, thus $p \in [q]_q$. □

COROLLARY 4.32. *Let $v \in \mathfrak{J}_u(\mathcal{F})$ and $p, q \in M$. If $q \in L(p)v$, then $p \in [q]_q$. (Recall from1.3that $\mathfrak{J}_u(\mathcal{F}) = \{v \in M : v^2 = v$ and $v \underset{\mathcal{F}}{=} u\}$, since $uv = v$).*

PROOF. Let $q \in L(p)v$. There exists $\alpha \in L(p)$ with $q = \alpha v$, so we obtain both $p \in [\alpha]_\alpha$ (by 4.31) and $\alpha \in F$ (by 4.29). Since by hypothesis $v \in \mathfrak{J}_u(\mathcal{F})$, we have that $q \in \mathfrak{E}_u(\mathcal{F})$, thus $[\alpha]_\alpha = [\alpha]_u = [\alpha]_q \overset{4.10}{=} [\alpha v]_q = [\alpha]_\alpha = [q]_q$, i.e. $p \in [q]_q$. □

COROLLARY 4.33. *We have the following inclusion,*

$$L(p)v \subseteq pQ \text{ for all } v \in \mathfrak{J}_u(\mathcal{F}).$$

PROOF. If $q \in L(p)v$, then by 4.32 $p \in [q]_q \overset{4.14}{\subseteq} qQ$. Thus, $q \in pQ$. □

COROLLARY 4.34. *We have,*

$$L(p)v_p \subseteq Hp \text{ for all } p \in \mathfrak{E}_u(\mathcal{F}).$$

PROOF. Observe that $L(p)v_p \subseteq (pQ) \cap Fv_p = Hp$. □

LEMMA 4.35. *If $L(p) \neq \emptyset$ and AH is a group, then $AL(p)v_p = AHp \in \mathfrak{E}_u(\mathcal{F})$ for all $p \in \mathfrak{E}_u(\mathcal{F})$.*

PROOF. By 4.34, $AHL(p)v \subseteq AHHp \subseteq AHAHp = AHp$, (since AH is a subgroup of F). Since $L(p) \neq \emptyset$, let $q \in L(p)$, therefore $qv \in L(p)v \subseteq AHL(p)v \subseteq AHp$. Thus, there exists $\alpha \in AH$ such that $q = \alpha p$. Take now an element $r \in AHp$. Then, there exists $\beta \in AH$ such that $r = \beta p$. Since AH is a group, $\beta\alpha^{-1} \in AH$ and $r = \beta\alpha^{-1}\alpha p = \beta\alpha^{-1}q \in AHL(p)v$. Hence, $AHp \subseteq AHL(p)v$, and since we have already proved the other inclusion, we are done. □

The property $L(p) \neq \emptyset$ will play a central role in what follows, so let's give it a name:

DEFINITION 4.36. A T-subalgebra $\mathcal{F}$ of $\mathfrak{A}(u)$ is said to be an L-algebra (with respect to u), or to satisfy the L condition, if $L(p) \neq \emptyset$ for all $p \in \mathfrak{E}_u(\mathcal{F})$.

A minimal flow (Y, T) is an L-flow, or it is said to satisfy the L condition if for all $y \in Y$ there exists $u \in J$ such that $yu = y$, and $al(Y, y)$ is an L-algebra with respect to u.

LEMMA 4.37. *We have,*

$$\alpha[q]_q = [\alpha q]_{\alpha q} \text{ for all } \alpha \in F \text{ and } q \in M.$$

PROOF. Obviously it suffices to show that $\alpha[q]_q \subseteq [\alpha q]_{\alpha q}$. But if $p \in [q]_q$, then there exists $\{p_i\}_i \subset \mathfrak{E}_q(\mathcal{F})$ and $\{t_i\}_i \subset T$ with $p_i \to p$, $p_i t_i \to q$ and $t_i \to v_q$. Hence, $\alpha p_i \to \alpha p$, $\alpha p_i t_i \to \alpha q$, and $\alpha p_i \underset{\mathcal{F}}{=} p_i \underset{\mathcal{F}}{=} q \underset{\alpha}{=} q$, thus, $\alpha p \in [\alpha q]_{\alpha q}$. □

LEMMA 4.38. *If $\mathcal{F}$ is an L-algebra and $\mathcal{F} \subseteq \mathcal{A}$ is full, then*

$$AF'p \subseteq A[p]_p \text{ for all } p \in \mathfrak{E}_u(\mathcal{F}).$$

PROOF. Let $p \in \mathfrak{E}_u(\mathcal{F})$ and $\alpha \in F'$. Then $p \in F'p = F'\alpha p \subseteq AF'\alpha p$. Since "full" implies that $AF' = AF^{\#} = AH$ (by 4.16, 4.17), we have $AF'\alpha p = AL(\alpha p)v(p)$ (by 4.35). Hence, there exists $\gamma \in A$, $q \in L(\alpha p)v(p)$ with $p = \gamma q$. Then $\gamma^{-1}p = q \in L(\alpha p)v(p)$, which by 4.32 implies

$$\alpha p \in [\gamma^{-1}p]_{\gamma^{-1}p}. \tag{9}$$

Since $\gamma \in A \subseteq F$, we have that $\gamma^{-1} \in F$, and so by 4.37,

$$[\gamma^{-1}p]_{\gamma^{-1}p} = \gamma^{-1}[p]_p \subseteq A[p]_p,$$

hence $\alpha p \in A[p]_p$ by (9). □

COROLLARY 4.39. *If $\mathcal{F}$ is an L-algebra and $\mathcal{F} \subseteq \mathcal{A}$ is full, then $Q^*(\mathcal{A} : \mathcal{F})$ is an equivalence relation on $\mathfrak{E}_u(\mathcal{F})$.*

PROOF. This follows from 4.38 and 4.25 □

THEOREM 4.40. *If $\pi : X \to Y$ is a full extension between minimal flows, and Y satisfies the L condition, then $Q(X, Y)$ is an equivalence relation.*

PROOF. Let $(x_1, x_2), (x_2, x_3) \in Q(X, Y)$. Let $y_0 = \pi(x_1)$ $(= \pi(x_i), i = 2, 3)$, u be an idempotent as in the definition of L for y_0 and $x_0 = \rho(u)$. Take $F, \mathcal{F}, A, \mathcal{A}$, etc, that correspond to these choices, and $p_i, i = 1, 2, 3$ with $p_i|_{\mathcal{A}} = x_i$, $i = 1, 2, 3$. With these choices, $p_i \in \mathfrak{E}_u(\mathcal{F})$ and $(p_1, p_2), (p_2, p_3) \in Q^*$. Since π is full, $F^{\#} \subseteq AF'$ and Q^* is an equivalence relation over $\mathfrak{E}_u(\mathcal{F})$ by 4.39, hence $(p_1, p_3) \in Q^*$ and $(x_1, x_3) \in Q(X, Y)$. □

REMARK 4.41. Observe that the condition of being full: $F^{\#} \subseteq AF'$; though it is a group condition on $\mathcal{A}$, it is NOT a group condition on $\mathcal{F}$, since $F^{\#}$ depends on $\mathcal{F}$ and not on F. In fact, the condition "$Q(\mathcal{A} : \mathcal{F})$ is an equivalence relation" is probably not a group condition on $\mathcal{F}$, because of the following:

PROPOSITION 4.42. *If $|\mathfrak{A}(F)|$ satisfies the L condition, then $Q(\mathcal{A} : \mathfrak{A}(F))$ is an equivalence relation for all $\mathcal{A} \supseteq \mathfrak{A}(F)$.*

PROOF. By 2.9, $F^{\#}$ in this case is equal to F'. Then certainly $F^{\#} \subseteq AF'$, no matter what A is. □

As mentioned above, probability this implies that the condition "$Q(\mathcal{A} : \mathcal{F})$ is an equivalence relation" is not a group condition, for suppose we can find $\mathcal{F}$ such that

a) $Q(A : \mathcal{F})$ is not an equivalence relation and,
b) $\mathfrak{A}(\mathcal{F})$ is an L-algebra,

then we would have two algebras $\mathcal{F}$ and $\mathfrak{A}(\mathcal{F})$ with the same group such that $Q(A : \mathfrak{A}(\mathcal{F}))$ is an equivalence relation while $Q(A : \mathcal{F})$ is not. These are several examples where condition (a) holds. Whether (b) also holds in these example remains to be examined.

COROLLARY 4.43. *Let $X \mapsto Y$ be an extension of minimal flows, with Y satisfying the L condition.*
If $P(X,Y)$ is closed, then $Q(X,Y)$ is an equivalence relation.

PROOF. If $P(X,Y)$ is closed, then it is an equivalence relation (3.15 or [**A1**]). In [**E2**], it is proved that if $P(X,Y)$ is a closed equivalence relation, then $F^* \subseteq A$. Thus, we have that $F^*F' \subseteq AF'$. Since $F^\# = F^*F'$ ([**E2**],14.14 or 2.8) we have $F^\# \subseteq AF'$ and so $X \mapsto Y$ is full and $Q(X,Y)$ is an equivalence relation. □

Let us recall that if $\mathcal{F}$ and $\mathcal{A}$ are T-subalgebras of $\mathfrak{A}(u)$, then

$$R(\mathcal{A} : \mathcal{F}) = \{(p|_{\mathcal{A}}, q|_{\mathcal{A}}) : p \underset{\mathcal{F}}{=} q\}.$$

COROLLARY 4.44. *If Y satisfies the L condition, then $\mathcal{F} \subseteq \mathcal{A}$ is full implies $Q(\mathcal{A} : \mathcal{F}) = R(\mathcal{A} : \mathcal{A} \cap \mathcal{F}^\#)$.*

PROOF. By 5.4 of [E3], $R(\mathcal{A} : \mathcal{A} \cap \mathcal{F}^\#) = P(\mathcal{A} : \mathcal{F})Q(\mathcal{A} : \mathcal{F})$. But

$$P(\mathcal{A} : \mathcal{F})Q(\mathcal{A} : \mathcal{F}) \subseteq Q(\mathcal{A} : \mathcal{F})Q(\mathcal{A} : \mathcal{F}) = Q(\mathcal{A} : \mathcal{F})$$

i.e., $R(\mathcal{A} : \mathcal{A} \cap \mathcal{F}^\#) \subseteq Q(\mathcal{A} : \mathcal{F})$. However $Q(\mathcal{A} : \mathcal{F}) \subseteq R(\mathcal{A} : \mathcal{A} \cap \mathcal{F}^\#)$, so we are done. □

COROLLARY 4.45. *A full extension of a flow with the L condition which has no equicontinuous factors is regionally proximal.*

PROOF. Let $\mathcal{F} \subseteq \mathcal{A}$ with no equicontinuous factors. Then $\mathcal{A} \cap \mathcal{F}^\# = \mathcal{F}$; so $Q(\mathcal{A} : \mathcal{F}) = R(\mathcal{A} : \mathcal{A} \cap \mathcal{F}^\#) = R(\mathcal{A} : \mathcal{F})$. □

REMARK 4.46. Observe that 6.3 of [**E3**] show that if the almost periodic points are dense in $R(\mathcal{A} : \mathcal{F})$, then the extension $\mathcal{F} \subseteq \mathcal{A}$ is full. Thus, if the almost periodic points are dense in $R(\mathcal{A} : \mathcal{F})$ and $\mathcal{F}$ is an L-algebra, (or, more generally, if it satisfies $F'p \subseteq [p]_p$), then $Q(\mathcal{A} : \mathcal{F})$ is an equivalence relation on $\mathfrak{E}_u(\mathcal{F})$. In particular, if $\mathcal{F} \subseteq \mathcal{A}$ is open and point distal, and $\mathcal{F}$ is an L-algebra, we have that $Q(\mathcal{A} : \mathcal{F})$ is an equivalence relation on $\mathfrak{E}_u(\mathcal{F})$.

Now, we will give a property that will ensure the $\mathcal{F}$-L condition, and we will show that the one point flow satisfies the L condition.

DEFINITION 4.47. If $\mathcal{F}$ is a T-subalgebra of $\mathfrak{A}(u)$, then u is said to be a *super $\mathcal{F}$-almost periodic point* ($\mathcal{F}$-sapp) if $u \in ap(F, M)$, and $ap(F, \mathfrak{E}_u(\mathcal{F}))$ is dense in $\mathfrak{E}_u(\mathcal{F})$.

Before proving the next lemma, let's recall another characterization of the τ-topology.

DEFINITION 4.48. If X is a topological space, define:

$$2^X = \{A \subseteq X : \emptyset \neq A = \overline{A}\}.$$

If X is T_1, the topology of X induces a topology on 2^X, called the finite or Vietoris topology. (see [**M**]). We are going to restrict ourselves to the case when X is compact, Hausdorff. In this case, 2^X is also compact, Hausdorf. When (X,T) is a flow, 2^X can also be made into a T-flow, by defining At as the set $\{at : a \in A\}$, $A \in 2^X, t \in T$. This action induces a (not necessarily continuous) action of βT, as with any flow. However, it is NOT true that the action of an element p of βT on an element A of 2^X is the set Ap ($\equiv \{ap : a \in A\}$). In order to distinguish these two sets, the set obtained by the action of p on A will be denoted $A \circ p$. Then, one can define , for any (closed or not) nonempty subset A of X, the set $A \circ p$ as simply $\overline{A} \circ p$.

A description of this set (see 1.3 of [EGS]) is given in the following.

PROPOSITION 4.49. *Let $\emptyset \neq A \subseteq X$, and $p \in \beta T$. Then, the following are equivalent :*

1. $x \in A \circ p$;
2. *there exists $\{a_i\}_i \subset A$ and $\{t_i\}_i \subset T$ with $t_i \to p$ and $a_i t_i \to x$;*
3. *there exists $\{a_i\}_i \subset A$ and $\{p_i\}_i \subset \beta T$ with $p_i \to p$ and $a_i p_i \to x$.*

This can of course also be applied to the flow $X = M$ or $X = \beta T$. In that case we have,

PROPERTY 4.50 ([**EGS**], 1.4). *For any non-empty $K \subseteq \beta T$ and $p \in \beta T$*

$$Kp \subseteq K \circ p.$$

REMARK 4.51. In the case of $X = M$, we have that if $A \subseteq G = Mu$ and $p \in M$, then $q \in A \circ p$ if and only if there exists $\{a_i\}_i \subset A$ and $\{p_i\}_i \subset M$ with $p_i \to p$ and $a_i p_i \to q$; (because we can find such a net with $p_i \in \beta T$). Then, $up_i \in M$, and since L_u is continuous, $up_i \to up = p$ and $a_i(up_i) = (a_i u)p_i = a_i p_i \to q$.

LEMMA 4.52. *Let $K \subseteq G$ then*

$$cls_\tau K = K \circ u \cap G = (K \circ u)u.$$

PROOF. The first equality is Lemma 2.4 of [**EGS**]. For the second, note that $K \circ u \cap G \subseteq (K \circ u)u$. But $(K \circ u)u \subseteq (K \circ u) \circ u = K \circ u^2 = K \circ u$, and $(K \circ u)u \subseteq G$. □

LEMMA 4.53. *Suppose that u is $\mathcal{F}$-sapp. Then $ic_\tau(h_u(V)u) \neq \emptyset$ for all $V \in p$ and $p \in \mathfrak{E}_u(\mathcal{F})$.*

PROOF. Let $V \in p$. Since $h_u(V) = h(V) \cap \mathfrak{E}_u(\mathcal{F})$ is open in $\mathfrak{E}_u(\mathcal{F})$ and, because u is $\mathcal{F}$-sapp, $\overline{ap(F, \mathfrak{E}_u(\mathcal{F}))} = \mathfrak{E}_u(\mathcal{F})$, we obtain that $h_u(V) \cap ap(F, \mathfrak{E}_u(\mathcal{F})) \neq \emptyset$. So, let $q \in ap(F, \mathfrak{E}_u(\mathcal{F}))$ with $q \in h_u(V)$ and take $w = v_q$. Since $q \in ap(F, \mathfrak{E}_u(\mathcal{F}))$, $\overline{Fq}$ is F-minimal. But $q \in \mathfrak{E}_u(\mathcal{F})$ implies $qu \in F$, so $Fq = Fqw = Fw$ and $\overline{Fw}$ is F-minimal. Let $\mathcal{O} = h(V) \cap \overline{Fw}$ $(= h_u(V) \cap \overline{Fw})$. Then $\mathcal{O}$ is a relatively open subset of $\overline{Fw}$, and since $q \in Fq = Fw \subseteq \overline{Fw}$ and $q \in h_u(V) \subseteq h(V)$, we have that $\mathcal{O} \neq \emptyset$. Since $\overline{Fw}$ is F-minimal, we have that $\overline{Fw} = F\mathcal{O} = \cup_{\alpha \in F} \alpha\mathcal{O}$, and, since $\overline{Fw}$ is compact in M, there exists $\{\alpha_1, \alpha_2, \ldots, \alpha_n\} \subseteq F$ such that $\overline{Fw} = \cup_{i=1}^n \alpha_i \mathcal{O}$. So, let $\mathcal{U} = \mathcal{O} \cap Fw$.

Claim: $Fw = \cup_{i=1}^n \alpha_i \mathcal{U}$.

Proof of the claim. Take $p \in Fw$. (Note then that $v_p = w$.). Then, there exist i such that $p \in \alpha_i \mathcal{O}$, so there exists $r \in \mathcal{O}$ with $p = \alpha_i r$. Hence $v_r = v_p = w$. Also, $r \in \mathcal{O} \subseteq h_u(V) \subseteq \mathfrak{E}_u(\mathcal{F}) \Rightarrow ru \in F$, therefore $r = ruw \in Fw$. Thus, $r \in \mathcal{O} \cap Fw = \mathcal{U}$ and $p \in \alpha_i \mathcal{U}$, so the claim is proved.

Continuing with the proof, we have:

$$\begin{aligned} Fw =& cls_{\tau_w} Fw = (Fw \circ w)w \\ =& ((\cup_{i=1}^n \alpha_i \mathcal{U}) \circ w)w \\ =& \cup_{i=1}^n ((\alpha_i \mathcal{U}) \circ w)w \\ =& \cup_{i=1}^n cls_{\tau_w} \alpha_i \mathcal{U}. \end{aligned}$$

Since $int_{\tau_w} Fw \neq \emptyset$, there exists an i such that $int_{\tau_w}(cls_{\tau_w} \alpha_i \mathcal{U}) \neq \emptyset$, and since L_{α_i} is τ-continuous, $int_{\tau_w} cls_{\tau_w}(\mathcal{U}) \neq \emptyset$ and since R_u is τ-continuous, $ic_\tau \mathcal{U}u \neq \emptyset$. But

$$\mathcal{U}u = (h_u(V) \cap Fw)u \subseteq (h_u(V)w)u = h_u(V)u \text{so } ic_\tau(h_u(V)u) \neq \emptyset.$$

□

LEMMA 4.54. *If u is $\mathcal{F}$-sapp, then it satisfies the $\mathcal{F}$-L condition.*

PROOF. By 4.53, $L(p) \neq \emptyset$ since (F, τ) is compact. □

COROLLARY 4.55. *The one point flow is an L-flow.*

PROOF. If $u \in ap(G, M)$ then it is always $\mathbb{R}$-sapp, since $u \in ap(G, M)$ implies that $ut \in ap(G, M)$ for all $t \in T$. Thus, the one point flow satisfies the L condition, since $ap(G, M) \cap J \neq \emptyset$. □

The following was proved in [**AEE**], 1.10 and 1.13. Let E denote the group of the universal equicontinuous flow, i.e., $E = \mathfrak{G}(\mathbb{R}^{\#})$.

COROLLARY 4.56. *If $E \subseteq AG'$, then $Q(X)$ is an equivalence relation.*

PROOF. The condition $E \subseteq AG'$ means that the extension $\mathbb{R} \subseteq \mathcal{A}$ is full. □

REMARK 4.57. Using 4.55 we see that 4.43 generalizes 1.15 of [**AEE**].

COROLLARY 4.58. *If T is nilpotent, then $Q(X)$ is always an equivalence relation.*

PROOF. If T is nilpotent, $\mathbb{R} = \mathfrak{A}(G)$, so the result follows from 4.42 and 4.55.□

REMARK 4.59. We have given two conditions on $\mathcal{F}$ alone that ensures that a full extension of $\mathcal{F}$ will make Q an equivalence relation. In 4.26 the condition is $F'p \subseteq [p]_p$, and in 4.39, the condition is $L(p) \neq \emptyset$. Here we pose a problem: Characterize those algebras $\mathcal{F}$ such that any full extension of $\mathcal{F}$ makes Q an equivalence relation.

We finish by giving a necessary and sufficient condition for Q to be an equivalence relation. Unfortunately, the conditions are heavily related to Q itself, not to the groups.

THEOREM 4.60. *The relation Q^* is an equivalence relation on $\mathfrak{E}_u(\mathcal{F})$ if and only if:*

$$I)\ AH \text{ is a group} \quad \text{and}$$

$$II)\ AHpv \subseteq pQ^* \text{for all } p \in \text{ and } v \in J.$$

PROOF. ($\Leftarrow$) Let (p_1,p_2), $(p_2,p_3) \in Q^* \cap \mathfrak{E}_u(\mathcal{F}) \times \mathfrak{E}(\mathcal{F})$. By 4.21, $p_i \in AHp_{i-1}v_{p_i}$, $i = 2,3$. Hence,

$$p_3 \in AHp_2v_{p_3} \subseteq AHAHp_1v_{p_2}v_{p_3} = AHAHp_1v_{p_3} \overset{I)}{=} AHp_1v_{p_3} \overset{II)}{\subseteq} p_1Q^*.$$

($\Rightarrow$) Let $\alpha, \beta \in AH$. By 4.21, $(\alpha, u), (\beta, u) \in Q^*$, so, since Q^* is an equivalence relation, $(\alpha, \beta) \in Q^*$, i.e. $(\alpha\beta^{-1}, u) \in Q^*$ (4.13ii). So, again by 4.21, $\alpha\beta^{-1} \in AH$, i.e. AH is a group.

Let now $q \in AHpv$. Hence, also by 4.21 $(q, pv) \in Q^*$. However, $(p, pv) \in P^* \subseteq Q^*$, so, since Q^* is an equivalence relation, $(q,p) \in Q^*$, i.e., $q \in pQ^*$ and $AHpv \subseteq pQ^*$. □

This allows us to give an alternate proof of 4.39 that bypasses 4.25.

COROLLARY 4.61. *If $L(p) \neq \emptyset$ and $\mathcal{F} \subseteq \mathcal{A}$ is full, then*

$$AHp = AL(p)v_p \text{ for all } p \in \mathfrak{E}_u(\mathcal{F}).$$

PROOF. Notice that,

$$AHp \overset{4.35}{=} AHL(p)v_p \overset{4.17}{=} AF'L(p)v_p \overset{4.30}{=} AL(p)v_p.$$

□

Another proof of Corollary 4.39: If $F^\# \subseteq AF'$, and $L(p) \neq \emptyset$; then $Q^*(\mathcal{A} : \mathcal{F})$ is an equivalence relation on $\mathfrak{E}_u(\mathcal{F})$. Let $p \in \mathfrak{E}_u(\mathcal{F})$ and $v \in J$. Then

$$AHpv \overset{4.61}{=} AL(p)v_pv = AL(p)v \overset{4.33}{\subseteq} A(pQ) \overset{4.13iv)}{\subseteq} pQ^*,$$

so we have II) of 4.60. Since $AH = AF'$, (by 4.20, 4.17), we also have that AH is a group, and by 4.60, Q^* is an equivalence relation on $\mathfrak{E}_u(\mathcal{F})$. □

References

[A1] J. Auslander, *On the proximal relation in topological dynamics*, Proc.Amer.Math.Soc. **11** (1960), 890–895.

[A2] ———, *Minimal flows and their extensions*, North Holland, Amsterdam, 1988.

[AEE] J. Auslander, D. Ellis and R. Ellis, *The regionally proximal relation*, Trans. Amer. Math. Soc. **11** (1995), n. 6, 2139–2146.

[E1] R. Ellis,, *Group-like extension of minimal sets*, Trans. Amer. Math. Soc. **127** (1967), 125-135.

[E2] ———, *Lectures in topological dynamics*, Benjamin, New York, 1969.

[E3] ———, *The Veech Structure Theorem*, Trans. Amer. Math. Soc. **186** (1973), 203-218.

[E4] ———, *Unpublished manuscript* (1993).

[EGS] R. Ellis, S. Glasner and L. Shapiro, *Proximal Isometric ($\mathcal{PI}$-) flows.*, Advances in Math. **17** (1975), 213-260.

[EK] R. Ellis and H. Keynes, *A Characterization of the equicontinuous structure relation*, Trans. Amer.Math. Soc. **161** (1971), 171-183.

[G] S. Glasner, *Proximal Flows*, Springer Verlag, Lecture Notes in Math. 517, 1976.
[M] E. Michael, *Topologies on spaces of subsets*, Trans. Amer. Math. Soc. **71** (1951), 152-182.
[S] L. Shapiro, *Proximality in minimal transformation groups*, Proc.Amer.Math.Soc. **20** (1970), 521-525.
[W] J.C.S.P. Woude, *Topological Dynamix*, Centrum voor Wiskunde en Informatica 22, Netherlands, 1986.

FACULTAD DE MATEMATICA, ASTRONOMIA Y FISICA, UNIVERSIDAD NACIONAL DE CORDOBA, CORDOBA, ARGENTINA
E-mail address: `penazzi@mate.uncor.edu`

Contemporary Mathematics
Volume **215**, 1998

Almost Equicontinuity and the Enveloping Semigroup

E. Akin, J. Auslander and K. Berg

To Bob Ellis on his retirement

The enveloping semigroup, defined by Bob Ellis in 1960, has proved to be one of the most important tools in topological dynamics. Dynamical properties can often be correlated with properties of the enveloping semigroup. For example, Ellis showed that a flow (X,T) is distal if and only if its enveloping semigroup $E(X)$ is a group, and that (X,T) is equicontinuous if and only if $E(X)$ is a group of homeomorphisms. The enveloping semigroup has also been used to obtain dynamical results. A notable instance of this is the proof of Furstenberg's structure theorem for distal minimal flows. See [**E**], [**E1**], [**E2**], [**E3**], [**F**].

This paper is a modest contribution to the program of relating the enveloping semigroup of a flow with dynamical properties. In an earlier paper [**AAB**], we considered topologically transitive non-equicontinuous maps which had points of equicontinuity. These systems are called almost equicontinuous. We are in the process of extending this work to arbitrary group actions [**AAB1**]. Here we develop a number of connections between the enveloping semigroup and almost equicontinuous flows.

We review basic concepts and establish our notation. A flow $(X,T,F) = (X,T)$ is a jointly continuous action $(t,x) \mapsto F^t x = tx$ of the topological group T on the compact metric space (X,d). For $x \in X$ the set $Tx \triangleq \{\, tx \,|\, t \in T\}$ is the orbit of x, and $\overline{Tx}$ is the orbit closure of x. A subset $A \subseteq X$ is invariant if $tA = A$ for all $t \in T$. The orbit of x and the orbit closure of x are invariant sets. A subset $A \subseteq X$ is minimal if it is the orbit closure of each of its points. The flow (X,T) is semisimple if every orbit closure is minimal. Equivalently, a flow is semisimple if x is in the orbit closure of y whenever y is in the orbit closure of x.

The space X^X consisting of all functions from X into X is a semigroup under composition. It is compact in the product topology. The enveloping semigroup E of the flow (X,T) is the closure of $\{\, F^t \,|\, t \in T\}$ in X^X. It is compact and, as the name suggests, it is a subsemigroup of X^X. Right multiplication $p \to pq$ by a fixed element $q \in E$ is continuous in p and left multiplication $p \to qp$ is continuous in p if the map $q : X \mapsto X$ is continuous. In particular $p \mapsto F^t p$ is continuous in p for each $t \in T$.

1991 *Mathematics Subject Classification.* 54H20.

If (X, d) is a metric space then $B(x, \delta) \triangleq \{\overline{x} \mid d(x, \overline{x}) < \delta\}$.

DEFINITIONS.

1. We say that $x \in X$ is an equicontinuity point of (X, T) if for every $\epsilon > 0$ there is a $\delta > 0$ such that if $d(x, \overline{x}) < \delta$ then $d(tx, t\overline{x}) < \epsilon$ for all $t \in T$. The set of equicontinuity points is denoted by $Equi$. If $Equi = X$ we say that (X, T) is equicontinuous. In that case, by compactness, the δ can be chosen to depend only on ϵ and not on X. That is, an equicontinuous flow is uniformly equicontinuous.
2. If $Equi$ is dense in X we say that (X, T) is almost equicontinuous.
3. We say x is a transitive point if its orbit closure is X. The set of transitive points is denoted by $Trans$. If $Trans \neq \emptyset$ then we say the action is topologically transitive.

NOTE. The condition for x to be an equicontinuity point has an equivalent formulation: For every $\epsilon > 0$ there is a $\delta > 0$ such that if x' and x'' are in $B(x, \delta)$ and $p \in E$ then $d(px', px'') \leq \epsilon$. Further, it is clear that $Ex \triangleq \{px \mid p \in E\}$ is the orbit closure of x, so $Trans = \{x \mid Ex = X\}$. Since E is a semigroup it follows that if $p \in E$, $x \in X$ and $px \in Trans$ then $x \in Trans$.

The map $\theta : E \times X \to X$ is defined by $\theta(p, x) = px$.

LEMMA 1. *Let $x \in X$. Then θ is continuous at (p, x) for all $p \in E$ if and only if $x \in Equi$.*

PROOF. Suppose $x \in Equi$ and $p \in E$. Let $\epsilon > 0$. Choose $\delta > 0$ such that if $d(x, x') < \delta$ then $d(qx, qx') < \epsilon$ for all $q \in E$. Suppose that $q \in E$ is in the neighborhood of p defined by $d(px, qx) < \epsilon$. Then if $d(x, x') < \delta$ it follows that $d(px, qx') < 2\epsilon$. This establishes continuity of θ at (p, x)

Suppose conversely, for some $x \in X$, that θ is continuous at (p, x) for every $p \in E$. Let $\epsilon > 0$. For each $p \in E$ there is an open neighborhood U_p of p and a $\delta_p > 0$ such that if $q \in U_p$ and $x' \in B(x, \delta_p)$ then $d(qx', px) < \frac{\epsilon}{2}$, $d(qx, px) < \frac{\epsilon}{2}$, and so $d(qx', qx) < \epsilon$. From the compactness of E it easily follows that there is a $\delta > 0$ such that if $q \in E$ and $d(x', x) < \delta$ then $d(qx', qx) < \epsilon$, so $x \in Equi$. □

This characterization of equicontinuity by the (joint) continuity of θ, which with suitably adjusted definitions holds in the non-metrizable case, can be difficult to apply. Our analysis will provide a stronger result in the metric case. Implementation of Lemma 1 is made easier by the following result.

LEMMA 2. *Let X' be a dense subspace of X and let θ' be the restriction of θ to $E \times X'$. Suppose θ' is jointly continuous at $(p, x') \in E \times X'$. Then θ is jointly continuous at (p, x').*

PROOF. Let $\epsilon > 0$. Let U be an open subset of E with $p \in U$ and let $\delta > 0$ be such that if $q \in U$ and $y \in B(x', \delta) \cap X'$ then $d(px', qy) < \epsilon$. In particular if $F^t \in U$ and $y \in B(x', \delta) \cap X'$ then $d(px', F^t y) < \epsilon$. Since each F^t is continuous and X' is dense in X, we have $d(px', F^t x) \leq \epsilon$ for all $F^t \in U$ and $x \in B(x', \delta)$. It follows that $d(px', qx) \leq \epsilon$ for all $x \in B(x', \delta)$ and $q \in U$. □

PROPOSITION 3.

1. *If x is in the orbit closure of y and if $x \in Equi$ then $y \in Equi$ and y is in the orbit closure of x.*

2. *Trans is an invariant* G_δ *and it is either dense or void.*
3. *Equi is an invariant* G_δ.
4. *If both Equi and Trans are non-void then they are equal.*
5. *If* (X,T) *is minimal then either* $Equi = \emptyset$ *or* (X,T) *is equicontinuous.*

PROOF.
The invariance of $Trans$ and $Equi$ is obvious.

1. We first show that $y \in Equi$. Let $\epsilon > 0$. Let $\delta_1 > 0$ be such that if $x',\ x'' \in B(x,\delta_1)$ then $d(tx',tx'') < \epsilon$ for all $t \in T$. Choose $t_0 \in T$ and $\delta > 0$ so that $t_0 B(y,\delta) \subseteq B(x,\delta_1)$. It follows that if $y' \in B(y,\delta)$ then $d(tt_0y', tt_0y) < \epsilon$ for all $t \in T$. Since T is a group, this establishes the equicontinuity of the action at y. Furthermore, since $t_0 y \in B(x,\delta_1)$ it follows that, with $t = t_0^{-1}$, we have $d(y,tx) < \epsilon$. This shows that y is in the orbit closure of x.
2. The void set is a G_δ. We now consider the topologically transitive case. Clearly $Trans$ is dense. Let $\{\, U_n \mid n \in \mathbf{N}\}$ be a countable base for the topology of the compact metric space X. Each set $TU_n = \bigcup_{t\in T} tU_n$ is open, and $Trans = \bigcap_{n\in\mathbf{N}} TU_n$.
3. For each $n \in \mathbf{N}$ let $V_n = \{\, x \in X \mid$ there is a $\delta_x > 0$ such that if $x',\ x'' \in B(x,\delta_x)$ then $d(tx',tx'') < \frac{1}{n},\ \forall t \in T\}$. The sets V_n are open and $Equi = \bigcap_{n\in\mathbf{N}} V_n$.
4. This follows immediately from (1).
5. This follows immediately from (4).

□

We recall some facts about Baire spaces.

NOTATION AND REMARKS. A subset R of a metric space X is residual if it contains a set R' which is a countable intersection of open dense sets. A topological space is called a Baire space (or has the property of Baire) if any countable union of closed subsets without interior has no interior. Equiuivalently every residual set is dense. Every G_δ subset of a complete metric space is a Baire space. It is an easy exercise to show that if a Baire space is written as the countable union of closed sets then the union of the interiors of the closed sets is dense.

The following technical result is crucial. A related result appears in [**M**].

LEMMA 4. *Let* K *be a compact topological space, let* B *be a separable Baire space, and let* (Z,d) *be a metric space. Suppose that* $\pi : K \times B \to Z$ *is continuous in each variable separately. Then there is a residual subset* $R \subseteq B$ *such that* π *is jointly continuous at each point of* $K \times R$.

PROOF. We first suppose that K is a metric space with metric ρ. For positive integers m and n set

$$B_{m,n} = \{\, b \in B \mid \text{ If } \ k,\ \overline{k} \in K \text{ and } \rho(k,\overline{k}) < \frac{1}{m} \text{ then } d(\pi(k,b),\pi(\overline{k},b)) \le \frac{1}{n}\}.$$

It follows from the continuity of π in b, and the continuity of d, that each $B_{m,n}$ is closed. It follows from the continuity, and thus uniform continuity, of π in k that for each n, $\bigcup_{m\in\mathbf{N}} B_{m,n} = B$. The set $R \triangleq \bigcap_{n\in\mathbf{N}}\bigcup_{m\in\mathbf{N}} \operatorname{int}(B_{m,n})$ is therefore residual in B. Let $k \in K$ and $b \in R$. We will show that π is jointly continuous at (k,b). Let $n \in \mathbf{N}$. By the construction of R, there is an $m \in \mathbf{N}$ such that $b \in \operatorname{int}(B_{m,n})$.

We choose an open set U such that $b \in U \subseteq B_{m,n}$ and such that if $\bar{b} \in U$ then $d(\pi(k,b),\pi(k,\bar{b})) < \frac{1}{n}$. For any such $\bar{b}$, if $\rho(k,\bar{k}) < \frac{1}{m}$ then $d(\pi(k,\bar{b}),\pi(\bar{k},\bar{b})) \le \frac{1}{n}$. So, if $\bar{b} \in U$ and $\rho(\bar{k},k) < \frac{1}{m}$ then, by the triangle inequality, $d(\pi(\bar{k},\bar{b}),\pi(k,b)) < \frac{2}{n}$. This proves that π is jointly continuous at (k,b).

Now we establish the result for general compact K by reducing to the metrizable case. Let A be a countable dense subset of B. For each $k \in K$, let $\phi(k) \in Z^A$ be defined by $\phi(k)_a = \pi(k,a)$. Clearly ϕ is continuous on K, so $\phi(K)$ is a compact subset of the metrizable product space Z^A. Therefore $\phi(K)$ is metrizable. If $\phi(k) = \phi(\bar{k})$ then $\pi(k,a) = \pi(\bar{k},a)$ for all $a \in A$. Since π is continuous in its second variable and since A is dense in B, $\pi(k,b) = \pi(\bar{k},b)$ for all $b \in B$. We define $\bar{\pi} : \phi(K) \times B \to Z$ by $\bar{\pi}(\phi(k),b) = \pi(k,b)$. It is clear that $\bar{\pi}$ is continuous in b. To see that it is continuous in the first variable, suppose that $\phi(k_\nu)$ converges to a point of $\phi(K)$. Taking a subnet, we may assume that k_ν converges to a point $k \in K$ and $\phi(k_\nu) \to \phi(k)$. Then $\bar{\pi}(\phi(k_\nu),b) = \pi(k_\nu,b) \to \pi(k,b) = \bar{\pi}(\phi(k),b)$. As we have seen, there is a residual set $R \subseteq B$ such that $\bar{\pi}$ is jointly continuous at each point of $\phi(K) \times R$. So $\pi = \bar{\pi} \circ (\phi \times 1)$ is jointly continuous. □

THEOREM 5. *Let (X,T) be a flow, with X compact metric. Then the following are equivalent:*

1. *The flow (X,T) is almost equicontinuous.*
2. *There is a dense subset Y of X with the Baire property such that the restriction to Y of each map $p \in E$ is continuous.*
3. *There is a dense subset D of X such that θ restricted to $E \times D$ is jointly continuous.*
4. *There is a residual subset R of X such that θ is jointly continuous at all points of $E \times R$.*

PROOF. We will show that (1) $\Longrightarrow$ (4) $\Longrightarrow$ (2) $\Longrightarrow$ (3) $\Longrightarrow$ (1).

(1) $\Longrightarrow$ (4): If (X,T) is almost equicontinuous then $R = Equi$ is dense by definition and is a G_δ set by Proposition 3, (3). The joint continuity of θ on $E \times R$ follows from Lemma 1.

(4) $\Longrightarrow$ (2): This is obvious.

(2) $\Longrightarrow$ (3): The map θ restricted to $E \times Y$ is separately continuous. By Lemma 4 there is a residual subset R of Y such that the restricted map is jointly continuous at each point of $E \times R$. Since Y is a Baire space, R is dense in Y and then, since Y is dense in X, R is dense in X.

(3) $\Longrightarrow$ (1): By Lemma 2 and Lemma 1, $D \subseteq Equi$ so (X,T) is almost equicontinuous.

□

The flow (X,T) is said to be weakly almost periodic if, for every $f \in C(X)$, the set $\{ f_t \mid t \in T\}$ is relatively compact in the weak topology on $C(X)$ (where $f_t \in C(X)$ is defined by $f_t(x) = f(tx)$). Ellis and Nerurkar [**EN**] showed that (X,T) is weakly almost periodic if and only if every $p \in E$ is a continuous map. This result, together with Theorem 5, immediately yield:

COROLLARY 6. *If (X,T) is weakly almost periodic, it is almost equicontinuous.*

Any factor of a weakly almost periodic flow is weakly almost periodic (see [**EN**]). On the other hand, there exist almost equicontinuous flows with minimal non-equicontinuous factors (see [**AAB**], [**GW**]). Thus the converse of Corollary 6 is false.

COROLLARY 7. *Suppose that (X,T) is topologically transitive. Then (X,T) is almost equicontinuous if and only if the restriction to $Trans$ of each map $p \in E$ is continuous. In this case, $\theta : E \times X \to X$ is (jointly) continuous at each point of $E \times Trans$.*

PROOF. Suppose each p restricted to $Trans$ is continuous. The set $Y = Trans$ is dense in X and, since it is a G_δ subset of a complete metric space, it has the Baire property. Theorem 5 shows that (X,T) is almost equicontinuous.

Conversely, suppose that (X,T) is almost equicontinuous. By Lemma 1, θ is jointly continuous at each point of $E \times Equi$ and, by Proposition 3, $Equi = Trans$. In particular, the restriction to $Trans$ of each element of E is clearly continuous.□

When the acting group T is abelian and the action is topologically transitive, we can get more detailed information concerning $Trans$ and its relation to the enveloping semigroup.

DEFINITION. Let $\Lambda = \{p \in E \,|\, p\,Trans \cap Trans \neq \emptyset\}$.

Since $x \in Trans$ if $px \in Trans$, $\Lambda = \{p \in E \,|\, pX \cap Trans \neq \emptyset\}$.

LEMMA 8. *Assume that T is abelian and let (X,T) be an almost equicontinuous, topologically transitive flow.*

1. *If $p \in \Lambda$ and $q \in E$ then $pq = qp$.*
2. *If $p, q \in \Lambda$ such that $px = qx$ for some $x \in Trans$, then $p = q$.*
3. *If $p \in E$ is such that $px = x$ for some $x \in Trans$ then $p = 1_X$.*
4. *$\Lambda = \{p \in E \,|\, p\,Trans \subseteq Trans\} = \{p \in E \,|\, p\,Trans = Trans\}$.*
5. *If p is an element of E then the following conditions are equivalent:*
   (a) *p is a surjective map, i.e. $pX = X$.*
   (b) *There exists $q \in E$ such that $pq = qp = 1_X$.*
   (c) *$p \in \Lambda$.*

PROOF.
1. Since $p \in \Lambda$ we can choose x so that x and px are in $Trans$. Let $w \in E$ and $z \in X$. Since T is abelian $twx = wtx$ for all $t \in T$. Since $\{F^t \,|\, t \in T\}$ is dense in E and w is continuous at the transitive point px we have $wpx = pwx$. Now let $r \in E$ such that $rx = z$. By the above result applied first with $w = r \in E$ and then with $w = qr \in E$, $qpz = qprx = qrpx = pqrx = pqz$.
2. Let $y \in X$ and $r \in E$ be such that $y = rx$. Then $py = prx = rpx = rqx = qrx = qy$.
3. Since $px = x$, $p \in \Lambda$. Apply (2) with $q = 1_X$ which is also in Λ.
4. Let $p \in \Lambda$ and $y \in Trans$. It suffices to show that $py \in Trans$ and $y \in pTrans$. Choose x and z in $Trans$ such that $px = z$. Let $r,\delta \in E$ such that $y = rx$ and $x = \delta y$. Then $x = \delta rx$ so $\delta r = 1_X$ by (3) and $py = prx = rpx$ by (1), so $\delta py = \delta rpx = px = z$. Since $z \in Trans$ it follows that $py \in Trans$. It follows that $y = qpy$ for some $q \in E$ and so by (1) $y = pqy$. Because $pqy \in Trans$, $qy \in Trans$ and so y is an element of $pTrans$.
5. $(a) \implies (c)$: Notice that $pX = X$ implies $pX \cap Trans \neq \emptyset$.
   $(c) \implies (b)$: If $px \in Trans$ then there exists $q \in E$ such that $qpx = x$.

Then $qp = 1_X$ by (3) and $pq = qp$ by (1).
$(b) \implies (a)$: This is obvious.

□

For an arbitrary flow (X, T) with enveloping semigroup E, an element $p \in E$ is a *unit* if there exists $q \in E$ such that $pq = qp = 1_X$, i.e. p is a bijection of X and its inverse map lies in E. The subset of such invertible elements forms a group called the *group of units* of E. The group of units always includes at least $\{ F^t \,|\, t \in T\}$.

THEOREM 9. *Assume that T is abelian and that (X, T) is an almost equicontinuous, topologically transitive flow with enveloping semigroup E. The set $\Lambda \triangleq \{p \in E \,|\, pTrans \cap Trans \neq \emptyset\}$ is the group of units of E. With the topology inherited from E, Λ is an abelian topological group. For any $x \in Trans$ the continuous map $\theta_x : E \to X$ defined by $\theta_x(p) = px$ restricts to a homeomorphism of Λ onto $Trans$.*

PROOF. By (5) of Lemma 8, Λ is the group of units in E. It is abelian by (1) of Lemma 8.

Before showing that Λ is a topological group we prove the last assertion of the theorem. For the remainder of our proof we suppose that x is some specific point of $Trans$.

The map θ_x is continuous on E and is surjective since $x \in Trans$. If $\theta_x(p) = px \in Trans$, then $p \in \Lambda$ by definition of Λ. Conversely, if $p \in \Lambda$ then $\theta_x(p) \in Trans$ by (4) of Lemma 8. Thus, θ_x restricts to a surjective map from Λ to $Trans$. The restriction is injective by (2) of Lemma 8.

We use the following compactness trick. Given a net q_n in E and $q \in E$. In order to prove that $q_n \to q$ we may, by going to a subnet, assume that $q_n \to \overline{q}$ and then prove $q = \overline{q}$.

First, suppose q_n is a net in Λ and $q \in \Lambda$ such that $\theta_x(q_n) \to \theta_x(q)$, i.e. $q_n x \to qx$ in $Trans$. If $q_n \to \overline{q} \in E$ then by continuity of θ_x on E, $\overline{q}x = qx$. Hence, $\overline{q} \in \Lambda$ and so $\overline{q} = q$ by (2) of Lemma 8, proving $q_n \to q$. This proves that the inverse map $\theta_x^{-1} : Trans \to \Lambda$ is continuous. Thus θ_x restricts to a homeomorphism from $\Lambda \to Trans$.

Now define $\mu : \Lambda \times \Lambda \to \Lambda$ by $\mu(p,q) = pq$. Observe that $\theta_x \circ \mu(p,q) = pqx = \theta(p, \theta_x(q))$. By Corollary 7, θ is continuous on $E \times Trans$ and by (4) of Lemma 8 $\theta(\Lambda \times Trans) = Trans$. As $\theta_x : \Lambda \to Trans$ is a homeomorphism, it follows that the multiplication map μ is continuous on $\Lambda \times \Lambda$.

Finally, we will show that inversion is continuous on Λ. Since multiplication is continuous, it is sufficient to show that inversion is continuous at 1_X. Assune $p_n \to 1_X \in \Lambda$ and let $q_n = p_n^{-1}$. We prove that $q_n \to 1_X$. Assume that $q_n \to \overline{q} \in E$. As θ is continuous on $E \times Trans$, $x = q_n p_n x = \theta(q_n, \theta_x(p_n)) \to \theta(\overline{q}, x) = \overline{q}x$. Hence, $\overline{q} = 1_X$ by (3) of Lemma 8, proving $q_n \to 1_X$. □

References

[AAB] E. Akin, J. Auslander and K. Berg, *When is a transitive map chaotic*, Convergence in Ergodic Theory and Probability (Bergelson/March/Rosenblatt, eds.), Walter de Gruyter & Co., Berlin, New York, 1996, pp. 25-40.

[AAB1] E. Akin, J. Auslander, and K. Berg, *Almost equicontinuous group actions*, In Preparation.

[E] R. Ellis, *Locally compact transformation groups*, Duke Math. J. **24** (1957), 119-125.

[E1] R. Ellis, *Distal transformation groups*, Pacific J. Math **8** (1958), 401-405.

[E2] R. Ellis, *A semigroup associated with a transformation group*, Trans. Amer. Math. Soc. **94** (1960), 272-281.

[E3] R. Ellis, *Lectures in Topological Dynamics*, Benjamin, 1969.

[EN] R. Ellis and M. Nerurkar, *Weakly almost periodic flows*, Trans. Amer. Math. Soc. **313** (1989), 103-119.

[F] H. Furstenberg, *The structure of distal flows*, Amer. J. Math **85** (1963), 477-515.

[GW] E. Glasner and B. Weiss, *Sensitive dependence on initial conditions*, Nonlinearity **6** (1993), 1067-1075.

[M] D. Montgomery, *Continuity in topological groups*, Bull. Amer. Math. Soc **42** (1936), 879-882.

Department of Mathematics, City College of New York, New York, NY 10031-9100, U. S.A.

E-mail address: ETKCC@CUNYVM.CUNY.EDU

Department of Mathematics, University of Maryland, College Park, MD 20742-0001, U.S.A.

E-mail address: jna@math.umd.edu

Department of Mathematics, University of Maryland, College Park, MD 20742-0001, U.S.A.

E-mail address: krb@math.umd.edu

Contemporary Mathematics
Volume **215**, 1998

Some Universal Constructions in Abstract Topological Dynamics

V. Pestov

Dedicated to Professor Robert Ellis on the occasion of his retirement

ABSTRACT. This small survey of basic universal constructions related to the actions of topological groups on compacta is centred around a new result — an intrinsic description of extremely amenable topological groups (i.e., those having a fixed point in each compactum they act upon), solving a 1967 problem posed by Granirer. Another old problem whose solution (in the negative) is noted here, is a 1957 conjecture by Teleman on irreducible representations of general topological groups. Our exposition covers the greatest ambit and the universal minimal flow, as well as closely related constructions and results from representation theory. We present in a simplified fashion the known examples of extremely amenable groups and discuss their relationship with a 1969 problem posed by Ellis. Open questions, including those from bordering disciplines, are reviewed along the way.

0. Introduction

The principal object of study in abstract topological dynamics is a G-flow (topological transformation group, topological dynamical system), that is, a triple $\mathfrak{X} = (G, X, \tau)$ consisting of a Hausdorff topological group G, a Tychonoff topological space X, and a continuous action $\tau: G \times X \to X$. The theory is best developed for compact phase spaces X, in which case τ can be thought of as a continuous homomorphism $G \to \operatorname{Homeo} X$, where the group of self-homeomorphisms is equipped with the compact-open topology. It is therefore most natural to begin a journey into the realm of abstract topological dynamics by asking: does *every* topological group G admit an effective action on a compact space, thereby allowing for a dynamically substantial theory?

This fundamental question was answered in the positive, forty years ago by Teleman [**T**] who produced a fine collection of results on the crossroads of topological dynamics with representation theory and theory of C^*-algebras. We present those results, leading to the concept of the greatest ambit, in Sections 1 and 4-5. Among the later corollaries, is a 1986 Uspenskiĭ's observation [**U**] of the existence

1991 *Mathematics Subject Classification.* 54H20, 43A07, 43A65.

of a universal Polish group, answering the 'right' version of Schreier–Ulam's question from the Scottish Book (Section 3). We also point out, possibly for the first time, that a 1957 conjecture of Teleman can be refuted through combining some developments in operator theory and in abstract harmonic analysis (Section 2).

Of all compact G-flows, minimal G-flows (that is, as name suggests, those having no proper compact G-subflows) are of particular importance. In Section 6 we introduce the universal minimal flow, $\mathcal{M}(G)$, for any topological group G. The next important question to address is that of the 'size' of the universal minimal flow. In particular, when is $\mathcal{M}(G)$ nontrivial, and when is the action of G upon $\mathcal{M}(G)$ effective?

A topological group G such that the universal (therefore, every) minimal G-flow is a singleton is called *extremely amenable* [**Gr1, Mi, Gr2**] (or else a group having the *fixed point on compacta property* [**G3**]).

The very existence of such groups was at first an open problem, though topological semigroups with a similar property were known to exist in abundance [**Gr1, Mi**]. By now there are known examples of extremely amenable topological groups of at least three different kinds [**HC, G3, P**]. In 1967 Granirer [**Gr1**] suggested the problem of describing extremely amenable topological groups in intrinsic terms. We propose such a description in Sections 7 and 8. Recall that a subset S of a topological group G is *left syndetic* if $KS = G$ for some compact $K \subseteq G$. We show that a topological group G is extremely amenable if and only if for every left syndetic subset $S \subseteq G$ the set SS^{-1} is everywhere dense in G. (One can also replace 'left syndetic' with 'big on the left,' that is, left syndetic in the discrete topology.)

Employing this new criterion, we sketch the known examples of extremely amenable groups. Historically the first such example, constructed by Herer and Christensen, utilises the existence of abelian topological groups without nontrivial unitary representations (see Section 13). Another source of examples, presented in Section 9, was discovered by Glasner and (independently) Furstenberg and Weiss, who built up on the work of Gromov and Milman, and it includes the (even monothetic) group of measurable maps from I to the unit circle, equipped with the L_1-metric. Finally, among the extremely amenable groups proposed by the present author (Section 10), is the group $\mathrm{Homeo}_+(I)$ of orientation-preserving homeomorphisms of the unit interval. We review a number of dynamical consequences for groups having importance in Analysis and observe that beyond the locally compact case, amenability is not inherited by closed subgroups (Section 10).

The existence of extremely amenable groups answers in the negative a 1969 question by Ellis from [**E3**], which was largely responsible for research in the area and can be reformulated in equivalent terms as follows: are points of the greatest ambit $\mathcal{S}(G)$ of a topological group G separated by continuous equivariant maps to minimal G-flows? This problem and related developments form the subject of Section 11.

Our criterion of extreme amenability is accompanied by a satellite result describing, also for the first time, those topological groups G admitting effective minimal actions on compacta (Section 12). A topological group G acts effectively on $\mathcal{M}(G)$ if and only if for every $g \in G$, $g \neq e$ there is a left syndetic set S with $x \notin \overline{SS^{-1}}$. We derive from this criterion an easy proof of the fact that every locally compact group acts effectively on the universal minimal flow. This is of course a corollary of a stronger theorem by Veech [**V2**] stating that every locally compact group acts *freely* on a compactum — the result of which a 'soft' proof has so far escaped us.

Finally, in Section 13 we discuss a few results and questions related to the old and still open problem of describing those topological groups admitting a complete system of strongly continuous unitary representations in Hilbert spaces.

Our usage of basic concepts from topology, functional analysis, topological algebra, representation theory, and topological dynamics is fairly standard, while our list of bibliography is kept to a bare minimum.

Acknowledgements. The present article has grown out of author's presentations of some of the results at the University of Maryland (September 1996), at the Semi-Annual Workshop on Dynamical Systems, Penn State University (October 1996), and at the Tsukuba Topological Conference, Tsukuba Science City, Japan (December 1996). The author is most thankul to Professors Joe Auslander and Dmitri Shakhmatov, as well as to the Organizers of the above Conferences, for making those presentations possible. Most stimulating e-mail discussions with Prof. Eli Glasner, useful advice from Dr. Arkady Leiderman and important bibliographical remarks by Prof. Alexander Kechris and Dr. Michael Megrelishvili are also acknowledged. Finally, a number of useful improvements were suggested by the referee of this article.

1. Flows and Representations

Topological dynamics is intimately linked to representation theory, as can be seen from the following observation. Let ρ be a representation of an abstract group G in a Banach space E by isometries, that is, $\rho: G \to \mathrm{Iso}(E)$ is a homomorphism to the group of all linear isometries of E onto itself. For every $g \in G$ the dual operator, $\rho_g^*: E^* \to E^*$, to the isometry $\rho_g: E \to E$ is an isometry again, and the restriction, ρ_g^B, of $\rho_{g^{-1}}^*$ to the unit ball, B_w^*, of E^*, equipped with the w^*-topology (the coarsest topology making each evaluation map $\hat{x}: E^* \to \mathbb{K}$, $\hat{x}(f) = f(x)$, continuous), is a self-homeomorphism. It is easy to check that the resulting map

$$\rho^B: G \ni g \mapsto \rho_g^B \in \mathrm{Homeo}(B_w^*)$$

is a group homomorphism, that is, an action of G on the compact space B_w^*.

Now assume that G is a topological group. Recall that a representation $\rho: G \to \mathrm{GL}(E)$ (in particular, a representation $\rho: G \to \mathrm{Iso}(E)$) is *strongly continuous* if it is continuous with respect to the topology on $\mathrm{GL}(E)$ inherited from the Tychonoff product topology on E^E. (This, the so-called *strong operator topology,* though not a group topology on $\mathrm{GL}(E)$, becomes a group topology when restricted to $\mathrm{Iso}(E)$.)

PROPOSITION 1.1 (Teleman [**T**]). (i) *If ρ is strongly continuous then ρ^B is continuous.*

(ii) *If ρ is faithful then ρ^B is effective.*

(iii) *If $\rho: G \hookrightarrow \mathrm{Iso}(E)$ is an embedding of topological groups, then so is $\rho^B: G \hookrightarrow \mathrm{Homeo}(B_w^*)$.*

PROOF. Let us compare standard basic neighbourhoods of identity in the topological group $\mathrm{Iso}(E)$,

$$\{u \in \mathrm{Iso}(E): \forall i = 1, 2, \ldots, n,\ \| u(x_i) - x_i \| < \varepsilon\},\ x_i \in E,\ \varepsilon > 0,$$

and in Homeo (B_w^*),

$$\{u \in \text{Homeo}\,(B_w^*) \colon \forall f \in B_w^*,\ (u(f), f) \in V\},\ V \in \mathcal{U}_X,$$

where $\mathcal{U}_X$ is the (unique) compatible uniformity on B_w^*. Applying the Hahn–Banach theorem in the first case and noticing that $\mathcal{U}_X$ is induced by the additive uniformity on E^* equipped with the weak* topology in the second case, we reduce both neighbourhoods to the same form:

$$\{u \colon \forall i = 1, 2, \ldots, n,\ \forall f \in B_w^*,\ |f(u(x_i)) - f(x_i)| < \varepsilon\},\ x_i \in E,\ \varepsilon > 0.$$

It implies that Iso (E) canonically embeds into Homeo (B_w^*) as a topological subgroup, whence all three statements follow. □

To conclude that every topological group G acts effectively on a compact space, it is enough to observe the following.

THEOREM 1.2 (Teleman [**T**]). *Every topological group G admits a faithful strongly continuous representation ρ by isometries in a suitable Banach space E. Moreover, $\rho \colon G \to \text{Iso}\,(E)$ is in this case a topological embedding.*

PROOF. We avoid using the terms 'left' and 'right' uniform structure altogether because there is a considerable confusion about their usage. Instead we denote, once and for all, by $\mathcal{U}_{\Rsh}(G)$ the uniform structure on G whose base is formed by the entourages of the diagonal

$$V_{\Rsh} = \{(x, y) \colon xy^{-1} \in V\},$$

where V is a neighbourhood of e_G. We say that a function $f \colon G \to \mathbb{C}$ is *$\mathcal{U}_{\Rsh}$-uniformly continuous*, or *$\mathcal{U}_{\Rsh}$-u.c.*, if

$$\forall \varepsilon > 0,\ \exists V \in \mathcal{N}(G),\ \forall x, y \in G,\ xy^{-1} \in V \Rightarrow |f(x) - f(y)| < \varepsilon \qquad (*)$$

The uniformity $\mathcal{U}_{\Lsh}$ and $\mathcal{U}_{\Lsh}$-uniformly continuous functions on a topological group are introduced in a similar way, though they are never used in this article.

The totality of all bounded complex-valued $\mathcal{U}_{\Rsh}$-u.c. functions on a topological group G equipped with the supremum norm and pointwise operations is easily checked to form a commutative C^*-algebra — indeed a C^*-subalgebra of $C(\beta G) \cong l_\infty(|G|)$ — which we denote by $C^*_{\Rsh}(G)$.

Every element $g \in G$ determines a left shift $L_g \colon C^*_{\Rsh}(G) \to C^*_{\Rsh}(G)$ via

$$L_g(f)(h) = f(g^{-1}h)$$

The emerging representation $L \colon G \to \text{Iso}\, C^*_{\Rsh}(G)$ is strongly continuous: if V is symmetric and as in $(*)$, then $\| L_g f - f \| < \varepsilon$ whenever $g \in V$. Moreover, L is a topological embedding: given a symmetric neighbourhood of identity, V, one can construct a $\mathcal{U}_{\Rsh}$-u.c. bounded $f \colon G \to \mathbb{C}$ with $f \equiv 0$ outside V and $f(e) = 1$; the set $W = \{u \in \text{Iso}\, C^*_{\Rsh}(G) \colon \| u(f) - f \| < 1\}$ is open in the strong topology and $e \in W \cap G \subseteq V$. □

It seems that Teleman's results later became a part of the folklore and were even sometimes reproduced with proofs without referring to the original.

2. Teleman's Conjecture on Irreducible Banach Representations

Having in view the celebrated 1943 Gelfand–Raĭkov theorem on the existence of a complete system of irreducible unitary representations for any locally compact group, Teleman asked in his 1957 paper [**T**] whether every topological group admits a complete system of strongly continuous irreducible representations in Banach spaces by isometries. A negative answer to this question can be deduced from more recent results obtained in operator theory and abstract harmonic analysis, but it is not apparent if it was ever noticed by anyone.

Recall that a topological group is called *minimally almost periodic* if it admits no nontrivial finite-dimensional unitary representations, and *monothetic* if it contains an everywhere dense cyclic subgroup.

PROPOSITION 2.1. *A monothetic minimally almost periodic topological group G admits no nontrivial irreducible strongly continuous representations in a complex Banach space by isometries.*

PROOF. Assume the existence of such a representation, ρ. Let x be a generator of an everywhere dense cyclic subgroup of G. The operator ρ_x is an invertible isometry of the complex Banach space of representation, E, and therefore, assuming $\dim E > 1$, there exists nontrival closed suspace F that is invariant for any operator commuting with ρ_x. (This fact follows, e.g. from Beauzamy's theorem 12.8.1 in [**B**] and remarks at the end of both pp.277 and 279.) Since F is then invariant for all operators ρ_g, $g \in G$, and ρ is, contrary to our assumption, reducible, one concludes that $\dim E = 1$ and the representation ρ, being a continuous character $G \to U(1)$, is trivial. □

Since there are numerous known examples of minimally almost periodic monothetic topological groups (see e.g. [**DPS**]), Teleman's conjecture does not survive.

QUESTIONS 2.2. (i) Does every topological group admit a complete system of strongly continuous irreducible representations in complex Banach spaces?

(ii) Assuming a topological group G admits a faithful strongly continuous representation in a Hilbert space, does it admit a complete system of irreducible representations in Hilbert spaces?

If answered positively, the second question, applied to the topological group from Glasner–Furstenberg–Weiss's example (see Section 9 below), would provide a solution to the Invariant Subspace Problem.

3. Universal Group of Countable Weight

It was asked by Schreier and Ulam in the Scottish Book (problem 103 in the 1981 edition [**Ma**]) whether or not there existed a universal separable topological group, that is, a separable topological group G such that every other such group is isomorphic to a topological subgroup of G. The answer is 'no' because of simple set-theoretic reasons: while a separable group can have at most $\mathfrak{c}$ pairwise distinct subgroups, there are $\exp \mathfrak{c}$ pairwise non-isomorphic separable topological groups, cf. the comments on pp. 184–186 in [**Ma**]. However, the following elegant result seems to answer in the positive the 'right' version of Schreier–Ulam's question as suggested by Kallman (*ibid.*).

THEOREM 3.1 (Uspenskiĭ [**U**]). *The group* Homeo$(I^{\aleph_0})$ *is a universal separable metrizable topological group. In other words, every separable metrizable topological group is isomorphic with a topological subgroup of* Homeo$(I^{\aleph_0})$.

PROOF. If G is separable metrizable, then its points are separated from closed subsets by elements of a countable subset — and therefore by those of a separable Banach subspace, say E — of $C^*_{\mathfrak{r}}(G)$, and such a subspace can be also assumed invariant under left translations. The action of G on E determines an isometric embedding $G \hookrightarrow \mathrm{Iso}(E)$, and Proposition 1.1 yields an embedding of topological groups $G \hookrightarrow \mathrm{Homeo}(B^*_w)$. Now it remains to apply Keller's theorem [**BP**]: all convex compact infinite-dimensional metrizable subsets of locally convex spaces are homeomorphic to the Hilbert cube $Q = I^{\aleph_0}$. □

4. Dynamics and C^*-Algebras

Let us take a closer look at Teleman's construction. So far we made no use of the fact that $C^*_{\mathfrak{r}}(G)$ is a commutative unital C^*-algebra rather than just a Banach space, and the shifts L_g are C^*-algebra automorphisms and not merely isometries.

Assume for a while that a topological group G acts strongly continuously on a commutative unital C^*-algebra A by automorphisms. The Gelfand space, $\mathrm{Spec}_{\mathbb{C}}A$, formed by all multiplicative linear functionals $f\colon A \to \mathbb{C}$, is a closed subspace of the unit ball B^*_w, and it is clearly invariant under the action ρ^B of G. Therefore, G acts on $\mathrm{Spec}_{\mathbb{C}}A$ continuously. Going in the opposite direction, if X is a compact G-flow, then G acts strongly continuously by left shifts on the commutative C^*-algebra $C(X)$ of all complex-valued continuous functions on X, and $X \cong \mathrm{Spec}_{\mathbb{C}}C(X)$ as G-spaces. The functorial correspondence $A \mapsto \mathrm{Spec}_{\mathbb{C}}A$ is an anti-equivalence of categories.

THEOREM 4.1. *Let G be a topological group. The category of all compact G-flows and G-equivariant continuous maps is anti-equivalent to the category of all strongly continuous representations of G in commutative unital C^*-algebras by automorphisms and intertwining unital C^*-algebra morphisms.* □

This simple observation helps to explain why theory of strongly continuous actions of topological groups on *arbitrary* unital C^*-algebras by automorphisms is universally accepted as the adequate version of *non-commutative dynamics*, cf. [**Gu**].

5. The Greatest Ambit

For a topological group G, denote the Gelfand space $\mathrm{Spec}_{\mathbb{C}}C_{\mathfrak{r}}(G)$ by $\mathcal{S}(G)$. Each element $g \in G$ determines an evaluation functional, $\hat{g} \in \mathcal{S}(G)$, via $\hat{g}(f) = f(g)$. Since $\mathcal{U}_{\mathfrak{r}}$-u.c. functions separate points and closed subsets in a topological group, the resulting canonical map $G \ni g \mapsto \hat{g} \in \mathcal{S}(G)$ is a homeomorphic embedding. It is also equivariant with respect to the action of G on itself by left translations: $L^*_{g^{-1}}(\hat{x}) = \widehat{gx}$ for all $g, x \in G$. For this reason, we will be identifying each element $g \in G$ with $\hat{g} \in \mathcal{S}(G)$. Notice that the G-flow $\mathcal{S}(G)$ has a distinguished point, e. The G-orbit of e is everywhere dense in $\mathcal{S}(G)$, since the latter space is a compactification of G; it is, in fact, the Samuel compactification (see e.g. [**En**]) of the uniform space $(G, \mathcal{U}_{\mathfrak{r}}(G))$.

DEFINITION 5.1. A compact G-flow with a distinguished point, (X,x), is called a *G-ambit* if the orbit of x is everywhere dense in X. A *morphism* between G-ambits (X,x) and (Y,y) is a continuous G-equivariant map $X \to Y$ taking x to y.

Let $X = (X,x)$ be an arbitrary G-ambit for a topological group G. Every $f \in C(X)$ determines a function $\tilde{f}: G \to \mathbb{C}$ via 'pullback': $\tilde{f}(g) = f(gx)$. It is easy to see that $\tilde{f}$ is bounded and $\mathcal{U}_{\mathrm{r}}$-u.c. The C^*-algebra monomorphism $C(X) \hookrightarrow C^*_{\mathrm{r}}(G)$, determined by the correspondence $f \mapsto \tilde{f}$, intertwines the representations of G in $C(X)$ and $C^*_{\mathrm{r}}(G)$ and therefore determines, according to our earlier observation, a dual morphism of G-flows $\mathcal{S}(G) \to X$, which takes e to x and is therefore onto. (The image of $\mathcal{S}(G)$ is compact and therefore closed, and is at the same time everywhere dense since it contains the orbit $G \cdot x$.) We come to the following.

THEOREM 5.2 (Teleman). *Every topological group G possesses the greatest ambit $\mathcal{S}(G) = (\mathcal{S}(G), e)$ in the sense that every G-ambit is the image of $\mathcal{S}(G)$ under a unique morphism of G-ambits. The greatest ambit contains G itself as an everywhere dense G-subspace.* □

As we will see later, the concept is very useful. It would be interesting to know the following.

QUESTION 5.3. Is there an analogue of the greatest ambit in noncommutative dynamics?

EXAMPLE 5.4. Let us describe two 'diametrically opposite' situations where the structure of the greatest ambit is relatively well understood.

(i) If G is precompact, then $\mathcal{S}(G)$ is simply the compact topological group, $\widehat{G}$, completion of G, equipped with the left action of the subgroup G.

(ii) If G is discrete, then every complex-valued function on G is $\mathcal{U}_{\mathrm{r}}$-u.c. and as a corollary, $\mathcal{S}(G)$ is βG, the Stone-Čech compactification of G, equipped with the natural action of G on elements of βG (which can be identified with all ultrafilters on the set G) by left translations: if $g \in G$ and $U \in \beta G$, then $gU = \{gA: A \in U\} \in \beta G$. (See [**A,E2,Vr**].)

The situation 'in between' is much less thoroughly understood. One significant recent advance is the description by Turek [**Tu**] of the greatest ambit, $\mathcal{S}(\mathbb{R})$, of the additive group of reals, $\mathbb{R}$, with the usual topology.

It only takes a slight modification of the arguments used in 1.1 and 1.2 to establish the following.

THEOREM 5.5. *The canonical action of a topological group G on its greatest ambit, $\mathcal{S}(G)$, determines an embedding of topological groups, $G \hookrightarrow$* Homeo $(\mathcal{S}(G))$.

The terminology we suggest using on such occasions is that the action is *topologically effective.*

Presently the standard reference to the concept of the greatest ambit is Brook's paper [**Br**].

6. The Universal Minimal Flow

Let G be a topological group. It is clear that a compact G-flow, X, is minimal, that is, admits no proper compact G-subflows, if and only if the orbit of every

$x \in X$ is everywhere dense in X. Since for an arbitrary compact G-flow the family of all compact non-empty G-subflows survives the intersections of chains, Zorn's lemma applies and yields the following result.

PROPOSITION 6.1. *Every compact G-flow contains a minimal G-subflow.*

A minimal G-subflow of the greatest ambit, $\mathcal{S}(G)$, is denoted by $\mathcal{M}(G)$ and called the *universal minimal G-flow*. The name is justified by the following result.

PROPOSITION 6.2. *For a topological group G every minimal G-flow, X, is the image of the universal minimal flow, $\mathcal{M}(G)$, under a surjective morphism of G-flows.*

PROOF. Let $x \in X$ be any point. The restriction to (an isomorphic copy of) $\mathcal{M}(G)$ of the unique morphism of G-ambits $\mathcal{S}(G) \to X$ sending e to x has a compact (therefore closed) image in X, which contains the everywhere dense G-orbit of x and thus coincides with X. □

The ambiguity in choosing an $x \in X$ in the above proof is responsible for the fact that there is, generally speaking, no *canonical* morphism from $\mathcal{M}(G)$ onto X, unlike the case of the greatest ambit. This is also the reason why the proof of uniqueness is slightly more involved, and we refer the reader to e.g. [**A**].

PROPOSITION 6.3. *For a topological group G, the universal minimal G-flow is unique up to an isomorphism.*

What is the 'size' of the universal minimal flow for a given topological group? In particular, when is $\mathcal{M}(G)$ nontrivial and when G acts upon $\mathcal{M}(G)$ effectively? By far the strongest result in this direction to date is the following.

THEOREM 6.4 (Veech [**V1**]). *Every locally compact group G acts freely on the universal minimal flow.*

The Veech's proof is rather technical. While we are unable to find a 'soft' version of it, we offer a simple proof of the weaker statement: every locally compact group acts *effectively* on the universal minimal flow. This will be achieved by means of a new description of topological groups admitting effective minimal flows.

7. Syndetic Sets and The Size of Universal Minimal Flow

A subset S of a topological group G is called *left syndetic* if for some compact $K \subseteq G$ one has $KS = G$. If a set S is left syndetic with respect to the discrete topology on a group (that is, $FS = G$ for some F finite), then S is called *discretely left syndetic* or *big on the left*. If either G is abelian or S is symmetric then 'left' in both definitions can be dropped.

Here is how syndetic sets emerge in topological dynamics. Suppose a topological group G acts on a minimal compact G-flow X and let $x \in V \subseteq X$, where V is open. Then the open set $\widetilde{V} = \{g \in G: gx \in V\}$ is big on the left in G. Indeed, since the translations hV, $h \in G$ form an open cover of X (assuming a $y \in X$ that is not in their union, the orbit of y would miss V, contrary to everywhere density in X), there is a finite subcover $\{fV: f \in F\}$, where $F \subseteq G$, $|F| < \infty$. It remains to notice that $\widetilde{fV} = f\widetilde{V}$ for each $f \in G$ and therefore $F\widetilde{V} = G$. (To see that $\widetilde{V}$ is not necessarily *right* syndetic, it is enough to study the full homeomorphism group acting on the circle $\mathbb{S}^1$.)

THEOREM 7.1. *Let G be a topological group and $g \in G$. The following are equivalent.*

(i) *The motion of the universal minimal flow $\mathcal{M}(G)$ determined by g is nontrivial, that is, $gx \neq x$ for some $x \in \mathcal{M}(G)$.*

(ii) *There is a left syndetic set $S \subseteq G$ with $g \notin \overline{SS^{-1}}$.*

(iii) *There is a set $S \subseteq G$ that is big on the left and such that $g \notin \overline{SS^{-1}}$.*

(iv) *There is an open set $S \subseteq G$ that is big on the left and such that $g \notin \overline{SS^{-1}}$.*

PROOF. (i) $\Rightarrow$ (iv): Let $\mathcal{U}_X$ denote the (unique) uniform structure on $X = \mathcal{M}(G)$. For every $E \in \mathcal{U}_X$, let $\tilde{E} = \{(g,h) \in G \times G\colon (gx, hx) \in E\}$. The collection $\{\tilde{E}\colon E \in \mathcal{U}_X\}$ forms a basis of a uniform structure, $\mathcal{V}$, on G contained in $\mathcal{U}_{r}(G)$.

Let $x \in X$ be such that $gx \neq x$. For some $E \in \mathcal{U}_X$ one has $E(x) \cap E(gx) = \emptyset$. Since $\mathcal{V} \subseteq \mathcal{U}_{r}(G)$, for a suitable symmetric $\mathcal{O} \in \mathcal{N}(G)$ and an $E_1 \in \mathcal{V}$, one has $\mathcal{O}_{r} \circ E_1 \subseteq E$ and therefore $\mathcal{O}E_1(x) \cap \mathcal{O}E_1(gx) = \emptyset$. Set $V = E_1(x) \cap g^{-1}E_1(gx)$; it is a neighbourhood of x and $\mathcal{O}V \cap gV \subseteq \mathcal{O}V \cap \mathcal{O}gV = \emptyset$. The set $S = \widetilde{V} \subseteq G$ is open, big on the left, and $\mathcal{O}\widetilde{V} = \widetilde{\mathcal{O}V}$, which implies that $\mathcal{O}S \cap gS = \emptyset$, that is, $g \notin \overline{SS^{-1}}$, as desired.

(iv) $\Rightarrow$ (iii) $\Rightarrow$ (ii): These implications are trivial.

(ii) $\Rightarrow$ (iv): Find a symmetric $V \in \mathcal{N}(G)$ with $\overline{SS^{-1}} \cap VgV = \emptyset$ and then a $W \in \mathcal{N}(G)$ such that $W^2 \subseteq V$. One has $WgW \cap (WS)(WS)^{-1} = \emptyset$. Choose a compact K with $KS = G$ and then a finite F with $FW \supseteq K$. Then $F(WS) = (FW)S \supseteq KS = G$, and the set WS is open, big on the left, and $g \notin \overline{(WS)(WS)^{-1}}$.

(iv) $\Rightarrow$ (i): Let $U \in \mathcal{N}(G)$ be symmetric and such that $UgU \cap SS^{-1} = \emptyset$. It means that $gUS \cap US = \emptyset$. Find a right-invariant continuous pseudometric ρ on G with $\rho \leq 1$ and $\mathcal{O}^{\rho}_1(e) \subseteq U$. For any $x \in G$, set $f(x) = \rho\text{-dist}\,(x, G \setminus US)$. Clearly, $0 \leq f(x) \leq 1$. Whenever $xy^{-1} \in \mathcal{O}^{\rho}_{\varepsilon}(e)$, one has by the triangle inequality $|f(x) - f(y)| \leq \rho(x,y) < \varepsilon$, that is, f is in $C^*_{r}(G)$. If $x \in S$ then $f(x) = 1$, and if $x \notin US$ then $f(x) = 0$. Therefore, $f|_{gS} \equiv 0$ and $f|_S \equiv 1$.

Choose a finite $F \subseteq G$ with $FS = G$. Denote by $[\cdot]$ the closure operator in $\mathcal{S}(G)$; since $G = \cup_{\kappa \in F} \kappa S$ and F is finite, one has $\mathcal{S}(G) = \cup_{\kappa \in F}[\kappa S]$. For each $\kappa \in F$, one clearly has $\overline{L_\kappa(f)}|_{[\kappa S]} \equiv 1$ and $\overline{L_\kappa(f)}|_{(\kappa g \kappa^{-1})[\kappa S]} \equiv 0$, where the horizontal bar denotes the unique continuous extension of a $\mathcal{U}_{r}(G)$-u.c. bounded function to $\mathcal{S}(G)$.

For at least one $\kappa \in F$ one has $[\kappa S] \cap \mathcal{M}(G) \neq \emptyset$, where $\mathcal{M}(G)$ is identified with a subflow of the greatest ambit. Choose an x from the latter intersection. The property from the previous paragraph implies that $(\kappa g \kappa^{-1})x \neq x$ (because these two points are separated by the function $\overline{L_\kappa(f)}$) and therefore $g(\kappa^{-1}x) \neq \kappa^{-1}x$, where $\kappa^{-1}x \in \mathcal{M}(G)$. □

8. Trivial Universal Minimal Flows

Let us now examine the issue of (non)triviality of the universal minimal flow. A topological semigroup S is called *extremely (left) amenable* [**Gr1, Mi, Gr2**] if every compact S-flow has a fixed point. For topological groups this amounts to saying that $\mathcal{M}(G) = \{*\}$ is a singleton. This property is a considerably strengthened version of amenability, as can be seen from a well-known characterization of amenable topological groups as those whose every continuous action upon a convex compact set by affine transformations has a fixed point [**G1**]. (There are many descriptions of extremely amenable topological semigroups, including even one in terms of geometry of Banach spaces [**Gr2**].) While examples of extremely amenable topological semigroups abound, the problem of existence of extremely amenable topological groups, explicitely stated by Mitchell in 1970 [**Mi**], happened to be much more difficult. The first such example was published in 1974 by Herer and Christensen [**HC**], and in 1996 more examples with different combinations of properties were published by Eli Glasner [**G3**], who kept his construction unpublished for a while and observed that it was obtained independently by Furstenberg and B. Weiss, and by the present author [**P**].

Granirer proposed in 1967 [**Gr1**] the problem of characterizing extremely amenable topological groups in intrinsic terms. We suggest the following solution, which is an immediate corollary (and in fact is an equivalent form of) Theorem 7.1.

THEOREM 8.1 (Extreme Amenability Criterion). *A topological group G is extremely amenable if and only if whenever $S \subseteq G$ is big on the left, SS^{-1} is everywhere dense in G.*

COROLLARY 8.2 (Glasner [**G3**]). *If there exists a monothetic minimally almost periodic topological group G that is not extremely amenable, then there exists a big subset $S \subseteq \mathbb{Z}$ such that $S - S$ is not a Bohr neighbourhood of zero.*

PROOF. Minimal almost periodicity of a topological group is obviously equivalent to the property: every neighbourhood of identity in the finest precompact topology on the underlying discrete group is everywhere dense in the original topology on G. Now it is enough to apply Theorem 8.1 to an everywhere dense copy of $\mathbb{Z}$. □

The question of existence of a monothetic minimally almost periodic non-extremely amenable topological group, suggested by Glasner, remains open, as is the classical question on the existence of a big set, S, of integers such that $S - S$ is not a Bohr neighbourhood of zero. (Cf. [**G3, V1**].)

9. Extreme Amenability of Levy Groups

The following is a slight extension of a definition from [**G3**], where we drop the metrizability restriction. We say that a topological group G is a *Levy group* if G contains an increasing chain of compact subgroups $G_i, i \in \mathbb{N}$ with the everywhere dense union and such that whenever $A_i \subseteq G_i$ have the property that $\liminf \mu_i(A_i) > 0$, then for every neighbourhood of identity, V, $\lim \mu_i(VA_i \cap G_i) = 1$, where μ_i denotes the normalized Haar measure on G_i.

An example of a Levy group (due to Glasner and, independently, Fusternberg and Weiss) is the group of all measurable $\mathbb{S}^1$-valued complex functions on a nonatomic Lebesgue measure space X equipped with the metric

$$d(f,g) = \int |f(x) - g(x)| d\mu(x)$$

Here one can choose as the compact subgroups G_i, tori of increasing dimension formed by step functions corresponding to a sequence of refining partitions of X. A subtle argument [**G3**] shows that this group is also monothetic.

The following was proved by Gromov and Milman [**GM**] under the additional condition of equicontinuity of the action, later removed by Glasner [**G3**] and (independently, unpublished) by Furstenberg and B. Weiss. (The assumed metrizability of the group is also redundant.)

THEOREM 9.1. *Every Levy group is extremely amenable.*

PROOF. Let S be any subset big on the left, and let F be finite with $FS = G$; one can assume that $F \subseteq \cup G_i$. Then $\liminf \mu_i(G_i \cap S) \geq 1/n$, where $n = |F|$. (Indeed, $\mu_i(G_i \cap S) \geq 1/n$ whenever $F \subseteq G_i$.) Since G is a Levy group, for any neighbourhood of identity, V, one has $\lim \mu_i(V(G_i \cap S) \cap G_i) = 1$. For all i large enough, $\mu_i((V(G_i \cap S) \cap G_i) \cdot (S^{-1} \cap G_i)) = 1$, since assuming the contrary implies that for a confinal set of i, a suitable translate of $G_i \cap S$ in G_i, having measure $\geq 1/n$, does not meet the set $V(G_i \cap S) \cap G_i$ of measure $> 1 - 1/n$- a contradiction. Since $VSS^{-1} \cap G_i \supseteq (V(G_i \cap S) \cap G_i) \cdot (S^{-1} \cap G_i))$, the set $VSS^{-1} \cap G_i$ has full measure and is everywhere dense in G_i starting from some i and thus VSS^{-1} is everywhere dense in G. Since V is arbitrary, SS^{-1} is everywhere dense in G, and Theorem 8.1 finishes the proof. □

10. Extreme Amenability of Groups of Order Automorphisms

Let us say that a group G of order automorphisms of a linearly ordered set X is *ω-transitive* if it takes any finite subset to any other subset of the same size.

THEOREM 10.1 (Pestov [**P**]). *An ω-transitive group of order automorphisms of an infinite linearly ordered set X, equipped with the topology of simple convergence on X, is extremely amenable.*

PROOF. Stabilizers, St_M, of finite subsets $M \subseteq X$ are open subgroups forming a neighbourhood basis at identity, and the right factor space G/St_M can be naturally identified with the collection $X^{(n)}$ of all ordered n-subsets of X, where $n = |M|$. Let $F \subseteq G$ be finite and such that $FS = G$. According to Infinite Ramsey's Theorem [**GRS**], for some infinite $A \subseteq X$ and some $f \in F$ one has $A^{(n)} \subseteq \pi_M(fS)$, where $\pi_M : G \to X^{(n)}$ is the factor-map. The set $B = f^{-1}(A)$ is also infinite and $B^{(n)} \subseteq \pi_M(S)$. In other words, $\mathrm{St}_M S$ contains all transformations taking M to a subset of B. Consequently, $S^{-1}\mathrm{St}_M$ contains all transformations such that the image of B contains M. Since $B \setminus M$ is infinite, any transformation from G can be represented as a composition of one from $S^{-1}\mathrm{St}_M$ and one from $\mathrm{St}_M S$, that is, $\mathrm{St}_M SS^{-1}\mathrm{St}_M = G$. Since M is arbitrary, it means that $G = \overline{SS^{-1}}$, as required. □

An example of such a topological group is the group $\mathrm{Aut}\,(\mathbb{Q})$ of all order automorphisms of the rational numbers, equipped with the topology of pointwise convergence on the discrete set $\mathbb{Q}$. Since this group naturally embeds as a topological subgroup into the unitary group $\mathrm{U}(l_2(|\mathbb{Q}|))$ with the strong operator topology, we conclude that the unitary group of an infinite-dimensional Hilbert space admits no free compact flow. The same conclusion and for the same reason holds for the full group of permutations, $\mathrm{Sym}\,(X)$, of an infinite set X equipped with the topology of simple convergence.

A further series of examples is obtained by observing that images of extremely amenable topological groups under continuous homomorphisms onto are extremely amenable. This is how one deduces extreme amenability of the groups of orientiation-preserving homeomorphisms endowed with the compact-open topology, Homeo $_+(I)$, Homeo $_+\mathbb{R}$, and the stabilizer of any point $s \in \mathbb{S}^1$ in Homeo $_+(\mathbb{S}^1)$.

From the latter fact it is easy to deduce that $\mathbb{S}^1$ forms the universal minimal flow for the topological group Homeo $_+(\mathbb{S}^1)$, and thus we arrive at an example showing that a topological group can act on the universal minimal flow effectively but not freely.

Since the free group on two generators, F_2, embeds into the multiplicative group of a countable linearly ordered skewfield [**N**], and any such skewfield is order isomorphic to $\mathbb{Q}$, one can easily deduce that F_2 embeds into Aut $(\mathbb{Q})$ as a closed discrete topological subgroup. It means that a closed topological subgroup of an amenable (even extremely amenable) topological group need not be amenable, unlike the locally compact case (cf. [**Gre**]).

CONJECTURE 10.2. *Every topological group is isomorphic to a topological subgroup of an extremely amenable topological group.*

All the above examples are presented in detail in [**P**].

11. Enveloping Semigroup of a Flow and Ellis' Problem

One of the guiding problems of topological dynamics is that of decomposing generic G-flows into some basic, simpler building blocks. Many important structure theorems known in topological dynamics serve this purpose. Here we show that a 1969 problem by Ellis on the enveloping semigroup of the universal minimal flow is of the same kind, and discuss its solution.

The following construction belongs to Ellis [**El, E3**]. Let G be a topological group, and let X be a compact G-flow. The *enveloping semigroup*, $\mathcal{E}(X)$, of X is a G-ambit defined as follows. Recall that every element $g \in G$ determines a self-homeomorphism, τ_g, of X, called the *g-motion.* The underlying topological space of $\mathcal{E}(X)$ is the closure of the collection of all g-motions, $g \in G$, in X^X equipped with the topology of simple convergence. Since X^X is compact, so is $\mathcal{E}(X)$. Since every element of $\mathcal{E}(X)$ is a (not necessarily continuous) map $X \to X$, a natural left action of G upon $\mathcal{E}(X)$ is determined by composing an $f \in \mathcal{E}(X)$ with the g-motion on the left:

$$\mathcal{E}(X) \ni f \stackrel{g}{\mapsto} \tau_g \circ f \in \mathcal{E}(X).$$

The correctness of the definition and the continuity of the action $G \times \mathcal{E}(X) \to \mathcal{E}(X)$ are checked easily. The identity map Id_X has an everywhere dense orbit in $\mathcal{E}(X)$ (it is simply the set of all motions), and therefore $\mathcal{E}(X) = (\mathcal{E}(X), \mathrm{Id}_X)$ is a G-ambit. Moreover, $\mathcal{E}(X)$ supports an obvious natural semigroup structure, but we are not interested in it.

The following explains the significance of the enveloping semigroup.

THEOREM 11.1. *The enveloping semigroup $\mathcal{E}(X)$ of a G-flow X is the greatest G-ambit with the property that morphisms into X separate points. In other words, morphisms of G-flows $\mathcal{E}(X) \to X$ separate points in $\mathcal{E}(X)$, and whenever (Z, z) is a G-ambit such that morphisms of G-flows $Z \to X$ separate points in Z, there exists a unique morphism of G-ambits $(\mathcal{E}(X), \mathrm{Id}_X) \to (Z, z)$.*

PROOF. The first statement is easy to verify: if $f, g \in \mathcal{E}(X)$ and $f \neq g$, then for some $x \in X$ one must have $f(x) \neq g(x)$ and therefore $\hat{x}(f) \neq \hat{x}(g)$, where

$\hat{x}: \mathcal{E}(X) \to X$ is the evaluation map at x. It remains to observe that $\hat{x}$ is continuous (by the very definition of the topology of simple convergence) and equivariant (by the definition of composition of mappings), that is, a morphism of G-flows.

Now assume that (Z, z) is a G-ambit whose points are separated by continuous equivariant maps to X. Denote by $\mathcal{F}$ the collection of all such maps $Z \to X$. Any $f \in \mathcal{F}$ is uniquely determined by the image of z, and thus the map

$$\mathcal{F} \ni f \mapsto f(z) \in X$$

is an injection, identifying $\mathcal{F}$ with an everywhere dense subset of X. To every $a \in Z$, associate an element of $X^{\mathcal{F}}$ of the form $f \mapsto f(a)$; the resulting mapping $Z \to X^{\mathcal{F}}$ is one-to-one and continuous and therefore (as X is compact) a homeomorphic embedding. Finally, it is easy to verify that the natural projection $X^X \to X^{\mathcal{F}}$, dual to the inclusion $\mathcal{F} \hookrightarrow X$, determines a continuous equivariant map from $\mathcal{E}(X) \subseteq X^X$ onto Z. □

EXAMPLES 11.2. **1.** The enveloping semigroup $\mathcal{E}(\mathcal{S}(G))$ of the greatest ambit of a topological group G is canonically isomorphic to the greatest ambit $\mathcal{S}(G)$ itself.

2. Here is a more interesting case where Theorem 11.1 applies. Let G be a discrete group. The *shift system* over G is topologically a (compact, zero-dimensional) Cantor cube $\mathbb{Z}_2^G$, upon which G acts by left translations. Such G-flows form the subject of study of *symbolic dynamics.* It is shown [**G2, Lemma 4.1**] that, rather remarkably, the enveloping semigroup $\mathcal{E}(\mathbb{Z}_2^G)$ is isomorphic to $\mathcal{S}(G)$ and therefore the shift system in a sense carries the complete dynamical information pertaining to a discrete group.

COROLLARY 11.3. *For a topological group G, the following are equivalent.*
(i) *The canonical morphism $\mathcal{S}(G) \to \mathcal{E}(\mathcal{M}(G))$ is an isomorphism of G-ambits.*
(ii) *Points of $\mathcal{S}(G)$ are separated by continuous equivariant mappings to minimal G-flows.*

PROOF. $\Rightarrow$: This follows from Theorem 11.1.
$\Leftarrow$: Let $x, y \in \mathcal{S}(G)$ be arbitrary, and let them be separated by a continuous equivariant mapping, f, to a minimal G-flow, X. Fix a morphism $i: \mathcal{M}(G) \to X$, which is necessarily onto, and choose any $a \in i^{-1}(f(e))$. There exists a unique morphism of G-flows $\varphi: \mathcal{S}(G) \to \mathcal{M}(G)$ sending e to a. Since $i(\varphi(e)) = f(e)$ and therefore $i \circ \varphi = f$, one has $\varphi(x) \neq \varphi(y)$, as required. □

Ellis asked in 1969 [**E3**] whether or not the condition (i) was true for every topological group G. In view of the above result, it amounts to asking whether or not the greatest ambit of a topological group can be reconstructed using only minimal G-flows, that is, is 'residually minimal.' Now that the existence of extremely amenable groups has been established, it is clear that in general Ellis' question is answered in the negative. Indeed, while the greatest ambit $\mathcal{S}(G)$ is always nontrivial whenever G is so, the universal minimal flow of an extremely amenable topological group is a singleton and maps to it do not separate points.

The answer to Ellis' question is positive for every precompact group G, because, as we observed earlier, in this case $\mathcal{M}(G)$ coincides with all of $\mathcal{S}(G)$ and is isomorphic, as a G-flow, to the compact group completion of G equipped with the natural left action of G by translations. Surprisingly, it remains the only known class of topological groups answering Ellis' question in the positive. The limitations of space prevent us from presenting a proof of the following recent result by Glasner [**G4**], since this observation is based on deep results [**F, GW**].

THEOREM 11.4. *For the group $\mathbb{Z}$ with the discrete topology, the points of the greatest ambit $\mathcal{S}(\mathbb{Z}) \cong \beta\mathbb{Z}$ are not separated by continuous equivariant maps to the universal minimal flow $\mathcal{M}(\mathbb{Z})$.* □

In view of this result, we put forward the following.

CONJECTURE 11.5. For a topological group G the following statements are equivalent:
(i) continuous equivariant maps to minimal G-flows separate points in the greatest G-ambit,
(ii) G is precompact.

12. Effective Minimal Flows

The following result, deduced at once from Theorem 7.1, is in fact an equivalent twin of Theorem 8.1, and can be obtained from it by applying the latter result to the image of a topological group G in Homeo $(\mathcal{M}(G))$ under the action homomorphism.

THEOREM 12.1. *A topological group G acts effectively on its universal minimal flow $\mathcal{M}(G)$ if and only if for every $g \in G$, $g \neq e$, there is a left syndetic subset $S \subseteq G$ such that $g \notin \overline{SS^{-1}}$.*

As the first application of this criterion, we are able to offer a simple proof of the following weakened form of Veech's theorem.

THEOREM 12.2. *A locally compact group acts effectively on its universal minimal flow.*

PROOF. Let G be a locally compact group, let $g \in G$ and $g \neq e$. We will verify the condition from Theorem 12.1. Fix a compact neighbourhood of g, say V. Assume $e \notin V$. Zorn's Lemma implies the existence of a maximal subset $A \subseteq G$ with $e \in A$ and $AA^{-1} \cap V = \emptyset$. We claim that $(V^{-1} \cup V)A = G$ and in particular A is left syndetic, which finishes the proof.

Assuming that this is not so, there is some $g \in G$ with

$$g \notin (V^{-1} \cup V)A \equiv V^{-1}A \cup VA,$$

which means that $Ag^{-1} \cap V = \emptyset = gA^{-1} \cap V$. Now we have

$$(A \cup \{g\})(A \cup \{g\})^{-1} = (AA^{-1} \cap V) \cup (Ag^{-1} \cap V) \cup (gA^{-1} \cap V) \cup (\{e\} \cap A) = \emptyset,$$

in contradiction with the assumed maximality of A. □

PROBLEM 12.3. Describe in intrinsic terms those topological groups acting *freely* on their universal minimal flows and use this description to obtain a 'soft' proof of Veech's theorem.

Remember that Homeo$_+(\mathbb{S}^1)$ provides an example of a topological group whose action on the universal minimal flow is effective but not free (Section 11).

13. On the Existence of Unitary Representations

Out of multitude of ways to define an amenable topological group [**G1, Gre**], the one most apt for abstract topological dynamics is this: a topological group G is amenable if and only if every compact G-flow X admits an invariant mean, that is, a positive linear functional $\varphi: C(X) \to \mathbb{C}$ of norm 1 such that $\varphi(L_g f) = \varphi(f)$ for all $f \in C(X)$, $g \in G$. (Cf. e.g. [**A**].)

The following is a somewhat strengthened form of Theorem 4 from [**HC**].

THEOREM 13.1. *Let an amenable topological group G act effectively on the universal minimal flow $\mathcal{M}(G)$. Then G admits a faithful strongly continuous representation in a Hilbert space.*

PROOF. Let $\varphi: C(\mathcal{M}(G)) \to \mathbb{C}$ be an invariant mean. Set for all $f, h \in C(\mathcal{M}(G))$: $\langle f, g \rangle = \varphi(f\bar{h})$. Clearly, $\langle , \rangle$ is an invariant sesquilinear form on $C(\mathcal{M}(G))$. Denote by $\mathcal{H} = \widehat{C(\mathcal{M}(G))}/\mathcal{N}_\varphi$ the associated Hilbert space, where $\mathcal{N}_\varphi = \{x \in C(\mathcal{M}(G)): \langle x, x \rangle = 0\}$. The group G acts on $\mathcal{H}$ by isometries. This action is strongly continuous. Indeed, so is the action of G on $C(\mathcal{M}(G))$ and since the pre-Hilbert topology on $C(\mathcal{M}(G))$ is coarser than the uniform topology, each orbit map $G \ni g \mapsto g \cdot f \in \mathcal{H}$ is continuous whenever $f \in C(\mathcal{M}(G))$. It means that the representation of G in $\mathcal{H}$ is continuous with respect to the topology of simple convergence on an everywhere dense subset of $\mathcal{H}$. But on the unitary group $U(\mathcal{H})$ any such topology coincides with the strong operator topology.

It remains to show that the representation is faithful. Let $g \in G$ be any element, and let $x \in \mathcal{M}(G)$ be such that $gx \neq x$. Fix a neighbourhood $V \ni x$ with $gV \cap V = \emptyset$. There exists an $f \in C(\mathcal{M}(G))$ with $0 \leq f \leq 1$, $\operatorname{supp} f \subset V$, and $\varphi(f) > 0$. Indeed, otherwise the support of the invariant mean φ would not meet any translate gV, and since finitely many of those cover $\mathcal{M}(G)$ by force of its minimality, the support of φ would be empty, in contradiction to $\| \varphi \| = 1$. Now it is easy to see that $\| f - g \cdot f \| = 2\| f \| > 0$, which means that g as an operator in $\mathcal{H}$ is different from identity. □

Inverting the above result, one concludes that an amenable topological group admitting no nontrivial strongly continuous unitary representations is extremely amenable. This observation (made in the particular case of abelian topological groups, which are all known to be amenable) was employed in [**HC**] by Herer and Christensen, who constructed an abelian topological group without nontrivial unitary representations and deduced its extreme amenability.

Combining Theorems 13.1 and 7.1 leads to the following new existence result for unitary representations.

COROLLARY 13.2. *Let G be an amenable topological group and let $g \in G$ be such that for some left syndetic subset $S \subseteq G$ one has $g \notin \overline{SS^{-1}}$. Then g is separated from identity by a strongly continuous unitary representation of G.*

PROOF. Denote the action of G on $\mathcal{M}(G)$ by τ. According to Theorem 7.1, $\tau(g) \neq e$ in the topological group $\tau(G) \subseteq \operatorname{Homeo}(\mathcal{M}(G))$. Since the latter group acts effectively and minimally on $\mathcal{M}(G)$, it admits a faithful unitary representation π by 13.1, and the unitary representation $\pi \circ \tau$ of G is strongly continuous and separates g from identity. □

Unfortunately, the above result falls short of a criterion, since even for amenable topological groups the existence of such a left syndetic set S is not necessary for the existence of a nontrivial unitary representation. Indeed, as observed in [**G4**], the monothetic extremely amenable group from Glasner–Furstenberg–Weiss example (Section 10) embeds as a topological subgroup into the unitary group of l_2 with the strong operator topology. Thus, the old problem of describing in intrinsic terms those topological groups admitting a complete system of strongly continuous unitary representations remains open. The importance of big sets in abstract harmonic analysis and representation theory is underlined by the following classical result.

THEOREM 13.3 (Cotlar and Ricabarra [**CR**]). *Let G be an abelian topological group. An element $g \in G$ is separated from identity by a continuous character if and only if there exists a big symmetric open set $S \subseteq G$ with $g \notin S^6$.*

Later Ellis and Keynes [**EK**] reduced the number 6 to 4.
In this connection, the following might be interesting.

CONJECTURE 13.4. *Let G be a topological group. An element $g \in G$ is separated from identity by a continuous finite-dimensional unitary representation if and only if there exists a big symmetric open set $S \subseteq G$ with $g \notin S^4$ (or S^6, etc.).*

Of course, the necessity ($\Rightarrow$) is always valid. Possibly, one extra condition to be imposed on S is that of being *invariant* under inner automorphisms of G.

References

[A] J. Auslander, *Minimal Flows and Their Extensions*, North-Holland Mathematics Studies **153**, North-Holland, Amsterdam–NY–London–Tokyo, 1988.

[B] B. Beauzamy, *Introduction to Operator Theory and Invariant Subspaces*, North-Holland Mathematics Library **42**, North-Holland, Amsterdam–NY–Oxford–Tokyo, 1988.

[BP] C. Bessaga and A. Pełczyński, *Selected Topics in Infinite-Dimensional Topology*, PWN, Warszawa, 1975.

[Br] R.B. Brook, *A constructions of the greatest ambit*, Math. Systems Theory **4** (1970), 243–248.

[CR] M. Cotlar and R. Ricabarra, *On the existence of characters in topological groups*, Amer. J. Math. **76** (1954), 375–388.

[DPS] D. Dikranjan, I. Prodanov, and L. Stoyanov, *Topological Groups. Characters, Dualities, and Minimal Group Topologies*, Monographs and Textbooks in Pure and Applied Mathematics **130**, Marcel Dekker, Inc., NY–Basel, 1989.

[E1] R. Ellis, *A semigroup associated with a transformation group*, Trans. Amer. Math. Soc. **94** (1960), 272–281.

[E2] ———, *Universal minimal sets*, Proc. Amer. Math. Soc. **11** (1960), 540–543.

[E3] ———, *Lectures on Topological Dynamics*, Mathematical Lecture Note Series, W.A. Benjamin Inc., New York, 1969.

[EK] R. Ellis and H.B. Keynes, *Bohr compactifications and a result of Følner*, Israel J. Math. **12** (1972), 314–330.

[En] R. Engelking, *General Topology*, PWN, Warczawa, 1977.

[F] H. Furstenberg, *Disjointness in Ergodic Theory, minimal sets and a problem of diophantine approximation*, Math. System Theory **1** (1967), 1-55.

[G1] S. Glasner, *Proximal Flows*, Lecture Notes in Math. **517**, Springer-Verlag, Berlin–Heidelberg–NY, 1976.

[G2] ———, *Divisibility properties and the Stone-Cech compactification*, Canadian J. Math **32** (1980), 993-1007.

[G3] ———, *Structure theory as a tool in topological dynamics*, Lectures given during the Descriptive Set Theory and Ergodic Theory Joint Workshop (Luminy, June 1996), Tel Aviv University preprint, 26 pp.

[G4] ———, *On minimal actions of Polish groups*, Tel Aviv University preprint, October 1996; to appear in Proc. 1996 Prague Topol. Symposium, 6 pp.

[GW] S. Glasner and B. Weiss, *Interpolation sets for subalgebras of* $l^\infty(\mathbb{Z})$, Israel J. Math. **44** (1983), 345-360.

[GRS] R.L. Graham, B.L. Rothschild, and J.H. Spencer, *Ramsey Theory*, Wiley-Interscience, NY, 1980.

[Gr1] E. Granirer, *Extremely amenable semigroups 2*, Math. Scand. **20** (1967), 93-113.

[Gr2] ———, *Functional analytic properties of extremely amenable semigroups*, Trans. Amer. Math. Soc. **137** (1969), 53–75.

[Gre] F.P. Greenleaf, *Invariant Means on Topological Groups*, Van Nostrand Mathematical Studies **16**, Van Nostrand – Reinhold Co., NY–Toronto–London–Melbourne, 1969.

[GM] M. Gromov and V.D. Milman, *A topological application of the isoperimetric inequality*, Amer. J. Math. **105** (1983), 843–854.

[Gu] A. Guichardet, *Systèmes dynamiques non commutatifs*, Asterisque **13-14**, 1974.

[HC] W. Herer and J.P.R. Christensen, *On the existence of pathological submeasures and the construction of exotic topological groups*, Math. Ann. **213** (1975), 203–210.

[Ma] R.D. Mauldin (ed.), *The Scottish Book*, Birkhäuser, Boston–Basel–Stuttgart, 1981.

[Mi] T. Mitchell, *Topological semigroups and fixed points*, Illinois J. Math.. **14** (1970), 630-641.

[N] B.H. Neumann, *On ordered division rings*, Trans. Amer. Math. Soc. **66** (1949), 202–252.

[P] V.G. Pestov, *On free actions, minimal flows, and a problem by Ellis*, Research report RP-96-193, Dept. Math., Victoria Univ. of Wellington, August 1996, 15 pp. (Available at `http://www.vuw.ac.nz/~vova/papers/ellis.tex`).

[T] S. Teleman, *Sur la représentation linéaire des groupes topologiques*, Ann. Sci. Ecole Norm. Sup. **74** (1957), 319–339.

[Tu] S. Turek, *Universal minimal dynamical system for reals*, Comment. Math. Univ. Carol. **36** (1995), 371–375.

[U] V.V. Uspenskiĭ, *A universal topological group with countable base*, Funct. Anal. Appl. **20** (1986), 160–161.

[V1] W.A. Veech, *The equicontinuous structure relation for minimal abelian transformation groups*, Amer. J. Math. **90** (1968), 723–732.

[V2] ———, *Topological dynamics*, Bull. Amer. Math. Soc. **83** (1977), 775–830.

[Vr] J. de Vries, *Elements of topological dynamics*, Mathematics and Its Applications **257**, Kluwer Academic Publ., Dordrecht–Boston–London, 1993.

School of Mathematical and Computing Sciences, Victoria University of Wellington, P.O. Box 600, Wellington, New Zealand

E-mail address: `vladimir.pestov@vuw.ac.nz`

Contemporary Mathematics
Volume **215**, 1998

Weakly Almost Periodic Flows and Hidden Eigenvalues

T. Downarowicz

ABSTRACT. We are searching for group-related structures behind the dynamics of a weakly almost periodic point transitive flow with trivial set of eigenvalues. The main idea is to consider the non-compact flow obtained either by discarding a closed invariant subset or by taking the set of all transitive points.

1. Introduction

The notion of strong and weak almost periodicity was introduced for Markov operators by K. Jacobs in the beginning of the 60's and was investigated by B. Jamison, M. Rosenblatt, R. Sine, and others. As an application of one of the main results on w.a.p. operators to topological dynamics, we obtain that every transitive w.a.p. cascade (**N**-action or **Z**-action) is uniquely ergodic.

The history of the study of weakly almost periodic mappings goes back to 1974 when R. Sine considered them as *W.A.P. deterministic Markov operators* ([S]). Among other things, using the DeLeeuw-Glicksberg decomposition, he proved that for every w.a.p. cascade the union M of all minimal sets is closed and the action is equicontinuous on M. Later, in 1980, J. Montgomery, R. Sine, and E. Thomas wrote a paper [MST] fully devoted to w.a.p. mappings. Their result that M is a retract of the phase space is of crucial significance. Interestingly, this result was obtained using operator theory, without applying the Ellis semigroup machinery.

With additional transitivity and recurrence assumption some implications of the topological structure of the phase space of a w.a.p. cascade have been obtained in [D1] and [D2] (1987) (in this case every **N**-action can be extended to a **Z**-action). In particular it has been proved that if the space is zero-dimensional or it is a one- or two-manifold then every such action is minimal (and hence equicontinuous). An effective example of a non-minimal w.a.p. transitive cascade has also been presented.

Weakly almost periodic group actions (flows) were studied by R. Ellis and M. Nerurkar in [EN] (1989). Using the Ellis semigroup theory they proved many results known for cascades, such as continuity of the limit maps, unique ergodicity, equicontinuity on minimal sets. Uniqueness of a minimal idempotent provides an analogue to the retract statement of [MST]. Some naturally appearing w.a.p. flows

---

1991 *Mathematics Subject Classification.* 28D05, 54H20.

Supported by KBN grant PB 666/2/91

in the theory of almost periodic functions and in the theory of measure preserving flows have been indicated.

The inspiration for this note comes from the peculiarities of the cascade in the explicit example presented in [D1]. It is visible that a p-adic adding machine G_p is strongly involved in the dynamics. But G_p is neither the maximal equicontinuous factor nor the minimal set of the cascade; both are singletons. It was natural to try to clarify the role of the group G_p in this example, and to search for similar "hidden" structures in the general case.

For illustration, consider a transitive cascade (X, T) obtained as a factor of another transitive cascade (Y, T) (rich in eigenvalues) by contracting all the points of an invariant subset $F \subset Y$ to a singleton. Clearly, the set of eigenvalues of (X, T) is trivial, however the action on the remaining part of X is "capable" of producing all the eigenvalues of (Y, T).

We present two different approaches in search for group-related behavior (richer than the maximal equicontinuous factor) behind the dynamics of a transitive flow: geometric and spectral. The first is described for w.a.p. flows, the second for any transitive cascade. In case of a w.a.p. transitive cascade the resulting structures are proved to coincide (by duality). In both methods we obtain a family of groups $\{M_F\}$ (or $\{\Lambda_F\}$) indexed by closed invariant subsets of the phase space. In the spectral method we define one more group, Λ_{h}, containing each Λ_F, determined by the set of the transitive points. Some applications of the notion of Λ_{h} are presented.

A naturally occurring class of w.a.p. transitive cascades is related to probability measures μ on the circle group $\mathbb{T}$. The group Λ_{h} for such a cascade corresponds to the dual to the $L^1(\mu)$-closure of the characters on $\mathbb{T}$. The cascade is recurrent if and only if μ is a Dirichlet measure.

For the cascade of the mentioned example of [D1] (which also turns out to be determined by a Dirichlet measure), we succeeded in effectively describing all the introduced groups. For each F, M_F is isomorphic to G_p, the group Λ_{h} is essentially larger than the group dual to G_p.

The author would like to thank Eli Glasner and Anzelm Iwanik, for inspiring discussions and suggestions.

2. Basics

Throughout this paper X will denote an infinite compact Hausdorff space and G a group acting on X, i.e., via a map $\pi : G \times X \to X$, continuous with respect to $x \in X$, satisfying

1) $\pi(e, x) = x$ for every $x \in X$ and the unity e of G, and
2) $\pi(g, \pi(h, x)) = \pi(gh, x)$ for each $x \in X$ and $g, h \in G$.

The pair (X, G) will be called a *flow*. Note that, unlike in many papers considering *transformation groups* (see, e.g.,[A], [EN]) we are using "left" (and not "right") action. The difference is, of course, inessential. The "left" action corresponds to composition of homeomorphisms, most often used for cascades. We establish the following notation: if $g \in G$ then $T_g : X \to X$ is the homeomorphism defined by $T_g x = \pi(g, x)$. We denote by G_X the group of corresponding homeomorphisms of X, $\{T_g : g \in G\}$, viewed as a subset of the family X^X of all transformations from X to itself, endowed with the (compact) product topology. (Thus, G_X is now a transformation group in the sense of [A].) The *Ellis semigroup* $E_{(X,G)}$ of (X, G) is the closure of G_X in X^X. The elements of G_X will be denoted by T, S, etc., while for denoting the elements of $E_{(X,G)}$ we will use Greek letters τ, σ, etc. It is a classical fact that $(E_{(X,G)}, \circ)$ with composition of transformations is a compact semitopological semigroup ($\tau \to \tau \circ \sigma$ is continuous for every σ). The group G_X (and hence G) acts in a natural continuous way on $E_{(X,G)}$ by $T\tau = T \circ \tau$.

Given a flow, we shall consider three types of sets: *invariant sets*, i.e., sets $F \neq \emptyset$ such that $T(F) = F$ for every $T \in G_X$, closed invariant sets (note that if F is closed and invariant then $\tau(F) \subset F$ for every $\tau \in E_{(X,G)}$), and *minimal sets*, i.e., closed invariant sets containing no proper closed invariant subsets. By Zorn's Lemma a minimal set occurs in every closed invariant set.

A flow (X, G) is called *point transitive* (briefly *transitive*) if there exists a point $x_0 \in X$ (called a *transitive point*) such that the *orbit* $G(x_0) = \{Tx_0 : T \in G_X\}$ of x_0 is dense in X. We will denote by X_{tr} the set of all transitive points of (X, T). In particular, a flow is minimal if and only if $X_{\mathrm{tr}} = X$. When discussing transitive flows, we will always denote by x_0 a fixed transitive point.

The flow (X, G) is *equicontinuous* if the homeomorphisms $T \in G_X$ are equicontinuous. Clearly, if a flow is both transitive and equicontinuous then it is minimal.

A factor map between two flows $\phi : (X, G) \to (Y, G)$ is a map satisfying $\phi \circ T = S \circ \phi$ for any pair of homeomorphisms $T \in G_X$ and $S \in G_Y$ corresponding to the same element of the (common) acting group G. The flow (Y, G) is then called a *factor* of (X, G). The flow (Y, G) is called the *maximal equicontinuous factor* of (X, G) if (Y, G) is an equicontinuous flow and every other equicontinuous factor of (X, G) is a factor of (Y, G). It is well known that the maximal equicontinuous factor exists for every flow. Transitivity, minimality, equicontinuity and weak almost periodicity (see next section) are properties which pass from (X, G) to every factor (Y, G).

Two flows (X, G) and (Y, G) are *isomorphic* if there exists an invertible factor map between them. We will identify isometric flows.

A large part of our investigation concerns the case where $G = \mathbf{Z}$ is the group of integers. The flow, in this case called a *cascade*, is then fully determined by a single homeomorphism T, so, instead of $(X, \mathbf{Z})$, it will be denoted by (X, T). Analogously, the corresponding Ellis semigroup will be denoted by E_T. A point x of a cascade (X, T) is *recurrent* if it is not isolated from the set $\{T^n x : n \neq 0\}$. In a transitive cascade (X, T) a transitive point x_0 is recurrent if and only if X has no isolated points. Therefore, in this case, all transitive points are recurrent. We will call such cascades *r-transitive*. Moreover, for metrizable X, r-transitivity implies *positive transitivity*, i.e., existence of a point whose positive orbit $\{T^n x : n \geq 1\}$ is dense in X.

Let E be a semitopological semigroup. A *semicharacter* on E is a continuous homomorphism from E to the circle $\mathbb{T} = \{z \in \mathbf{C} : |z| = 1\}$ viewed as a multiplicative group. The family $\widehat{E}$ of all semicharacters on E is also a multiplicative group.

Return to the cascade (X, T). If f is some function defined on X then we denote $fT = f \circ T$. For $\tau \in E_T$, $f\tau$ is defined analogously. An *eigenfunction* pertaining to an *eigenvalue* $\lambda \in \mathbf{C}$ is a nonzero continuous complex-valued function f on X such that $fT = \lambda f$. If (X, T) is transitive then every eigenfunction f satisfies $|f| = \mathit{const}$, and the set Λ of all eigenvalues is a subgroup of $\mathbb{T}$. Moreover, for each $\lambda \in \Lambda$ there exists a unique eigenfunction f such that $f(x_0) = 1$. As easily verified, this function determines a semicharacter γ on E_T by $\gamma(\tau) = f\tau(x_0)$.

Suppose now that (X, T) is both minimal and equicontinuous. By the Halmos-von Neumann Theorem, E_T is a compact monothetic group and the cascades (X, T) and (E_T, T) are topologically isomorphic. As noticed above, every eigenvalue λ determines a character on this group. Also, in this case, every character γ is an eigenfunction pertaining to the eigenvalue $\lambda = \gamma(T)$. Thus Λ is algebraically isomorphic to the character group (*dual*) $\widehat{E_T}$ of E_T. Conversely, by the Pontryagin duality theorem, the group E_T can be identified with $\widehat{\Lambda}$, where Λ is viewed as a discrete group. In this setting, T corresponds to multiplying the elements of $\widehat{\Lambda}$ by the identity character γ_1 on Λ (defined by $\gamma_1(\lambda) = \lambda$).

Recall the definition of a group G_p of *p-adic integers*. The base is an increasing sequence $p = (p_m)$ of natural numbers, such that, for each $m \geq 1$, p_{m+1} is a multiple of p_m. The group is then defined as the set of all sequences $g = (g(m))$ of natural numbers, such that

$$0 \leq g(m) < p_m, \text{ and } g(m+1) = g(m) \pmod{p_m},$$

for each $m \geq 1$. Addition in G_p is defined coordinatewise as addition modulo p_m. This group can be viewed as a compactification of $(\mathbf{Z}, +)$ by identifying each g which is eventually constant (j) with the integer j and each g which is eventually equal to $(p_m - j)$, with the integer $-j$. Hence G_p is monothetic. The character group $\widehat{G}_p$ can be identified with the union (over m) of groups of roots of unity of degree p_m.

3. Weakly Almost Periodic Transitive Flows

We define weak almost periodicity by condition that is most convenient for us. Our definition and the original one are equivalent, see [EN].

DEFINITION 1. A flow (X, G) is called *weakly almost periodic* (*w.a.p.*) if every transformation $\tau \in E_{(X,G)}$ is continuous on X.

For example every equicontinuous flow is w.a.p. Some other classical facts from the theory of w.a.p. flows will be needed. We state them (without proofs) in the transitive case. We refer the reader to [MST], [D1], and [EN] for more information on w.a.p. flows. Recall that a *retract map* is a continuous transformation $\tau : X \to X$ such that $\tau \circ \tau = \tau$.

FACT 1. *Let (X, G) be a w.a.p. transitive flow. Then*
a) there exists a unique minimal subset M of X;
b) the flow (M, G) is equicontinuous, in particular, M is a compact group;
c) there exists a unique retract factor map $\tau_M \in E_{(X,G)}$ from (X, G) onto (M, G);
d) τ_M commutes with each $\tau \in E_{(X,G)}$;
e) (M, G) is the maximal equicontinuous factor of (X, G).
f) The group of semicharacters $\widehat{E}_{(X,G)}$ is algebraically isomorphic to the group of characters $\widehat{M}$

The last statement follows immediately from the fact that every semicharacter γ on $E_{(X,G)}$ satisfies $\gamma(\tau_M) = \gamma(\tau_M^2) = 1$.

In the commutative transitive case we have an equivalent formulation of the w.a.p. property:

FACT 2. *(cf. [D1, Lemma 3]) If G is commutative and (X, G) transitive then the following conditions are equivalent:*
a) (X, G) is w.a.p.,
b) $E_{(X,G)}$ is commutative.
The flows $(E_{(X,G)}, G)$ and (X, G) are then topologically isomorphic.

PROOF. Let $\tau \in E_{(X,G)}$. Since τ is a pointwise limit of elements of G_X, it commutes with every $T \in G$. Thus, if τ is continuous, it also commutes with each $\sigma \in E_{(X,G)}$. Therefore a) $\Longrightarrow$ b) without transitivity assumption.

Conversely, let $\tau \in E_{(X,G)}$ and let (x_ι) be a convergent net in X. Let $x = \lim x_\iota$. By transitivity, there are σ_ι and σ in $E_{(X,G)}$ such that $\sigma_\iota x_0 = x_\iota$ and $\sigma x_0 = x$. For each subnet of the indices (ι) there exists a sub-subnet (ι') such that $\sigma_{\iota'}$ converge to some σ' in $E_{(X,G)}$. Of course, $\sigma' x_0 = x$. By commutativity, $\tau x_{\iota'} = \tau \sigma_{\iota'} x_0 = \sigma_{\iota'} \tau x_0 \to \sigma' \tau x_0 = \tau \sigma' x_0 = \tau x$, hence τ is continuous.

For the last part of the assertion, observe that $\phi : E_{(X,G)} \to X$ defined by $\phi(\tau) = \tau x_0$ is continuous, onto, and preserves the action of G. To see that it is 1-1, note, by commutativity, that if $\tau x_0 = \sigma x_0$ then τ and σ coincide on the orbit of x_0. By continuity, these maps are then equal. □

Consider an arbitrary compact semitopological semigroup E containing some group G. Then G acts on E in a natural way, as left multipliers. The following statements are straightforward:

FACT 3. *For every compact semitopological commutative semigroup E containing a group G the following holds: $E_{(E,G)}$ is the closure $\overline{G}$ of G in E, and the flow (E, G) is w.a.p. Further,*
a) (E, G) is transitive if and only if $\overline{G} = E$, a transitive point is the unit e of G (and hence of E).
If transitivity holds then:
b) the set $E_{\rm tr}$ of all all transitive points of (E, G) is equal to the subgroup of all invertible elements of E,
c) (E, G) is equicontinuous if and only if (E, G) is minimal if and only if E is a group. □

Moreover, if $G = \mathbf{Z}$ and E is metrizable with no isolated points then the condition equivalent to transitivity in a) can be weakened. A (multiplicative) semigroup E is *monothetic* if there exists an element $T \in E$ (called a *generator* of E) such that the set $\{T, T^2, T^3, ...\}$ is dense in E.

FACT 4. *(cf. [D1, Lemma 1]) A non-minimal cascade (X, T) with X metric, is w.a.p. r-transitive if and only if it is isomorphic to a rotation by a generator of a compact metric monothetic semitopological commutative semigroup E that is not a group and that has no isolated points.*

PROOF. Let E be as above. There exists a sequence $n_k \geq 2$ of natural numbers, such that $T^{n_k} \to T$. The limits of any convergent subsequences of $T^{n_k - 1}$ and $T^{n_k - 2}$ are, respectively, the unit e of E and an element T^{-1} inverse to T. Therefore E contains the group of all integer iterates of T. By Fact 3, (E, T) is the desired cascade.

Conversely, let the cascade (X, T) be given. By Facts 2, 3a) and c), we have all the desired properties for $E = E_T$ except for monotheticity. By r-transitivity and metrizability, there exists a forward transitive point in X, which, along with weak almost periodicity, implies that the forward orbit under T of the identity e is dense in E. □

The w.a.p. cascades described in Fact 4 are of special interest to us. We call them *Dirichlet cascades because they are analogous to Dirichlet measures (see later in this paper). Some more topological information about the phase space of a Dirichlet cascade can be derived (cf. [D1, Lemmas 1 and 2]):*

FACT 5. *Let (X,T) be a Dirichlet cascade. Then every point $x \in X$ is positively recurrent and it cannot be disconnected from its image $\tau_M x$ in the minimal set M.*

PROOF. By Fact 4, there exists a recurrent positively transitive point x_0, hence there exists a sequence (n_k) of natural numbers, such that $T^{n_k}x_0 \to x_0$. By the last statement of Fact 2, the sequence T^{n_k} converges pointwise to identity e in E_T. This proves positive recurrence of each $x \in X$. Further, fix an $x \in X \setminus M$ and suppose there exists a closed and open set U containing x and disjoint with M. For every natural k there exists n_k such that $T^{n_k}x$ is so close to $\tau_M x$ that $T^{n_k+i}x \notin U$ for each natural $i \leq k$. Since x is positively recurrent, we can pick a sequence m_k of "farewell visits" in U, i.e., such that $T^{m_k}x \in U$ and $T^{m_k+i}x \notin U$ for each natural $i \leq k$. The limit point of any convergent subsequence of $T^{m_k}x$ is isolated from its positive orbit, a contradiction to positive recurrence. We have proved that x cannot be disconnected from M. By compactness of M, the intersection C of all closed and open sets containing x has a representative y in M. Suppose $\tau_M x \notin C$. Let $U \ni x$ be a closed and open set not containing $\tau_M x$, and denote by U^c the complement of U. We have $y \in U$. Consider the sets $\tau_M^{-1}(U)$ and $\tau_M^{-1}(U^c)$. They are disjoint, both closed and open, the first containing y, the second containing x, a contradiction to the definition of y. □

Dirichlet cascades admit plenty of invariant subsets:

FACT 6. *Let (X,T) be a Dirichlet cascade. Denote the metric on X by d, and let $r = \sup\{d(x,M) : x \in X\}$. For every $\epsilon \in [0,r]$ there exists a closed invariant set F_ϵ such that $\sup\{d(x,M) : x \in F_\epsilon\} = \epsilon$.*

PROOF. The statement is trivial for $\epsilon = 0$ and for $\epsilon = r$. Consider $0 < \epsilon < r$. Let x_r be a point realizing the maximal distance r from M. Let $T^{m_k}x_r$ be the sequence of "farewell visits" of the orbit of the point x_r in U^c, the complement of the open ϵ-ball U around M, i.e., such that $T^{m_k}x_r \in U^c$ and $T^{m_k+i}x_r \in U$ for each natural $i \leq k$. The limit point x of any convergent subsequence of $T^{m_k}x_r$ is in U^c, but its positive orbit is in the closure $\overline{U}$. By positive recurrence of x, the closed invariant set F generated by x is in $\overline{U}$. Thus $x \in U^c \cap \overline{U}$, hence $d(x,M) = \epsilon$. □

4. Dirichlet Measures

In this section we present a classical construction of a semitopological commutative semigroup using a measure, adapted to our particular needs. The connections of such semigroups with the notion of weak almost periodicity have been known since a long time (see [BH], [DR]).

Let μ be a probability measure on the unit circle $\mathbb{T}$. By the Banach-Alaoglu theorem, the unit ball B in $L^\infty(\mu)$ (complex) is compact and metrizable in the weak* topology. More explicitly, we consider

$$B = \{f \in L^\infty(\mu) : |f| \le 1\},$$

and we say that $f_n \to f$ in B if

$$\int f_n g \, d\mu \to \int fg \, d\mu,$$

for every $g \in L^1(\mu)$. Obviously, B is closed under multiplication. Moreover, if $f_n \to f$ in B and $f' \in B$ then

$$\int f_n f' g \, d\mu \to \int f f' g \, d\mu,$$

for every $g \in L^1(\mu)$, because $f'g \in L^1(\mu)$. This proves that B (with multiplication) is a compact semitopological commutative semigroup with unity $\mathbf{1}$ (the constant function). We select in B the compact subsemigroup X_μ generated by the characters of the circle group $\mathbb{T}$, i.e.,

$$X_\mu = \overline{\{\gamma_n : n \in \mathbf{Z}\}} \subset B,$$

where $\gamma_n(z) = z^n$, $(z \in \mathbb{T})$. By Fact 3, (X_μ, T), where T denotes multiplying by γ_1, is a w.a.p. transitive cascade. The above cascade (X_μ, T) is isomorphic to the subsystem (Y_μ, S) of $(\mathbb{D}^{\mathbf{Z}}, S)$ (the countable product of the unit discs with the shift transformation) obtained as the orbit-closure of the sequence $\widehat{\mu}$ representing the Fourier transform of μ. In fact, it is clear that the map sending each complex-valued function f to its Fourier transform $\widehat{f} = (\widehat{f}(n)) = (\int \gamma_n f d\mu)$ is a 1-1 factor map from the cascade (B, T) into $(\mathbb{D}^{\mathbf{Z}}, S)$ in which the image of (X_μ, T) is (Y_μ, S).

The Wiener lemma (see, e.g., [K, p.42]) applied to the nonatomic part μ_0 of μ says that $|\widehat{\mu}_0(n)|$ tend in average to zero. Thus the sequence $(..., 0, 0, 0, ...)$ belongs to the shift orbit-closure of $\widehat{\mu}_0$. Therefore there exist in X_μ functions equal to zero mod μ_0, i.e., supported only by the atoms of μ (if there are any atoms). The collection M of these functions coincides, on the set of the atoms of μ, with the pointwise closure of $\{\gamma_n : n \in \mathbf{Z}\}$, so it is the group dual to the discrete (countable) subgroup of $\mathbb{T}$ generated by the atoms of μ. It is now seen that M is the unique minimal subset of X_μ. In the case of μ nonatomic, we have $M = \{\mathbf{0}\}$.

The transitive point $\mathbf{1}$ is recurrent in (X_μ, T) if and only if it can be weak* approximated by the characters, i.e., when the measure μ is a so called *Dirichlet measure* (see [HMP]). In other words, each Dirichlet measure defines a Dirichlet cascade.

An explicit example of a Dirichlet cascade (in the form of a subshift on the Hilbert cube) was presented in [D1]. It turns out that this cascade (slightly modified, the modification being a topological isomorphism) falls in the class described above, i.e., the transitive point is a Fourier transform of a non-atomic Dirichlet measure. Since we intend to conduct further study of this particular cascade, we will present the construction in full details.

EXAMPLE 1.
Let (w_m) be a sequence of natural numbers, such that

$$\sum_{m=1}^{\infty} \frac{1}{w_m} < \infty.$$

Denote $p_1 = 2w_1$, $p_m = 2w_m p_{m-1}$ for $m > 1$, and let $\alpha_m = \frac{1}{p_m}$. Let $a_m = (a_m(n))$ be the p_m-periodic sequence such that

$$a_m[0, p_m - 1] = 0, \alpha_m, 2\alpha_m, 3\alpha_m, ..., \frac{p_m}{2}\alpha_m, (\frac{p_m}{2} - 1)\alpha_m, (\frac{p_m}{2} - 2)\alpha_m, ..., \alpha_m.$$

Observe that each a_m is a subadditive and symmetric function, i.e., for any two integers i and j we have

$$a_m(i+j) \leq a_m(i) + a_m(j) \text{ and } a_m(-i) = a_m(i).$$

Let S denote the shift transformation on $I^{\mathbf{Z}}$, where $I = [0, \infty]$ (the one-point compactification of nonnegative real numbers), i.e.,

$$Sx(n) = x(n+1) \text{ for } x = (x(n)) \in I^{\mathbf{Z}}.$$

In the sequel we will several times use the following observation:

(1) Let $s_k = \Sigma_{m=k+1}^{\infty} S^{n_{k,m}} a_m$, where $(n_{k,m})$ is an arbitrary double sequence of integers. Suppose that s_k converges coordinatewise as $k \to \infty$. Then the limit is a constant sequence $(..., r, r, r, ...)$, $r \in I$. If $n_{k,m}$ does not depend on k then either $r = 0$ or $r = \infty$.

The first statement holds since the distance between the values at any two neighbouring coordinates in s_k is estimated by $\Sigma_{m=k+1}^{\infty} \alpha_m$ which tends to 0. Now, the second statement is immediate since, in this case, $s_k(0)$ is the tail of a fixed infinite series of nonnegative numbers.

We define the space X as a subset of $I^{\mathbf{Z}}$ (with the product topology) consisting of elements of the form

$$x = x_{(r,g)} = r + \sum_{m=1}^{\infty} S^{g(m)} a_m,$$

where r stands for the constant sequence $(..., r, r, r, ...) \in I^{\mathbf{Z}}$ and $g = (g(m))$ is an element of the group G_p of p-adic integers with the base $p = (p_m)$. We will prove that X is compact. Let $x_{(r_k, g_k)}$ be a sequence of elements of X. By compactness of I and using a diagonal argument, we can choose a subsequence of the indices (also denoted by (k)) such that

(2) $$r_k \to r_0 \in I \text{ and } g_k(m) = g_m(m) \pmod{p_m},$$

for each $m \geq 1$ and $k \geq m$. Let $g_0 = (g_m(m)) \in G_p$. Now, using (1) and (2), we obtain

$$x_{(r_k, g_k)} = r_k + \sum_{m=1}^{k} S^{g_m(m)} a_m + \sum_{m=k+1}^{\infty} S^{g_k(m)} a_m \to$$

$$r_0 + \sum_{m=1}^{\infty} S^{g_0(m)} a_m + r' = x_{(r_0 + r', g_0)},$$

with some $r' \in I$, which completes the proof of compactness.

Using the second statement of (1) it can be proved that for each $x = x_{(r,g)} \in X$ either $x = x_{(\infty)} = (..., \infty, \infty, \infty, ...)$ or it is a real-valued sequence with infimum r. Moreover, for each $x \neq x_{(\infty)}$, g is also uniquely determined. In fact, by observing the intervals of linearity of $(x(n))$ one can determine $g(1)$, then subtracting $S^{g(1)}a_1$ from x and observing intervals of linearity one finds $g(2)$, etc. Also, it can be verified, that the projection $x_{(r,g)} \to g$ of X to G_p is continuous on $X \setminus \{x_{(\infty)}\}$, which is not true for the projection $x_{(r,g)} \to r$ to I (we will see later that $\{x_{(0,g)}\}$ is dense in X). Let

$$G = \{g \in G_p : x_{(0,g)}(0) < \infty\}.$$

It is easily seen that $G \neq G_p$ (it can be proved that G is of Haar measure zero in G_p). We have

$$X = \{x_{(r,g)} : r \in [0,\infty), g \in G\} \cup \{x_{(\infty)}\}.$$

We will show that G is a subgroup of G_p. We have $x_{(0,0)}(0) = 0$, so $0 \in G$. Choose $g_1, g_2 \in G$. Then

$$x_{(0,g_1+g_2)}(0) \leq x_{(0,g_1)}(0) + x_{(0,g_2)}(0) < \infty,$$

by subadditivity of each a_m and by the definition of G. This proves that $g_1+g_2 \in G$. Finally, by symmetry of the sequences a_m it is seen that $g \in G$ implies $-g \in G$.

We will now endow X with a structure of a semitopological commutative semigroup. For this we let

$$x_{(\infty)} \oplus x = x \oplus x_{(\infty)} = x_{(\infty)} \text{ for any } x \in X,$$

and

$$x_{(r_1,g_1)} \oplus x_{(r_2,g_2)} = x_{(r_1+r_2,g_1+g_2)},$$

for $r_1, r_2 < \infty$ and $g_1, g_2 \in G$. (It is worth noticing that the map $(r,g) \to x_{(r,g)}$ is a homomorphism onto X of $I \times G$, but not of $I \times G_p$. This can be seen by applying it to $(0,g)$ and $(0,-g)$ with some $g \notin G$). We need to check continuity of the operation $\oplus$ with respect to one (say the first) variable. This holds trivially if the second variable is fixed at $x_{(\infty)}$. Thus, consider an element $x_{(r,g)} \neq x_{(\infty)}$ in X. Let $x_{(r_k,g_k)}$ be a convergent sequence of elements of $X \setminus \{x_{(\infty)}\}$ satisfying (2) (by compactness, it suffices to consider only such sequences). The limit is $x_{(r_0+r',g_0)}$, as in the proof of compactness of X (here $x_{(\infty)}$ is admitted). If $x_{(r_0+r',g_0)} \neq x_{(\infty)}$ then

$$x_{(r_0+r',g_0)} \oplus x_{(r,g)} = x_{(r_0+r'+r,g_0+g)}.$$

If $x_{(r_0+r',g_0)} = x_{(\infty)}$ then either $r_0 + r' = \infty$ or $g_0 \notin G$, the last implying $g_0 + g \notin G$. Thus, in both cases $x_{(r_0+r'+r,g_0+g)} = x_{(\infty)}$, so the displayed formula for $x_{(r_0+r',g_0)} \oplus x_{(r,g)}$ is also valid. On the other hand

$$\begin{aligned} x_{(r_k,g_k)} \oplus x_{(r,g)} &= x_{(r_k+r,g_k+g)} \\ &= r_k + r + \sum_{m=1}^{k} S^{g_m(m)+g(m)} a_m + \sum_{m=k+1}^{\infty} S^{g_k(m)+g(m)} a_m \\ &\to r_0 + r + \sum_{m=1}^{\infty} S^{g_0(m)+g(m)} a_m + r'' = x_{(r_0+r+r'',g_0+g)}, \end{aligned}$$

for some $r'' \in I$ (by (1)). It suffices to show that $r' = r''$. Recall that

$$r' = \lim_k \sum_{m=k+1}^{\infty} S^{g_k(m)} a_m(0) \text{ and } r'' = \lim_k \sum_{m=k+1}^{\infty} S^{g_k(m)+g(m)} a_m(0).$$

By subadditivity of each a_m we have

$$\sum_{m=k+1}^{\infty} S^{g_k(m)+g(m)}a_m(0) \leq \sum_{m=k+1}^{\infty} S^{g_k(m)}a_m(0) + \sum_{m=k+1}^{\infty} S^{g(m)}a_m(0)$$

and

$$\sum_{m=k+1}^{\infty} S^{g_k(m)}a_m(0) \leq \sum_{m=k+1}^{\infty} S^{g_k(m)+g(m)}a_m(0) + \sum_{m=k+1}^{\infty} S^{-g(m)}a_m(0).$$

Applying the second part of (1), and the fact that g (and hence $-g$) is in G, we obtain that $r'' \leq r' \leq r''$, as desired.

Finally we shall observe that the compact semitopological commutative semigroup $(X, \oplus)$ is monothetic, has no isolated points, and that it is not a group. In fact, it is seen that for every $x \in X$, $x \oplus x_{(0,1)} = Sx$, in particular

$$S^n x_{(0,0)} = x_{(0,n)}.$$

Further, for every $(r,g) \in [0,\infty) \times G_p$, $x_{(r,g)}$ can be approximated by $x_{(0,n)}$ with $n \in \mathbf{N}$. For this observe that for every $\epsilon > 0$ and $k \in \mathbf{N}$ there exist natural n and k' such that

$$n = g(m) \pmod{p_m} \quad \text{for each } m \leq k,$$

$$\left|\sum_{m=k+1}^{k'} S^n a_m(0) - r\right| < \epsilon, \quad \text{and} \quad \sum_{m=k'+1}^{\infty} S^n a_m(0) < \epsilon.$$

This implies that the positive multiples (with respect to the operation $\oplus$) of $x_{(0,1)}$ are dense in X. In particular, the neutral element $x_{(0,0)}$ of X is not isolated, and so is every point of X. Finally the existence of another idempotent $x_{(\infty)}$ shows that X is not a group. By Fact 6, (X, S) is a Dirichlet cascade with the transitive point $x_{(0,0)}$. It is immediate that the unique minimal subset M of X is $\{x_{(\infty)}\}$.

Clearly, the set of all invertible (and hence transitive, see Fact 3b)) points of X is equal to $\{x_{(0,g)} : g \in G\}$, so algebraically X_{tr} is the same as G. However, the topology of X_{tr} is stronger: every character on G can be extended to one on G_p, (a general fact concerning subgroups of compact abelian groups). In the sequel we will show that X_{tr} admits essentially more characters than does G_p.

REMARK 1. The above cascade is determined by a Dirichlet measure as described earlier. Namely, the exponential function $y(x) = e^{-x}$ is a homeomorphism between $[0,1]$ and I, translating sums into products. Thus, the flow (X, S) is topologically isomorphic to the shift orbit-closure (Y, S) of the sequence $y_0 = \prod_m e^{-a_m}$ in the Hilbert cube $[0,1]^{\mathbf{Z}}$. Since the finite products $\prod_{m=1}^k e^{-a_m}$, converge pointwise to y_0, it suffices to show that $e^{-a_m} = \widehat{\mu}_m$ for each m, where μ_m is some probability measure on $\mathbb{T}$. Then the convolutions $\mu_1 * \mu_2 * ... * \mu_k$ (weak*) converge to a probability measure μ on $\mathbb{T}$, and $y_0 = \widehat{\mu}$. By the proved properties of the cascade (X, S), it will follow that μ is a non-atomic Dirichlet measure.

Fix an $m \geq 1$. Consider the group $H \equiv H_m = \{1, z, z^2, ..., z^{p_m - 1}\}$ of p_mth roots of unity ($z = e^{\frac{2\pi i}{p_m}}$), and let f be the function on H defined by $f(z^j) = e^{-a_m(j)}$,

$j = 0, 1, ..., p_m - 1$. Since $\widehat{H} = H$, the Fourier transform of f is another function on H and its values are

$$q(j) = \widehat{f}(z^j) = \sum_{l=0}^{p_m-1} e^{-a_m(l)} z^{jl}.$$

Below we prove that each $q(j)$ is positive. Let $p = p_m$, $\beta = e^{-\alpha_m}$, and $\kappa = \frac{2j\pi}{p}$. We have

$$\begin{aligned} q(j) &= \sum_{l=0}^{\frac{p}{2}-1} (\beta z^j)^l + (\beta z^j)^{\frac{p}{2}} + \sum_{l=\frac{p}{2}+1}^{p-1} \beta^{p-l} z^{jl} \\ &= (-1)^j \beta^{\frac{p}{2}} - 1 + 2\mathrm{Re}\left(\sum_{l=0}^{\frac{p}{2}-1} (\beta z^j)^l \right), \end{aligned}$$

since $z^{\frac{p}{2}} = -1$, and since the first sum and second sum enlarged by 1 are mutually conjugate. Further, the last term equals

$$\begin{aligned} 2\mathrm{Re}\left(\frac{1 - (\beta z^j)^{\frac{p}{2}}}{1 - \beta z^j} \right) &= 2(1 - (-1)^j \beta^{\frac{p}{2}})\mathrm{Re}\left(\frac{1 - \beta z^{-j}}{|1 - \beta z^j|^2} \right) \\ &= 2\frac{(1 - (-1)^j \beta^{\frac{p}{2}})(1 - \beta \cos \kappa)}{1 - 2\beta \cos \kappa + \beta^2}. \end{aligned}$$

Finally,

$$\begin{aligned} q(j) &= 2\frac{(1 - (-1)^j \beta^{\frac{p}{2}})(1 - \beta \cos \kappa)}{1 - 2\beta \cos \kappa + \beta^2} - (1 - (-1)^j \beta^{\frac{p}{2}}) \\ &= (1 - (-1)^j \beta^{\frac{p}{2}}) \frac{1 - \beta^2}{|1 - \beta z^j|^2}. \end{aligned}$$

Since $\beta < 1$, $q(j)$ is seen to be positive, as desired. We conclude, by duality, that f is the Fourier transform of the probability measure μ_m on H with the atoms $cq(j)$ at z^j (c is the normalizing factor). This implies that e^{-a_m} is the Fourier transform of the same measure μ_m viewed as an atomic measure on $\mathbb{T}$.

5. The Maximal Weakly Almost Periodic Factor

Let E be an arbitrary semigroup of (not necessarily continuous) transformations of a compact Hausdorff space X. Consider the following family of continuous complex-valued functions on X:

$$\mathcal{A} = \{f \in C(X) : \forall_{T \in E} \; fT \in C(X)\}.$$

It is obvious that $\mathcal{A}$ is an E-invariant C^*-subalgebra of $CB(X)$ closed in the uniform norm. Defining on X an equivalence relation

$$x \sim_{\mathcal{A}} y \text{ whenever } f(x) = f(y) \text{ for each } f \in \mathcal{A},$$

and letting $Y = X/\sim_{\mathcal{A}}$ (Y is compact, Hausdorff), we can identify $\mathcal{A}$ with $C(Y)$. The transformations from E preserve equivalence classes in X, hence they define transformations of Y. All these transformations are continuous on Y.

DEFINITION 2. For a flow (X, G), the *maximal weakly almost periodic factor* is defined as the flow (Y, G), where Y is constructed as above for the Ellis semigroup $E = E_{(X,G)}$.

Of course, the system (Y, G) is w.a.p. We omit verifying that (Y, G) is in fact a factor of (X, G). The standard proof of the following statement will also be omitted.

FACT 7. *Let (Y', G) be a w.a.p. factor of a compact flow (X, G). Then (Y', G) is a factor of the maximal w.a.p. factor (Y, G) of (X, G), and the factor map $\phi' : X \to Y'$ is the composition $\psi \circ \phi$ of the factor maps $\phi : X \to Y$ and $\psi : Y \to Y'$.*

Recall that every transitive flow with the action of G, is a factor of the *universal transitive flow* $(\beta G, G)$ provided by the Čech-Stone compactification βG of G.

DEFINITION 3. The maximal w.a.p. factor of the universal transitive flow $(\beta G, G)$ is called *the universal w.a.p. transitive flow.*

Using Fact 7 we obtain

FACT 8. *Every w.a.p. transitive flow (X, G) is a factor of the universal w.a.p. transitive flow.*

If the group G is topological then the subalgebra of those bounded continuous functions on G which extend to continuous functions on the universal w.a.p. transitive flow, with G-action, consists of the so called *weakly almost periodic functions* on G. These have been investigated by many authors (see e.g. [BH] and [EN]).

6. The F-replacement in W.A.P. Transitive Flows

Let (X, G) be a w.a.p. transitive flow. As noticed in Fact 1e) and f), the three naturally occurring groups: the unique minimal set, the maximal equicontinuous factor and the dual to the discrete group of semicharacters of the Ellis semigroup, all coincide. The construction presented below is one of the possible ways in search of group related behavior inside a w.a.p. transitive flow, richer than the above trio.

Let F denote a proper closed invariant subset of X. We will replace F by another set M_F, such that the action of G on M_F is minimal, the whole new flow remains w.a.p., and M_F is maximal in the sense that for any other set M'_F satisfying the above, the flow (M'_F, G) is a factor of (M_F, G). We proceed in the following way.

1) The set $F^c = X \setminus F$ is open, dense, and G-invariant. Let $(\beta(F^c), G)$ be the extension of the G-action to the Čech-Stone compactification of F^c (in which F^c is again open and dense).

2) Let (Y', G) be the maximal w.a.p. factor of $(\beta(F^c), G)$. Since (X, G) is a w.a.p. compactification of (F^c, G), it is a factor of (Y', G) and hence all factor maps in Fact 7 applied to $(\beta(F^c), G)$, (Y', G), and (X, G) are identities on F^c. So, we can think of (Y', G) as (X, G) with the closed invariant subset F "replaced" by another closed invariant set F', of which F is a factor. Of course, the flow (Y', G) remains transitive.

3) Let M_F be the unique minimal subset of (Y', G). Obviously, $M_F \subset F'$. We want to replace F' by M_F. Let $\tau'_{M_F} : Y' \to M_F$ denote the unique retract factor map of the w.a.p. flow (Y', G) onto its minimal set. Consider the following family of functions:

$$\mathcal{B} = \{f \in C(Y') : f = f\tau'_{M_F} \text{ on } F'\}.$$

Using Fact 1d) one obtains that $\mathcal{B}$ is an $E_{(Y',G)}$-invariant algebra closed in the uniform norm, so it can be represented as $C(Y)$ (with $Y = Y'/{\sim_{\mathcal{B}}}$), so that (Y,G) is a w.a.p. factor of (Y',G). Further, the corresponding factor map is identity on $F^c \cup M_F$ since the algebra $\mathcal{B}$ separates points of this set. Each point y of the remaining set $F' \setminus M_F$ is sent to $\tau'_{M_F} y \in M_F$. This means that Y can be identified with $F^c \cup M_F$ endowed with a new compact topology in which M_F becomes a boundary of F^c.

DEFINITION 4. For a given w.a.p. transitive flow (X,G) and a proper closed invariant subset F of X, the set M_F constructed above will be called the *F-replacement.*

The proof of the following statement, based on comparing subalgebras of functions, will be omitted.

FACT 9. *Let (X,G) be a w.a.p. transitive flow and F its proper closed invariant subset. Then for every w.a.p. compactification of F^c by a set M' on which the action is minimal, (M',G) is a factor of (M_F,G).*

7. Hidden Eigenvalues

Let (X,T) be an arbitrary transitive cascade and let F be a proper closed invariant subset of X. Only the eigenvalues of the cascade (F,T) are candidates to be eigenvalues of (X,T). As before, we are interested in discovering a group-related structure carried by the cascade, free from the restrictions forced by the proper closed invariant subsets.

DEFINITION 5. Let (X,T) be a transitive cascade, and D a dense invariant subset of X. A complex number $\lambda \in \mathbb{T}$ will be called a *D-eigenvalue* of (X,T) if there exists on D a continuous nonzero function f such that $Tf = \lambda f$. The function f will be called a *D-eigenfunction.*

Clearly, the set of D-eigenvalues is a subgroup of $\mathbb{T}$. Mostly, we will be interested in the following two cases:
1) $D = F^c$ is the complement of a proper closed invariant set $F \subset X$.
The set of all F^c-eigenvalues will be denoted by Λ_F, and by Λ_{inv} we denote the group generated by the union $\bigcup \Lambda_F$ over all proper closed invariant subsets F of X.
2) $D = X_{\text{tr}}$.
In this case the D-eigenvalues will be called the *hidden eigenvalues* of the cascade (X,T). The set of all hidden eigenvalues will be denoted by Λ_{h}.

Clearly, $\Lambda \subset \Lambda_F \subset \Lambda_{\text{inv}} \subset \Lambda_{\text{h}}$ (since $X_{\text{tr}} \subset F^c$) for each proper closed invariant $F \subset X$. We will show that Λ_{h} can be determined by observing the orbit of an arbitrary transitive point x_0.

THEOREM 1. *If x_0 is a transitive point, $D = \mathbf{Z}(x_0)$, and f is a D-eigenfunction, then f can be extended to an X_{tr}-eigenfunction.*

PROOF. It suffices to show that f can be extended continuously to X_{tr}. Let $x \in X$, and suppose λ_1, λ_2 are two different limit values of f at x. Thus, at every point of the orbit of x there are two limit values of f at the same distance from each other as λ_1, λ_2. This property passes to all points of the orbit-closure of x. Therefore x cannot be a transitive point, because f admits a unique limit value at x_0. □

It follows from the above theorem, that for minimal cascades, the notion of D-eigenvalues coincides with the ordinary notion of eigenvalues, regardless of the choice of the invariant set D.

In the metrizable case the following characterization follows immediately from Theorem 1:

THEOREM 2. *Let (X,T) be a transitive cascade and x_0 a transitive point. Then $\lambda \in \Lambda_{\mathrm{h}}$ if and only if $\lambda^{n_k} \to 1$ for every sequence (n_k) for which $T^{n_k}x_0 \to x_0$.*

It is now obvious that if (X,T) is not r-transitive then $\Lambda_{\mathrm{h}} = \mathbb{T}$, and in the opposite case Λ_{h} is of Lebesgue measure zero.

REMARK 2. Spectral invariants similar to hidden eigenvalues have been considered in ergodic theory. This generalization of eigenvalues is defined by the condition $\lambda^{n_k} \to 1$ along sequences (n_k) of integers being so called "rigidity times". Seemingly, not much is known about these invariants. The author was unable to find an appropriate reference. Some analogue to hidden eigenvalues can also be found in [KS] and in [I], where additional assumptions are made on the speed of the above convergence.

The rest of this section will be devoted to presenting some simple applications of the notion of hidden eigenvalues. First, for transitive cascades we recall the notion of weak disjointness (see [P]). For two cascades (X,T) and (Y,S) by the direct product we mean the cascade $(X \times Y, T \times S)$, where $T \times S(x,y) = (Tx, Sy)$.

DEFINITION 6. Two transitive cascades are *weakly disjoint* if their direct product is transitive.

Note that for minimal cascades our condition is weaker than the classical notion of topological disjointness.

THEOREM 3. *Let (X,T) and (Y,S) be transitive cascades. If (X,T) and (Y,S) have a common hidden eigenvalue $\lambda \neq 1$ then they are not weakly disjoint. If (Y,S) is a rotation of a compact monothetic (additive) group G by a generator* 1 *then the converse implication is true.*

PROOF. Suppose there exists a transitive point (x_0, y_0) in the product $(X \times Y, T \times S)$ and suppose λ is a common hidden eigenvalue. Let f and g be corresponding eigenfunctions on $\mathbf{Z}(x_0)$ and $\mathbf{Z}(y_0)$ respectively. The function $h(x,y) = \frac{f(x)}{g(y)}$ is defined and continuous on the product $\mathbf{Z}(x_0) \times \mathbf{Z}(y_0)$. But it is constant on $\mathbf{Z}((x_0,y_0))$, a dense subset of the domain. This implies that $f(x_0) = fT(x_0)$, hence $\lambda = 1$.

Conversely, let $(X \times G, T \times 1)$ be not transitive. Denote by P the orbit closure of $(x_0, 0)$ (here 0 is the neutral element of G). The set $H = \{g \in G : (x_0, g) \in P\}$ is a closed proper subset of G. By an easy approximation argument, this set is a closed subsemigroup, hence a closed subgroup. Every nontrivial character γ of the quotient group G/H defines an eigenvalue for $(G,1)$ by $\lambda = \gamma(H+1) \neq 1$. Now, since P is closed and invariant, whenever $T^{n_k}x_0 \to x_0$, then also $H + n_k \to H$, and hence $\lambda^{n_k} \to 1$. By Theorem 2 we obtain $\lambda \in \Lambda_{\mathrm{h}}$. □

Recall that a cascade (X,T) is *topologically weakly mixing* if every invariant set in the direct product of (X,T) with itself is either nowhere dense or dense. In the metrizable case, topological weak mixing can be expressed in terms of weak disjointness: (X,T) is topologically weakly mixing if and only if it is weakly disjoint from itself. An example of a topologically weak mixing transitive cascade is the full shift on two symbols.

It is well known (see, e.g., [A]) that for minimal cascades topological weak mixing is equivalent to not having eigenvalues other than 1. For transitive cascades, as a consequence of Theorem 3, we obtain

THEOREM 4. *Every topologically weakly mixing transitive cascade has no hidden eigenvalues other than* 1.

FACT 10. *Any metrizable w.a.p. transitive cascade* (X,T) *is not topologically weakly mixing.*

PROOF. Suppose (x_0, x) is a transitive point in the product $(X \times X, T \times T)$. Since x_0 is a transitive point in (X,T), $T^{n_k}x_0 \to x_0$ implies $T^{n_k}x \to x$. Thus the orbit closure of (x_0, x) in $X \times X$ contains no (x_0, y) with $y \neq x$, a contradiction.

We will show in further discussion that not having hidden eigenvalues other than 1 is essentially weaker than the topological weak mixing condition.

8. The Case of a W.A.P. Transitive Cascade

In the case of a w.a.p. transitive cascade (X,T) there is a strong relation between the F-replacement and Λ_F.

THEOREM 5. *Let* (X,T) *be a w.a.p. transitive cascade and* F *a proper closed invariant subset of* X. *Then the set* Λ_F *is identical with the set of all eigenvalues of the cascade* (M_F, T). *In other words* $M_F = \widehat{\Lambda}_F$.

PROOF. Let λ be an eigenvalue of (M_F, T), and let f_1 be a corresponding eigenfunction. Let τ_{M_F} denote the retract factor map of the w.a.p. compactification of F^c by M_F onto M_F. Then $f = f_1 \tau_{M_F}$ (restricted to F^c) is an F^c-eigenfunction pertaining to λ, hence $\lambda \in \Lambda_F$.

For the converse consider the following set of functions

$$\mathcal{B} = \overline{\{f \in CB(F^c) : f = f_0|_{F^c} + f_1, \quad f_0 \in C(X), \quad f_0 = 0 \text{ on } F, \quad \text{and } f_1 \in \mathcal{F}\}},$$

where $\mathcal{F}$ is the uniformly closed subalgebra of $CB(F^c)$ spanned by the F^c-eigenfunctions. By w.a.p. assumption, it can be verified that $\mathcal{B}$ is an E_T-invariant subalgebra of $CB(F^c)$ separating the points of F^c. So it has the form $C(Y)$, where (Y,T) is some w.a.p. compactification of (F^c, T). For each F^c-eigenvalue λ the function $0 + f$, where f is an F^c-eigenfunction pertaining to λ, is in $\mathcal{B}$, hence it extends to an eigenfunction $\tilde{f}$ on (Y,T). Next, the restriction of $\tilde{f}$ is an eigenfunction on the unique minimal subcascade of (Y,T). But, by Fact 9, the latter is a factor of the cascade (M_F, T) on the F-replacement. Eventually we obtain that the set of all eigenvalues of the cascade (M_F, T) contains Λ_F. □

Recall that, by Facts 2 and 3b), X_{tr} has a group structure, the same as $E_{\mathrm{tr}} \equiv (E_T)_{\mathrm{tr}}$.

THEOREM 6. *Let* (X,T) *be a w.a.p. transitive cascade. Then the group* Λ_{h} *of hidden eigenvalues is isomorphic to the dual* $\widehat{X}_{\mathrm{tr}}$ *to the group of the transitive points in* X.

PROOF. Let λ be a hidden eigenvalue and f the corresponding eigenfunction on $\mathbf{Z}(x_0)$, satisfying $f(x_0) = 1$. By Theorem 1, f can be extended to X_{tr}, hence it defines a character on X_{tr}. The converse is clear: every character γ on X_{tr} is an X_{tr}-eigenfunction pertaining to the hidden eigenvalue $\lambda = \gamma(T)$. We omit proving that the above map from $\widehat{X}_{\mathrm{tr}}$ onto Λ_{h} is continuous and open. □

9. Dirichlet Measures – Continuation

Return to the w.a.p. transitive cascade (X_μ, T) obtained as the weak* closure in $L^\infty(\mu)$ of the characters γ_n on $\mathbb{T}$, where μ is some probability measure on $\mathbb{T}$. It is obvious that f is an invertible element of the semigroup X_μ if and only if $|f| = 1$. Within the group of unimodular functions the weak* convergence coincides with the $L^1(\mu)$ convergence. Thus $X_{\mu\mathrm{tr}}$ can be identified with the closure of the characters in $L^1(\mu)$. This group has been effectively described only in few particular cases (see [HMP]). We are interested in describing Λ_{h}, i.e., the characters on $X_{\mu\mathrm{tr}}$. Of course, only Dirichlet measures are interesting here – otherwise $\Lambda_{\mathrm{h}} = \mathbb{T}$.

THEOREM 7. *If λ is contained in the intersection of all subgroups H of $\mathbb{T}$ satisfying $\mu(H) = 1$, then it is a hidden eigenvalue of the flow (X_μ, T).*

PROOF. By Theorem 2, it suffices to show that $\lambda^{n_k} \to 1$, whenever $\gamma_{n_k} \to \mathbf{1}$ in $L^1(\mu)$. But in this case, every subsequence of n_k has a sub-subsequence n'_k such that $\gamma_{n'_k} \to \mathbf{1}$ μ-almost everywhere, i.e., on a set A with $\mu(A) = 1$. The convergence holds also on the group H generated by A (finite products of the elements of $A \cup A^{-1}$). But $\lambda \in H$ since $\mu(H) = 1$, which ends the proof. □

Obtaining some more information about Λ_{h} and (especially) Λ_{inv} and their relation with the support of μ seems to be difficult. In two extreme cases Theorem 7 can be reversed:

1) When μ has an absolutely continuous part (with respect to the Lebesgue measure). Then $X_{\mu\mathrm{tr}} = \{\gamma_n : n \in \mathbf{Z}\}$ is discrete, so $\Lambda_{\mathrm{h}} = \mathbb{T}$. On the other hand, the only subgroup of full measure μ is $\mathbb{T}$. Such measures, however, being far from Dirichlet measures are not of our main interest.

2) When μ is a so called a *Kronecker measure.* Then $X_{\mu\mathrm{tr}}$ is equal to the group of all unimodular functions in $L_1(\mu)$, a classical example of a Polish group with no nontrivial characters. This means that there are no hidden eigenvalues of (X, T) other than 1. On the other hand, for every $\lambda \neq 1$ we can find a subgroup of full measure μ, not containing λ. We omit the proofs.

The latter example along with Fact 10 show (as announced) that the condition $\Lambda_{\mathrm{h}} = \{1\}$ is essentially weaker than the topological weak mixing condition.

In Example 1 we are able to fully describe the sets Λ_{inv} and Λ_{h}. Interestingly, the first is countable and equal to Λ_F for any proper closed invariant subset F, while the second is uncountable. Unfortunately, we do not know much about the support of the corresponding Dirichlet measure.

PROPOSITION 1. *In Example 1, for any proper closed invariant set F, we have $\Lambda_F = \widehat{G}_p$. In other words, the action on the F-replacement, (M_F, T), is isomorphic to $(G_p, 1)$.*

PROOF. Let F be a proper closed invariant subset of X, and choose $x_{(r,g)} \in F$. Then $x_{(r',h)} = x_{(r,g)} \oplus x_{(r'-r,h-g)}$ also belongs to F, for each $r' \geq r$ and $h \in G_p$. Let $r_F = \inf\{r : x_{(r,g)} \in F\}$. We have $r_F > 0$ (because F is proper), and $F = \{x_{(r,g)} : r \geq r_F,\ g \in G_p\}$. Recall that $X = \{x_{(r,g)} : r \in [0,\infty), g \in G\} \cup \{x_{(\infty)}\}$, so $F^c = \{x_{(r,g)} : r < r_F, g \in G\}$. Let λ be an eigenvalue of $(G_p, 1)$, and let $\tilde{f}$ be a corresponding eigenfunction. We define $f(x_{(r,g)}) = \tilde{f}(g)$ on F^c. Clearly, f is an F^c-eigenfunction pertaining to λ, so $\lambda \in \Lambda_F$.

Conversely, by symmetry of the sequence $x_{(0,0)}$, for any F^c-eigenfunction f satisfying $f(x_{(0,0)}) = 1$, we have $f(x_{(r,g)}) = f(x_{(r,-g)})^{-1}$. Thus $f(x_{(r,0)}) = 1$ or -1, the latter possibility being excluded by continuity of f on F^c . We have proved that

$$f(x_{(r,0)}) = 1 \quad \text{for every } r < r_F. \tag{3}$$

Suppose λ is an F^c-eigenvalue other than those arising from $(G_p, 1)$. We will view the multiplicative group $\mathbb{T}$ as the interval $[0, 1)$ with addition modulo 1. In this setting, λ (as well as any number from $[0, 1)$) can be represented as a series

$$\lambda = \sum_{m=1}^{\infty} \frac{i_m}{p_m},$$

where, for each m, i_m is an integer with $0 \leq i_m < 2w_m$. Moreover, both the sets $\{m : i_m > 0\}$ and $\{m : i_m < 2w_m - 1\}$ are infinite (otherwise λ is an eigenvalue of $(G_p, 1)$). Equivalently, there are infinitely many indices m which satisfy the alternative: either
(a) $0 < i_{m+1} < 2w_{m+1} - 1$, or
(b) $i_{m+1} = 2w_{m+1} - 1$ and $i_{m+2} = 0$.
Let $m = k$ satisfy the above alternative. We have

$$p_k\lambda + N = p_k \sum_{m=k+1}^{\infty} \frac{i_m}{p_m} = \frac{i_{k+1}}{2w_{k+1}} + p_k \sum_{m=k+2}^{\infty} \frac{i_m}{p_m},$$

(N denotes an integer). The last sum is at most $\frac{1}{p_{k+1}}$, which multiplied by p_k equals $\frac{1}{2w_{k+1}}$. Thus, in case (a),

$$\frac{1}{2w_{k+1}} \leq p_k\lambda + N \leq 1 - \frac{1}{2w_{k+1}}.$$

In case (b) the last sum is at most $\frac{1}{p_{k+2}}$, which multiplied by p_k equals $\frac{1}{4w_{k+1}w_{k+2}}$. In any case, we can write

$$\frac{1}{4w_{k+1}} \leq p_k\lambda + N \leq 1 - \frac{1}{4w_{k+1}}.$$

Let $\delta = \min\{1/4, r_F/4\}$. By the above inequalities, there exists a positive integer t such that $tp_k\lambda$ is at least δ apart from the nearest integer and this t can be chosen from among $1, 2, ..., \min\{w_{k+1}, [w_{k+1}r_F] + 1\}$. Returning to the multiplicative representation of $\mathbb{T}$, we have obtained that λ^{tp_k} is 4δ-apart from 1. Let f be the eigenfunction on F^c, pertaining to the eigenvalue λ, satisfying $f(x_{(0,0)}) = 1$.

Consider the point $x = S^{tp_k}x_{(0,0)} = x_{(0,tp_k)}$. The value of f at x is λ^{tp_k} proved to be apart from 1. On the other hand, we have

$$x = \sum_{m=1}^{k} a_m + S^{tp_k}a_{k+1} + \sum_{m=k+2}^{\infty} S^{tp_k}a_m.$$

Observe x at the positions between $-p_k$ and p_k. The first of the above sum differs there by at most $p_k \sum_{m=k+1}^{\infty} \alpha_m \leq \sum_{m=k+1}^{\infty} \frac{1}{2w_m}$ from $x_{(0,0)}$. Similarly, since $0 < tp_k < \frac{p_{k+1}}{2}$, the last series differs from zero at each of these positions by at most $\sum_{m=k+2}^{\infty} \frac{1}{2w_m}$. The remaining term, $S^{tp_k}a_{k+1}$, differs there by at most $\frac{1}{2w_{k+1}}$ from the constant $r_k = tp_k\alpha_{k+1}$, which, by the choice of t, is less than $\frac{r_F}{2} + \frac{1}{2w_{k+1}}$. Thus, we can say that x is close to $x_{(r_k,0)}$. Increasing the index k and choosing a subsequence we will obtain a sequence of points on the orbit of $x_{(0,0)}$ convergent to some $x_{(r,0)} \in F^c$, with the values of f not converging to 1, contradicting, in view of (3), continuity of f on F^c. □

As we expect, the set of all hidden eigenvalues in this example contains other numbers, too. The hidden eigenvalues are the numbers which admit sufficiently fast approximation by the rational eigenvalues of $(G_p, 1)$:

PROPOSITION 2. *In Example 1, $\lambda \in \mathbb{T}$ is a hidden eigenvalue of the cascade (X,T) if and only if there exists a bounded sequence of integers (i_m) such that*

$$\lambda = \sum_{m=1}^{\infty} \frac{i_m}{p_m} \ (mod\ 1),$$

in the additive representation of the circle group.

PROOF. To prove that every such λ is a hidden eigenvalue we need to show that $n_k\lambda$ converge to 0 in $\mathbb{T}$ for (a subsequence of) every sequence n_k of integers, such that $S^{n_k}x_{(0,0)} \to x_{(0,0)}$ (see Theorem 2). By symmetry, it suffices to examine positive sequences (n_k). Since $n_k \to 0$ in G_p, we can assume that each n_k is a multiple of p_k. Further, we can represent this multiple as

$$n_k = tp_k = t_0p_k + t_1p_{k+1} + t_2p_{k+2} + \dots + t_lp_{k+l},$$

where l is natural such that $n_k < p_{k+l+1}$, and $0 \leq t_j < 2w_{k+j+1}$, for each $j = 0, 1, \dots, l$. We have

$$S^{n_k}x_{(0,0)} = \sum_{m=1}^{k} a_m + \sum_{m=k+1}^{\infty} S^{n_k}a_m \to x_{(0,0)} + r = x_{(r,0)},$$

where r is a constant limit of the second series. By our assumption $r = 0$, hence, in particular

$$\sum_{m=k+1}^{\infty} a_m(n_k) \to 0,$$

as $k \to \infty$. By the definition of each a_m and the formula for n_k, for $j = 1, 2, \dots, l+1$, we have

$$a_{k+j}(n_k) = \begin{cases} \frac{t_{j-1}}{2w_{k+j}} + \epsilon_j & \text{or} \\ 1 - \frac{t_{j-1}}{2w_{k+j}} - \epsilon_j \end{cases}$$

depending on whether the first value is $\leq 1/2$ or not. Here

$$\epsilon_j = \frac{t_0 p_k + t_1 p_{k+1} + \ldots + t_{j-2} p_{k+j-2}}{p_{k+j}} \leq \frac{1}{2w_{k+j}}.$$

We have obtained that the following sums converge to zero in k:

$$\sum_{j=1}^{l+1} \left\| \frac{t_{j-1}}{2w_{k+j}} \right\|, \tag{4}$$

where $\|\cdot\|$ denotes the distance from 0 in $\mathbb{T}$ (l depends on k).

On the other hand

$$n_k \lambda = i_{k+1} \frac{t_0 p_k}{p_{k+1}} + i_{k+2} \frac{t_0 p_k + t_1 p_{k+1}}{p_{k+2}} + \ldots +$$

$$i_{k+l+1} \frac{t_0 p_k + t_1 p_{k+1} + \ldots + t_l p_{k+l}}{p_{k+l+1}} + n_k \sum_{m=k+l+2}^{\infty} \frac{i_m}{p_m} + N.$$

Since N is an integer, it can be omitted, and the preceding sum (jointly with the coefficient n_k) is estimated by

$$p_{k+l+1} \sum_{m=k+l+2}^{\infty} \frac{c}{p_m} \leq c \sum_{m=k+l+2}^{\infty} \frac{1}{2w_m}$$

($c = \sup_m |i_m|$), which decreases to zero as $k + l \to \infty$. The remaining sum is at most

$$c \sum_{j=1}^{l+1} \frac{1 + t_{j-1}}{2w_{k+j}}.$$

It suffices to show that the sums

$$\sum_{j=1}^{l+1} \frac{t_{j-1}}{2w_{k+j}}$$

converge to 0 in $\mathbb{T}$, as $k \to \infty$. But this follows, since (4) estimates the distance of the above sums from 0.

Finally, we show that numbers other than those described above are not hidden eigenvalues in this cascade. Consider such a λ. Then, (similarly as in the proof of Proposition 1 (case a)), for every $n \in \mathbf{N}$, we can find a $k(n)$ such that

$$\frac{n}{2w_{k(n)+1}} \leq p_{k(n)} \lambda + N(n) \leq 1 - \frac{n}{2w_{k(n)+1}} \tag{5}$$

(the integers $N(n)$ can be neglected). For appropriately chosen positive integers $t(n)$ satisfying $0 < t(n) \leq [\frac{w_{k(n)+1}}{2n}] + 1$, we have

$$\frac{1}{4} \leq \frac{n t(n)}{2w_{k(n)+1}} \leq \frac{3}{4} \text{ for each } n, \quad \text{and} \quad \frac{t(n)}{2w_{k(n)+1}} \to 0 \text{ as } n \to \infty. \tag{6}$$

Passing to the multiplicative representation, the first estimates along with (5) imply that

$$\lambda^{t(n) p_{k(n)}} \not\to 1.$$

From the convergence in (6) (using the same method of estimation as in the proof of Proposition 1), we obtain that

$$S^{t(n)p_{k(n)}}x_{(0,0)} \longrightarrow x_{(0,0)},$$

implying that $\lambda \notin \Lambda_{\mathrm{h}}$. □

We conclude this note by formulating some related questions. Below, (X, G) ((X, T)) always denotes a w.a.p. transitive flow (cascade), F a proper closed invariant subset of X, and M the unique minimal subset. By a *measure* we will mean a probability measure on $\mathbb{T}$.

QESTION 1. What is the characterization of the metrizable cascades (X, T) which are topologically isomorphic to those defined by measures? The same question can be asked for Dirichlet cascades and measures.

QUESTION 2. In what class of flows is the F-replacement of each F metrizable?

QUESTION 3. Describe the class of flows where all F-replacements are identical.

QUESTION 4. Suppose that $M_M = M$. What are the consequences?

QUESTION 5. Finally we repeat the question about the connections between the support of a measure and the sets Λ_{h} and Λ_{inv} for the resulting w.a.p. cascade.

References

[A] J. Auslander, *Minimal Flows and Their Extensions*, North Holland Mathematics Studies, 1988.

[BH] J. Berglund and H. Hofmann, *Compact Semitopological Semigroups and Weakly Almost Periodic Functions*, Springer Verlag, Berlin, 1967.

[D1] T. Downarowicz, *Some properties of weakly almost periodic mappings*, Colloq. Math. **54** (1987), 241–252.

[D2] T. Downarowicz, *Weakly almost periodic mappings on one and two-manifolds*, Colloq. Math. **54** (1987), 253–259.

[DR] C. Dunkl and D. Ramirez, *Representations of Commutative Semitopological Semigroups*, Springer Verlag, Berlin, 1975.

[EN] R. Ellis and M. Nerurkar, *Weakly almost periodic flows*, Trans. Amer. Math. Soc. **313** (1989), 103–120.

[HMP] B. Host, J. Méla and F. Parreau, *Analyse Harmonique des Mesures*, Astérisque, Société Mathématique de France, 1986.

[HR] E. Hewitt and K. Ross, *Abstract Harmonic Analysis*, vol. I, Springer Verlag, 1963.

[I] A. Iwanik, *Approximation by periodic transformations and diophantine approximation of the spectrum*, Ergodic Theory of $\mathbf{Z}^d$-actions, Proc. of the Warwick Symposium 1993-4, Cambridge University Press, 1996, pp. 387–401.

[KS] A. Katok and A. Stepin, *Approximations in ergodic theory*, Uspekhi Mat. Nauk **22** (1967), 81–106.

[K] Y. Katznelson, *An Introduction to Harmonic Analysis*, New York, 1968.

[MST] J. Montgomery, R. Sine and E. Thomas, *Some topological properties of weakly almost periodic mappings*, Topology and its Applications **11** (1980), 69–85.

[P] R. Peleg, *Some extensions of weakly mixing flows*, Israel J. Math. **8** (1971), 330–336.

[S] R. Sine, *Convergence theorems for weakly almost periodic Markov operators*, Israel J. Math. **19** (1974), 246–255.

Institute of Mathematics, Technical University, Wybrzeże Wyspiańskiego 27, 50-370 Wrocław, Poland.

E-mail address: downar@graf.im.pwr.wroc.pl

Contemporary Mathematics
Volume **215**, 1998

Enveloping Linear Maps

E. Akin

ABSTRACT. We present a direct construction for and generalization of the enveloping semigroup for the action of the general linear group on the sphere.

For a topological dynamicist a diffeomorphism of a compact manifold is a homeomorphism of a compact space together with the extra structure given by the tangent map acting on various vector bundles. By using the induced maps on sphere bundles or projective bundles we get compactness for the bundle maps as well. When you then consider Ellis' enveloping semigroup construction you discover that on the fibers the maps are rather special as they are limits of maps induced by linear isomorphisms. This motivates the study of the action of the general linear group on the sphere or projective space. For applications it is the projective actions which are wanted and these are what Ellis investigated, see [E2] and the notes of D. Ellis [DE]. However, the theory itself develops a bit better for the action on spheres as we will see below.

Our functions act on the left, reversing Ellis' notation. In a volume dedicated to Ellis the left action choice requires some justification.

It seems to me that it is now clear that the convention long ago adopted to write function notation as $y = f(x)$ (and to say "f of x") was in many ways the wrong choice. Especially once the arrow notation is adopted, $xf = y$ (and read "x by f") has many advantages. For example, composition of $f : X_1 \to X_2$ followed by $g : X_2 \to X_3$ is then fg with $x(fg) = (xf)g$, instead of gf. In several ways, Ellis is favoring the right over might. Instead, I bow to the unwisdom of tradition.

* * *

We consider real, normed vector spaces of finite, positive dimension, denoted V, V_1 etc. By a *subspace* W of V we will mean a nontrivial linear subspace, with norm inherited from V. We denote by $L(V_1; V_2)$ the vector space of linear maps with the operator norm, and by $Li(V_1; V_2)$ the open subset of injective linear maps. The set Li is nonempty iff $\dim V_1 \leq \dim V_2$.

For a nonzero vector x in V we let $\langle x \rangle$ denote the normalized vector $x/|x|$. For $A \subset V$, let $\langle A \rangle$ denote $\{\langle x \rangle : x \in A \backslash 0\}$ so that $\langle A \rangle$ is a nonempty set of unit vectors

1991 *Mathematics Subject Classification.* Primary 54H20; Secondary 34C35.

unless $A \subset \{0\}$. In particular, $\langle V \rangle$ is the unit sphere in V. If $T \in L(V_1; V_2)$ and $S \in L(W; V_1)$ with $T \circ S \neq 0$ then we define

$$\Phi(T, S) = \Phi^T(S) = \Phi_S(T) \equiv \langle T \circ S \rangle. \tag{1}$$

Thus, Φ is a continuous map from the open subset $\{(T, S) \in L(V_1; V_2) \times L(W; V_1) : T \circ S \neq 0\}$ to the unit sphere $\langle L(W; V_2) \rangle$. When W is a subspace of V_1 and S is the inclusion map, we denote Φ_S by Φ_W. Thus, Φ_W is defined and $\Phi_W(T) = \langle T|W \rangle$ on the set $\{T \in L(V_1; V_2) : W \backslash Ker\ T \neq \emptyset\}$. On the other hand, associating $x \in V_1$ to the linear map $S_x \in L(\mathbf{R}; V_1)$ such that $S_x(1) = x$, we obtain a natural isometric identification between $L(\mathbf{R}; V_1)$ and V_1. So with $W = \mathbf{R}$, we obtain for $T \in L(V_1; V_2)$ the continuous map $\Phi^T : V_1 \backslash Ker\ T \to \langle V_2 \rangle$, associating $x \mapsto \langle T(x) \rangle$.

LEMMA 1. *Assume $T_1, T_2 \in L(V_1; V_2) \backslash 0$. If $\langle T_1 \rangle = \langle T_2 \rangle$ then Ker T_1 = Ker T_2 and $\Phi^{T_1} = \Phi^{T_2}$ on $V_1 \backslash Ker\ T_1$. Conversely, if Ker T_1 = Ker T_2 and $\Phi^{T_1} = \Phi^{T_2}$ on $\langle W \rangle$, where W is a subspace of V_1 complementary to Ker T_1, then $\langle T_1 \rangle = \langle T_2 \rangle$.*

PROOF. The first statement is obvious. For the second it suffices to find a positive constant α such that $T_1 = \alpha T_2$ on W.

By homogeneity $\Phi^{T_1} = \Phi^{T_2}$ on all of $W \backslash 0$ and so for every $x \in W \backslash 0$ there is a unique positive constant α_x such that $T_1(x) = \alpha_x T_2(x)$, i.e. $\alpha_x = |T_1(x)|/|T_2(x)|$. We prove α_x is independent of x.

If, instead, $\alpha_x \neq \alpha_y$ for some $x, y \in W \backslash 0$ then $\{x, y\}$ is linearly independent and so $\{T_2(x), T_2(y)\}$ is linearly independent because T_2 is injective on W. Consequently, $\{T_2(x) + T_2(y), \alpha_x T_2(x) + \alpha_y T_2(y)\}$ is linearly independent. Thus, $\{T_2(x + y), T_1(x + y)\}$ is linearly independent, a contradiction. □

A *filtration of length* k for V_1 is a strictly decreasing sequence of subspaces: $V_1 = W_1 \supset W_2 \ldots \supset W_k$. So V_1 admits filtrations of length at most dim V_1, also note that $V_1 = W_1$ is the unique filtration of length 1. A *pennant* from V_1 to V_2 associated with the filtration $\{W_1, \ldots, W_k\}$ is a sequence of linear maps, $\mathbf{L} = \{L_1, \ldots, L_k\}$ satisfying

$$L_i \in \langle L(W_i; V_2) \rangle,$$

$$\text{Ker } L_i = W_{i+1} \quad i = 1, \ldots, k, \tag{2}$$

where W_{k+1} denotes the trivial subspace $\{0\}$ so that

$$L_k \in \langle Li(W_k; V_2) \rangle. \tag{3}$$

We denote by $\mathcal{L}(V_1; V_2)$ the set of all pennants from V_1 to V_2 with topology defined later. An injective linear map of norm one is exactly a pennant associated with the length 1 filtration. Thus,

$$\langle Li(V_1; V_2) \rangle \subset \mathcal{L}(V_1; V_2). \tag{4}$$

If $\{W_1, \ldots, W_k\}$ is a filtration for V_1 and $A \subset V_1$ with $A \backslash 0 \neq \emptyset$, then the *carrier* of A is the smallest filtration element which contains A. If W_i is the carrier of A then

(5) $$A \subset W_i \quad and \quad A \backslash W_{i+1} \neq \emptyset.$$

Now we extend the previous definition of the composition map Φ to define

$$F : \mathcal{L}(V_1; V_2) \times (L(W; V_1) \backslash 0) \to \langle L(W; V_2) \rangle$$

(6) $$F(\mathbf{L}, S) = F^{\mathbf{L}}(S) = F_S(\mathbf{L}) \equiv \langle L_i \circ S \rangle,$$

where W_i is the carrier of the image of S for the filtration associated with $\mathbf{L}$. Observe that, $L_i \circ S$ is defined because $S(W) \subset W_i$ and is not zero because $S(W) \backslash W_{i+1} \neq \emptyset$.

As before when W is a subspace of V_1 and S is its inclusion map, we denote F_S by F_W. For example, if W_i is a subspace in the filtration for $\mathbf{L}$ then

(7) $$F_{W_i}(\mathbf{L}) = L_i.$$

On the other hand, using $W = \mathbf{R}$ we obtain

$$F : \mathcal{L}(V_1; V_2) \times (V_1 \backslash 0) \to \langle V_2 \rangle$$

(8) $$F^{\mathbf{L}}(x) = \langle L_i(x) \rangle,$$

where W_i is the carrier of x. In particular, each pennant $\mathbf{L}$ from V_1 to V_2 induces the map $F^{\mathbf{L}}$ from the unit sphere $\langle V_1 \rangle$ to the unit sphere $\langle V_2 \rangle$.

LEMMA 2. *The map* $F^{\mathbf{L}} : \langle V_1 \rangle \to \langle V_2 \rangle$ *is continuous at* $x \in \langle V_1 \rangle$ *iff* $x \notin W_2$, *i.e. iff the carrier of* x *is the largest subspace in the filtration.*

PROOF. On the open set $V_1 \backslash W_2$, $F^{\mathbf{L}}$ is the continuous map Φ^{L_1} defined by (1). On the other hand, for $x \in W_2$ choose $y \in V_1 \backslash W_2$ and define the path $\langle x + ty \rangle = x_t$ in $\langle V_1 \rangle$ for $|t|$ small. For $t \neq 0$, $x_t \notin W_2$ and so

(9) $$F^{\mathbf{L}}(x_t) = \langle t L_1(y) \rangle = \pm \langle L_1(y) \rangle$$

with the initial sign that of t. As the path x_t passes through x at $t = 0$, $F^{\mathbf{L}}(x_t)$ reverses direction and so $F^{\mathbf{L}}$ is not continuous at x. □

REMARKS. (a) This is the key sphere result whose projective analogue is false. It allows us to detect the subset $\langle W_2 \rangle$ using the map F^L and so to recover the entire filtration (see Proposition 7 below).

(b) Notice that (9) implies that the sphere map $F^{\mathbf{L}}$ is not injective unless $W_2 = 0$, in which case the filtration length is one and $\mathbf{L} \in \langle Li(V_1; V_2) \rangle$ in the sense of the inclusion (4).

Given a finite or infinite non-increasing sequence of subspaces of V_1 (including the trivial subspace) we obtain a filtration of V_1 by *packing*. This procedure consists of introducing V_1 at the beginning of the sequence, eliminating repeats and any copies of the zero subspace, and then renumbering the resulting strictly decreasing sequence. For example, if $\{W_1, \dots, W_k\}$ is a filtration of V_1 and $S \in Li(W; V_1)$ then the sequence $\{S^{-1}(W_1), \dots, S^{-1}(W_k)\}$ is non-increasing and by packing we get a filtration $\{\tilde{W}_1, \dots, \tilde{W}_l\}$ of W called the filtration induced by S from $\{W_1, \dots, W_k\}$. For $t = 0, \dots, l$ we see that

$$W_i \textit{ is the carrier of } S(\tilde{W}_t) \Leftrightarrow$$

$$S^{-1}(W_i) = \tilde{W}_t \textit{ and } \tilde{W}_t \backslash S^{-1}(W_{i+1}) \neq \emptyset, \tag{10}$$

and so if W_i is the carrier of $S(\tilde{W}_t)$ then

$$S^{-1}(W_{i+1}) = \tilde{W}_{t+1}, \tag{11}$$

the first proper subspace of $S^{-1}(W_i)$ in the sequence $\{S^{-1}(W_1), \dots, S^{-1}(W_k), S^{-1}(W_{k+1})\}$.

Now for $\mathbf{L} \in \mathcal{L}(V_1; V_2)$ a pennant with filtration $\{W_1, \dots, W_k\}$ and $S \in Li(W; V_1)$ inducing the filtration $\{\tilde{W}_1, \dots, \tilde{W}_l\}$ of W, define for $t = 0, \dots, l$

$$\tilde{L}_{\tilde{i}} = F(\mathbf{L}, S|\tilde{W}_{\tilde{\mathbf{L}}}) = \langle L_i \circ S|\tilde{W}_{\tilde{i}} \rangle, \tag{12}$$

where W_i is the carrier of $S(\tilde{W}_t)$. Condition (11) implies that

$$Ker \tilde{L}_t = S^{-1}(\text{Ker } L_i) = \tilde{W}_{t+1}.$$

So $\tilde{\mathbf{L}}$ is a pennant from W to V_2 with filtration $\{\tilde{W}_1, \dots, \tilde{W}_l\}$. Thus, the association $(\mathbf{L}, S) \mapsto \tilde{\mathbf{L}}$ defines a map

$$\tilde{F} : \mathcal{L}(V_1; V_2) \times Li(W; V_1) \to \mathcal{L}(W; V_2). \tag{13}$$

Since $\tilde{W}_1 = W$, the first map of the $\tilde{\mathbf{L}}$ pennant sequence is given by

$$\tilde{F}(\mathbf{L}, S)_1 = \tilde{L}_1 = F(L, S). \tag{14}$$

As usual, if W is a subspace of V_1 and S is its inclusion map, we denote by $\tilde{F}_W$ the restriction map from $\mathcal{L}(V_1; V_2)$ to $\mathcal{L}(W; V_2)$. For example, if $\{W_1, \dots, W_k\}$ is the filtration for $\mathbf{L}$ and S is the inclusion of W_i $(1 \leq i \leq k)$ then $\{W_i, W_{i+1}, \dots, W_k\}$ (renumbered) is the induced filtration and

$$\tilde{F}_{W_i}(\mathbf{L}) = \{L_i, \dots, L_k\}. \tag{15}$$

We now relate these definitions to a simple limit procedure.

A series $\sum_{j=0}^{\infty} \epsilon^j T_j$ is called an *infinitesimally injective sum*, an IIS, in $L(V_1; V_2)$ if ϵ is a positive real variable and $\{T_0, T_1, \dots\}$ is a sequence in $L(V_1; V_2)$ such that $T_j = 0$ for j sufficiently large and

$$\cap_j \text{ Ker } T_j = 0. \tag{16}$$

Condition (16) says that distinct vectors in V_1 can be distinguished if not by T_0 then by some T_j whose coefficient we think of as an infinitesimal of order j.

For an IIS $\sum_j \epsilon^j T_j$ in $L(V_1; V_2)$ and $S \in L(W; V_1)\backslash 0$ we define the limit composition

$$(\textstyle\sum_j \epsilon^j T_j) * S \equiv Lim_{\epsilon \to 0^+} \langle \textstyle\sum_j \epsilon^j (T_j \circ S) \rangle \in \langle L(W; V_2) \rangle. \tag{17}$$

Observe that if j_S is the smallest value of j such that $T_j \circ S \neq 0$, which exists by (16), then

$$(\textstyle\sum_j \epsilon^j T_j) * S = \langle T_{j_S} \circ S\rangle. \tag{18}$$

The sequence $\{Ker\ T_0 \cap \ldots \cap\ Ker\ T_j : j = 0, 1, \ldots\}$ is non-increasing and so by packing we obtain the *filtration* $\{W_1, \ldots, W_k\}$ *associated with the* IIS. For $i = 1, \ldots, k$ define

$$j_i = \min\{j : T_j | W_i \neq 0\},$$

$$L_i = \langle T_{j_i} | W_i\rangle. \tag{19}$$

PROPOSITION 3. *(a) For an IIS* $\sum_j \epsilon^j T_j$ *in* $L(V_1; V_2)$*, the sequence* $\mathbf{L} = \{L_1, \ldots, L_k\}$ *is a pennant with filtration* $\{W_1, \ldots, W_k\}$ *called the pennant associated with the IIS. If* $S \in L(W; V_1)\backslash 0$ *then*

$$(\textstyle\sum_j \epsilon^j T_j) * S = F(\mathbf{L}, S). \tag{20}$$

If $S \in Li(W; V_1)$ *then* $\sum_j \epsilon^j (T_j \circ S)$ *is an IIS in* $L(W; V_2)$ *with associated pennant* $\tilde{F}(\mathbf{L}, S)$.

(b) For $\mathbf{L} = \{L_1, \ldots, L_k\}$ *in* $\mathcal{L}(V_1; V_2)$*, let* $T_j \in L(V_1; V_2)$ *with* $T_j | W_{j+1} = L_{j+1}$ *for* $j = 0, \ldots, k-1$ *and* $T_j = 0$ *for* $j \geq k$*. Then* $\sum_j \epsilon^j T_j$ *is an IIS in* $L(V_1; V_2)$ *with associated pennant* $\mathbf{L}$.

PROOF. (a) By (19) we have

$$j < j_i \Leftrightarrow W_i \subset\ Ker\ T_j \tag{21}$$

and so W_i is the intersection of the set $\{Ker\ T_j : 0 \leq j < j_i\}$, e.g. for $i = 1$, $W_1 = V_1$ and if $T_0 \neq 0$ then $j_1 = 0$ and this set is empty. Since $T_{j_i} | W_i$ is nonzero, its kernel is a proper subspace of W_i. Hence,

$$Ker\ L_i = \cap_{j=0}^{j_i}\ Ker\ T_j = W_{i+1}. \tag{22}$$

Thus, $\mathbf{L}$ is a pennant with filtration $\{W_1, \ldots, W_k\}$.

For $S \in Li(W; V_1)$ the filtration $\{\tilde{W}_1, \ldots, \tilde{W}_l\}$ induced from $\{W_1, \ldots, W_k\}$, is the same as that obtained by packing the sequence $\{Ker(T_0 \circ S) \cap \ldots \cap\ Ker(T_j \circ S) = S^{-1}(Ker\ T_0 \cap \ldots \cap\ Ker\ T_j)\}$. Thus, $\tilde{\mathbf{L}} = \tilde{F}(L, S)$ and the pennant $\hat{\mathbf{L}}$ associated with $\sum_j \epsilon^j (T_j \circ S)$ have the same filtration. For $t = 0, \ldots, l$, let W_i be the carrier of $S(\tilde{W}_t)$.

By (19) $\hat{L}_t = \langle T_{j_t} \circ S | \tilde{W}_t\rangle$ where j_t is the minimum index j such that $T_j \circ S | \tilde{W}_t \neq 0$. By (12) and (19), $\tilde{L}_t = \langle L_i \circ S | \tilde{W}_t\rangle = \langle T_{j_i} \circ S | \tilde{W}_t\rangle$. So we must prove that $j_i = j_t$ where i is the carrier of $S(\tilde{W}_t)$.

By (21) $j < j_i$ implies $T_j | W_i = 0$ and so by (10) $T_j \circ S | \tilde{W}_t = 0$ which implies $j < j_t$. Hence, $j_i \leq j_t$. By (22) the kernel of $T_{j_i} | W_i$ is W_{i+1} and so by (10) $T_{j_i} \circ S | \tilde{W}_t \neq 0$. So it is false that $j_i < j_t$, proving $j_i = j_t$.

For $S \in L(W; V_1)\backslash 0$ the proof of (20) uses (18). We then proceed as above to prove that $j_S = j_i$ when W_i is the carrier of $S(V_1)$, using (5) with $A = S(V_1)$ in place of (10).

(b) Observe that

$$W_{i+2} = \cap_{j=0}^{i} \operatorname{Ker} T_j \tag{23}$$

for $i = 0, 1, \ldots$ with $W_j \equiv 0$ for $j > k$. So (16) holds and the filtration associated with the resulting IIS is $\{W_1, \ldots, W_k\}$. The associated pennant is the original $\mathbf{L}$ by (19). □

Notice that the result in (b) is independent of the choice of extension T_j for L_{j+1}. In effect, the value of T_j on a vector only matters when the values of the lower order infinitesimal terms are zero.

COROLLARY 4. *If* $\mathbf{L} \in \mathcal{L}(V_1; V_2)$, $S_2 \in Li(W_2; V_1)$ *and* $S_1 \in Li(W_1; W_2)$ *then*

$$\tilde{F}_{S_1}(\tilde{F}_{S_2}(\mathbf{L})) = \tilde{F}_{S_2 \circ S_1}(\mathbf{L}) \in \mathcal{L}(W_1; V_2). \tag{24}$$

PROOF. If $\sum_j \epsilon^j T_j$ is an IIS in $L(V_1; V_2)$ with associated pennant $\mathbf{L}$, then the two sides are the pennants associated with

$$\sum_j \epsilon^j ((T_j \circ S_2) \circ S_1) = \sum_j \epsilon^j (T_j \circ (S_2 \circ S_1)).$$

□

COROLLARY 5. *If* $\mathbf{L} \in \mathcal{L}(V_1; V_2)$ *and* $S \in Li(W; V_1)$ *then with* $\tilde{\mathbf{L}} = \tilde{F}(\mathbf{L}, S)$ *the following diagram commutes.*

(25)

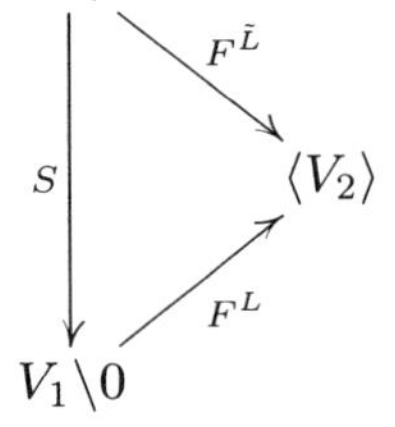

PROOF. With $x \in W\backslash 0$ apply Corollary 4 with $W_2 = W$, $S = S_2$ and $W_1 = \mathbf{R}$, $S_x = S_1$. Note that $S_2 \circ S_1$ is $S_{S(x)}$. Then (25) follows from (24) and (14). □

COROLLARY 6. *If W_1, W are subspaces of V_1 with $W_1 \subset W$ then the following diagram commutes.*

$$\begin{array}{ccc} \mathcal{L}(V_1;V_2) & & \\ \Big\downarrow \tilde{F}_W & \searrow^{F_{W_1}} & \\ & & \langle \mathcal{L}(W_1;V_2)\rangle \\ & \nearrow_{F_{W_1}} & \\ \mathcal{L}(W;V_2) & & \end{array} \tag{26}$$

PROOF. Apply Corollary 4 with $W_2 = W$, S_2 its inclusion and S_1 the inclusion of W_1 into W_2. Then (26) follows from (24) and (14). □

PROPOSITION 7. *Two pennants agree if they induce the same map of unit spheres. That is, for $\mathbf{L}$, $\tilde{\mathbf{L}} \in \mathcal{L}(V_1;V_2)$ if $F^{\mathbf{L}} = F^{\tilde{\mathbf{L}}} : \langle V_1\rangle \to \langle V_2\rangle$ then $\mathbf{L} = \tilde{\mathbf{L}}$.*

PROOF. The proof is given by induction on the length of the filtration $\{W_1, \dots, W_k\}$ for $\mathbf{L} = \{L_1, \dots, L_k\}$. Let $\{\tilde{W}_1, \dots, \tilde{W}_{\tilde{k}}\}$ be the filtration for $\tilde{\mathbf{L}} = \{\tilde{L}_1, \dots, \tilde{L}_{\tilde{k}}\}$.

On $\langle W_1 \backslash W_2\rangle$, $F^{\mathbf{L}}$ is the continuous map Φ^{L_1} and on $\langle \tilde{W}_1 \backslash \tilde{W}_2\rangle$ $F^{\tilde{\mathbf{L}}}$ is $\Phi^{\tilde{\mathbf{L}}_1}$. By Lemma 2, $F^{\mathbf{L}}$ is discontinuous at points of $\langle W_2\rangle$ and $F^{\tilde{\mathbf{L}}}$ is discontinuous at points of $\langle \tilde{W}_2\rangle$. So $F^{\mathbf{L}} = F^{\tilde{\mathbf{L}}}$ implies the subspaces W_2 and $\tilde{W}_2$ agree . Then Lemma 1 implies the maps of unit norm L_1 and $\tilde{L}_1$ agree on $W_1 = \tilde{W}_1 = V_1$.

Since $W_2 = \tilde{W}_2$, (15) implies $\tilde{F}_{W_2}(\mathbf{L}) = \{L_2, \dots, L_k\}$ and

$$\tilde{F}_{W_2}(\tilde{\mathbf{L}}) = \{\tilde{L}_2, \dots, \tilde{L}_{\tilde{k}}\}.$$

The commutative diagram of Corollary 5 implies that on $\langle W_2\rangle$ these pennants induce the restriction of $F^{\mathbf{L}}$ and $F^{\tilde{\mathbf{L}}}$ and so are equal. By induction hypothesis $\{L_2, \dots, L_k\} = \{\tilde{L}_2, \dots, \tilde{L}_{\tilde{k}}\}$. Since $L_1 = \tilde{L}_1$ as well, it follows that $\mathbf{L} = \tilde{\mathbf{L}}$. □

REMARK. As we will later see, the analogue of this result is not true in the projective case.

Now define the product map

$$F_\pi \equiv \Pi_W F_W : \mathcal{L}(V_1;V_2) \to \Pi_W \langle L(W;V_2)\rangle \tag{27}$$

where W varies over all (nontrivial) subspaces of V_1. For each $\mathbf{L} \in \mathcal{L}(V_1;V_2)$ the map of spheres $F^{\mathbf{L}}$ is in the space of maps $\langle V_2\rangle^{\langle V_1\rangle}$ which we regard as a product of copies of $\langle V_2\rangle$. So $\mathbf{L} \mapsto F^{\mathbf{L}}$ defines the map

$$F^{\#} : \mathcal{L}(V_1;V_2) \to \langle V_2\rangle^{\langle V_1\rangle}. \tag{28}$$

Proposition 7 says that $F^{\#}$ is injective. For each $x \in \langle V_1\rangle$, we can use $S_x : \mathbf{R} \to V_1$ to identify x with the one dimensional subspace $W = S_x(\mathbf{R})$. Thus, F_π is certainly injective. We define the topology on $\mathcal{L}(V_1;V_2)$ to be that induced from the product space by F_π. That is, we use the topology generated by the family of maps F_W from $\mathcal{L}(V_1; V_2)$ to the compact spaces $\langle L(W, V_2)\rangle$.

PROPOSITION 8. *If W is a subspace of V_1 then $\tilde{F}_W : \mathcal{L}(V_1; V_2) \to \mathcal{L}(W; V_2)$ is a continuous surjection.*

PROOF. If W_1 is a subspace of W then $F_{W_1} \circ \tilde{F}_W = F_{W_1}$ on $\mathcal{L}(V_1; V_2)$ by Corollary 6. So $\tilde{F}_W$ is continuous because the topology on $\mathcal{L}(W; V_2)$ is generated by such F_{W_1}'s.

If $\tilde{\mathbf{L}} = \{\tilde{L}_1, \dots, \tilde{L}_t\}$ is a pennant on W, then we can define a decreasing sequence of subspaces of V_1, $\{W_1, \dots, W_{i-1}, W_i\}$ with $W_1 = V_1$ and $W_i = W$, and linear maps $\{L_1, \dots, L_{i-1}\}$ such that $L_j \in \langle(L(W_j; V_2)\rangle$ with Ker $L_j = W_{j+1}$. Clearly, $\mathbf{L} = \{L_1, \dots, L_i, \tilde{L}_1, \dots, \tilde{L}_t\}$ is a pennant from V_1 to V_2 and by (15), $\tilde{F}_W(\mathbf{L}) = \tilde{\mathbf{L}}$. Thus, $\tilde{F}_W$ is surjective. □

REMARK. If dim $V_1 \leq$ dim V_2 then we can construct $\mathbf{L}$ from $\tilde{\mathbf{L}}$ by attaching a single linear map, i.e. $i = 2$. If, however, dim V_1 is large then a larger i might be required.

THOEREM 9. *The space of pennants $\mathcal{L}(V_1; V_2)$ is compact and satisfies the first axiom of countability. In fact, if $\mathbf{L} = \{L_1, \dots, L_k\}$ is a pennant with filtration $\{W_1, \dots, W_k\}$ and $\{\mathbf{L}^\alpha\}$ is a net in $\mathcal{L}(V_1; V_2)$ then $\{\mathbf{L}^\alpha\}$ converges to $\mathbf{L}$ in $\mathcal{L}(V_1; V_2)$ iff for $i = 1, \dots, k$, $\{F_{W_i}(\mathbf{L}^\alpha)\}$ converges to $F_{W_i}(\mathbf{L}) = L_i$ in $\langle L(W_i, V_2)\rangle$.*

The map $F^\# : \mathcal{L}(V_1; V_2) \to \langle V_2\rangle^{\langle V_1\rangle}$ is an embedding, i.e. it is a homeomorphism onto its image.

If $\dim V_1 \leq \dim V_2$ then $\langle Li(V_1; V_2)\rangle$ is a nonempty open sbset of $\langle L(V_1; V_2)\rangle$. The inclusion map (4) of $\langle Li(V_1; V_2)\rangle$ into $\mathcal{L}(V_1; V_2)$ as the set of pennants of length one, is a homeomorphism onto an open, dense subset of $\mathcal{L}(V_1; V_2)$.

PROOF. We first prove compactness and the convergence condition by induction on dim V_1.

For a net $\{L^\alpha\}$ in $\mathcal{L}(V_1; V_2)$ we can go to a subnet so that $\{L_1^\alpha\}$ converges to L_1 in the compact space $\langle L(V_1; V_2)\rangle$. Let $W_2 = Ker\ L_1$ be of dimension smaller than dim V_1. So we can apply the induction hypothesis to the net $\{\tilde{F}_{W_2}(\mathbf{L}^\alpha)\}$ in $\mathcal{L}(W_2; V_2)$. Thus, by going to a subnet we can find a pennant $\tilde{\mathbf{L}} = \{L_2, \dots, L_k\}$ in $\mathcal{L}(W_2; V_2)$ with filtration $\{W_2, \dots, W_k\}$ such that for $i = 2, \dots, k$, $\{F_{W_i}(\tilde{F}_{W_2}(\mathbf{L}^\alpha))\}$ converges to L_i in $\langle L(W_i; V_2)\rangle$. Clearly, $\mathbf{L} = \{L_1, L_2, \dots, L_k\}$ is a pennant in $\mathcal{L}(V_1; V_2)$ with filtration $\{W_1, \dots, W_k\}$ and $\{\mathbf{L}_1^\alpha = F_{W_1}(\mathbf{L}^\alpha)\}$ converges to L_1 by definition of L_1. On the other hand, by Corollary 6 $\{F_{W_i}(\tilde{F}_{W_2}(\mathbf{L}^\alpha)) = F_{W_i}(\mathbf{L}^\alpha)\}$ and so this net converges to L_i. Now we show that this condition implies that the net $\{\mathbf{L}^\alpha\}$ converges to $\mathbf{L}$ in $\mathcal{L}(V_1; V_2)$.

Again the induction hypothesis is applied to $\mathcal{L}(W_2; V_2)$ this time to conclude $\{\tilde{F}_{W_2}(\mathbf{L}^\alpha)\}$ converges to $\tilde{L}$ in $\mathcal{L}(W_2; V_2)$. By definition of the topology in $\mathcal{L}(V_1; V_2)$ we must show that $\{F_W(\mathbf{L}^\alpha)\}$ converges to $F_W(\mathbf{L})$ for every subspace W of V_1.

CASE 1. $W \subset W_2$: By Corollary 6 $\{F_W(\mathbf{L}^\alpha) = F_W(\tilde{F}_{W_2}(\mathbf{L}^\alpha))\}$ which converges to $F_W(\tilde{\mathbf{L}}) = F_W(F_{W_2}(\mathbf{L})) = F_W(\mathbf{L})$ by induction hypothesis and Corollary 6 again.

CASE 2. $W \backslash W_2 \neq \emptyset$: $L_1|W \neq 0$ and $\{L_1^\alpha\}$ converges to L_1 in $\langle L(V_1; V_2)\rangle$. So eventually $L_1^\alpha|W \neq 0$. Then by (6) and (1) $F_W(\mathbf{L}^\alpha) = \Phi_W(L_1^\alpha)$ which converges to $\Phi_W(L_1) = F_W(\mathbf{L})$ by continuity of Φ.

This completes the inductive proof of compactness and the convergence criterion.

Now for any subspace W of V_1 define the pseudo-metric d_W on $\mathcal{L}(V_1; V_2)$ by

$$(29) \qquad d_W(\mathbf{L}, \tilde{\mathbf{L}}) = |F_W(\mathbf{L}) - F_W(\tilde{\mathbf{L}})|,$$

i.e. pull back the norm distance on the sphere $\langle L(W; V_2)\rangle$. For $\mathbf{L} \in \mathcal{L}(V_1; V_2)$ with filtration $\{W_1, \dots, W_k\}$ the open sets

$$(30) \qquad U_n^{\mathbf{L}} = \{\tilde{\mathbf{L}} : d_{W_i}(\mathbf{L}, \tilde{\mathbf{L}}) < 1/n \text{ for } i = 1, \dots, k\}$$

define a countable base for the neighborhood system of $\mathbf{L}$ by the convergence criterion.

Now if $\mathbf{L} \in \mathcal{L}(V_1; V_2)$ and $x \in \langle V_1\rangle$ then let W be the one dimensional subspace $\mathbf{R}x$. Since $|x| = 1$, $S_x : \mathbf{R} \to W$ is an isometry and so $L(W; V_2) \simeq L(\mathbf{R}; V_2) \simeq V_2$ identifies $F_W(\mathbf{L})$ in $\langle L(W; V_2)\rangle$ with $F^{\mathbf{L}}(x)$ in $\langle V_2\rangle$. As $\mathbf{L} \mapsto F_W(\mathbf{L})$ is continuous, it follows that $\mathbf{L} \mapsto F^{\mathbf{L}}(x)$ is continuous and so $F^\#$ is continuous to the compact product space $\langle V_2\rangle^{\langle V_1\rangle}$, $F^\#$ is injective by Proposition 7 and $\mathcal{L}(V_1; V_2)$ is compact. So $F^\#$ is a homeomorphism onto its image.

Now assume dim $V_1 \leq \dim V_2$ so that $Li(V_1; V_2)$ is nonempty. Assume $L \in \langle Li(V_1; V_2)\rangle$. If $\{L^\alpha\}$ is a net in $\langle Li(V_1; V_2)\rangle$ converging in norm to L then regarding these as pennants with filtration $\{V_1\}$ the convergence criterion implies $\{L^\alpha\}$ converges to L in the topology of $\mathcal{L}(V_1; V_2)$. So the inclusion is continuous. If $\{L^\alpha\}$ is a net in $\mathcal{L}(V_1; V_2)$ converging there to L then $\{L_1^\alpha = F_{W_1}(\mathbf{L}^\alpha)\}$ converges to L in $\langle L(V_1; V_2)\rangle$ but $\langle Li(V_1; V_2)\rangle$ is open in $\langle L(V_1; V_2)\rangle$ and so eventually $L_1^\alpha \in Li$. Hence Ker $L_1^\alpha = 0$ eventually and so the filtrations of $\{\mathbf{L}^\alpha\}$ are eventually of length 1. Thus, the image of $\langle Li(V_1; V_2)\rangle$ is an open subset of $\mathcal{L}(V_1; V_2)$ and the inclusion is a homeomorphism.

Let $\mathbf{L} = \{L_1, \dots, L_k\}$ be an arbitrary element of $\mathcal{L}(V_1; V_2)$ with filtration $\{W_1, \dots, W_k\}$. Apply Proposition 3b extending each L_{j+1} to define T_j for $j = 0, \dots, k-1$ so that $\sum_j \epsilon^j T_j$ is an IIS with associated pennant $\mathbf{L}$. Recall from (3) that L_k is injective. Since $\dim V_1 \leq \dim V_2$ we can choose the extension $T_{k-1} \in Li(V_1; V_2)$. Then choose $P \in L(V_2; V_1)$ a left inverse for T_{k-1}, i.e. $P \circ T_{k-1} = 1_{V_1}$. As ϵ tends to infinity

$$(31) \qquad \epsilon^{1-k}(\textstyle\sum_j \epsilon^j P \circ T_j) \to 1_{V_1}.$$

So the polynomial in ϵ, the determinant of $\sum_j \epsilon^j P \circ T_j$ is not identically zero. There exists $\epsilon_0 > 0$ so that this polynomial has no roots in the open interval $(0, \epsilon_0)$. Hence, for ϵ in this interval $\Sigma_j \epsilon^j T_j \in Li(V_1; V_2)$. Let $L^\epsilon = \langle \sum_j \epsilon^j T_j\rangle \in \langle Li(V_1; V_2)\rangle$.

For any subspace W of V_1, let S be the inclusion map.

$$(32) \qquad F_W(L^\epsilon) = \Phi_W(L^\epsilon) = \langle \textstyle\sum_j \epsilon^j T_j \circ S\rangle$$

and as ϵ tends to 0^+ this converges to $F_W(\mathbf{L})$ by (17) and (20). Thus, in $\mathcal{L}(V_1; V_2)$

$$\mathit{Lim}_{\epsilon\to 0^+}\{L^\epsilon\} = \mathbf{L}. \tag{33}$$

Hence $\langle Li(V_1;V_2)\rangle$ is dense. □

COROLLARY 10. *For* $\mathbf{L}\in\mathcal{L}(V_1;V_2)$ *and* $\tilde{\mathbf{L}}\in\mathcal{L}(W;V_1)$ *the composition* $\mathbf{L}{*}\tilde{\mathbf{L}}$ *is defined as the unique element of* $\mathcal{L}(W;V_2)$ *such that for any* W_1 *and* $S\in L(W_1;W)\backslash 0$

$$F^{\mathbf{L}*\tilde{\mathbf{L}}}(S) = F^{\mathbf{L}}(F^{\tilde{\mathbf{L}}}(S)). \tag{34}$$

In particular, $F^{\mathbf{L}*\tilde{\mathbf{L}}} : \langle W\rangle \to \langle V_2\rangle$ *is the composition of sphere maps* $F^{\mathbf{L}}\circ F^{\tilde{\mathbf{L}}}$.

If $S\in\langle Li(W;V_1)\rangle$ *then*

$$\mathbf{L}{*}S \equiv \tilde{F}(\mathbf{L},S). \tag{35}$$

Composition is continuous in the left variable. That is, for $\tilde{\mathbf{L}}\in\mathcal{L}(W;V_1)$ $\mathbf{L}\mapsto\mathbf{L}{*}\tilde{\mathbf{L}}$ *defines a continuous map from* $\mathcal{L}(V_1;V_2)$ *to* $\mathcal{L}(W;V_2)$.

PROOF. We first construct some $\hat{\mathbf{L}}\in\mathcal{L}(W;V_2)$ so that $F^{\hat{\mathbf{L}}}(S) = F^{\mathbf{L}}(F^{\tilde{\mathbf{L}}}(S))$ for all W_1 and $S\in L(W_1;W)\backslash 0$.

Let $\sum_j \epsilon^j T_j$ be an IIS in $L(V_1;V_2)$ with associated pennant $\mathbf{L}$. Assume that $T_j = 0$ for $j\ge k$. Let $\sum_t \epsilon^t\tilde{T}_t$ be an IIS in $L(W;V_1)$ with associated pennant $\tilde{\mathbf{L}}$. Now for $p = j+tk$ $(j = 0,1,\dots,k-1)$ define

$$\hat{T}_p = T_j\circ\tilde{T}_t. \tag{36}$$

Clearly,

$$\textstyle\sum_p \epsilon^p\hat{T}_p = \sum_{j=1,t}^{k} \epsilon^{j+tk} T_j\circ\tilde{T}_t \tag{37}$$

is an IIS in $L(W;V_2)$. We denote its associated pennant by $\hat{\mathbf{L}}$.

For $S\in L(W_l;W)\backslash 0$, the index p_S, the smallest index such that $\hat{T}_p\circ S\neq 0$ is $j+tk$ where $t = t_S$, the smallest index such that $\tilde{T}_t\circ S\neq 0$ and then j the smallest index such that $T_j\circ(\tilde{T}_{t_S}\circ S)\neq 0$. So by (18) and (20)

$$F^{\hat{\mathbf{L}}}(S) = \langle\hat{T}_{p_S}\circ S\rangle = \langle T_j\circ(\tilde{T}_{t_S}\circ S)\rangle = F^{\mathbf{L}}(F^{\tilde{\mathbf{L}}}(S)). \tag{38}$$

In particular, $F^{\hat{\mathbf{L}}} : \langle W\rangle \to \langle V_2\rangle$ is the composition $F^{\mathbf{L}}\circ F^{\tilde{\mathbf{L}}}$ (as usual, use $S = S_x$ for $x\in\langle W\rangle$). But $F^{\#}$ from $\mathcal{L}(W;V_2)\to\langle V_2\rangle^{\langle W\rangle}$ is injective. So $\hat{\mathbf{L}}\equiv\mathbf{L}{*}\tilde{\mathbf{L}}$ is uniquely defined by this condition.

If $\tilde{\mathbf{L}}$ is the injective map S then use the IIS defined by $\tilde{T}_0 = S$ and $\tilde{T}_t = 0$ for $t>0$. Then $\hat{\mathbf{L}} = \tilde{F}(\mathbf{L},S)$ by Proposition 3a.

Composition is continuous in the left variable because this is true for the space of sphere maps and the $F^{\#}$'s are homeomorphisms. □

COROLLARY 11. *The embedding* $F^{\#} : \mathcal{L}(V;V) \to \langle V\rangle^{\langle V\rangle}$ *identifies* $\mathcal{L}(V;V)$ *with the enveloping semigroup of the topological group* $\langle Li(V;V)\rangle$ *acting on the sphere* $\langle V\rangle$. *For* $\mathbf{L} \in \mathcal{L}(V;V)$ *the following are equivalent.*

1. $\mathbf{L} \in \langle Li(V;V)\rangle$.
2. $F^{\mathbf{L}}$ *is injective on* $\langle V\rangle$.
3. $F^{\mathbf{L}}$ *is continuous on* $\langle V\rangle$.
4. $F^{\mathbf{L}}$ *is a homeomorphism of* $\langle V\rangle$.

PROOF. The enveloping semigroup results follow from Theorem 9 and Corollary 10. Clearly, (1) $\Rightarrow$ (4) and (4) $\Rightarrow$ (2),(3). On the other hand, by Lemma 2 and the Remarks thereafter either (2) or (3) implies the filtration for $\mathbf{L}$ has length one and so $\mathbf{L} \in \langle Li(V;V)\rangle$. □

If $\mathbf{L} = \{L_1, \dots, L_k\} \in \mathcal{L}(V_1;V_2)$ then $F^{\mathbf{L}} : \langle V_1\rangle \to \langle V_2\rangle$ satisfies

$$F^{\mathbf{L}}(-x) = -F^{\mathbf{L}}(x) \tag{39}$$

and so $F^{\mathbf{L}}$ induces a map between projective spaces. Clearly all 2^k pennants $\{\pm L_1, \dots, \pm L_k\}$ induce the same map of projective spaces. But it can happen that a much larger set will induce the same projective map. This is why the enveloping semigroup of the projective action appears not to be everywhere first countable.

For example, the projective space of $\mathbf{R}$ is a single point so all of $\mathcal{L}(\mathbf{R}^2;\mathbf{R})$ induces the same trivial projective map. But $\mathcal{L}(\mathbf{R}^2;\mathbf{R})$ itself, homeomorphic by $F^{\#}$ to a space of maps from the circle $\langle \mathbf{R}^2\rangle$ to the pair of points $\langle \mathbf{R}\rangle$, is an interesting space. It is, in fact a version of Ellis' "two-circle" example (see [E1] page 44). We leave this amusing exercise and the computation of its generalizations $\mathcal{L}(\mathbf{R}^n;\mathbf{R})$ to the reader. By including $\mathbf{R}$ as a subspace of $\mathbf{R}^n$ we obtain an uncountable subset of $\mathcal{L}(\mathbf{R}^n;\mathbf{R}^n)$, with uncountably many distinct filtrations, all members of which induce the same constant map at the projective level.

References

[E1] R. Ellis, *Lectures on topological dynamics*, W.A. Benjamin, New York, 1969.

[E2] R. Ellis, *The enveloping semigroup of projective flows*, Ergod. Th. & Dynam, Sys. **13** (1993), 635-660.

[DE] D. Ellis, *Enveloping semigroup of projective space*, unpublished notes.

[F] H. Furstenberg, *Rigidity and cocycles for ergodic actions of semisimple groups (after G.A. Margulis and R. Zimmer)*, Séminaire Bourbaki **559** (1979-1980).

Mathematics Department, The City College, 137 Street & Convent Avenue, New York, NY 10031

E-mail address: etkcc@cunyvm.cuny.edu

Contemporary Mathematics
Volume **215**, 1998

An Overview of the Construction of Suspension Flows Using Continuous Cocycles

H. Keynes, K. Madden, N. Markley, and M. Sears

1. Introduction

This paper provides an overview of our results about continuous $\mathbf{Z}^n$ cocycles, their continuous suspension flows and their role as models for more general $\mathbf{R}^n$ flows on compact metric spaces. Cocycles occur in a variety of dynamical settings. Before explaining the main themes of this survey paper, we describe the abstract context in which cocycles usually appear.

Suppose a group T acts on a space X. If we try to extend the action of T on the first co-ordinate of $X \times G$, where G is another space, to all of $X \times G$, we need a device which, given x, tells us how to act on (x, g) as g ranges over G. When G is a group the simplest approach is to multiply g by another group element which depends on x and t. In other words the natural device is a continuous function $h : X \times T \to G$, but this procedure will only define an action of T on $X \times G$ when h satisfies a condition called the it cocycle equation. This equation has arisen in various contexts in addition to its roots in homological algebra. If G is compact this new flow is the well studied skew extension flow. In the general case, and under certain conditions, by factoring by an appropriate equivalence relation we can obtain a flow on a compact space, called a *suspension flow*. The well known *flow under a function* is a special case of this.

Much is known about cocycles and suspension flows in the measure theoretic case (every multiparameter action is a suspension up to measure isomorphism), and there is a body of results in the case of differentiable actions. Robert Ellis [E2] produced a comprehensive study in the setting of abstract continuous actions using the Ellis algebras and groups. Less is known about the case of actions under connected groups other than the reals.

We have been interested in the situation where the acting groups are $\mathbf{Z}^n$ and $\mathbf{R}^n$. The subject is beginning to be developed and some understanding of these cocycles and suspensions has resulted. The aim of this paper is mainly expository: to put the area in context and to illuminate the ideas without the details of technical proofs. In some cases the basic ideas behind a proof will be given where we feel it helps understanding, but formal details will simply be referenced. Our attempt is to provide a more cohesive picture of our overall approach, progress to date, and directions for future investigation.

1991 *Mathematics Subject Classifications*: Primary 54H20.

Using the model of a flow built under a function, we first develop the basic theory of $\mathbf{R}^n$ suspensions. The study of suspensions under arbitrary acting groups has its roots in Robert Ellis' paper [E1] and our debt to his work is clearly apparent. Our first investigation concerns topological properties of the resulting suspension flows. It is noted that the space being Hausdorff is essentially equivalent to a uniform growth property of the cocycle. These so called *covering* cocycles become the central focus of the remainder of the paper. The relation between the compactness of the space and linear growth of the cocycle is illustrated, as well as the existence of nice suspension models. The notions of fundamental domains and associated tilings of $\mathbf{R}^n$ are also briefly surveyed.

We next turn to the space of cocycles and look at questions about the pervasiveness of covering cocycles. It turns out that using certain ergodic averages provides a tool to show that the density of covering cocycles is equivalent to a notion of unique ergodicity in this space, and to a nice characterization of the space of cocycles as the direct product of the linear cocycles and the closure of the coboundaries.

The easiest suspensions to analyze are the constant suspensions obtained from the identity linear map and their time changes. A central issue is the unfolding problem: when is a given covering cocycle essentially equivalent to such a cocycle? Partial answers are developed; one consequence is that in the case of a uniquely ergodic base flow this is always true.

The final section addresses the issue of how far from a linear cocycle a covering cocycle can be in the minimal flow situation. An important example called *the almost periodic spiral* and its exotic properties are sketched, including some highly nonlinear behavior. Some natural questions for further investigation are highlighted at the end.

2. Origins

The usual setting for continuous abstract dynamical systems, or simply *flows*, is a compact metric space X on which a non-compact abelian group T acts via a continuous function $\phi : X \times T \to X$ satisfying the following conditions:

1. $\phi(x, s+t) = \phi(\phi(x,s),t)$,
2. $\phi(x,0) = x$.

for all x in X and s, t in T. For convenience we write $\phi(x,t) = xt$ and denote the flow by (X,T). In the same vein given $E \subseteq X$ and $F \subseteq T$, let

$$EF = \{xt : x \in E \text{ and } t \in F\}.$$

To avoid certain pathologies, flows will always be assumed to have a free orbit ($xt \neq x$ when $t \neq 0$) which is dense (the closure of $\{xt : t \in T\}$ is X). The integers and reals will be denoted by $\mathbf{Z}$ and $\mathbf{R}$ respectively. Although we are primarily interested in the multidimensional cases when $T = \mathbf{R}^m$ or $T = \mathbf{Z}^m$, we will use the classical cases $T = \mathbf{R}$ and $T = \mathbf{Z}$ to motivate the theory. A single homeomorphism ϕ of X onto itself determines a $\mathbf{Z}$ action on X simply by $\phi(x,n) = \phi^n(x)$, and the study of $\mathbf{R}$ actions is akin to the qualitative study of solutions of differential equations.

Poincaré and others exploited the idea of a local section to study the recurrent behavior of solutions of ordinary differential equations in $\mathbf{R}^m$ and on differentiable manifolds. The concept of a local section from ordinary differential equations can be extended to actions of $\mathbf{R}$ on metric spaces. Given a flow (X, T), a closed subset S of X is a *local section* provided there exists an $\epsilon > 0$ satisfying:

1. the flow maps $S \times [-\epsilon, \epsilon]$ homeomorphically into X,
2. the interior of $\phi(S \times [-\epsilon, \epsilon])$ is non-empty.

If S is a local section, then it is easy to verify that there exists a subset S^o of S such that the interior of $\phi(S \times [-\epsilon, \epsilon])$ is $\phi(S^o \times (-\epsilon, \epsilon))$. If x is not a fixed point ($xt \neq x$ for some t), then there exists a *local section* S *at* x which means that $x \in S^o$. A proof of this fact can be found in Nemytskii and Stepanov [NS] on page 333. This result was extended to $\mathbf{R}^m$ actions when the isotropic subgroup $\{t : xt = x\}$ is a discrete subgroup of $\mathbf{R}^m$ by Zippin and a detailed proof was given by Thomas [T] in his thesis.

Let S be a local section at x_0 and suppose $x_0 t \in S^o$ for some $t > 0$. By continuity the set

$$U = \{x \in S^o : xt \in S^o \text{ for some } t > 0\}$$

is open in S^o, and for $x \in U$ let

$$\tau(x) = \min\{t > 0 : xt \in S^o\}.$$

Then $\psi : U \to S^o$ defined by $\psi(x) = x\tau(x)$ is just the Poincaré first return map and both $\tau(x)$ and $\psi(x)$ are continuous on U. Although a local section and first return map can provide useful information about the dynamics at x_0, they generally do not completely describe the dynamics of the flow $(X, \mathbf{R})$.

A local section S is a *global section* provided:

1. $S^0 = S$
2. for all $x \in X$ there exists t such that $xt \in S$.

When S is a global section the first return map ψ is defined on S and is a homeomorphism of S onto S. Now the dynamics of $(X, \mathbf{R})$ are largely dictated by the dynamics of the flow $(S, \mathbf{Z})$ defined by the homeomorphism ψ. For example, $(S, \mathbf{Z})$ is minimal if and only if $(X, \mathbf{R})$ is minimal. (A flow is *minimal* if each of its orbits is dense in the space. A point whose orbit closure has this property is called *almost periodic.*) However not all dynamical properties are preserved. (For example it is possible that $(S, \mathbf{Z})$ is equicontinuous and $(X, \mathbf{R})$ is weak mixing. [NS, p.299]) Conversely, given a flow $(S, \mathbf{Z})$ and a continuous positive real valued function $\tau(\cdot)$ on S, a flow $(X, \mathbf{R})$ can be constructed containing S as global section with τ as the first return time and ψ as the first return map. Form the space $S_\tau = \{(x, u) : 0 \leq u \leq \tau(x), x \in S\}/\sim$ where $\sim$ is the equivalence relation $(x, \tau(x)) \sim (\psi(x), 0)$ and define real orbits over each base point $(x, 0)$ by $(x, 0)t = (x, t)$ if $0 \leq t < \tau(x)$, by $(x, 0)t = (\psi^n(x), t - \sum_{k=0}^{n-1} \tau(\psi^k x))$ if $\tau(x) \leq t$, and similarly for $t < 0$. Then S_τ is a compact metric space and the above definition gives a genuine real action. This construction is the classical *flow under a function* . Notice that the "axis" $\{(x, 0) : x \in S\}$ is embedded in the new flow as a global section.

Following the ideas of Ellis we will subsume the construction of a flow under a function in the more general construction of a suspension flow $(X, \mathbf{R}^m)$ constructed from a $\mathbf{Z}^m$ action on a section. Ellis [E1] generalized the idea of a global section S to a topological group T acting on a topological space X by using a second group G acting on $S \times T$. The G action on $S \times T$ will be written on the left. There is also a T action on $S \times T$, given by $(y, s)t = (y, s + t)$, and this action will be written on the right. Specifically, Ellis required that the G and T actions on $S \times T$ commute and that the map $(y,t) \to yt$ of $S \times T$ onto X induce a homeomorphism of the G-orbit space $S \times T/G$ onto X. He then used this concept to obtain necessary conditions for the fundamental group of X to be non-trivial. We will use the same idea but from an external constructive point of view instead of from Ellis' internal structural point of view. The equivalence of these two points of view is essentially the content of Theorem 1 in Ellis' paper. In what follows, $\mathbf{Z}^m$ will play the role of G and $\mathbf{R}^m$ the role of T.

Given a flow $(X, \mathbf{Z}^m)$, we will now write the action on the left as ax when a is in $\mathbf{Z}^m$ and x is in X, but we will continue to write $\mathbf{R}^m$ actions on the right. A *cocycle* for $(X, \mathbf{Z}^m)$ is a continuous function $h : X \times \mathbf{Z}^m \to \mathbf{R}^n$, satisfying

$$h(x, a+b) = h(x,a) + h(ax, b)$$

for all x in X and a, b in $\mathbf{Z}^m$. Note that $h(x, 0) = x$ and that the set of $\mathbf{R}^n$ valued cocycles for $(X, \mathbf{Z}^m)$ is a vector space over $\mathbf{R}$.

There are two elementary kinds of cocycles that are easy to construct. First, if $T : \mathbf{R}^m \to \mathbf{R}^n$ is a linear map, then setting $h(x,a) = T(a)$ defines a cocycle called a *constant cocycle*. Second, a *coboundary* is obtained by setting $h(x,a) = f(ax) - f(x)$ for any continuous function $f : X \to \mathbf{R}^n$. There is also a basic equivalence relation of cocycles for $(X, \mathbf{Z}^m)$; two cocycles h and g are *cohomologous*, written $h \sim g$, if $h - g$ is a coboundary. The existence of cocycles which are not cohomologous to a constant cocycle has been observed by a number of people and is written down in [KMS3].

Given an $\mathbf{R}^n$ valued cocycle h for the flow $(X, \mathbf{Z}^m)$, for each a in $\mathbf{Z}^m$ define $T_a : X \times \mathbf{R}^n \to X \times \mathbf{R}^n$ by

$$T_a(x, v) = \big(ax, v - h(x,a)\big).$$

It is easy to see that $T_a \circ T_b = T_{a+b}$ and T_a is a homeomorphism with $(T_a)^{-1} = T_{-a}$. In particular, $\{T_a : a \in \mathbf{Z}^m\}$ is a group of homeomorphisms and thus $\mathbf{Z}^m$ acts on $X \times \mathbf{R}^n$. Consequently we can form the orbit space of this action by identifying (x, v) and $T_a(x,v)$ for a in $\mathbf{Z}^m$ and denote it by X_h. In other words, X_h is the quotient space

$$X \times \mathbf{R}^n / \{T_a : a \in \mathbf{Z}^m\}.$$

There is also a natural $\mathbf{R}^n$ action on $X \times \mathbf{R}^n$ given by $(x,v)w = (x, v+w)$. Since

$$T_a\big((x,v)w\big) = \big(T_a(x,v)\big)w,$$

on X_h, there is an induced $\mathbf{R}^n$ action defining a flow $(X_h, \mathbf{R}^n)$, called the *suspension flow*, which is the principle object of interest in this paper.

The space X_h can be rather pathetic. We will illustrate how bad it can be with a classical example. Let $\mathbf{K}$ be the unit circle and let $\rho = e^{2\pi\alpha i}$ where α is irrational. Thus $z \to \rho z$ is an irrational rotation of $\mathbf{K}$ and defines a $\mathbf{Z}$ action on $\mathbf{K}$. A. S. Besicovitch [B1], [B2] studied homeomorphisms $T : \mathbf{K} \times \mathbf{R} \to \mathbf{K} \times \mathbf{R}$ of the form $T(z,s) = \big(\rho z, s + f(z)\big)$ where $f : \mathbf{K} \to \mathbf{R}$ is continuous. Clearly

$$T^n(z,s) = \left(\rho^n z, s + \sum_{k=0}^{n-1} f(\rho^k z)\right).$$

Setting

$$h(z,n) = \begin{cases} -\sum_{k=0}^{n-1} f(\rho^k z) & n > 0 \\ 0 & n = 0 \\ \sum_{k=1}^{-n} f(\rho^{-k} z) & n < 0 \end{cases}$$

defines a cocycle for the irrational rotation $(\mathbf{K}, \mathbf{Z})$ determined by ρ. In fact, the above formula, with $\rho^k z$ replaced by $\phi^k(x)$, will always produce a cocycle from a continuous function $f : X \to \mathbf{R}$ and a homeomorphism $\phi : X \to X$. Furthermore, the corresponding T_n is just T^n.

Noting that the equivalence relation generating X_h is essentially the same as the one on the graph of τ giving S_τ, we have, in fact, exactly recovered the flow under a function situation as a special case of the general construction when $m = n = 1$ (and f is positive).

Besicovitch [B2] constructed a function f such that for some (z_0, s_0) the orbit $\{T^n(z_o, s_0) : n \in \mathbf{Z}\}$ was dense in $\mathbf{K} \times \mathbf{R}$. Since this orbit determines a point in X_h, the space X_h has a dense point and is not even close to being Hausdorff. Consequently determining when X_h is at least Hausdorff is a very basic question.

3. Uniform Growth

At the very least we want to restrict our attention to cocycles h for $(X, \mathbf{Z}^m)$ such that the space X_h is Hausdorff. The canonical projection $\pi : X \times \mathbf{R}^n \to X_h$ is also closely tied to the topological properties of X_h because π is obviously an open map. In particular, it is relatively easy to show that X_h is Hausdorff when $h(x,a)$ goes to infinity uniformly in x as a goes to infinity in $\mathbf{Z}^m$. The uniform growth of h also implies that π is a local homeomorphism. The following theorem from [FKMS] shows that uniform growth can be characterized topologically, identifying a natural class of cocycles for study.

THEOREM 3.1. *Let $h : X \times \mathbf{Z}^m \to \mathbf{R}^n$ be a cocycle. The space X_h is a Hausdorff space and π is a local homeomorphism if and only if $h(x,a)$ goes to infinity uniformly in x as a goes to infinity in $\mathbf{Z}^m$. (That is, given any $R > 0$ there exists an $r > 0$ such that $|h(x,a)| > R$ for all $x \in X$ and $a \in \mathbf{Z}^m$ with $|a| > r$.)*

One way of interpreting this theorem is that uniform growth of h means $\pi : X \times \mathbf{R}^n \to X_h$ is similar to a covering map, and cocycles satisfying this growth condition are called *covering cocycles.* When h is a covering cocycle, it readily follows that X_h is a locally compact metric space.

When $m = n = 1$, a cocycle h is completely determined by the continuous real valued function $h(x,1)$, and when $h(x,1)$ is positive for all t, the suspension flow $(X_h, \mathbf{R})$ is the flow under the function $h(x,1)$. However, h can be covering without $h(x,1)$ being positive for all x. In this context there is a simple test in [FKMS] for covering:

THEOREM 3.2. *A cocycle $h : X \times \mathbf{Z} \to \mathbf{R}$ is covering if and only if $\int_X h(x,1)d\mu \neq 0$ for every invariant probability measure μ for $(X, \mathbf{Z})$.*

In particular, if h is a covering cocycle there can exist x such that $h(x,1) = 0$.

Returning to the general cocycle $h : X \times \mathbf{Z}^m \to \mathbf{R}^n$, suppose $h(x,a) = 0$ and $ax \neq x$. It follows that

$$\pi(ax, 0) = \pi\big(ax, 0 - h(x,a)\big) = \pi(x, 0).$$

Hence the canonical projection π is not one-to-one on $X \times \{0\}$ and does not embed $X \times \{0\}$ in X_h. In particular, X does not naturally sit in X_h as a global section. Conversely a covering cocycle is called an *embedding cocycle* if $h(x,a) \neq 0$ for all $a \neq 0$. If h is an embedding cocycle, there exists $\epsilon > 0$ such that π is one-to-one, and hence a homeomorphism, on $X \times \{v : \|v\| < \epsilon\}$. In this case X is a global section of $(X_h, \mathbf{R}^n)$ in Ellis' sense.

An interesting open question is whether or not every covering cocycle is cohomologous to an embedding cocycle. To understand the significance of this question we need the following elementary fact: If h and g are cohomologous, then their suspension flows are isomorphic. The map $\Psi : X \times \mathbf{R}^n \to X \times \mathbf{R}^n$ defined by $\Psi(x,v) = \big(x, v - f(x)\big)$ induces an isomorphism of flows from $(X_h, \mathbf{R}^n)$ to $(X_g, \mathbf{R}^n)$ where $h(x,a) - g(x,a) = f(ax) - f(x)$. Consequently, if h is a covering cocycle, g is an embedding cocycle, and g is cohomologous to h, then X appears as a global section of $(X_h, \mathbf{R}^n)$ but not in the form of $\pi_h(X \times \{0\})$. Thus a positive answer to the question would prove that X could always be imbedded in X_h as a global section of the suspension flow $(X_h, \mathbf{R}^n)$ when h is covering.

The growth rate of a covering cocycle will play a significant role in the next section. We will end this section with a general discussion of this issue. It is convenient to use the norm $|v| = \sum_{i=1}^{n} |v_i|$ on $\mathbf{R}^n$. For a cocycle h set

$$\|h\| = \sup\{|h(x, e_i)| : x \in X \text{ and } 1 \leq i \leq n\}$$

where $e_1, \dots, e_m$ are the standard generators of $\mathbf{Z}^m$. It follows immediately that

$$|h(x,a)| \leq \|h\| |a|$$

and linear growth is the most we can expect of a covering cocycle.

Do all covering cocycles eventually grow linearly; that is, do there exist positive constants A, B, and C such that

$$A|a| \leq |h(x,a)| \leq B|a|$$

when $C \leq |a|$? In general the answer is no, but for one important class of covering cocycles, namely those for which $m = n$, the answer is yes. (See Theorem 4.3.)

In [M2] an example of a covering cocycle which did not grow linearly was constructed using the geometric substitution ideas of Dekking [D]. A few comments about the properties of this example will set the stage for the next section. In this example $m = 1$ and $n = 2$ and $(X, \mathbf{Z})$ is a uniquely ergodic substitution minimal set. We can write $h(x,k) = \big(h_1(x,k), h_2(x,k)\big)$ where the h_i are real valued cocycles. It turns out that if either h_1 or h_2 was covering, then h would grow linearly. So the trick is to find the real valued cocycles h_1 and h_2 such that neither one goes to infinity uniformly in x as k goes to infinity, but together, as a point in $\mathbf{R}^2$, the uniform growth required for h to be a covering cocycle is achieved. Thus the integrals of both h_i must be zero. If one connects the points $h(x,k)$ and $h(x,k+1)$ with a line segment, then the resulting curve looks like a von Koch curve that has been stretched to infinity.

4. Compactness and Tilings

In the case of a linear map (i.e. a constant cocycle) it is not hard to see that the map is covering if and only if the kernel is trivial. In an intuitive way cocycles behave like linear maps. For example, a cocycle $h : X \times \mathbf{Z}^2 \to \mathbf{R}$ need not have a one dimensional kernel as a linear map $T : \mathbf{R}^2 \to \mathbf{R}$ would have, but given $R > 0$, there must exist a in $\mathbf{Z}^2$ such that $|a| \geq R$ and h maps (x, a) within a fixed bounded distance from the origin. Hence we would not expect such a cocycle to be covering. This situation is covered in the following general result from [FKMS]:

THEOREM 4.1. *If $h : X \times \mathbf{Z}^m \to \mathbf{R}^n$ is a covering cocycle, then $n \geq m$.*

To prove this result one first notes that by replacing $\mathbf{Z}^m$ with a subgroup of finite index, we can assume without loss of generality that h is an embedding cocycle. Hence there exists $\epsilon > 0$ such that, given $x \in X$, the balls, $B_\epsilon(h(x, a))$, of radius ϵ around $h(x, a)$ in $\mathbf{R}^n$ are all disjoint. Asymptotically the number of points from $\mathbf{Z}^m$ in $B_R(0) \subset \mathbf{R}^m$ approaches a number proportional to the volume of an m dimensional sphere of radius R. Thus, the number of disjoint balls of the form $B_\epsilon(h(x, a))$ in $B_{R\|h\|+\epsilon}(0)$ also approaches a number proportional to the volume of an m dimensional sphere of radius R. As R goes to infinity there is just not enough room in $B_{R\|h\|+\epsilon}(0)$ when $n < m$.

The case when $m = n$ is special:

THEOREM 4.2. *Let $h : X \times \mathbf{Z}^m \to \mathbf{R}^n$ be a covering cocycle. The space X_h is compact if and only if $m = n$.*

A key step in the proof of the above theorem is that when $m = n$, covering cocycles must grow linearly:

THEOREM 4.3. *A cocycle $h : X \times \mathbf{Z}^m \to \mathbf{R}^m$ is a covering cocycle if and only if there exist constants A, B, and C such that*

$$A|a| \leq |h(x, a)| \leq B|a|$$

for all x in X when $|a| \geq C$.

As a consequence of Theorem 3.2, the suspension flow $(X_h, \mathbf{R}^m)$ given by a covering cocycle $h : X \times \mathbf{Z}^m \to \mathbf{R}^m$ is the natural higher dimensional analog of a flow under a function. Conversely Rudolph [R1] has shown the following for both the measure theoretic and continuous settings:

THEOREM 4.4. *If the flow $(Y, \mathbf{R}^m)$ has a free dense orbit, then there exists a suspension flow $(X_h, \mathbf{R}^m)$ given by an embedding cocycle h and a homomorphism $\theta : (X_h, \mathbf{R}^m) \to (Y, \mathbf{R}^m)$ which is one-to-one on a residual set of Y. In this case $(X_h, \mathbf{R}^m)$ is called an almost one-to-one extension of $(Y, \mathbf{R}^m)$.*

This result is the best possible because it is known that the horocycle flow of a compact surface of constant negative curvature does not contain a global section.

The values of the cocycles in Rudolph's theorem are specific vertices of tiles in a tiling of $\mathbf{R}^m$ by a finite set of rectangular tiles with faces perpendicular to the coordinate axes. It is quite evident from Rudolph's proof that these cocycles are always embedding cocycles. Consequently it is natural to ask if there is a method for constructing tilings of $\mathbf{R}^m$ from any embedding cocycle. The answer is yes and the construction is linked to the analog of a fundamental domain for a covering cocycle.

Every covering cocycle with $m = n$ has a fundamental domain in the following sense:

THEOREM 4.5. *If* $h : X \times \mathbf{Z}^m \to \mathbf{R}^m$ *is a covering cocycle, then there exists a compact set* F *contained in* $X \times \mathbf{R}^m$ *such that the following conditions are satisfied:*

1. $\pi(F) = X_h$,
2. π *is one-to-one on the interior of* F,
3. F *equals the closure of its interior.*

Notice that in the example of a flow under a function, a fundamental domain is simply given by the closed region between the axis and the graph of the suspending function. The proof of the theorem is a relatively simple Zorn's lemma argument. In general if F is a fundamental domain in the sense of the above theorem, the sets $F_x = \{v \in \mathbf{R}^m : (x, v) \in F\}$ can be very complicated. By forcing these sets to be "nice" and to fit together continuously to form F, we can characterize embedding cocycles.

THEOREM 4.6. *The cocycle* $h : X \times \mathbf{Z}^m \to \mathbf{R}^m$ *is embedding if and only if there exists a compact subset* F *of* $X \times \mathbf{R}^m$ *satisfying the following conditions:*

(a) $\pi(F) = X_h$,
(b) π *is one-to-one on the interior of* F,
(c) F *equals the closure of its interior,*
(d) F_x *is convex for all* x *in* X,
(e) $X \times \{0\}$ *is contained in* F,
(f) *the function* $g : X \times S^{m-1} \to \mathbf{R}$ *defined by*

$$g(x, u) = \sup\{t \geq 0 : (x, tu) \in F\}$$

is continuous where $S^{m-1} = \{u \in \mathbf{R}^m : \|u\| = 1\}$.

Again the graph of the suspending function satisfies all these properties except item (f) in the case $m = n = 1$. However the missing requirement can be obtained by shifting the global section up slightly from the axis of the graph. Turning to the proof of the theorem, necessity uses a standard geometric construction of a fundamental domain for covering spaces. Specifically, set

$$F = \{(x, v) : \|v\| \leq \|h(x, a) - v\| \text{ for all } a \in \mathbf{Z}^m\}$$

where $\| \; \|$ is the Euclidean norm. The sufficiency makes use of all the conditions to show that X_h is Hausdorff and π is one-to-one on $X \times B_\epsilon(0)$ for some $\epsilon > 0$.

Since $-h(ax, -a) = h(x, a)$, it follows that

$$T_{-a}(\{ax\} \times F_{ax}) = \{x\} \times (F_{ax} + h(x, a))$$

and for a fixed x, the sets $F_{ax} + h(x, a)$ tile $\mathbf{R}^m$. When F is constructed as above, this tiling is just the tiling by Voronoi polygons given by the set $\{h(x, a) : a \in \mathbf{Z}^m\}$ in $\mathbf{R}^m$. In general the properties of F in Theorem 4.6 guarantee that the tiles are polytopes and we have the following corollary.

COROLLARY 4.1. *Let* F *satisfy (a) through (f) as above. Then for all* x *in* X

$$\{F_{ax} + h(x, a) : a \in \mathbf{Z}^m\}$$

is a tiling of $\mathbf{R}^m$ *by polytopes.*

5. Cocycle Averages

Let $\mathcal{C}_n(X, \mathbf{Z}^m)$ denote the vector space of cocycles $h : X \times \mathbf{Z}^m \to \mathbf{R}^n$ for the flow $(X, \mathbf{Z}^m)$. The function

$$\|h\| = \sup\{|h(x, e_i)| : x \in X \text{ and } i = 1, \ldots, n\}$$

defines a norm on $\mathcal{C}_n(X, \mathbf{Z}^m)$ and it is easy to check that it is a Banach space. There is also a natural $\mathbf{Z}^m$ action on $\mathcal{C}_n(X, \mathbf{Z}^m)$ defined by

$$ah(x, b) = h(ax, b) = h(x, a + b) - h(x, a).$$

When $n = m$, the covering cocycles and the embedding cocycles form open subsets of $\mathcal{C}_m(X, \mathbf{Z}^m)$. A broad question is to better understand the structure of these open sets which are vaguely analogous to the invertible matrices as subsets of $\mathbf{R}^{m \times m}$. For example, we know that the set of covering cocycles has at least two components but can have more than two.

To motivate the ideas in this section we return to the $m = n = 1$ case again. An easy consequence of the integral characterization of covering given in Section 1 is the following: the covering cocycles are dense in $\mathcal{C}_1(X, \mathbf{Z}^1)$ if and only if $(X, \mathbf{Z})$ is uniquely ergodic. The space $C(X)$ is isometrically isomorphic to $\mathcal{C}_1(X, \mathbf{Z}^1)$ and it is well known that unique ergodicity is equivalent to the ergodic averages

$$\frac{1}{N} \sum_{k=0}^{N-1} f(kx)$$

converging uniformly in x to a constant as N goes to infinity for all $f \in C(X)$. In fact, for h in $\mathcal{C}_1(X, \mathbf{Z}^1)$

$$\frac{h(x, N)}{N} = \frac{1}{N} \sum_{k=0}^{N-1} h(kx, 1)$$

by the cocycle equation. Thus it is natural to investigate the cocycle averages of the form

$$\frac{h(x, a)}{|a|}.$$

These averages were studied in [KMS2] and also in the measurable case by Boivin and Derriennic in [BD].

The vector space $\mathcal{L}(m, n)$ of linear maps $T : \mathbf{R}^m \to \mathbf{R}^n$ can be embedded in $\mathcal{C}_n(X, \mathbf{Z}^m)$ as the constant cocycles. Since h in $\mathcal{C}_n(X, \mathbf{Z}^m)$ can be written as $h = (h_1, \ldots, h_n)$ with h_j in $\mathcal{C}_1(X, \mathbf{Z}^m)$, we consider real valued cocycles first.

An *invariant cocycle integral* is a linear function

$$\sigma : \mathcal{C}_1(X, \mathbf{Z}^m) \to \mathcal{L}(m, 1)$$

satisfying the following conditions:

1. $\|\sigma\| = 1$,
2. $\sigma(ah) = \sigma(h)$ for all $a \in \mathbf{Z}^m$,
3. $\sigma(T) = T$ for all $T \in \mathcal{L}(m, 1)$.

The set of all invariant cocycle integrals will be denoted by $\mathcal{I}$. Notice that when $m = 1$ the invariant cocycle integrals are given by the ordinary integrals with invariant measures regarded as linear maps on $\mathbf{R}^1$, since every continuous function generates a cocycle.

Letting invariant cocycle integrals play the role of invariant measure, the following result was obtained in [KMS2]. It forms the basis for the rest of the discussion of cocycle integrals in this section.

THEOREM 5.1. *Let h be a cocycle in $\mathcal{C}_1(X, \mathbf{Z}^m)$. The cocycle average $h(x,a)/|a|$ goes to 0 uniformly in x as $|a|$ goes to infinity if and only if $\sigma(h) = 0$ for all σ in $\mathcal{I}$.*

Because $\sigma(h-T) = 0$ if and only if $\sigma(h) = T$, the next theorem is an immediate consequence of this result.

THEOREM 5.2. *The flow $(X, \mathbf{Z}^m)$ has a unique cocycle integral if and only if for every h in $\mathcal{C}_1(X, \mathbf{Z}^m)$ there exists T_h in $\mathcal{L}(m,1)$ such that*

$$\frac{|h(x,a) - T_h(a)|}{|a|}$$

goes to 0 uniformly as $|a|$ goes to infinity.

All invariant cocycle integrals have the form

$$\sigma(h) = \sum_{j=1}^{m} \int_X h(x, e_j) d\mu_j \, e_j^*$$

where $\mu_1, \dots, \mu_m$ are invariant Borel measures for $(X, \mathbf{Z}^m)$ and $\{e_1^*, \dots, e_m^*\}$ is the dual basis of the standard basis $\{e_1, \dots, e_m\}$ in $\mathbf{R}^m$. Hence unique ergodicity implies there is a unique invariant cocycle integral, but the converse is a tantalizing open question. In fact $\mathcal{I}$ and its extreme points are not well understood.

The coboundaries defined in Section 1 also form a subspace of $\mathcal{C}_n(X, \mathbf{Z}^m)$ denoted by $\mathcal{B}_n$. Obviously $h = (h_1, \dots, h_n)$ is in $\mathcal{B}_n$ if and only if h_j is in $\mathcal{B}_1$ for $j = 1, \dots, n$. The invariant cocycle integrals determine $\overline{\mathcal{B}}_1$ in the sense that h lies in $\overline{\mathcal{B}}_1$ if and only if $\sigma(h) = 0$ for all σ in $\mathcal{I}$. It is also true in general that

$$\overline{\mathcal{B}}_n \cap \mathcal{L}(m,n) = 0.$$

The set of *close to linear cocycles*, $\mathcal{D}_n$, is defined by

$$\mathcal{D}_n = \overline{\mathcal{B}}_n \oplus \mathcal{L}(m,n).$$

The close to linear cocycles are studied in some detail in [KMS3]. One of their properties is that if $h = g + T$ with g in $\overline{\mathcal{B}}_n$ and T in $\mathcal{L}(m,n)$, then $\sigma(h_i) = T_i, i = 1, \dots, n$ for all σ in $\mathcal{I}$ and

$$\lim_{|a| \to \infty} \frac{|h(x,a) - T(a)|}{|a|} = 0.$$

Observe that this implies that T must be invertible if h is covering. It is also easy to prove that one can find a sequence of cocycles h_n, all cohomologous to h, with $h_n \to T$. This justifies the name "close to linear". It turns out that all cocycles are close to linear exactly when there is a unique cocycle integral. It was also shown in [KMS3] that the covering cocycles are dense in $\mathcal{D}_m$ ($n = m$ case), allowing us to end the section with a generalization of our opening remark.

THEOREM 5.3. *The set of covering cocycles is dense in $\mathcal{C}_m(X, \mathbf{Z}^m)$ if and only if the flow $(X, \mathbf{Z}^m)$ has a unique invariant cocycle integral.*

6. Furstenberg's Unfolding Problem

Given a cocycle $h : X \times \mathbf{Z}^m \to \mathbf{R}^m$, i.e. with $n = m$, the only kind we will consider in this section, the image $h(x, \mathbf{Z}^m)$ can be thought of as a $\mathbf{Z}^m$ net cast in $\mathbf{R}^m$. The nicest situation is when an invertible linear map casts the net because then $a \to h(x, a)$ extends to a linear homeomorphism of $\mathbf{R}^m$ onto itself. The worst case is a coboundary which leaves the entire net folded up near the origin in a hopeless tangle. Even when h is covering and $h(x, \mathbf{Z}^m)$ is more uniformly cast because there are constants A, B, and C such that

$$A|a-b| \leq |h(x,a) - h(x,b)| \leq B|a-b|$$

when $|a - b| \geq C$, there could be a lot of local folding. Can we unfold these local problems by adding a coboundary and then continuously extend the cohomologous cocycle to a homeomorphism of $\mathbf{R}^m$ onto itself for all x? We need some additional definitions to state the problem precisely and explain its relevance.

A cocycle h in $\mathcal{C}_m(X, \mathbf{Z}^m)$ is *invertible* if there exists a continuous function $H : X \times \mathbf{R}^m \to \mathbf{R}^m$ satisfying the following conditions for all x in X, a in $\mathbf{Z}^m$, and v in $\mathbf{R}^m$:

(a) $H(x, a) = h(x, a)$,
(b) $H(x, a + v) = h(x, a) + H(ax, v)$,
(c) the map $v \to H(x, v)$ is a homeomorphism of $\mathbf{R}^m$ onto itself.

Since $\mathbf{Z}^m$ is the set of vertices for a triangulation of $\mathbf{R}^m$ which is invariant under $\mathbf{Z}^m$ translation, the piecewise linear extension of $h(x, a)$ provides a map H satisfying conditions (a) and (b). The stumbling block is (c). When h is covering, the Borsuk-Ulam Theorem can be applied to show that any continuous H satisfying conditions (a) and (b) must be onto. Consequently the problem of showing a covering cocycle is invertible is reduced to the question of whether or not there exists a one-to-one extension H satisfying conditions (a) and (b) in the definition.

The unfolding problem can be stated as: Is every covering cocycle cohomologous to an invertible cocycle? To understand this problem from a different perspective, let I denote the identity cocycle given by $I(x, a) = a$ for any $(X, \mathbf{Z}^m)$. If h is invertible, then H induces a homeomorphism of X_I onto X_h which preserves orbits. In other words, the suspension flow $(X_h, \mathbf{R}^m)$ is a time change of $(X_I, \mathbf{R}^m)$. From this point of view the unfolding problem asks whether suspension flows of covering cocycles are all just time changes of $(X_I, \mathbf{R}^m)$.

When $m = n = 1$, h is invertible if and only if $h(x, 1)$ is either a positive or a negative function. It can be shown that every covering cocycle is cohomologous to a positive cocycle and therefore is invertible. The key to proving this and the other theorems in this section is the following observation:

LEMMA 6.1. *A cocycle $h : X \times \mathbf{Z}^m \to \mathbf{R}^m$ is cohomologous to the cocycle defined on the generators $e_1, \dots, e_m$ of $\mathbf{Z}^m$ by the formula*

$$h_N(x, e_i) = \frac{1}{N^m} \sum \{h(ax, e_i) : 1 \leq a_j \leq N, 1 \leq j \leq m\}.$$

This lemma is proved by using the cocycle equation to show that

$$h_N(x, e_i) - h(x, e_i) = f(e_i x) - f(x)$$

where $a = (a_1, \dots, a_m)$ and

$$f(x) = \frac{1}{N^m} \sum \{h(x,a) : 1 \leq a_j \leq N, 1 \leq j \leq m\}.$$

To see how this lemma is used, note that in the $m = n = 1$ case, when $\int_X h(x,1)d\mu > 0$ for every invariant measure μ, it follows that $h_N(x,1)$ is positive for large N and thus h is cohomologous to a positive cocycle.

Note that a constant cocycle T in $\mathcal{L}(m,m)$ is invertible as a cocycle if and only if T is invertible as a linear map. In [KMS3] the next result was obtained using piecewise linear topology.

THEOREM 6.1. *If T is an invertible constant cocycle, then T is in the interior of the invertible cocycles.*

Consider a covering cocycle h in $\mathcal{D}_m$. As we observed in the previous section, we can find a linear map T and a cocycle cohomologous to h as close to T as we please. Also by those remarks T must be invertible. Now Theorem 5.1 can be applied to finish the proof of the next result.

THEOREM 6.2. *Every close to linear covering cocycle is cohomologous to an invertible cocycle. In particular, every covering cocycle is cohomologous to an invertible one when there is a unique invariant cocycle integral.*

A larger class of covering cocycles than the close to linear covering cocycles is the class of nonsingular cocycles. A cocycle h is *nonsingular* if given invariant Borel measures $\mu_1, \dots, \mu_m$ for $(X, \mathbf{Z}^m)$ the matrix

$$\begin{pmatrix} \int_X h_1(x,e_1)d\mu_1 & \cdots & \int_X h_1(x,e_m)d\mu_m \\ \vdots & & \vdots \\ \int_X h_m(x,e_1)d\mu_1 & \cdots & \int_X h_m(x,e_m)d\mu_m \end{pmatrix}$$

is nonsingular in the usual sense. Note that the rows of this matrix are the matrices of $\sigma(h_1), \dots, \sigma(h_m)$ for a fixed σ in $\mathcal{I}$. The formal functional analytic relationship between the above matrices and $\mathcal{I}$ was worked out in [M1]. Nonsingular cocycles are of interest because of the following result proved in [MM], again using piecewise linear topology:

THEOREM 6.3. *Every nonsingular cocycle is cohomologous to an invertible cocycle.*

This theorem allows us to prove a very general theorem about $\mathbf{R}^m$ actions. It is not difficult to verify that the cocycles Rudolph constructs to prove Theorem 4.4 are all nonsingular. Thus, as a consequence of Theorems 4.4 and 6.3, we now have:

THEOREM 6.4. *If $(Y, \mathbf{R}^m)$ is a flow with a free dense orbit, then there exists a flow $(X, \mathbf{Z}^m)$ such that some time change of $(X_I, \mathbf{R}^m)$ is an almost one-to-one extension of $(Y, \mathbf{R}^m)$.*

7. The Almost Periodic Spiral

A question left open by the theory described in the first five sections of this paper is how far from a linear map can a covering cocycle be? The answer is not completely known. In this section we will describe the properties of one exotic example constructed in [M1]. It is the only known example of an embedding cocycle for a minimal flow which is singular and hence not close to linear, and it is counter-intuitive in the way covering cocycles can behave directionally.

First we need to describe a method of simultaneously constructing a cocycle and a dynamical system. The idea is to start with the desired values of the cocycle h for a single x with a dense orbit and build from there. Specifically, let

$$\mathcal{S} = \{\nu : \mathbf{Z}^m \to \mathbf{R}^m : \nu(0) = 0\}$$

be given the topology of pointwise convergence. There is a natural $\mathbf{Z}^m$ action on $\mathcal{S}$ defined by

$$a\nu(b) = \nu(a+b) - \nu(a)$$

and an evaluation cocycle $e : \mathcal{S} \times \mathbf{Z}^m \to \mathbf{R}^m$ defined by $e(\nu, a) = \nu(a)$.

A *net* is a function ν in $\mathcal{S}$ satisfying

$$\sup \left\{ \frac{|\nu(a) - \nu(b)|}{|a-b|} : a \neq b \right\} < \infty.$$

If ν is a net, then the orbit closure of ν, denoted by $\overline{\mathcal{O}(\nu)}$ is a compact metric space and by restricting the evaluation cocycle to $\overline{\mathcal{O}(\nu)}$ we get a cocycle $h_\nu : \overline{\mathcal{O}(\nu)} \times \mathbf{Z}^m \to \mathbf{R}^m$ such that $h_\nu(\nu, a) = \nu(a)$. In some sense $(\overline{\mathcal{O}(\nu)}, \mathbf{Z}^m)$ is the smallest flow on which $a \to \nu(a)$ can occur as a cocycle restricted to some $\{x\} \times \mathbf{Z}^m$. Note that if $\nu = T$, a linear map in $\mathcal{L}(m,n)$, then $\overline{\mathcal{O}(\nu)}$ is trivial because the linear maps are the fixed points of the $\mathbf{Z}^m$ action on $\mathcal{S}$.

The *almost periodic spiral* is constructed using the following spiraling Lipschitz function of $\mathbf{C}$ into itself as a starting point:

$$\lambda(z) = \begin{cases} z & \text{when } 0 \leq r \leq \exp(e^{2\pi}) \\ z e^{i \log\log r} & \text{when } \exp(e^{2\pi}) < r \end{cases}$$

where as usual $z = re^{i\theta}$. The restriction ν of λ to $\mathbf{Z}^2 = \{p + qi \in \mathbf{C} : p, q \in \mathbf{Z}\}$ is a net. It is not hard to see that the limit points of ν are the rotations on $\mathbf{Z}^2$ and thus fixed points of $(\mathcal{S}, \mathbf{Z}^2)$ because rotations are linear maps. The slightly more surprising fact is that h_ν is a covering cocycle. However, the suspension flow associated with h_ν is not minimal, and the construction of the almost periodic spiral requires further work.

Covering cocycles must grow linearly in the $m = n$ case, but how do the different directions contribute to their growth? What can be seen from the behavior of the spiraling net is that the "sufficient" growth required to produce a covering cocycle can be the result of the combined contributions of all the component cocycles with no single component having the uniform growth itself in any direction. To make this more precise let $T : \mathbf{R}^2 \to \mathbf{R}$ be a linear map. Then $T \circ h$ is a cocycle when h is one. If we fix a in $\mathbf{Z}^2$, then $x \to ax$ defines a $\mathbf{Z}$ action. Thus

$$g(x,k) = T\big(h(x, ka)\big)$$

is a cocycle of the $g : X \times \mathbf{Z} \to \mathbf{R}$ type with $m = n = 1$. The spiraling behavior of h_ν implies that for all T and a as above

$$g(x,k) = T\big(h_\nu(x, ka)\big)$$

is not covering. Since a cocycle h' cohomologous to h would have the same behavior, this property is an invariant of the cohomology class. In contrast, if h is close to linear, for every a there exists T such that $T(h(x,a))$ is covering. Thus h_ν is an example of a covering cocycle which is not close to linear.

The underlying dynamics of h_ν, that is the dynamics of the $\mathbf{Z}^2$ flow on the orbit closure of ν, are simple, consisting of a circle of fixed points and a single dense orbit. However, by carefully modifying the function λ, an almost periodic net can be constructed without destroying the spiraling behavior. Although this construction carried out in [M1] is technically very demanding, the underlying idea is easily described. The constructive process must syndetically insert copies of larger and larger pieces of the central behavior of the function including all earlier modifications. The latter point is the difficult aspect of the construction. The result is a minimal flow and a cocycle, called the almost periodic spiral, with the exotic properties delineated in our final theorem.

THEOREM 7.1. *There exists a minimal flow $(X, \mathbf{Z}^2)$ and an embedding (and hence covering) cocycle $h : X \times \mathbf{Z}^2 \to \mathbf{R}^2$ with the following properties:*

1. *for every linear map T in $\mathcal{L}(2,1)$ and every non-zero a in $\mathbf{Z}^2$, the cocycle $g(x,k) = T\big(h(x,ka)\big)$ does not grow uniformly in x and is not covering,*
2. *the cocycle h is singular and therefore not close to linear,*
3. *the cocycle h is invertible,*
4. *the cocycle h is not in either of the components of the covering cocycles containing the linear maps in $\mathcal{L}(2,2)$ with positive determinants or negative determinants,*
5. *when μ is an ergodic invariant measure for $(X, \mathbf{Z}^2)$ the set of matrices*

$$\begin{pmatrix} \int_X h_1(x,e_1)d\mu & \int_X h_1(x,e_2)d\mu \\ \int_X h_2(x,e_1)d\mu & \int_X h_2(x,e_2)d\mu \end{pmatrix}$$

   coincides with the group of rotations of $\mathbf{R}^2$,

6. *when μ_1 and μ_2 are invariant measures for $(X, \mathbf{Z}^2)$ the set of matrices*

$$\begin{pmatrix} \int_X h_1(x,e_1)d\mu_1 & \int_X h_1(x,e_2)d\mu_2 \\ \int_X h_2(x,e_1)d\mu_1 & \int_X h_2(x,e_2)d\mu_2 \end{pmatrix}$$

   coincides with the set of matrices

$$\left\{ \begin{pmatrix} a & b \\ c & d \end{pmatrix} : a^2 + c^2 \le 1, b^2 + d^2 \le 1 \right\}.$$

The original goal which led to the almost periodic spiral was to construct a covering cocycle for a minimal flow where the distance between its cohomology class and the linear maps was positive. In [KMS3] it was subsequently shown that a cocycle h is close to linear if and only if the distance between its cohomology class and the set of linear maps is zero. Consequently the almost periodic spiral is not close to linear. However, the nonlinear behavior of the almost periodic spiral is even more dramatic and is better reflected by part (4) of Theorem 7.1.

Since the almost periodic spiral is invertible, there exists an extension $H : X \times \mathbf{R}^2 \to \mathbf{R}^2$ of it as described in Section 6. Fix x and consider the function $\psi(t) = H(x, te_1)$ of $\mathbf{R}$ into $\mathbf{R}^2$. If the almost periodic spiral had any linear behavior beyond that given by Theorem 4.3, one would expect $\psi(\mathbf{R})$ to bear some slight resemblance to a straight line. It does not. Writing $\psi(t) = r(t)e^{i\theta(t)}$, it can be shown that $\lim_{t\to\pm\infty} \theta(t) = \pm\infty$. In other words, the spiraling behavior of the underlying function $\lambda(z)$ is not lost in the process of making the net almost periodic. In fact, this is precisely the behavior of $\theta(t)$ that is used to prove part (4) of Theorem 7.1.

Covering cocycles can be characterized using invariant measures when $m = n = 1$ as stated in Theorem 3.2. Furthermore, when the flow $(X, \mathbf{Z}^m)$ has a unique invariant measure μ the cocycle h is covering if and only if

$$\begin{pmatrix} \int_X h_1(x, e_1) d\mu & \cdots & \int_X h_1(x, e_m) d\mu \\ \vdots & & \vdots \\ \int_X h_m(x, e_1) d\mu & \cdots & \int_X h_m(x, e_m) d\mu \end{pmatrix}$$

is a nonsingular matrix. These two special results make the idea of using invariant measures to characterize covering cocycles very appealing. But the almost periodic spiral exposes some of the fundamental difficulties in obtaining such a result.

The obvious approach is to study the set of matrices

$$\begin{pmatrix} \int_X h_1(x, e_1) d\mu & \cdots & \int_X h_1(x, e_m) d\mu \\ \vdots & & \vdots \\ \int_X h_m(x, e_1) d\mu & \cdots & \int_X h_m(x, e_m) d\mu \end{pmatrix}$$

where μ varies over the set of invariant measures. From part (5) of Theorem 7.1 we know that for the almost periodic spiral the ergodic measures give us the group of rotations which includes I and $-I$. Because the set of invariant measures is convex, some invariant measure produces the zero matrix. Unfortunately, it is also not enough to require that the above matrices are nonsingular just for the ergodic measures (even when $m = n = 1$). Consequently the best hope is to try using the sign of the determinant function, but whether or not this will work is still an open question.

One of the motivations for developing the cocycle integrals was to characterize covering. Here one easily shows that a cocycle h is covering when the matrices

$$\begin{pmatrix} \int_X h_1(x, e_1) d\mu_1 & \cdots & \int_X h_1(x, e_m) d\mu_m \\ \vdots & & \vdots \\ \int_X h_m(x, e_1) d\mu_1 & \cdots & \int_X h_m(x, e_m) d\mu_m \end{pmatrix}$$

are all nonsingular. It is now evident from part (6) of Theorem 7.1 that the converse is false even when the measures are ergodic. Of course, it is still possible that some property of the cocycle integrals of a cocycle will characterize covering.

For our final observation we consider the suspension $(X_{h_\nu}, \mathbf{R}^2)$ constructed using the cocycle h_ν of the almost periodic spiral. From Rudolph's theorem (Theorem 4.4) there exists a suspension, $(Y_g, \mathbf{R}^2)$, constructed using an embedding, nonsingular cocycle $g : Y \times \mathbf{Z}^2 \to \mathbf{R}^2$ which is an almost one-to-one extension of $(X_{h_\nu}, \mathbf{R}^2)$. It is a small step (see [KMS1]) to verify that $(Y_g, \mathbf{R}^2)$ is isomorphic to a suspension flow $(U_h, \mathbf{R}^2)$ where the cocycle $h : U \times \mathbf{Z}^2 \to \mathbf{R}^2$ takes on the same values as h_ν and has the same invariant cocycle integrals. Thus dramatically different cocycles can give rise to isomorphic suspensions. One can view $(Y, \mathbf{Z}^2)$ and $(U, \mathbf{Z}^2)$ as being Kakutani equivalent in a topological sense. Whether or not there is a topological variant of Kakutani equivalence providing a useful coarse classification of lattice actions of compact metric spaces is unknown.

References

[B1] A. Besicovitch, *A Problem on Topological Transformations of the Plane*, Fundam. Math. **28** (1937), 61-65.

[B2] A. Besicovitch, *A Problem on Topological Transformations of the Plane II*, Proc. Camb. Phil. Soc. **41** (1951), 38-45.

[BD] D. Boivin and Y. Derriennic, *The Ergodic Theorem for Additive Cocycles of Z^d or R^d*, Ergodic Theory and Dynamical Systems 11 No 1 (1991), 19-39.

[D] F. Dekking, *Recurrent Sets*, Adv.in Math. **44** (1982), 78-104.

[E1] R. Ellis, *Global Sections of Transformation Groups*, Illinois J. of Math. **8 No 3** (1964), 380-394.

[E2] R. Ellis, *Cocycles in Topological Dynamics*, Topology **17** (1978), 111-130.

[FKMS] H. Furstenberg, H. Keynes, N. Markley and M. Sears, *Topological Properties of R^n Suspensions and Growth Properties of Z^n Cocycles*, Proc. London Math. Society **(3) 66** (1993), 431-448.

[GH] W. Gottschalk and G. Hedlund, *Topological Dynamics*, Am. Math. Soc., Colloquium Publications **36** (1955).

[Ka] A. Katok, *The Special Representation Theorem for Multidimensional Group Actions*, Astérisque **49** (1977), 117-140.

[KMS1] H. Keynes, N. Markley and M. Sears, *The Structure of Minimal R^n Actions*, Quaestiones Mathematicae **16 No 1** (1993), 81-102.

[KMS2] H. Keynes, N. Markley and M. Sears, *Ergodic Averages and Integrals of Cocycles*, Acta Math. Univ. Comenianae **64 No 1** (1995), 123-139.

[KMS3] H. Keynes, N. Markley and M. Sears, *Close to Linear Cocycles*, Proc. Am. Math. Soc. **124 No 6** (1996), 1923- 1931.

[M1] K. Madden, *The Existence and Consequences of Exotic Cocycles, Ph.D. Thesis, University of Maryland, College Park,* (1994.).

[M2] K. Madden, *A Covering Cocycle Which Does Not Grow Linearly*, Trans. Am. Math. Soc. **347 6** (1995), 2225-2234.

[MM] K. Madden and N. Markley, *Nonsingular Cocycles and Piecewise Linear Time Changes*, Acta Math. Univ. Comenianae (to appear).

[NS] V. Nemytskii And V. Stepanov, *Qualitative Theory of Differential Equations,* (1960), Princeton University Press, Princeton, New Jersey.

[R1] D. Rudolph, *Rectangular Tilings of R^n and Free R^n Actions,*, Dynamical Systems Proceedings, University of Maryland (1986-1987), 342-353.

[R2] D. Rudolph, *Markov Tilings of R^n and Representations of R^n Actions*, Contemporary Math. **94** (1988), 271- 290.

[T] R. Thomas, *Commuting Continuous Flows, Ph.D. Thesis, University of Southampton* (1968).

School of Mathematicsm, University of Minnesota, Minneapolis, MN 55455
E-mail address: keynes@math.umn.edu

Department of Mathematics, Lafayette College, Easton, PA 18042
E-mail address: MaddenK@lafvax.lafayette.edu

Department of Mathematics, University of Maryland, College Park, MD 20742
E-mail address: ngm@glue.umde.edu

Department of Computational and Applied Mathematics, University of Witwatersrand, Johannesburg 2050, South Africa
E-mail address: 036mis@cosmos.wits.ac.za

Contemporary Mathematics
Volume **215**, 1998

Suspensions, Inheritance, and Flows on Homogeneous Spaces

D. Ellis

ABSTRACT. Let S be a subgroup of T. Given an S-flow, (X, S), one can form its suspension $(X \underset{S}{\times} T, T)$, thus obtaining a T-flow. On the other hand if (Z, T) is a T-flow then (Z, S) is an S-flow. In this paper we consider the following questions: What is the relationship between the dynamical properties of (X, S) and those of $X \underset{S}{\times} T$? Which T-transformation groups are suspensions of S-transformation groups? and Which dynamical properties of (Z, T) are inherited by (Z, S)? We show that when T is locally compact and S is a closed syndetic subgroup, answers to these questions are closely related to the dynamical properties of the natural actions of S and T on the homogeneous space T/S.

1. Introduction

Let T be a locally compact topological group and $S \subset T$ a closed syndetic subgroup given the induced topology. (Recall that S is a syndetic subgroup of T if there exists a compact subset $K \subset T$ with $SK = T$). Given an S-flow, (X, S), one can form its suspension $(X \underset{S}{\times} T, T)$, thus obtaining a T-flow. On the other hand if (Z, T) is a T-flow then (Z, S) is an S-flow. Three natural questions arise:

1. What is the relationship between the dynamical properties of (X, S) and those of $X \underset{S}{\times} T$?
2. Which T-transformation groups are suspensions of S-transformation groups?
3. Which dynamical properties of (Z, T) are inherited by (Z, S)?

It turns out that answers to these questions are closely related to the dynamical properties of the natural actions of S and T on the homogeneous space T/S. This paper is divided into three sections. Each section deals with one of these questions.

In the first section we show that a pointed T-flow (Z, z_0) is the suspension of the S-flow $(X, S) = (\overline{z_0 S}, S)$ if and only if $(T/S, T)$ is a factor of (Z, T). In fact, in general we show that $(X \underset{S}{\times} T, T)$ is isomorphic to the supremum $(Z \bigvee (T/S), T)$ of the flows (Z, T) and $(T/S, T)$. These results generalize corollary 4.5 of [**DE**], where the assumption is made that S is normal in T.

1991 *Mathematics Subject Classifications*: Primary 54H20.

In the second section we show that the suspension of every distal S-flow is a distal T-flow if and only if $(T/S, T)$ is distal. We also give necessary and sufficient conditions for the suspension of every equicontinuous S-flow to be an equicontinuous T-flow. These results generalize lemma 2.3 and proposition 5.6 of [**DE**].

In the third section we consider the question of inheritance. It is well known (see for example [**GH**]) that when S is normal in T, a flow (Z, T) is pointwise almost periodic under T if and only if it is pointwise almost periodic under S. We generalize this by showing that pointwise almost periodicity is inherited in this sense if and only if the S-flow $(T/S, S)$ is pointwise almost periodic, which in turn holds if and only if $(T/S, T)$ is distal. We also show that every minimal T-flow (Z, T) has exactly one S-minimal set if and only if $(T/S, T)$ is proximal.

In this paper the phase spaces of all flows are assumed to be compact and pointed.

2. Realizing T-flows as Suspensions

We begin by recalling the definition of the suspension:

DEFINITION 2.1. Let $S \subset T$ be a closed syndetic subgroup of T and (X, S) an S-flow. We define an action of S on $X \times T$ by

$$s(x,t) = (xs^{-1}, st)$$

Then the *suspension* of (X, S) is given by

$$X \underset{S}{\times} T = X \times T \Big/ \sim$$

where $(x_1, t_1) \sim (x_2, t_2)$ iff $s(x_1, t_1) = (x_2, t_2)$ *for some* $s \in S$.

We will denote the equivalence class of (x, t) by $[x, t]$. We observe that $X \underset{S}{\times} T$ is a compact Hausdorff space because S is closed and syndetic. The map

$$\begin{array}{ccc} (X \underset{S}{\times} T) \times T & \to & X \underset{S}{\times} T \\ ([x,t], t_1) & \to & [x, tt_1] \end{array}$$

defines a continuous action of T on $X \underset{S}{\times} T$.

The following proposition characterizes those T-flows which can be realized as suspensions of S-flows. It is important to keep in mind that we are dealing with pointed flows so that homomorphisms of flows are unique.

PROPOSITION 2.2. *Let S be a closed syndetic subgroup of T and (Z, z_0) a pointed T-flow. Set $X = \overline{z_0 S}$. Then $Z \cong X \underset{S}{\times} T$ if and only if $(T/S, T)$ is a factor of (Z, T).*

PROOF. Clearly the map

$$\begin{array}{ccc} X \underset{S}{\times} T & \overset{\psi}{\to} & T/S \\ [x,t] & \to & St \end{array}$$

is a T-homomorphism; hence T/S is a factor of $X \underset{S}{\times} T$.

Now suppose that T/S is a factor of (Z,T), and consider the T-homomorphism

$$\begin{array}{ccc} X\underset{S}{\times}T & \overset{\varphi}{\to} & Z \\ [x,t] & \to & xt. \end{array}$$

If $\varphi([x_1,t_1]) = \varphi([x_2,t_2])$, then $\psi([x_1,t_1]) = \psi([x_2,t_2])$ so

$$x_1t_1 = x_2t_2 \quad \text{and} \quad St_1 = St_2.$$

This implies that $[x_1,t_1] = [x_1,t_1t_2^{-1}t_2] = [x_1t_1t_2^{-1},t_2] = [x_2,t_2]$. Thus φ is an isomorphism, and the proof is complete. □

In order to state a useful consequence of Proposition 2.2 we recall that the *supremum* $(Z_1 \bigvee Z_2, T)$, of two pointed T-flows (Z_1,T) and (Z_2,T) is the pointed T-flow which has both flows as factors and is minimal with respect to that property.

PROPOSITION 2.3. *Let S be a closed syndetic subgroup of T and (Z,z_0) a pointed T-flow. Set $X = \overline{z_0S}$. Then $X\underset{S}{\times}T$ is the supremum $Z\bigvee(T/S)$ of the flows (Z,T) and $(T/S,T)$.*

PROOF. As was seen in the proof of 2.2, $X\underset{S}{\times}T$ is an extension of (Z,T) and $(T/S,T)$. Hence it is an extension of $Z\bigvee(T/S)$. On the other hand T/S is a factor of $Z\bigvee(T/S)$, so by 2.2 $Z\bigvee(T/S)$ is a suspension of $\hat{X} = \overline{z_0S} \subset Z\bigvee(T/S)$. But (X,S) is a factor of $(\hat{X},S)$, so $X\underset{S}{\times}T$ is a factor of $\hat{X}\underset{S}{\times}T \cong Z\bigvee(T/S)$. □

COROLLARY 2.4. *Let S be a closed syndetic subgroup of T with $(T/S,T)$ distal (proximal, almost periodic). Let (Z,z_0) be a pointed T-flow. Set $X = \overline{z_0S}$. Then $X\underset{S}{\times}T$ is a distal (proximal, almost periodic) extension of (Z,T).*

PROOF. This follows immediately from the fact that if (Y,T) is distal (proximal, uniformly almost periodic), then $(Z\bigvee Y,T)$ is a distal (proximal, almost periodic) extension of (Z,T). □

3. Distality and Equicontinuity under Suspension

PROPOSITION 3.1. *Let S be a closed syndetic subgroup of T and assume that (Y,S) is a distal extension of (X,S). Then $Y\underset{S}{\times}T$ is a distal extension of $X\underset{S}{\times}T$.*

PROOF. Let $\varphi : Y \to X$ denote the S-homomorphism of pointed flows, and

$$\begin{array}{cccc} \hat{\varphi}: & Y\underset{S}{\times}T & \to & X\underset{S}{\times}T \\ & [y,t] & \to & [\varphi(y),t] \end{array}$$

the associated homomorphism of suspensions. Assume that $\hat{\varphi}([y_1,t_1]) = \hat{\varphi}([y_2,t_2])$. Then $t_1t_2^{-1} \in S$ and $(\varphi(y_1))(t_1t_2^{-1}) = \varphi(y_1(t_1t_2^{-1})) = \varphi(y_2)$. If $[y_1,t_1],[y_2,t_2]$ are proximal, then there exists a net $\{r_i\} \subset T$ with

$$\lim[y_1(t_1t_2^{-1}),e]t_2r_i = \lim[y_1(t_1t_2^{-1}),t_2]r_i = \lim[y_1,t_1]r_i = \lim[y_2,e]t_2r_i.$$

Since S is syndetic in T, $T = SK$ for some compact set $K \subset T$, and we can write $t_2 r_i = s_i k_i$ where $s_i \in S$ and $k_i \in K$. It then follows that

$$\begin{aligned}\lim[y_1(t_1t_2^{-1})s_i, k_i] &= \lim[y_1(t_1t_2^{-1}), s_ik_i] = \lim[y_1(t_1t_2^{-1}), t_2r_i] \\ &= \lim[y_2, e]t_2r_i = \lim[y_2s_i, k_i].\end{aligned}$$

By passing to subnets if necessary we may assume that $y_1(t_1t_2^{-1})s_i \to z_1$, $y_2s_i \to z_2$, and $k_i \to k \in K$. Then $[z_1, k] = [z_2, k]$ which implies that $z_1 = z_2$ and hence $y_1(t_1t_2^{-1})$ is proximal to y_2. Since φ is distal and we know that $\varphi(y_1(t_1t_2^{-1})) = \varphi(y_2)$, we have shown that $y_1(t_1t_2^{-1}) = y_2$, from which it follows that $[y_1, t_1] = [y_2, t_2]$. □

COROLLARY 3.2. *Let S be a closed syndetic subgroup of T and assume that T acts distally on T/S. Let (X, S) be a distal S-flow. Then $X \underset{S}{\times} T$ is a distal T-flow.*

PROOF. We apply Proposition 3.1 to the distal extension $X \to \{*\}$. We conclude that $X \underset{S}{\times} T$ is a distal extension of T/S. (Note that the flow $(T/S, T)$ is the suspension of the trivial flow.) Since a distal extension of a distal flow is distal, the proof is complete. □

REMARKS 3.3. 1) When T is given the discrete topology, T/S is finite so T always acts distally on T/S, (indeed $(T/S, T)$ is an equicontiuous T-flow). Hence the suspension of every distal S-flow is a distal T-flow. This is the content of lemma 2.3(3) of [**DE**]. We will show that equicontiuity is also preserved under suspension in this case.

2) For a general topology on T it is possible for T to act distally but not equicontinuously on T/S. In this case the suspension of a distal flow is distal, but since every suspension has T/S as a factor, no suspension is equicontinuous. Thus the suspension of any equicontinuous S-flow will be distal but not equicontinuous. A specific example is provided by:

$$S = \left\{ \begin{bmatrix} 1 & m & n \\ 0 & 1 & r \\ 0 & 0 & 1 \end{bmatrix} \mid m, n, r \in \mathbf{Z} \right\} \subset \left\{ \begin{bmatrix} 1 & x & y \\ 0 & 1 & z \\ 0 & 0 & 1 \end{bmatrix} \mid x, y, z \in \mathbf{R} \right\} = T \subset SL(3, \mathbf{R});$$

(see [**AGH**] pages 46-47 for further examples and details).

3) Even when S is a closed syndetic normal subgroup of T (so that T/S is equicontinuous and hence distal) there are examples where (X, S) is equicontinuous but $(X \underset{S}{\times} T, T)$ is not (of course it is distal). See for instance [**DE**] example 2.6.

4) There are of course many examples where T does not act distally on T/S. Indeed, when $T = SL(2, \mathbb{R})$ and $S \subset T$ is the subgroup of upper triangular matrices, $(T/S, T)$ is proximal. In this case no suspension of a distal flow is distal.

In order to describe necessary and sufficient conditions for the suspension of every equicontinuous flow to be equicontinuous we will need to review some notation, (for details see [**E2**] and [**V**]).

NOTATION 3.4. Let βT denote the Stone-Cech compactification of the group T when provided with the discrete topology. The algebra of continuous functions from βT to $\mathbf{R}$ will be denoted $\mathcal{C}(\beta T)$. We write $\Re_T$ for the T-subalgebra of $\mathcal{C}(\beta T)$ consisting of all bounded right-uniformly continuous functions on T. We write $|\Re_T|$ for the space whose algebra of continuous functions is $\Re_T$.

The space $|\Re_T|$ is a T-flow; indeed it has some "universal" properties. We state these in the following lemmas which appear as lemmas 3.2 and 3.3 of [**DE**].

LEMMA 3.5. *Let T be a locally compact topological group and (X,T) a pointed T-transformation group. Then*

1. *there exists a unique homomorphism of T-transformation groups $(|\Re_T|,T) \to (X,T)$,*
2. *$|\Re_T|$ admits a semigroup structure in which left multiplication by p is continuous for all $p \in |\Re_T|$ and right multiplication by t is continuous for all $t \in T$; moreover the canonical T-homomorphism $\pi : \beta T \to |\Re_T|$ is a homomorphism of semigroups,*
3. *the natural inclusion $T \to \beta T \xrightarrow{\pi} |\Re_T|$ gives an isomorphism of T into $|\Re_T|$ with $\overline{T} = |\Re_T|$.*

PROOF. (1) is proven in [**E2**], (2) follows immediately, and (3) is proven in [**V**] (see his appendix pp. 823-824) □

LEMMA 3.6. *Let $S \subset T$ be a closed syndetic subgroup and denote by $\overline{S}$ the closure of S in $|\Re_T|$. Then*

1. *the product map $\overline{S} \times T \to |\Re_T|$ induces an isomorphism $(\overline{S}\underset{S}{\times}T,T) \to (|\Re_T|,T)$ of T-transformation groups.*
2. *$(\overline{S},S)$ is universal for continuous pointed S-transformation groups. (Here S is provided with the induced topology).*

PROOF. (1) follows immediately from 3.5 and 2.2. See [**DE**] for a proof of (2). □

We need some further notation from [**DE**]:

NOTATION 3.7. Let M be a fixed minimal ideal in $|\Re_T|$. Fix an idempotent

$$u^2 = u = u_S \in M_S = M \cap \overline{S} \subset M.$$

We set

$$G = Mu \qquad \text{and} \qquad G_S = M_S u.$$

Then G_S is a subgroup of the group G; also the pointed flows (M_S,u) and (M,u) are universal for pointed minimal S and T flows respectively. Now let (Z,T) and (X,S) be pointed minimal flows; we define

$$G(Z,T) = \{\alpha \in G \mid z_0\alpha = z_0\} \quad \text{and} \quad G(X,S) = \{\alpha \in G_S \mid x_0\alpha = x_0\}.$$

In the following proposition $B(S)$ and $B(T)$ denote the Bohr compactifications (maximal equicontinuous flows) of S and T respectively. (See page 98 of [**G**])

PROPOSITION 3.8. *Let S be a closed syndetic subgroup of T with $(T/S,T)$ distal. Assume that $G(B(T),T) = G(B(S),S)$. Then the suspension of every equicontinuous minimal S-flow is equicontinuous.*

PROOF. Consider the T-homomorphism

$$\varphi : B(S)\underset{S}{\times}T \to B(T).$$

It is shown in [**DE**] that for any S-flow (X,S), $G(X,S) = G(X\underset{S}{\times}T,T)$. Thus the assumption that $G(B(T),T) = G(B(S),S)$ guarantees that φ is a proximal extension. (The proof of this fact is the same as that for the discrete topology on T; see [**E2**] for details) On the other hand since $(T/S,T)$ is distal, 3.2 implies that $B(S)\underset{S}{\times}T$ is distal. Therefore φ is an isomorphism.

Now suppose that (X,S) is equicontinuous. Then (X,S) is a factor of $(B(S),S)$, and hence $X\underset{S}{\times}T$ is a factor of $B(S)\underset{S}{\times}T \cong B(T)$. Thus $(X\underset{S}{\times}T,T)$ is equicontinuous. □

REMARKS 3.9. 1) It can be shown that the condition

$$G(B(T),T) = G(B(S),S)$$

always holds when T is given the discrete topology. Therefore in this case the suspension of every equicontinuous flow is equicontinuous.

2) Proposition 3.8 is a generalization of proposition 5.6 of [**DE**], which is just 3.8 with the stronger assumption that S is normal in T. (In that case of course T/S is a compact topological group and hence is equicontinuous)

3) Since T/S is the suspension of the trivial flow, it follows that under the assumptions of 3.8, T/S is equicontinuous. It is not true however that the conclusion of 3.8 follows from the assumption that T/S is equicontinuous (compare remark 3.3(4)).

4) To see that the first assumption in 3.8 does not follow from the second consider the following example:

Let $T = SL(n,\mathbf{R})$ with $n \geq 3$, and

$$S = \left\{ \begin{pmatrix} A & w \\ 0 & 1 \end{pmatrix} \mid A \in SL(n-1,\mathbf{R}), w \in \mathbf{R}^{n-1} \right\}.$$

The linear action of T on $\mathbf{R}^n$ (on the right) induces an action of T on real projective space $\mathbb{R}P(n-1)$. Moreover S is the stabilizer of the point corresponding to the vector $(0,\dots,0,1)$. Indeed T acts proximally on $T/S \cong \mathbb{R}P(n-1)$. On the other hand in this case both $B(S)$ and $B(T)$ are trivial. Hence for this example $B(S)\underset{S}{\times}T \cong T/S \to B(T)$ is a non trivial proximal extension.

4. Inheritance

In this section we examine which properties of a flow (Z,T) are inherited by the flow (Z,S). As in the previous sections the dynamics of the homogeneous space T/S plays a key role.

We begin with a generalization of the classical inheritance theorem for closed normal syndetic subgroups. Recall that it states that when S is syndetic and normal in T any almost periodic point of the flow (Z,T) is an almost periodic point for the flow (Z,S). Note that when S is syndetic and normal in T, $(T/S,T)$ is equicontinuous and hence distal.

PROPOSITION 4.1. *Let* S *be a closed syndetic subgroup of* T*. Then the following are equivalent:*

1. *If* $u^2 = u \in |\Re_T|$ *is an idempotent, then* $u \in \overline{S}$.
2. *For every* (Z,T), $z \in Z$ *is* T*-almost periodic if and only if* z *is* S*-almost periodic.*
3. *The action of* S *on* T/S *is pointwise almost periodic.*
4. St *is proximal to* S *if and only if* $t \in S$.
5. $(T/S,T)$ *is distal.*

PROOF. (1) $\Rightarrow$ (2) Recall from 3.7 that the pointed flow (M,u) is universal for pointed minimal T-flows where $u^2 = u \in M$ is an idempotent, and $M \subset |\Re_T|$ is a minimal ideal. Now let $z \in Z$ be T-almost periodic. Then $(\overline{zT}, z)$ is a pointed minimal T-flow so there exists a T-homomorphism

$$(M,u) \to (\overline{zT}, z).$$

Thus $\overline{zS}$ is the image of $\overline{uS} = M_S$. But (M_S, S) is minimal, hence $(\overline{zS}, S)$ is minimal and z is S-almost periodic.

(2) $\Rightarrow$ (3) If (2) holds then every point in T/S is S-almost periodic since $(T/S,T)$ is minimal.

(3) $\Rightarrow$ (4) If (3) holds and St is proximal to S, then

$$S = Sp = Stp \in St\overline{S} \quad \text{for some} \quad p \in \overline{S},$$

and hence $S = St$.

(4) $\Rightarrow$ (1) Assume that (4) holds and let $u^2 = u$ be an idempotent. It follows from the fact that S is syndetic in T (or 3.6(1)) that we can write

$$u = pt \quad \text{with} \quad p \in \overline{S}.$$

Now Su is proximal to S, but $Su = Spt = St$. Thus by our assumption $t \in S$, and $u \in \overline{S}$.

(5) $\Rightarrow$ (4) This is immediate.

(1) $\Rightarrow$ (5) Assume that (1) holds and let $u^2 = u \in |\Re_T|$ be an idempotent. Then for every $t \in T$, tut^{-1} is also an idempotent. Hence

$$tut^{-1} \in \overline{S}$$

for every $t \in T$. Thus in $(T/S,T)$:

$$St = S(tut^{-1})t = (St)u,$$

in other words u restricted to the enveloping semigroup of T/S is the identity. It follows that the enveloping semigroup $E(T/S,T)$ is a group. Hence $(T/S,T)$ is distal. (see [**E2**] proposition 5.3) □

The next proposition is the analog of 4.1 when $(T/S,T)$ is proximal.

PROPOSITION 4.2. *Let* S *be a closed syndetic subgroup of* T *and* $M \subset |\Re_T|$ *a minimal ideal. Then the following are equivalent:*

1. *For any* S*-flow* (X,S)*, every point of* $X \underset{S}{\times} T$ *is* S*-proximal to a point in* X.
2. *Every minimal* T*-flow has exactly one* S*-minimal set.*

3. *For every* $t \in T, tM_S \subset \overline{S}$.
4. *The flow* $(T/S, T)$ *is proximal.*
5. *Every point of* T/S *is proximal to* S.
6. *The singleton set* $\{S\}$ *is the only* S*-minimal set in* T/S.

PROOF. (1) $\Rightarrow$ (2) Let (Z,T) be a minimal T-flow and $z_0 \in Z$ be an S-almost periodic point. Set $X = \overline{z_0 S}$. We claim that X is the only S-minimal subset of Z. To see this recall that the product map induces a T-homomorphism

$$\pi : X \underset{S}{\times} T \to Z.$$

For any S-minimal subset $Y \subset Z$, consider the S-invariant set

$$\hat{Y} = \pi^{-1}(Y) \subset X \underset{S}{\times} T.$$

Let $\hat{X} \subset \hat{Y}$ be an S-minimal set. Then

$$\pi(\hat{X}) = Y$$

because Y is S-minimal. On the other hand by our assumption (1), every point of $\hat{X}$ is S-proximal to a point of X. In particular

$$X \cap \hat{X} \neq \emptyset \quad \text{and hence} \quad X = \hat{X}.$$

It follows that $Y = \pi(\hat{X}) = \pi(X) = X$.

(2) $\Rightarrow$ (3) Let $M_S = M \cap \overline{S} \subset M \subset |\Re_T|$ be as in 3.7. Then M is minimal under T and M_S is minimal under S. It follows that for every $t \in T$, tM is minimal under T and tM_S is minimal under S. Thus tM_S and $tM \cap \overline{S}$ are both S-minimal subsets of the T-minimal flow tM. Hence $tM_S = tM \cap \overline{S} \subset \overline{S}$.

(3) $\Rightarrow$ (4) Assume that $tM_S \subset \overline{S}$ for every $t \in T$, and let $St_1, St_2 \in T/S$. Then for any $p \in M_S$ we have

$$(St_1)p = S(t_1 p) = S = S(t_2 p) = (St_2)p.$$

Thus St_1 is proximal to St_2.

(4) $\Rightarrow$ (5) This is immediate.

(5) $\Rightarrow$ (6) Let $St \in T/S$ be S-almost periodic. Assuming (5) there exists $q \in M$ with $Stq = Sq$. The fact that (M,u) is universal for pointed minimal T-flows together with proposition 2.2 implies that

$$\begin{array}{ccc} M_S \underset{S}{\times} T & \to & M \\ [p,t] & \to & pt \end{array}$$

is an isomorphism. It follows that we can write $q = pr$ with $p \in M_S$ and $r \in T$. Then

$$S = Sp = Sqr^{-1} = Stqr^{-1} = Stp \in (St)\overline{S},$$

whence $St \in (St)\overline{S} = \{S\}$ and S is the only S-almost periodic point.

(6) $\Rightarrow$ (1) Let $[x,t] \in X \underset{S}{\times} T$ and $u^2 = u \in M$ be a minimal idempotent. Then the point $Stu \in T/S$ is an almost periodic point. Hence by our assumption (6) we must have $Stu = S$. It follows that $tu \in \overline{S}$, whence

$$[x,t]u = [x,e](tu) = [x,e](tu)u = [x(tu),e]u.$$

Thus $[x, t]$ is proximal to $[xtu, e]$. □

COROLLARY 4.3. *Let S be a closed syndetic subgroup of T and suppose that $(T/S, T)$ is proximal. Let (Z, T) be any distal minimal T-flow. Then (Z, S) is minimal.*

PROOF. Let $X \subset Z$ be the S-minimal subset of Z. Then every point of Z is proximal to a point of X. Since Z is distal this implies that $Z = X$. □

COROLLARY 4.4. *Let S be a closed syndetic subgroup of T and suppose that $(T/S, T)$ is proximal. Then the canonical map $\iota : B(S) \to B(T)$ is onto, (similarly for the maximal distal flows $D(S) \to D(T)$).*

PROOF. Note that $B(T)$ is distal and minimal under T; moreover $\iota(B(S)) \subset B(T)$ is invariant under S. The result now follows immediately from 4.3. □

References

[AGH] L. Auslander, L. Green and F. Hahn, *Flows on Homogeneous Spaces,*, Annals of Math. Studies, Princeton University **53** (1963).

[DE] D. Ellis, *Suspensions of Topological Transformation Groups*, Ergod. Th. & Dynam. Sys. **10** (1990), 101-117.

[E1] R. Ellis, *Group-Like Extensions of Minimal Sets*, Amer. Math. Soc. **127** (1967), 125-135.

[E2] R. Ellis, *Lectures on Topological Dynamics*, Benjamin, New York, 1969.

[G] S. Glasner, *Proximal Flows*, Springer-Verlag, Lecture Notes in Mathematics 517, Berlin, Heidelberg, New York, 1976.

[GH] W. Gottschalk and G. Hedlund, *Topological Dynamics*, Am. Math. Soc. Coll. Pub. **XXVI** (1955).

[V] W. Veech, *Topological Dynamics*, Bull. Amer. Math. Soc. **83** (1977), 775-830.

DEPARTMENT OF MATHEMATICS AND COMPUTER SCIENCE, BELOIT COLLEGE, BELOIT WI 53511, USA

E-mail address: ellis@beloit.edu

Contemporary Mathematics
Volume **215**, 1998

On the Lifting of Transformation Semigroups

H. Ikeshoji and T. Wu

ABSTRACT. Let (X, S) be a transformation semigroup, which we call a flow, with compact Hausdorff phase space X. Since S is not necessarily a group, there is no guarantee that S acts on X as homeomorphisms. We call a flow (W, S) a lifting of (X, S) if S acts on W as homeomorphisms and if there is a homomorphism $\mu : (W, S) \to (X, S)$. In this paper, we show that if liftings exist for (X, S), then a minimal lifting exists in the product space X^T, where T is the group generated by S, and we give some sufficient conditions for the existence of liftings.

Introduction

Let S be a semigroup and consider a transformation semigroup, which we simply call a flow, (X, S) with compact Hausdorff phase space X. Since S is not necessarily a group, there is no guarantee that S acts on X as homeomorphisms. So we want to relate (X, S) with some flow $(\widetilde{X}, S)$ where S acts as homeomorphisms. If (X, S) is embedded in such a flow $(\widetilde{X}, S)$, then S acts on X as homeomorphisms and there is no problem in this case. If (X, S) is a homomorphic image of $(\widetilde{X}, S)$, then we can deduce many properties of the action of S on X based on the fact that S acts on $\widetilde{X}$ as homeomorphisms. In this paper we consider the second case and discuss conditions for the existence of such $(\widetilde{X}, S)$ and a homomorphism $\widetilde{\pi} : (\widetilde{X}, S) \to (X, S)$. We will call such a flow $(\widetilde{X}, S)$ a lifting of (X, S). That is, a flow $(\widetilde{X}, S)$ is called a lifting of a flow (X, S) if S acts on $\widetilde{X}$ as homeomorphisms and there exists a (surjective) homomorphism $\widetilde{\pi} : (\widetilde{X}, S) \to (X, S)$. In this paper we identify some important subflows related to the original flow (X, S) and clarify conditions for the existence of liftings in terms of these flows.

In the following we assume that S is embeddable in a group. In fact, if S acts on $\widetilde{X}$ as homeomorphisms and the action is effective, i.e., if $w \neq z$ for $w, z \in \widetilde{X}$, there exists some $s \in S$ such that $ws \neq zs$, then we may assume that S is a subsemigroup of the group $H(\widetilde{X})$ of all homeomorphisms on $\widetilde{X}$. We denote the group generated by S in a group by T. It should be noted that such a T is essentially uniquely determined.

In this paper we follow the notations and conventions of [A], [E]. For example, actions are written on the left of points in the phase space while homomorphisms on the left. Also we assume that all the actions are effective and discrete, i.e., the semigroup S (or T) is endowed with discrete topology.

1991 *Mathematics Subject Classifications*: Primary 54H20.

1. The Coordinate Actions and Conditions for Liftings

Let S be a discrete semigroup embeddable in a group and T the group generated by S. Let (X, S) be a flow with compact Hausdorff phase space X. A standard way to find an S-flow with homeomorphism actions related to (X, S) is to consider the "co-ordinate" actions. Let D be any nonempty set and consider the product space X^D. For each $z \in X^D$, let z_i be the value of ith co-ordinate of z, i.e., the image of z under the projection $\pi_i : X^D \to X$. If S acts on D from left, as a set map without topological consideration, then we can define an S-flow on X^D as follows: For each $s \in S$, let $\phi_s : X^D \to X^D$ be the mapping defined by $(z\phi_s)_i = z_{si}$ for each $i \in D$. Since we consider the product topology on X^D, each ϕ_s is continuous and, by the assumption that S acts on D from left, we have $(z\phi_s\phi_t)_i = (z\phi_s)_{ti} = z_{s(ti)} = z_{(st)i} = (z\phi_{st})_i$ for any $i \in D$. So $\phi_s : X^D \to X^D$ satisfies $\phi_s\phi_t = \phi_{st}$. Thus we can define a discrete flow on X^D. We call this action of S the co-ordinate action as opposed to the usual product action and we will denote this action using "ϕ". For example, we write the flow on X^D under this co-ordinate action by $(X^D, \phi(S))$. The importance of co-ordinate action is that it easily offers actions of homeomorphisms. Indeed, if S acts on D from left bijectively, then each $\phi_s : X^D \to X^D$ is a homeomorphism. For such a set D we can always choose the most natural one, i.e., T, the group generated by S. But the co-ordinate flow has some defects in the sense that the most natural mappings associated with X^D, i.e., the projections π_i are in general not homomorphisms. In fact, $\pi_i(z\phi_s) = (z\phi_s)_i = z_{si}$ is, in general, not equal to $\pi_i(z)s = z_i s$. But co-ordinate actions have many homomorphisms among them. Let C be a subset of D which is closed under the action of S. Then, clearly, the restriction of $z \in X^D$ to C, i.e., $\pi_C z = z|C$ defines a homomorphism $\pi_C : (X^D, \phi(S)) \to (X^C, \phi(S))$. (Note that $C = \{i\}$ is usually not closed under S-action on D.) So, if we can find a copy of (X, S) in $(X^C, \phi(S))$, then we can relate the flow (X, S) to $(X^D, \phi(S))$ and circumvent the difficulties due to the fact that π_i are not homomorphisms. Under these preliminary considerations, we proceed formally as follows.

Let $D = T$ and $C = S$. Define $\Psi : X \to X^S$ by $\Psi(x)_s = xs$. It is easily checked that Ψ defines a homomorphism $\Psi : (X, S) \to (X^S, \phi(S))$, and by the assumption that S acts on X effectively, Ψ is injective. Since Ψ is clearly continuous and X is compact, $\Psi(X)$ is closed in X^S and so the closed subflow $(\Psi(X), \phi(S))$ is an isomorphic copy of (X, S) in $(X^S, \phi(S))$. Let $\pi_S : X^T \to X^S$ be the restriction to the co-ordinate S and let $\widehat{X} = \pi_S^{-1}(\Psi(X))$. Clearly $\widehat{X}$ is closed. Since $\pi_S : (X^T, \phi(S)) \to (X^S, \phi(S))$ is a homomorphism, $(\widehat{X}, \phi(S))$ is a subflow of $(X^T, \phi(S))$ and π_S, restricted to $\widehat{X}$, is a surjective homomorphism from $(\widehat{X}, \phi(S))$ to the isomorphic copy of (X, S), i.e., $(\Psi(X), \phi(S))$. So, in order to find a lifting, the only thing we need to check is whether each $\phi_s : \widehat{X} \to \widehat{X}$ is a homeomorphism or not. Since $\phi_s : X^T \to X^T$ is a homeomorphism, ϕ_s, restricted to $\widehat{X}$, is injective. But it is, in general, not surjective. For example, let $z \in \widehat{X}, s \in S$. To show the surjectivity of $\phi_s : X^T \to X^T$, we need to find $w \in \widehat{X}$ such that $w\phi_s = z$. Assume that there exists such $w \in \widehat{X}$ and let $x \in X$ be such that $\pi_S(w) = \Psi(x)$. If there exists $t \in T \setminus S$ such that $st \in S$, then we must have $z_t = (w\phi_s)_t = w_{st} = (\pi_S w)_{st} = \Psi(x)_{st} = x(st)$. Now pick any $z' \in X^T$ such that $z'_{t'} = z_{t'}$ if $t' \neq t$ and $z'_t \neq x(st)$. Then it is clear that $z' \in \widehat{X}$, but there is no

$w \in \widehat{X}$ such that $w\phi_s = z$. This suggests that we need to restrict $\widehat{X}$ to a smaller $\phi(S)$-invariant subset whose elements behave more regularly.

In spite of this fact, $(\widehat{X}, \phi(S))$ is the "first approximation" of a lifting of (X, S) and it has many important properties. For example, $\pi_e : X^T \to X$, the projection to the co-ordinate of identity e of T, is a homomorphism when it is restricted to $\widehat{X}$. To see this, note that π_e, restricted to $\widehat{X}$, is equal to $\Psi^{-1}\pi_S$ and that $\Psi : (X, S) \to (X^S, \phi(S))$ is an isomorphism. So π_e is a homomorphism which is clearly surjective. In fact, $(\widehat{X}, \phi(S))$ is the largest subflow in$(X^T, \phi(S))$ on which π_e is a homomorphism. Let $(Z, \phi(S))$ be a subflow of $(X^T, \phi(S))$ such that $\pi_e : Z \to X$ is a homomorphism. Then for each $z \in Z$, $\pi_e(z\phi_s) = \pi_e(z)s$ for each $s \in S$. But this implies that $z_s = z_{se} = z_e s = \Psi(z_e)_s$. So $\pi_S(z) \in \Psi(X)$ and $Z \subset \widehat{X}$. Summarizing, we have the following proposition, when T has an identity.

PROPOSITION 1.1. *Let $\widehat{X} = \pi_S^{-1}(\Psi(X))$. Then $(\widehat{X}, \phi(S))$ is the largest subflow of $(X^T, \phi(S))$ on which $\pi_e : X^T \to X$, the projection to the co-ordinate $e \in T$, is a homomorphism. Moreover, $\pi_e : (\widehat{X}, \phi(S)) \to (X, S)$ is surjective.*

Another important subflow of $(X^T, \phi(S))$ is defined as follows. Let $\widetilde{X}$ be the set of all elements z in X^T satisfying $z_{ts} = z_t s$ for all $t \in T$ and $s \in S$. $\widetilde{X}$ is closed because $\widetilde{X} = \bigcap_{t \in T, s \in S}\{z \in X^T | \, z_{ts} = z_t s\}$ and each $\{z \in X^T | \, z_{ts} = z_t s\}$ is clearly closed. It is also easily checked that $\widetilde{X}$ is a $\phi(S)$-invariant subset of $\widehat{X}$. Actually, it is $\phi(T)$-invariant. In fact, let $z \in \widetilde{X}$ and $t \in T$. Then $(z\phi_t)_{us} = z_{t(us)} = z_{(tu)s} = z_{tu}s = (z\phi_t)_u s$ holds for any $u \in T$ and $s \in S$. So $z\phi_t \in \widetilde{X}$ and $\widetilde{X}$ is $\phi(T)$-invariant. Also, $(\pi_S z)_s = z_s = z_{es} = z_e s = \Psi(z_e)_s$ shows that $\widetilde{X} \subset \widehat{X}$. The importance of the subflow $(\widetilde{X}, \phi(S))$ for the problems of liftings is expressed in the following proposition.

PROPOSITION 1.2. *The flow $(\widetilde{X}, \phi(S))$ is a subflow of $(X^T, \phi(S))$ where S acts as homeomorphisms and $\pi_e : (\widetilde{X}, \phi(S)) \to (X, S)$ is a homomorphism. Let $X^* = \bigcap_{s \in S} Xs$. Then, X^* is an S-invariant closed subset of X and $\pi_e(\widetilde{X}) \subset X^*$. In particular, $\widetilde{X}$ is actually a subset of $(X^*)^T$. If (W, S) is any flow where S acts as homeomorphisms and if there exists a homomorphism $\mu : (W, S) \to (X, S)$, then there exists a unique homomorphism $\lambda : (W, S) \to (\widetilde{X}, \phi(S))$ such that $\mu = \pi_e\lambda$.*

PROOF. Since $\widetilde{X}$ is $\phi(T)$-invariant, it is clear that each $\phi_s : \widetilde{X} \to \widetilde{X}$ is a homeomorphism. Also it is clear that $\pi_e : \widetilde{X} \to X$ is a homomorphism because $\widetilde{X} \subset \widehat{X}$. Since $\phi_s : \widetilde{X} \to \widetilde{X}$ is a homeomorphism, $\widetilde{X}\phi_s = \widetilde{X}$. So, $\pi_e(\widetilde{X}) = \pi_e(\widetilde{X}\phi_s) = \pi_e(\widetilde{X})s \subset Xs$ for any $s \in S$. So, $\pi_e(\widetilde{X}) \subset \bigcap_{s \in S} Xs = X^*$. It is clear that $X^* = \bigcap_{s \in S} Xs$ is S-invariant and, since Xs is closed for any $s \in S$ because X is compact Hausdorff, X^* is also closed. For any $t \in T$, and $z \in \widetilde{X}$, $z_t = z_{te} = (z\phi_t)_e = \pi_e(z\phi_t)$. Since $z\phi_t \in \widetilde{X}$, $z_t \in \pi_e(\widetilde{X}) \subset X^*$. Thus $\widetilde{X} \subset (X^*)^T$. Assume that S acts on W as homeomorphism and let $H(W)$ be the group of all homeomorphisms on W. Note that S may be regarded as a subsemigroup of $H(W)$ and the action of S on W can be extended to that of T on W. Using this extended T-action, define $\lambda : W \to X^T$ by $\lambda(w)_t = \mu(wt)$. Then $\lambda(w) \in \widetilde{X}$. In fact, $\lambda(w)_{ts} = \mu(w(ts)) = \mu((wt)s) = \mu(wt)s = \lambda(w)_t s$ for any $t \in T$ and $s \in S$. So, $\lambda(w) \in \widetilde{X}$. Also, $(\lambda(w)\phi_s)_t = \lambda(w)_{st} = \mu(w(st)) = \mu((ws)t) = \lambda(ws)_t$ shows that $\lambda(ws) = \lambda(w)\phi_s$, and so $\lambda : (W, S) \to (\widetilde{X}, \phi(S))$ is a homomorphism.

Clearly, $\pi_e\lambda(w) = \lambda(w)_e = \mu(we) = \mu(w)$ for any $w \in W$. So, $\mu = \pi_e\lambda$. To show that μ is factored uniquely, let $\lambda, \lambda' : (W, S) \to (\widetilde{X}, \phi(S))$ be homomorphisms satisfying $\mu = \pi_e\lambda = \pi_e\lambda'$. Then, for any $w \in W$, $\lambda(w)_e (= \mu(w)) = \lambda'(w)_e$. Note that S acts on $W, \widetilde{X}$ as homeomorphisms and so λ, λ' can be extended to T- homomorphisms by $\lambda(wu) = \lambda(w)\phi_u, \lambda'(wu) = \lambda'(w)\phi_u$ for each $u \in T$. So, $\lambda(w)_u = \lambda(w)_{ue} = (\lambda(w)\phi_u)_e = (\lambda(wu))_e$ and similarly $\lambda'(w)_u = (\lambda'(wu))_e$. So, $\lambda(w)_u = \lambda'(w)_u$ for any $u \in T$. Thus, $\lambda(w) = \lambda'(w)$ and, since $w \in W$ was arbitrary, $\lambda' = \lambda$. □

We will call $(\widetilde{X}, S)$ the minimal lifting of (X, S). It is clear that if there exists a flow (W', S) with homeomorphism that satisfies the unique factorization property in Proposition 1.2, then $(\widetilde{X}, S)$ and (W', S) are isomorphic. It is also clear from Proposition 1.2 that there is no surjective homomorphism $\mu : (W, S) \to (X, S)$ if S acts on W as homeomorphism unless $X^* = X$, i.e., $(\forall s \in S)(Xs = X)$. So we will call a flow (W, S) a lifting of (X, S) if S acts on W as homeomorphisms and if there exists a homomorphism $\mu : (W, S) \to (X, S)$ such that $\mu(W) = X^*$. From Proposition 1.2 it is clear that a flow (X, S) has a lifting in this sense if $\pi_e(\widetilde{X}) = X^*$. Incidentally, note that $\pi_t(\widetilde{X}) = \pi_e(\widetilde{X})$ for any $t \in T$, since $\widetilde{X}$ is $\phi(T)$-invariant.

We consider sufficient conditions for the existence of liftings in the next Section.

2. Sufficient Conditions for the Existence of Liftings

As shown in Section 1, the subflow $(\widetilde{X}, \phi(S))$ plays an important role when we consider the problem of liftings. In this section we give two sufficient conditions for the existence of liftings, i.e., $\pi_e(\widetilde{X}) = X^*$. First, we note that $\widetilde{X}$ can be identified as the limit of an inverse system. So a natural condition for $\widetilde{X} \neq \emptyset$ is that the preorder defining the inverse system is directed. It turns out that the condition of directedness is sufficient to guarantee that $\pi_e(\widetilde{X}) = X^*$. As we will see below, this directedness condition is a rather common condition in semigroup theory. Another sufficient condition considered is that S is a free semigroup.

Let S, T be as defined in Section 1 and (X, S) a flow where X is compact Hausdorff. Define a preorder $\prec$ in T by $t' \prec t \Leftrightarrow t' = ts$ for some $s \in S$ or $t' = t$. Clearly "$\prec$" defines a preorder in T. For each t', t with $t' \prec t$, define $\xi^{t',t} : X \to X$ by $\xi^{t',t}(x) = xs$ if $t' = ts$ and $\xi^{t',t}(x) = x$ if $t' = t$. (Note that we assume the existence of the identity in S.) Also, let $X_t = X$ for each $t \in T$. Then $\{X_t, \xi^{t',t}\}$ defines an inverse system and it is clear that $\widetilde{X}$ is the inverse limit of $\{X_t, \xi^{t',t}\}$. In fact, the defining condition of $\widetilde{X}$ was given by $z_{ts} = z_t s$ for $t \in T, s \in S$. So, using $\{\xi^{t',t}\}$, this condition on $z \in X^T$ is rewritten as $\pi_{ts}z = \xi^{ts,t}(\pi_t z)$. So, including the case where $t' = t$, the condition is rewritten as $\pi_{t'}z = \xi^{t',t}(\pi_t z)$, and so $\widetilde{X} = \varprojlim \{X_t, \xi^{t',t}\}$. Therefore, if T is directed in this preorder, $\widetilde{X} \neq \emptyset$.

Before we show, in this case, that $\pi_e(\widetilde{X}) = X^*$ and that λ in Proposition 1.2 is uniquely determined, we discuss the condition of directedness. Note that the set following $u \in T$, i.e., $\{t \in T | u \prec t\}$, is given by $uS^{-1} \cup \{u\}$. So, the condition that T is directed is given by (a) $(\forall u, v \in T)(uS^{-1} \cap vS^{-1} \neq \emptyset)$.

In fact, if (a) holds, then any $t \in uS^{-1} \cap vS^{-1}$ follows both u and v. On the other hand, assume that u and v both are followed by a common $t \in T$. If $t \neq u, v$, then $t \in uS^{-1} \cap vS^{-1}$. If t is equal to one of u, v, say $t = u$, then $v \prec u$ and in this case we have $uS^{-1} \subset vS^{-1}$. Clearly this implies $uS^{-1} \cap vS^{-1} \neq \emptyset$. Thus, the directedness condition of T can be expressed by (a). But the condition (a) is related to a rather common condition in semigroup theory ([CP]). We first give a following proposition related to the directedness of T. As the following proposition shows, the condition is more natural than it appears.

PROPOSITION 2.1. *Assume that S is embeddable into a group and let T be the group generated by S. Then the following conditions are equivalent:*

(a) $(\forall u, v \in T)(uS^{-1} \cap vS^{-1} \neq \emptyset)$;
(b) $(\forall s, t \in S)(Ss \cap St \neq \emptyset)$;
(c) $T = S^{-1}S$.

PROOF. *(a) ⇒ (b)* Assume (a) and let $s, t \in S$. Since $s^{-1}S^{-1} \cap t^{-1}S^{-1} \neq \emptyset$, there are $p, q \in S$ such that $s^{-1}p^{-1} = t^{-1}q^{-1}$. Thus $ps = qt$ and so $Ss \cap St \neq \emptyset$.

(b) *(b) ⇒ (c)* See p.36, [CP].

(c) ⇒ *(a)* Assume (c) and let $u, v \in T$. Then $v^{-1}u \in T = S^{-1}S$, and so $v^{-1}u = s^{-1}t$ for some $s, t \in S$. So $ut^{-1} = vs^{-1}$. Thus $uS^{-1} \cap vS^{-1} \neq \emptyset$. □

The condition (b) is called right invertibility ([CP]) and it is well known that if S is cancellative and satisfies (b), then S is embeddable into a group (and the group T that S generates is of the form $S^{-1}S$ as above). When S is cancellative and abelian, then in the usual way S can be embedded in the quotient group T and the conditions in Proposition 2.1 are satisfied. If $(\forall u \in T)(Su \cap S \neq \emptyset)$, then clearly $T = S^{-1}S$ and S satisfies the condition (a). For example, let $GL(n, \mathbf{R})$ be the group of of all invertible $n \times n$ matrices with real entries, $M_n(\mathbf{Z})$ the semigroup of all $n \times n$ matrices with integer entries. If $S = GL(n, \mathbf{R}) \cap M_n(\mathbf{Z})$, then $T = GL(n, \mathbf{Q})$ and it is clear, e.g., by multiplying u by kI_n with an appropriate integer $k \in \mathbf{Z}$, that $Su \cap S \neq \emptyset$ for any $u \in T$.

Under the condition on S as above we can add the following to Proposition 1.2.

PROPOSITION 2.2. *Let (X, S) be a flow where S separates points in X. Assume that S is embeddable into a group and satisfies one of the conditions in Proposition 2.1. Then, $\pi_e(\widetilde{X}) = X^*$. In particular, X^* is the largest subset in X on which each $s \in S$ acts surjectively.*

PROOF. It was shown in Proposition 1.2 that $\pi_e(\widetilde{X}) \subset X^*$. To show the converse inclusion, let $a \in X^*$. We show that there exists some $z \in \widetilde{X}$ such that $\pi_e(z) = a$. For each $u \in T$, let $F(u) = \{z \in X^T \mid \pi_e(z) = a, z_{us} = z_u s \text{ for all } s \in S\}$. Since $F(u) = \pi_e^{-1}(a) \cap (\bigcap_{s \in S}\{z \mid z_{us} = z_u s\})$ it is clear that $F(u)$ is closed. Clearly $F(u) \supset F(v)$ if $u \prec v$. In fact, assume that $u \prec v$ and $u \neq v$. Then $u = vt$ for some $t \in S$. If $z \in F(v)$, then $z_{us} = z_{(vt)s} = z_{v(ts)} = z_v(ts) = (z_v t)s = z_{vt}s = z_u s$ for any $s \in S$. So, $z \in F(u)$ and $F(v) \subset F(u)$. So, if $F(u) \neq \emptyset$ for each $u \in T$, then $\{F(u)\}_{u \in T}$ is a family of closed sets with the finite intersection property in X^T. So $\bigcap_{u \in T} F(u) \neq \emptyset$ and it is easily checked that any $z \in \bigcap_{u \in T} F(u)$ is an element in $\widetilde{X}$ with $\pi_e(z) = a$. To show $F(u) \neq \emptyset$ for each $u \in T$, we first show that $F(t^{-1}) \neq \emptyset$ for each $t \in S$. Let $t \in S$. Since $a \in X^* \subset Xt$, there exists some $b \in X$ with $bt = a$. Choose any $w \in X^T$ such that $\pi_S(w) = \Psi(b)$ and $w_e = b$, and let $z = w\phi_t$. Then, $z_{t^{-1}s} = (z\phi_{t^{-1}})_s = (w\phi_t\phi_{t^{-1}})_s = w_s = bs$. Note that $z_{t^{-1}} = (w\phi_t)_{t^{-1}} = w_e = b$.

So $z_{t^{-1}s}(=bs) = z_{t^{-1}}s$ for any $s \in S$. Thus $F(t^{-1}) \neq \emptyset$ for any $t \in S$. Now let $u \in T$ be arbitrary. Since $T = S^{-1}S$ ((c) in Proposition 2.1), $u = t^{-1}s$ for some $s, t \in S$. Then $u \prec t^{-1}$ and so $F(t^{-1}) \subset F(u)$ as shown above. So $F(u) \neq \emptyset$. So, for each $a \in X^*$, there exists some $z \in \widetilde{X}$ such that $\pi_e(z) = a$. Thus, $\pi_e(\widetilde{X}) \supset X^*$ and so $\pi_e(\widetilde{X}) = X^*$. In particular, $X^*s = \pi_e(\widetilde{X})s = \pi_e(\widetilde{X}\phi_s) = \pi_e(\widetilde{X}) = X^*$ for each $s \in S$. □

To show $\pi_e(\widetilde{X}) = X^*$, we could have used results of inverse limits (e.g., p.435, Problem 4, [D]) since $\widetilde{X} = \varprojlim \{X_t, \xi^{t',t}\}$.

Next we will show that if S is free, a flow (X, S) has liftings. For a technical reason explained after Proposition 2.4, we assume $(\forall s \in S)(Xs = X)$ in the remaining of this section. We first give a lemma.

LEMMA 2.3. *Let (X, S) be a flow where S acts surjectively, i.e., $(\forall s \in S)(Xs = X)$. Then, (X, S) has a lifting if and only if there exists a function $F : X \times T \to X$ satisfying the following conditions;*

(i) *$F(x, e) = x$ for any $x \in X$,*
(ii) *$F(x, us) = F(x, u)s$ for any $x \in X, u \in T, s \in S$.*

PROOF. Sufficiency: Assume that there exists a function F satisfying (i) and (ii). For each $x \in X$, let $\lambda(x) \in X^T$ be defined by $\lambda(x)_t = F(x, t)$. Then (ii) shows that $\lambda(x) \in \widetilde{X}$ and (i) shows that $\pi_e : \widetilde{X} \to X$ is surjective.

Necessity: Assume (X, S) has a lifting. Then $\pi_e : \widetilde{X} \to X$ is surjective. For each $x \in X$, choose $z \in \widetilde{X}$ satisfying $\pi_e(z) = x$. Let $z = \lambda(x)$ and define $F : X \times T \to X$ by $F(x, t) = \lambda(x)_t$. Then, clearly, F satisfies (i) and (ii). □

Note that in Lemma 2.3 the continuity of F is irrelevant.

Now assume that S is a free semigroup. Then the group T generated by S consists of reduced words. For each $s \in S$ and $x \in X$, pick any $y \in X$ such that $ys = x$. Denote this y by $\nu(x, s)$. By induction on the length of reduced words, define $F : X \times T \to X$ as follows: First, set $F(x, e) = x$ for all $x \in X$. Assume that F is defined for all $x \in X$ and $u \in T$ whose length is equal to or less than n. Any word $v \in T$ of length $n + 1$ can be written as $v = us$ or $v = us^{-1}$ with $s \in S$ and $u \in T$ of length n. Define $F(x, v)$ by $F(x, v) = F(x, u)s$ if $v = us$ and $F(x, v) = \nu(F(x, u), s)$ if $v = us^{-1}$. It is routinely checked that $F : X \times T \to X$ is unambiguously defined and it satisfies the conditions (i) and (ii). Thus, by Lemma 2.3, we have the following proposition.

PROPOSITION 2.4. *Assume that S is a free semigroup and let (X, S) be a flow where S acts surjectively. Then, (X, S) has a lifting.*

The surjectivity condition $(\forall s \in S)(Xs = X)$ was used to define $\nu : X \times S \to X$. (Otherwise, ν would be a partial function.) If X' is a subset of X satisfying $(\forall s \in S)(X's = X')$, then we can define $\nu : X' \times S \to X'$ as above and the same argument shows that $\pi_e(\widetilde{X}) \supset X'$ (in particular, $\widetilde{X} \neq \emptyset$). But we do not know whether S acts on X^* surjectively or not. (When S satisfies the condition of Proposition 2.1, $(\forall s \in S)(X^*s = X^*)$ was a result of $\pi_e(\widetilde{X}) = X^*$.)

3. Minimal Liftings as a Functor

As shown in Sections 1,2, the existence of liftings for a flow (X, S) heavily depends on the property of S and once an appropriate S is chosen the construction of the minimal lifting $(\widetilde{X}, S)$ from (X, S) is standardized and applicable in a uniform way to a class of S-flows. In this section we clarify the "uniformity" of the construction of $(\widetilde{X}, S)$ as a functor. In the following we assume that S is a semigroup such that for any flow (X, S), $\pi_e^X(\widetilde{X}) = X^*$, where π_e^X is the projection to the co-ordinate of the identity e. For simplicity all the flows (X, S) considered in this section are assumed to satisfy $X^* = X$, i.e., $(\forall s \in S)(Xs = X)$.

Let $(X, S), (Y, S)$ be flows and let $f : X \to Y$ be arbitrary. Define $f^T : X^T \to Y^T$ by $(f^T(z))_t = f(z_t)$. Then it is easily checked that f^T is a homomorphism from X^T to Y^T with respect to the co-ordinate actions as well as the product actions. So, since $\widetilde{X}, \widetilde{Y}$ are $\phi(S)$-invariant, f^T restricted to $\widetilde{X}$ is a homomorphism from $(\widetilde{X}, \phi(S))$ to $(\widetilde{Y}, \phi(S))$ if $f^T(\widetilde{X}) \subset \widetilde{Y}$. But this condition is not always satisfied by an arbitrary f. Actually this condition characterizes the homomorphisms from (X, S) to (Y, S).

LEMMA 3.1. *Let $Hom(X, Y)$ be the set of all homomorphisms from (X, S) to (Y, S) and let $f : X \to Y$ be arbitrary. Then, $f \in Hom(X, Y)$ if and only if $f^T(\widetilde{X}) \subset \widetilde{Y}$.*

PROOF. Sufficiency: Assume $f^T(\widetilde{X}) \subset \widetilde{Y}$ and let $z \in \widetilde{X}, t \in T, s \in S$ be arbitrary. Then $f^T(z)_{ts} = f^T(z)_t s = f(z_t)s$ because $f^T(z) \in \widetilde{Y}$. On the other hand, $f^T(z)_{ts} = f(z_{ts}) = f(z_t s)$ because $z \in \widetilde{X}$. So $f(z_t s) = f(z_t)s$. By assumption $\{z_t | z \in \widetilde{X}, t \in T\} = X$. So, $f \in Hom(X, Y)$.

Necessity: Assume $f \in Hom(X, Y)$ and let $z \in \widetilde{X}$. Then, $f^T(z)_{ts} = f(z_{ts}) = f(z_t s) = f(z_t)s = (f^T(z)_t)s$. So, $f^T(z) \in \widetilde{Y}$. Thus, $f^T(\widetilde{X}) \subset \widetilde{Y}$. □

Now, for each $f \in Hom(X, Y)$ denote $f^T|\widetilde{X} : \widetilde{X} \to \widetilde{Y}$ by $\widetilde{f}$. Then $\widetilde{f}$ clearly satisfies $f\pi_e^X = \pi_e^Y \widetilde{f}$. By Proposition 1.2, $\widetilde{f} : \widetilde{X} \to \widetilde{Y}$ is the unique homomorphism satisfying this condition. Let $f \in Hom(X, Y), g \in Hom(Y, Z)$. Then $\pi_e^Z \widetilde{g \circ f} = (gf)\pi_e^X = g \circ \pi_e^Y \widetilde{f} = \pi_e^z(\widetilde{g} \circ \widetilde{f})$. So, again by Proposition 1.2, $\widetilde{g \circ f} = \widetilde{g} \circ \widetilde{f}$. This shows that $(X, S) \mapsto (\widetilde{X}, S), f \mapsto \widetilde{f}$ defines a functor. Thus we have the following proposition, whose trivial proof is omitted.

PROPOSITION 3.2. *The construction of the minimal lifting $(\widetilde{X}, S)$ from (X, S) together with $f \mapsto \widetilde{f} : Hom(X, Y) \to Hom(\widetilde{X}, \widetilde{Y})$ defines a functor from the category of flows of surjective S-actions to its subcategory of T-extendable flows. The projections $\{\pi_e^{(\cdot)}\}$ is a natural transformation from this functor to the identity functor.*

References

[A] J. Auslander, *Minimal flows and their extensions*, North-Holland.

[CP] A. Clifford and G Preston, *The algebraic theory of semigroups*, Amer. Math. Soc., Rhode Island **.1** (1961).

[D] J. Dugundji, *Topology*, Allyn and Bacon, Boston, 1966.

[E] R. Ellis, *Lectures on topological dynamics*, Benjamin, New York.

[IW] H. Ikeshoji and T. Wu, *Notes on transformation semigroups, submitted.*

DEPARMENT OF INDUSTRIAL MANAGEMENT ENGINEERING, HIROSHIMA INSTITUTE OF TECHNOLOGY, ITSUKAICHI, HIROSHIMA 731-51, JAPAN
E-mail address: ikeshoji@cc.it-hiroshima.ac.jp

DEPARTMENT OF MATHEMATICS, CASE WESTERN RESERVE UNIVERSITY, CLEVELAND, OH 44106
E-mail address: txw@po.cwru.edu

Part II.
Applications and Other Dynamical Results

Contemporary Mathematics
Volume **215**, 1998

Idempotent Measures Associated to a Locally Compact Topological Group

D. Dokken

This paper is dedicated to Professor Robert Ellis in honor of his retirement.

Abstract. Let T be a locally compact topological group and μ a regular Borel probability measure on T. A function f on T is said to be μ-harmonic if $\int f(t't)\,d\mu(t') = f(t)$. Let the μ-harmonic functions on T be denoted by $\mathcal{H}_\mu$. For a semi-simple Lie group T, Furstenberg proved that there exists a compact Hausdorff space B, with the property that $C(B)$ is isometrically isomorphic to $\mathcal{H}_\mu$ as a Banach space. In addition he proved that the space B was a compact homogeneous space associated with the group if μ was absolutely continuous with respect to a Haar measure on T. This proof was probabilistic in nature. We prove Furstenberg's theorem for $SL(2,\mathbf{R})$. The proof given here uses ideas from Ellis' abstract topological dynamics and is nonprobabilistic. In particular it uses the measure calculus on the universal pointed flow of the group. We discuss the measure calculus and also indicate how it can be used to study additive cocycles on the group.

0. Introduction

Let T be a locally compact topological group and $M(T)$ be the set of regular Borel probability measures on T. Let (X,T) be a flow, where X is a compact Hausdorff space and denote the set of regular Borel probability measures on X by $M(X)$. Let $\mathcal{R}$ the set of bounded real valued right uniformly continuous functions on T; i.e. $f : T \to \mathbf{R}$ is in $\mathcal{R}$ iff f is bounded and $\lim_{t\to e}||tf - f|| = 0$. The set $\mathcal{R}$ is a Banach algebra with respect to pointwise multiplication and the *sup* norm. Then there exists a compact hausdorff space P such that $\mathcal{R} = C(P)$. Using this identification we may identify right uniformly continuous functions on T with continuous functions on P. The group acts on the space P to give a flow (P,T). The flow is called the *universal pointed flow* of T.

Let $\mu \in M(P)$. A bounded continuous function on P is said to be μ-*harmonic* if

$$\int f(pt)\,d\mu(p) = f(t).$$

1991 *Mathematics Subject Classification.* Primary 54H20; Secondary 28C10, 46J10, 58F25.

The μ-harmonic functions on P form a Banach space and we will denote them by $\mathcal{H}_\mu$. Note that if $f \in \mathcal{R}$ and $\mu \in M(T)$ then the above definition of μ-harmonic can be written as

$$\int f(t't)\, d\mu(t') = f(t),$$

and we may think of the μ -harmonic functions as functions on T.

In [**F1**] Furstenberg proved that given $\mu \in M(T)$ there exists a compact Hausdorff space B (called the Poisson flow) and a right T-action on B such that the right uniformly continuous μ-harmonic functions on T are given by

$$h(t) = \int_B f(xt)\, d\omega(x) \equiv \mathcal{R}_\omega(f)(t),$$

where $\omega \in M(B)$, $f \in C(B)$, and the mapping

$$\begin{aligned} \mathcal{R}_\omega : C(B) &\to \mathcal{H}_\mu \\ f &\to \mathcal{R}_\omega(f) \end{aligned}$$

is an isometry. In the case of $T = SL(2, \mathbf{R})$ and for an appropriately chosen measure, this correspondence is just the classical Poisson integral formula.

The methods used in the construction of B and the proof of its structure use probabilistic techniques. In [**D**] an alternative construction of the Poisson flow is given. Let $\mu \in M(T)$. Then one finds (via Kakutani's fixed point theorem) a stationary measure ν, for μ in $M(P)$, with the property that $\nu\mu = \nu$ (stationarity) and $\int f(xt)\, d\nu(x) = f(t)$ if and only if $f \in \mathcal{H}_\mu$, that is $\mathcal{H}_\nu = \mathcal{H}_\mu$. Then $B = \overline{ex}\,(\overline{cn\nu}(\nu T))$ where bar "—" indicates closure with respect to the weak-topology on $M(P)$. The representing measure ω in $M(B)$ is given by $\omega \in M(B)$ such that $b(\omega) = \nu$ where $b : M(B) \to \overline{cn\nu}(\overline{\nu T})$ is the barycenter mapping.

The purpose of this paper is to construct the Poisson flow using abstract topological dynamics machinery developed by Bob Ellis, and show how other related ideas can be studied in this context as well.

In Section 1 we review the properties of the flow $(\beta T, T)$.

In Section 2 we prove the existence the unversal pointed flow (P, T) and give some of its relevant properties.

In Section 3 we introduce the measure calculus and prove some if its properties.

In Section 4 we introduce the Poisson flow and construct it using the measure calculus.

In Section 5 we use the measure calculus to compute the Poisson flow for the group $SL(2, \mathbf{R})$.

In Section 6 we consider the case of a continuous addtive cocycle on a $X \times T$ to $\mathbf{R}$. We show that the cocycle can be extended to a continuous function on the universal pointed flow and given $\mu \in M(T)$, under certain conditions on X and T, that there is a $\nu \in M(P)$ such that $\forall x \in X$, $\rho(x, p)$ equals $+\infty$ for all p in the support ν.

Finally I would like to thank the referees and the editors for patient reading and important suggestions on improving the text.

1. The Flow $(\beta T, T)$

Let $(T, \mathcal{T})$ be a locally compact topological group; then we denote by $M(T, \mathcal{T})$ the set of *regular Borel probability measures* on $(T, \mathcal{T})$.

Let X be a compact Hausdorff space; then we denote by $M(X)$ the set of *regular Borel probability measures* on X.

DEFINITION 1.1. A *flow* is a pair (X,T) with X and T as above, and a map $\pi : X \times T \to X$, $(x,t) \to xt = \pi(x,t)$, satisfying the following properties:

1. $xe = x$ $(x \in X)$, where e is the identity in T,
2. $x(t_1 t_2) = (xt_1)t_2$ for all $x \in X$, $t_1, t_2 \in T$.

In general the letters T,π will be suppressed and the flow (X,T,π) denoted simply by X.

DEFINITION 1.2. Let X be a flow , $f : X \to \mathbf{R}$ and $t \in T$. Then $tf : X \to \mathbf{R}$ is the map such that $(tf)(x) = f(xt)$ $(x \in X)$.

PROPOSITION 1.3. *Let T be a group provided with the discrete topology and βT its Stone-Cech or β-compactification. Then there exists a map $\pi : \beta T \times T \to \beta T$ such that $(\beta T, T, \pi)$ is a flow.*

The proof of 1.3 is by now standard and so are the following remarks.

REMARKS 1.4.

1. The group T may be identified with an open everywhere dense subset of βT.
2. The space βT may be provided with a semigroup structure such that:

(a) T is a subgroup of βT,

(b) The maps $p \to qp : \beta T \to \beta T$ are continuous $(q \in \beta T)$.

(c) The maps $p \to pt : \beta T \to \beta T$ are continuous $(t \in T)$.

(d) The point $\pi(p,t)$ equals the product of p and t $(p \in \beta T, t \in T)$.

(e) Let $f : T \to Y$, Y be compact T_2. Then there exists a unique continuous function $\overline{f} : \beta T \to Y$ such that $\overline{f}(t) = f(t)$ $(t \in T)$. Thus the set of bounded functions on T may be identified with $C(\beta T)$- the set of continuous functions on βT.

(f) Let (X,T) be any flow for which there exists a dense orbit, that is there exists a point $x_0 \in X$ such that $\overline{x_0 T} = X$. Then there exists a flow homomorphism from $\beta T \to X$ such that the identity element $e \in T$ maps to $x_0 \in X$. The flow $(\beta T, T)$ is called the universal pointed flow for the group T with the discrete topology.

2. The Flow (P,T)

Let $(T,\mathcal{T})$ be a locally compact group T with topology $\mathcal{T}$. The set

$$\mathcal{R}(T,\mathcal{T}) \equiv \mathcal{R}$$

of right uniformly continuous functions on T may be identified with a uniformly closed subalgebra of $C(\beta T)$ such that $tf \in \mathcal{R}$ $(t \in T, f \in \mathcal{R})$.

PROPOSITION 2.1. *Let $R = \{(p,q) | f(p) = f(q), f \in \mathcal{R}\}$. Then R is a closed T-invariant equivalence relation on βT such that (P,T) is a flow where $P = \beta T/R$.*

PROOF. The proof that R is a closed invariant equivalence relation is obvious. Hence $P = \beta T/R$ is a compact Hausdorff space on which T acts. Using the right uniform continuity of $f \in \mathcal{R}$ one sees that (P,T) is a flow.

□

PROPOSITION 2.2. *Let $\pi : \beta T \to P$ be the canonical map. Then*
(i) $\pi|_T$ *is one to one;*
(ii) $\overline{\pi(T)} = P$;
(iii) $\pi(T)$ *is open.*

Thus T may be identified with a dense subset of P.

PROOF. (i) The functions with compact support are right uniformly continuous and seperate points therefore $\pi|_T$ is one to one.

(ii) Since R is a closed equivalence relation the map π is closed, therefore $\pi(\overline{T}) = \overline{\pi(T)}$.

(iii) Note the right uniformly continuous functions with compact support on T, with the locally compact topology, are continuous on T with the discrete topology as well, and they seperate points in βT from T. Therefore $\pi^{-1}(T) = T$. As $\pi : \beta T \to P = \beta T/R$ is a closed map and T is open in βT, we have $\pi^{-1}(T) = T$ is open in βT. Therefore T is open in P.

□

3. The Measure Calculus

DEFINITION 3.1. Let X be a flow, α a Borel probability measure on X, and $f \in C(X)$. Then

$$< f, \alpha > = \int f(x)\, d\alpha(x)$$

and $f\alpha : T \to R$ is given by

$$(f\alpha)(t) = < tf, \alpha > = \int f(xt) d\alpha(x) \ (t \in T).$$

We use notation similar to that defined above for *convolution of functions and measures* on $(T, \mathcal{T})$. Let $\mu \in M(T, \mathcal{T})$ and $f \in C(T, \mathcal{T})$. Then $\mu f = < f, \mu > = \int f(t)\, d\mu(t)$ and $(f\mu)(t) = \int f(t't)\, d\mu(t')$.

PROPOSITION 3.2. *Let X be a flow and α a Borel probability measure on X. Then*
(i) $f\alpha \in \mathcal{R}$ $(f \in C(X))$,
(ii) *the map $\phi : f \to f\alpha$ is linear and $\phi(tf) = t\phi(f)$, $(f \in C(X), t \in T)$.*

PROOF. Clearly ϕ maps $C(X)$ into $C(T)$ and note that ϕ is a positive linear operator such that $\phi(1) = 1$. Also

$$||\phi f|| = \sup_t \left| \int f(xt)\, d\alpha(x) \right| \le ||f|| \sup_t \int d\alpha(x) = ||f||,$$

hence $||\phi f||/||f|| \le 1$ implies $||\phi|| \le 1$. Hence $||\phi|| = 1$.

To see $\phi(f) \in \mathcal{R}$, let (t_β) be a net such that t_β converges to t, then

$$||t_\beta(\phi f) - t(\phi f)|| = ||\phi(t_\beta f) - \phi(tf)|| \le ||\phi|| \cdot ||t_\beta f - tf||$$

converges to 0 as t_β converges to t.

□

REMARK 3.3. In 3.2 set $\alpha = \delta_x$ for some $x \in X$. Then

1. $(f\alpha)(t) = < tf, \delta_x > = f(x)$ $(f \in C(X)$,
2. $(f\delta_x)(t) = < tf, \delta_x > = (tf)(x) = f(xt)$,
3. ϕ is an algebra homomorphism,
4. if $\overline{xT} = X$, the map ϕ is injective and so induces a continuous function ψ from P onto X such that $\psi(e) = x$ and $\psi(pt) = \psi(p)t$ $(p \in P, t \in T)$.

DEFINITION 3.4. Proposition 3.2 allows us to define the *convolution of a measure on a flow* (X,T) *and a measure on* P. Let $\alpha \in M(X)$ and $\beta \in M(P)$. Then let $\alpha\beta$ be the measure on X defined by

$$< f, \alpha\beta > = \int f(x)\, d(\alpha\beta)(x) = \int_P \left[\int_X f(xt)\, d\alpha(x) \right] d\beta(t)$$

where $f \in C(X)$ and $\int_X f(xt)\, d\alpha(x)$ is the extension of $\int_X f(xt)\, d\alpha(x)$ to P. If we take X equal to P, then we see that the space of regular Borel probability measures on P, $M(P)$, has an analogous semi-group structure with convolution of measures as multiplication. Thus let $\alpha, \beta \in M(P)$ and $f \in C(P)$. Then by 3.2, $f\alpha \in \mathcal{R} = C(P)$ whence $< f\alpha, \beta >$ exists. We define $< f, \alpha\beta >$ to be $< f\alpha, \beta >$.

PROPOSITION 3.5. *With the notation introduced above, we have*

(i) $(\alpha\beta)\gamma = \alpha(\beta\gamma)$, $(\alpha, \beta, \gamma \in M(P))$.

(ii) *The map* $\beta \to \alpha\beta : M(P) \to M(P)$ *is continuous for each* $\alpha \in M(P)$;

(iii) $(a\alpha + (1-a)\beta)\gamma = a\alpha\gamma + (1-a)\beta\gamma$ *and* $\gamma(a\alpha + (1-a)\beta) = a\gamma\alpha + (1-a)\gamma\beta$ *if* $\alpha, \beta, \gamma \in M(P)$ *and* $0 \le a \le 1$.

(iv) *The map* $x \to \delta_x : P \to M(P)$ *is a semigroup homomophism onto a closed subset of* $M(P)$. *Thus* P *may be identified with a closed subsemigroup of* $M(P)$.

The proof of the above result is standard,see [**D**] .

REMARK 3.6. Let $M(T)$ denote the set of Borel probability measures on T.

1. If $\mu \in M(T)$, then $f \to \mu(f) : \mathcal{R} = C(P) \to \mathbf{R}$ is a positive linear functional on $C(P)$ and so may be veiwed as an element $\hat{\mu}$ of $M(P)$.
2. Recall that if $\mu, \nu \in M(T)$, we may form the convolution $\mu * \nu$ where $< f, \mu * \nu >= \int_T \left[\int_T f(st)\, d\mu(s) \right] d\nu(t)$. Then $*$ defines a semigroup structure on $M(T)$. The map $\mu \to \hat{\mu} : M(T) \to M(P)$ defines a semigroup monomorphism. In the future we suppress the $*$ and write $\mu\nu$ for $\mu * \nu$.
3. Let $\alpha \in M(P)$. Then $\alpha = \hat{\mu}$ for some $\mu \in M(T)$ if and only if $\alpha(T) = 1$.
4. The map $\alpha \to \alpha\hat{\mu} : M(P) \to M(P)$ is continuous $(\mu \in M(T))$.
5. Let (X,T) be a flow. The map $\alpha \to \alpha\hat{\mu} : M(X) \to M(X)$ is continuous $(\mu \in M(T))$.

Henceforth we will identify μ with $\hat{\mu}$ $(\mu \in M(T))$.

4. The Poisson Flow

In this chapter we give a proof of a result which allows one to identify the μ-harmonic functions on T with the continuous functions on a compact hausdorff space B, on which the group T acts. The space B is called the Poisson flow.

In [**A**], [**F3**], and [**G**] it is shown that the μ-harmonic functions on $(T, \mathcal{T})$ have boundary values in the following sense:

THEOREM 4.1. *There exists a compact Hausdorff space B and a measure $\omega \in M(B)$ such that the mapping $\mathcal{R}_\omega : C(B) \to \mathcal{H}_\mu$ given by $f \to \int_B f(xt)\, d\omega(x)$ is an isometry onto. The compact Hausdorff space B is a flow for $(T, \mathcal{T})$ and (B, ω) is called the Poisson flow.*

In [**D**] an alternative approach to the Poisson flow is given.

Let $\mu \in M(T, \mathcal{T})$. Then we may consider μ as a measure on P as follows: We identify $C(P)$ with $\mathcal{R}(T, \mathcal{T})$, say $\overline{f} \in C(P)$ is identified with $f \in \mathcal{R}(T, \mathcal{T})$, then define $\overline{\mu} \in M(P)$ by $\overline{\mu}\overline{f} = \mu f$.

Using the Kakutani fixed point theorem there is a fixed point, ν, for r_μ-the right convolution by μ, in the closed convex hull of μ^n, $\overline{cn\nu}\{\mu^n : n \in \mathbf{N}\}$ in $M(P)$. It can be shown that measure ν is an idempotent measure, that is $\nu^2 = \nu$.

Define $B = \overline{ex}\left(\overline{cn\nu}(\overline{\nu T})\right)$, the bar "—" indicating closure taken in $M(P)$. Recall that the barycenter mapping $b : M(B) \to \overline{cn\nu}(\overline{\nu T})$ is onto. Let $\overline{\nu}$ be any measure in $M(B)$ such that $b(\overline{\nu}) = \nu$. It is easily seen that (B, T) is a flow. I claim the Poisson flow for $\mu \in M(T, \mathcal{T})$ is (B, T, ω) where $\omega \in M(B)$ is such that $\omega = \overline{\nu}\nu$. Using the idempotence of the measure ν and the fact that the barycenter mapping is a flow homomorphism, it is easy to see $b(\omega) = \nu$.

To prove that $\mathcal{R}_\omega$ maps $C(B)$ isometrically onto $\mathcal{H}_\mu$ we need the following definition.

DEFINITION 4.2. Let (X, T) be a flow and $\nu \in M(X)$. Then we say ν is *contractible* if $f \to f\nu : C(X) \to R(T, \mathcal{T})$ is an isometry. If ν is contractible we will say the flow (X, T, ν) is a *contractible flow.*

LEMMA 4.3. *Let (X, T) be a flow and $\nu \in M(X)$. Then ν is contractible if and only if $\{\delta_x : x \in X\} \subseteq \overline{\nu T}$.*

For a proof of this result see [**A**],[**D**] or [**G**].

Recall that the barycenter mapping, $b : M(B) \to \overline{cn\nu}(\overline{\nu T})$ maps $b(\omega) = \nu$ and is one to one on extreme points. Therefore by the definition of B, the space (B, ω) is a contractible flow. Hence $\mathcal{R}_\omega$ is an isometry into.

To see that $\mathcal{R}_\omega$ is an isometry onto we show that every $g \in C(B)$ has the form $\mathcal{A}(f)|_B$ for some $f \in \mathcal{H}_\mu$, where $\mathcal{A}(f)$ is the affine extension of f to $M(P)$ and $\mathcal{A}(f)|_B$ is the restriction of $\mathcal{A}(f)$ to $B = \overline{ex}\left(\overline{cn\nu}(\overline{\nu T})\right)$. In fact

$$\mathcal{R}_\omega(g) = \int_B g(xt)\, d\omega(x) = \int_B \mathcal{A}(f)|_B(xt)\, d\omega(x),$$

and by definition of the barycenter mapping,

$$\int_B \mathcal{A}(f)|_B(xt)\, d\omega(x) = \int_P f(pt)\, d\nu(p).$$

Hence as

$$\int_P f(pt)\, d\nu(p) = f(t),\ (f \in \mathcal{H}_\mu = \mathcal{H}_\nu),$$

we see

$$\mathcal{R}_\omega(g) = f.$$

Since $\mathcal{R}_\omega$ maps $C(B)$ isometrically onto $\mathcal{H}_\mu$.

THEOREM 4.4. *The cardinality of $b^{-1}(\nu)$ is 1.*

PROOF. Note from the above dicussion that every continuous function on B is the restiction of a continuous affine extension of a μ harmonic function on $\overline{cn\nu}(\nu\overline{T})$. That is if $f \in C(B)$, then $\exists h \in \mathcal{H}_\mu$ such that $\mathcal{A}(h)|_B = f \in C(B)$ and

$$\mathcal{R}_\omega \mathcal{A}(h)|_B = h = \mathcal{R}_\omega f.$$

Let $K \subseteq C(B)$ be such that $f \in K$ if $\mathcal{A}(h)|_B = f$ for some $h \in \mathcal{H}_\mu$. Assume that $b(\alpha) = b(\omega)$. Then $f\alpha = h = f\omega$. Since $\mathcal{R}_\omega$ is an isometry, $C(B) \subseteq K$. Therefore $\alpha = \omega$. □

THEOREM 4.5. *The barycenter mapping* $b : M(B) \to \overline{cn\nu}(\nu\overline{T})$ *is one to one, onto and therefore* $\overline{cn\nu}(\nu\overline{T})$ *is a Bauer simplex.*

PROOF. Let α and $\beta \in M(B)$. Now

$$l_\omega : M(P) \to M(B) : \alpha \to \omega\alpha$$

is onto therefore $\alpha = \omega\eta$ and $\beta = \omega\gamma$. Hence $b(\alpha) = \nu\eta$ and $b(\beta) = \nu\gamma$. This implies that for $f \in C(B)$

$$f\alpha = (f\omega) \cdot \eta = (f\omega) \cdot b(\alpha)$$

and

$$f\beta = (f\omega) \cdot \gamma = (f\omega) \cdot b(\beta).$$

Therefore if $b(\alpha) = b(\beta)$, then $f\alpha = f\beta$, for every $f \in C(B)$, hence $\alpha = \beta$.

□

In the next chapter we show how the measure calculus can be used to give alternative proofs of several results of Furstenberg and others in boundary theory. The arguments given here are topological dynamical in nature.

5. Semisimple Lie Groups

In this chapter we prove a result of Furstenberg [**F2**]. It states that if T is a semi-simple Lie group with finite center and no-compact factors and $\mu \in M(T)$ is absolutely continuous with respect to Haar measure on T, then T acts transitively on the Poisson flow (B, ω) of μ.

THEOREM 5.1 (FURSTENBERG). *Let T be a connected semi-simple Lie group with no compact factors and let $\mu \in M(T)$ be absolutely continuous with respect to Haar measure on T. If $T = SK$ (where S is solvable and K is a compact subgroup) is the Iwasawa decomposition of T then K acts transitively on the Poisson flow B, of μ.*

REMARK 5.2. The group $SL(2, \mathbf{R})$ is a semi-simple Lie group with no-compact factors and finite center. We prove the above result for the case of $SL(2, \mathbf{R})$. Let S be the set of lower triangular matrices with positive elements on the diagonal in $SL(2, \mathbf{R})$, let A be the subgroup of diagonal matrices in $SL(2, \mathbf{R})$, $K = SO(2, \mathbf{R})$ and let N^+ and N^- be the upper triangular and lower triangular matrices respectively with ones on the diagonal. Then we have the following decomposition

$$\begin{aligned} SL(2, \mathbf{R}) &= S\ SO(2, \mathbf{R}) \\ &= SO(2, \mathbf{R})\ A\ SO(2, \mathbf{R}) \\ &= SO(2, \mathbf{R})\ AN^- = SO(2, \mathbf{R})\ AN^+. \end{aligned}$$

REMARK 5.3. We show that the group $SO(2,\mathbf{R})$ acts transitively on the space B. The basic idea is to transfer a geometric estimate of Azencott (Lemma 5.11) on $T = SL(2,\mathbf{R})$ to an estimate on P, then to an estimate on B. We consider a metric flow (X,T) (i.e. a flow for which X is metrizible) which is an image of the flow (B,T). We will show that $SO(2,\mathbf{R})$ acts transitively on the (X,T).

The metric images of the flow (B,T) seperate points of B. To see this let $x, y \in B$ and $f \in C(B)$ such that $f(x) \neq f(y)$. Consider the equivalence relation on B defined by

$$R = \{(w,z) | f(wt) = f(zt), \quad \forall t \in T\}.$$

Then R is a closed T invariant equivalence relation on B. Therefore $(B/R,T)$ is a factor of the flow (B,T), and as $SL(2,\mathbf{R})$ is second countable, the flow $(B/R,T)$ is metrizible. Under the flow homomorphism $\pi : B \to B/R$ we have that $\pi(x) \neq \pi(y)$. Therefore the metric images seperate the points of the flow (B,T). From this it will follow that $SO(2,\mathbf{R})$ acts transitively on B as well.

DEFINITION 5.4. Let $M = \{k \in K : kak^{-1} = a, a \in A\}$. In other words M is the *centralizer* of A in K. Let M_0 be the connected component of the identity in M. Let $P = NAM$.

Let X be a locally compact space.

DEFINITION 5.5. Let $M^+(X)$ be the *set of bounded positive Borel measures on* X. Let $M^+(T)$ be the *set of bounded positive Borel measures on* T.

DEFINITION 5.6.

1. If $\alpha, \beta \in M^+(X)$ $(M^+(T), M^+(P))$, we say $\alpha < \beta$ if $f\alpha < f\beta$ for every $f \in C_c(X)$ $(C_c(T), C(P))$, $f \geq 0$ (where C_c is the set of continuous functions with compact support.
2. If α and β are as above, then $\alpha \wedge \beta$ is the *sup of all measures* γ *such that* $\alpha > \gamma$ and $\beta > \gamma$.
3. If α is a measure on X (or on T or P), we define $\|\alpha\|_X = \alpha(X)$ (or $\|\alpha\|_T = \alpha(T)$ or $\|\alpha\|_P = \alpha(P)$, respectively).

In this Section we prove a result of Azencott (Theorem 5.14) which was motivated by the result of Furstenberg. Our proof is topological in nature. Here the idea is to transfer a result from the group to P and then to B.

Let U be an open subset of T and α and $\beta \in M(T)$. We say α *dominates* β on U if there exists $c > 0$ such that for every $f \in L^1(T)$ satisfying $f \geq 0$ and $supp(f) \subseteq U$, then we have $\int f\, d\alpha > c \int f\, d\beta$.

REMARK 5.7. We will assume μ is *étalée*; that is, some power μ^n dominates a multiple of Haar measure on a nonempty open subset of T.

DEFINITION 5.8. Let T be a locally compact group with a countable base and let $\mu \in M(T)$. We denote by S_μ the set of $t \in T$ such that we can find a Haar measure m on T and an integer q such that the restriction of m to an open neighborhood of t is dominated by μ^q.

REMARK 5.9. The set S_μ is an open sub-semigroup of T.

LEMMA 5.10 (AZENCOTT). *Let U be an open relatively compact subset of T, m a right Haar measure on T, η a measure dominating m on an open neighborhood of $\overline{U}$. Then there exists an $\epsilon > 0$ such that, for any a and $b \in T$ satisfying $Ua \cap Ub \neq \emptyset$, one has $\epsilon \leq \|\eta^2 a \wedge \eta^2 b\|$.*

PROOF. Let Q be compact neighborhood of $\overline{U}$ and Q^o the interior of Q. Let η be a measure dominating $m1_Q$; then η^2 dominates the measure $m * f$, where $f = 1_Q * 1_Q$. One can easily show that f is a continuous strictly positive function on Q^{o2}, and therefore on the compact set $\overline{U}^2$. It follows that there exists a number $\epsilon > 0$ such that η^2 dominates ϵm on U^2. If a and b in T are such that $Ua \cap Ub$ is nonempty, then there exists $t \in T$ such that $Ut \subset U^2a \cap U^2b$. Let θ be the restriction of m to U^2; the restriction of m to Ut is dominated by θa and θb. It follows that $m(U) - m(Ut) \leq \|\theta a \wedge \theta b\|$. As $\theta\epsilon \leq \eta^2$, we obtain $\epsilon < m(U) \leq \|\eta^2 a \wedge \eta^2 b\|$.

□

LEMMA 5.11 (AZENCOTT). *Let $h \in T$, $(t_\alpha) \subseteq T$ a net of elements in T, μ an étalée measure on T, S_μ the open semigroup defined above. If the net $t_\alpha h t_\alpha^{-1}$ converges to an element of $S_\mu^{-1}S_\mu$, then there exists a number $\epsilon > 0$ and a measure θ that is a convex combination of a finite number of measures of the form μ^n such that, for every α sufficiently large one has*

$$\epsilon \leq \|\theta t_\alpha \wedge \theta t_\alpha h\|.$$

PROOF. Suppose that $g \in S_\mu^{-1}S_\mu$ such that $\lim_\alpha t_\alpha h t_\alpha^{-1} = g$. One can write $g = t^{-1}s$, where s and t are in S_μ. There exists open relatively compact neighbourhoods V of s and W of t and integers n_s and n_t such that μ^{n_s} (μ^{n_t}) dominates right Haar measure m on the neighborhood of $\overline{V}$ (respectively $\overline{W}$). If $U = V \cup W$ one has $g \in U^{-1}U$. Therefore for α sufficiently large, one has $t_\alpha h t_\alpha^{-1} \in U^{-1}U$. This is equivalent to saying $Ut_\alpha h \cap Ut_\alpha$ is nonempty. Let $\eta = \frac{1}{2}(\mu^{n_s} + \mu^{n_t})$; the measure η dominates $\frac{1}{2}m$ on a neighborhood of $\overline{U}$. Applying Lemma 3.5, there exists $\epsilon > 0$ such that for sufficiently large α one has $\epsilon \leq \|\eta^2 t_\alpha h \wedge \eta^2 t_\alpha h\|$. Then one has

$$\eta^2 = \frac{1}{4}\left(\mu^{2n_s} + \mu^{2n_t} + 2\mu^{(n_t+n_s)}\right).$$

This proves the lemma.

□

THEOREM 5.12. *Let T be a locally compact group and $\mu \in M(T)$ be étalée and $h \in T$. If for $x \in P$ such that $\nu x \in B$, there is a net $t_\alpha \to x$ such that the net $t_\alpha h t_\alpha^{-1}$ has a limit point in $S_\mu^{-1}S_\mu$, then $\nu x h = \nu x$.*

PROOF. Let t_α and x be as in the theorem. Then by Lemma 5.11 there is an $\epsilon > 0$ and θ a convex combination of measures of the form μ^n, such that

$$\epsilon \leq \|\theta t_\alpha h \wedge \theta t_\alpha\|.$$

If ν is the idempotent associated to μ, then (by going to a subnet of t_α) we may assume $\theta t_\alpha h \wedge \theta t_\alpha$ converges and $\theta x h \geq \lim_\alpha(\theta t_\alpha h \wedge \theta t_\alpha)$ which implies

$$\theta x h \wedge \theta x \geq \lim_\alpha(\theta t_\alpha h \wedge \theta t_\alpha)$$

and hence as $\|\theta t_\alpha h \wedge \theta t_\alpha\| \geq \epsilon$ for every α sufficiently large, we see that

$$\|\theta x h \wedge \theta x\| \geq \epsilon.$$

Next we show

$$\|\nu x h \wedge \nu x\| \geq \epsilon.$$

First note

$$\nu xh = \nu\theta xh \geq \nu(\theta xh \wedge \theta x)$$

and

$$\nu x = \nu\theta x \geq \nu(\theta xh \wedge \theta x).$$

This implies

$$\nu h \wedge \nu x \geq \nu(\theta xh \wedge \theta x).$$

Hence $\|\nu xh \wedge \nu x\| \geq \|\nu(\theta xh \wedge \theta x)\| \geq \epsilon$ as ν is a Borel probability measure.

Now we show $\nu xh = \nu x$. This will follow if we can show $\|\delta_{(\nu x)}h \wedge \delta_{(\nu x)}\| \geq \epsilon$, since $supp(\delta_{(\nu x)}h) \cap supp\delta_{(\nu x)} \neq \emptyset$ and as $supp(\delta_{(\nu x)}h) = \nu xh$ and $supp\delta_{(\nu x)} = \nu x$, we obtain $\nu x = \nu xh$.

Note first that $b(\delta_{(\nu x)}) = \nu x$ and $b(\delta_{(\nu x)}h) = b(\delta_{(\nu x)})h = \nu xh$. For every $f \in C(P)$ such that $f \geq 0$, $\mathcal{A}(f)|_B(\delta_{(\nu x)}h) > f(\nu xh \wedge \nu x)$ and $\mathcal{A}(f)|_B(\delta_{(\nu x)}) = f(\nu x) \geq f(\nu xh \wedge \nu x)$. Hence $\mathcal{A}(f)|_B(\delta_{(\nu x)}h \wedge \delta_{(\nu x)}) \geq f(\nu xh \wedge \nu x)$ (because $\delta_{(\nu x)}h \wedge \delta_{\nu x}$ is the supremum of all measures less than $\delta_{(\nu x)}h$ and less than $\delta_{(\nu x)}$). Also evaluation of f at $\nu xh \wedge \nu x$ defines a bounded linear functional on $C(B) = \{\mathcal{A}(f)|_B : \mathcal{H}_\nu = \mathcal{H}_\mu\}$. Therefore $\|\delta_{(\nu x)}h \wedge \delta_{(\nu x)}\| \geq \|\nu xh \wedge \nu x\| \geq \epsilon$. And hence as noted above, it follows that $\nu xh = \nu x$. □

REMARK 5.13. Since the barycenter mapping from $b : M(B) \to \overline{cn\nu}(\overline{\nu T})$ maps $b(\omega) = \nu$ and is one to one and onto the above result implies the following.

THEOREM 5.14 (AZENCOTT). *Let T be a locally compact group and $\mu \in M(T)$ such that μ is étalée and $h \in T$. If for $x \in P$ such that $\omega x \in B$, there is a net $t_\alpha \to x$ such that the net $t_\alpha h t_\alpha^{-1}$ has a limit point in $S_\mu^{-1}S_\mu$, then $\omega xh = \omega x$.*

LEMMA 5.15. *There exists sequences $(k_n) \in SO(2,\mathbf{R})$ and $(a_n) \in A$ such that $\omega k_n a_n \to x_0 \in X$ with $\overline{x_0 T} = X$.*

PROOF. Let $(t_n) \subseteq T$ be a sequence in T such that $\omega t_n \to x \in X$ and $\overline{x_0 T} = X$. Let $t_n = k_n a_n l_n$ where (k_n) and $(l_n) \subseteq SO(2,\mathbf{R})$ and $(a_n) \in A$. We may assume $k_n \to k$ and $l_n \to l$ in $SO(2,\mathbf{R})$, and $\omega k_n a_n \to \rho \in M(X)$. Then $\omega t_n = \omega k_n a_n l_n \to \rho l$. Whence $\lim \omega k_n a_n = \rho = x l^{-1} \in X$. Then

$$\overline{x_0 T} = \overline{x_0 l^{-1} T} = \overline{xT} = X.$$

□

REMARK 5.16. Note that since μ is étalée, ${S_\mu}^{-1}S_\mu$ contains an open niegh-borhood O of e in T. Fix a sequence $(a_i) \subseteq A$, $(k_i) \subseteq K$ such that $\omega k_i a_i \to x_0 \in X$ with $\overline{x_0 T} = X$ and $k_i \to k \in K$. I claim that if

$$a_i = \begin{bmatrix} s_i & 0 \\ 0 & s_i^{-1} \end{bmatrix}$$

then $s_i \to +\infty$ or $-\infty$. This follows from the fact that $SL(2,\mathbf{R})$ is non-amenable and therefore the minimal right ideals in $M(P)$ are nontrivial. Hence the Poisson flow is non trivial as it is the set of extreme points of an ideal in $M(P)$. Therefore $k_n a_n$ cannot converge to a point in T. Hence (by going to subnet if necesarry) k_i converges to a point in K and we must have that a_i converges to a point in P, therefore $s_i \to +\infty$ or $-\infty$. If we assume that $s_i \to \infty$ then ${s_i}^{-1} \to 0$, and choose

$$n = \begin{bmatrix} 1 & 0 \\ t & 1 \end{bmatrix} \in N^-, \text{then}$$

$$a_i\ n\ a_i^{-1} = \begin{bmatrix} s_i & 0 \\ 0 & s_i^{-1} \end{bmatrix} \begin{bmatrix} 1 & 0 \\ t & 1 \end{bmatrix} \begin{bmatrix} s_i & 0 \\ 0 & s_i^{-1} \end{bmatrix}^{-1}$$

$$= \begin{bmatrix} 1 & 0 \\ s_i^{-2}t & 1 \end{bmatrix} \to \begin{bmatrix} 1 & 0 \\ 0 & 1 \end{bmatrix} = e.$$

If we assume that $s_i \to 0$ then ${s_i}^{-1} \to \infty$ and chose

$$n = \begin{bmatrix} 1 & t \\ 0 & 1 \end{bmatrix} \in N^+,$$

then we can show that $a_i\ n\ a_i^{-1} \to e$.

REMARK 5.17. In the arguments that follow we will assume that $s_i \to \infty$ as $i \to \infty$ and therefore that $s_i^{-1} \to 0$ as $i \to \infty$. The other case is left to the reader.

LEMMA 5.18. *There exists a neighbourhood U_A of e in A such that $x_0 U_A = x_0$.*

PROOF. Let O be as in Remark 5.16. Let $k_i \to k$. Then $ad(k_i) \to ad(k)$ where $[ad(k)](t) = ktk^{-1}$. Now $ad(k)$ is continuous on T and $ad(k)(e) = e$. Therefore there exists a neighborhood U of e in T such that $[ad(k)](U) \subseteq O$ and $U = U_{SO(2,\mathbf{R})} U_A U_{N^-}$ where $U_{SO(2,\mathbf{R})}$, U_A, and U_{N^-} are open neighborhoods of e in $SO(2,\mathbf{R})$, A, and N^- respectively. Then $kU_Ak^{-1} \subseteq O$. Now since A is abelian,

$$k_i a_i a = k_i a a_i = k_i a {k_i}^{-1} k_i a_i.$$

Hence if $a \in U_A$ and we take $(t_i) = (k_i a_i)$ then

$$(k_i a_i) a (k_i a_i)^{-1} = k_i a {k_i}^{-1} k_i a_i {a_i}^{-1} {k_i}^{-1} = k_i a {k_i}^{-1} \to kak^{-1} \in O.$$

Therefore by Theorem 5.14 $\omega pa = \omega p$.

□

LEMMA 5.19. *There exists a neighborhood U_M of e in M, such that $x_0 l = x_0$ $(l \in U_M)$, whence $x_0 l = x_0$ $(l \in M_0)$.*

PROOF. As in the preceding proposition let $k_i \to k$. Then $ad(k)$ is continuous on T, hence there exists a neighborhood U_M of e in M such that $ad(k)(U_M) \subseteq O$. If $l \in M$,

$$k_i a_i l = k_i l a_i = k_i l {k_i}^{-1} k_i a_i.$$

Now $k_i l {k_i}^{-1} \to klk^{-1}$. Hence if $l \in U_M$ and we take $(t_i) = (ka_i)$ then

$$(k_i a_i) l (k_i a_i)^{-1} = k_i l {k_i}^{-1} k_i a_i {a_i}^{-1} {k_i}^{-1} = k_i l {k_i}^{-1} \to klk^{-1} \in O.$$

Hence $\omega pl = \omega p$.

□

THEOREM 5.20. *Let $i \in N^-$. Then $x_o i = x_0$.*

PROOF. Consider the sequence of mappings $\pi_i : \pi_i(n) = a_i n a_i^{-1}$. As $i \to \infty$, this sequence converges to the constant mapping $\phi : N^- \to N^- : n \to I$, in the compact-open topology. Hence the mapping $\psi = ad(k) \circ \phi$ is continuous. Consequently there exists an open set $U_{N^-} \subseteq O$ containing e, such that $\psi(U_{N^-}) \subseteq O$. Hence setting $(t_n) = (k_n a_n)$ in Theorem 5.14 we have that, $x_0 t = x_0$, $t \in U_{N^-}$. □

REMARK 5.21. Note $x_0 s = x_0$ for all $s \in A \cup N^-$.

Finally we prove that $SO(2, \mathbf{R})$ acts transitively on the Poisson flow. This is immediate from the previous lemmas. Note that

$$x_0 t = x_0 ank = x_0 k.$$

Therefore as any metric image of the Poisson flow is point transitive ([**G**]), we see that

$$\overline{x_0 T} = \overline{x_0 SO(2, \mathbf{R})} = x_0 SO(2, \mathbf{R}).$$

Therefore $SO(2, \mathbf{R})$ acts transitively on X. Hence, by Remark 5.3, $SO(2, \mathbf{R})$ acts transitively on the Poisson flow (B, T).

□

6. Cocycles

Let T be a locally compact group and let (X, T) be a flow on a compact space X.

(i) Let $\sigma : X \times T \to \mathbf{R}$ is a continuous additive cocycle (i.e. if $x \in X$ and $t_1, t_2 \in T$, then σ satisfies the following properties: $\sigma(x, t_1 t_2) = \sigma(x, t_1) + \sigma(x t_1, t_2)$ and $\sigma(x, e) = 0$);

(ii) $\omega \in M(X)$ and $\mu \in M(T)$ satisfy $\omega\mu = \omega$ where $\omega\mu$ is the convolution of μ with ω induced by the flow (X, T);

(iii) $\nu \in M(P)$ and $\nu\mu = \nu$, where P is the universal pointed flow of the group T.

The function $t \to \sigma(x, t) : T \to \mathbf{R}$ is right uniformly continuous $(x \in X)$. Hence it can be extended to a continuous (possibly $+-\infty$ valued) function on the universal poinited flow P of T. Under certain conditions on X and σ the following can be shown: Let $\alpha_\mu \equiv \int_T \int_X \sigma(x, t)\, d\omega(x)\, d\mu(t)$, if $\alpha_\mu > 0$, then the function $p \to \sigma(x, p) : P \to \mathcal{R}$ is equal to $+\infty$ on the support of ν; if $\alpha_\mu < 0$, then $p \to \sigma(x, p) : P \to \mathcal{R}$ is equal to $-\infty$ on the support of ν. The behavior of the function $p \to \sigma(x, p)$ influences the behavior of the limit of nets on T which converge to points in the support of ν.

THEOREM 6.1. *The function $t \to \sigma(x, t) : T \to \mathbf{R}$ is right uniformly continuous $(x \in X)$. Hence it can be extended to a continuous (possibly $+-\infty$ valued) function on the universal poinited flow P of T.*

PROOF. Let $\sigma : X \times T \to \mathbf{R}$ be a continuous additive cocycle such that $\sigma(x, e) = 0$, for all $x \in X$. Then

$$|\sigma(x, t_1 t_2) - \sigma(x, t_1)| \leq |\sigma(x t_1, t_2)|.$$

Since σ is continuous we can find a covering of X by open sets so that $|\sigma(x,t_2)| \leq \epsilon$, for all $x \in X$ if t_2 is in a fixed sufficiently small neighborhood of e. Hence

$$|\sigma(x,t_1t_2) - \sigma(x,t_1)| \to 0$$

as $t_2 \to e$. Hence $\sigma_x(t) = \sigma(x,t)$ is right uniformly continuous on T.

If p and q in βT are such that $f(p) = f(q)$ for every $f \in \mathcal{R}$, but $\sigma_x(p) \neq \sigma_x(q)$. Assume that $\sigma_x(p) < \sigma_x(q)$. Let $r < s$, where $r, s \in \mathbf{R}$ be such that

$$\sigma_x(p) < r < s < \sigma_x(p).$$

Then for every $p' \in \beta T$ define

$$\overline{\sigma_x}(p') = \begin{cases} s, & \text{if } \sigma_x(p') \geq s \\ \sigma_x(p'), & \text{if } r \leq \sigma_x(p') \leq s \\ r, & \text{if } r \geq \sigma_x(p'). \end{cases}$$

Then $\overline{\sigma_x} \in \mathcal{R}$, hence

$$r = \overline{\sigma_x}(p) = \overline{\sigma_x}(q) = s.$$

But this is a contradiction. Therefore $\overline{\sigma_x}$ is defined on P and continuous on P.

□

THEOREM 6.2. *Let (X,T) be the Poisson flow for $\mu \in M(T)$ and let $\alpha_\mu \equiv \int_T \int_X \sigma(x,t)\, d\omega(x)\, d\mu(t)$. If $\alpha_\mu > 0$, then a.e.ω, $\sigma(x,p)$ extends to $+\infty$ for every p in the support of ν; if $\alpha_\mu < 0$, then a.e.ω, $\sigma(x,p)$ extends to $-\infty$ for every p in the support of ν.*

PROOF. The proof follows from the ergodic theorem. □

THEOREM 6.3. *Let $\alpha_\mu > 0$, then for every $x \in X$, $\sigma(x,p) = +\infty$ for all $p \in supp(\nu)$. Let $\alpha_\mu < 0$, then for every $x \in X$, $\sigma(x,p) = -\infty$ for all $p \in supp(\nu)$.*

PROOF. Let $\sigma : X \times T \to \mathbf{R}$ be a continuous additive cocycle. If $\sigma(x,p) = \infty$ a.e. X (with respect to ω) and for all $p \in supp(\nu)$ then we have for every $L \in \mathbf{R}_+$ and a.e. ω there exists an open set $O_{x,L}$ containing x and $t_{x,L} \in T$ such that $t_{x,L}$ is close to p and

$$\sigma(y, t_{x,L}) \geq L \text{ for all } y \in O_{x,L}.$$

Hence by compactness of B there is a finite set $\{O_{x_i,L}\}_{i=1,\dots,n}$, which covers X and such that $\sigma(b', t_{x_i,L}) \geq L$ for every $b' \in O_{x_i,L}$. Now as $\sigma(x,t_\alpha) \to \infty$, a.e. ω, as $t_\alpha \to p \in supp(\nu)$, we may choose L arbitrarily large. Therefore for every $x \in X$, $\sigma(x,t_\alpha) \to \infty$, as $t_\alpha \to p \in supp(\nu)$. □

PROPOSITION 6.4. *Fix $x \in X$. The set of $p \in P$ for which $\sigma(x,p) = \infty$ is an ideal in P. Similar statement holds for the set of points p for which $\sigma(x,p) = -\infty$.*

PROOF. This follows from the cocycle property. □

QUESTION 6.5. And important question is: How does the function σ_x extend to the set $supp(\nu)$ if $\alpha_\mu = 0$? I would conjecture that there exist two sets of positive ν measure, A^+ and A^- contained in P, such that for every $p \in A^+$ we have $\sigma(x,p) = +\infty$ and for every $p \in A^-$ we have $\sigma(x,p) = -\infty$.

REMARK 6.6. The results above can be used to study the limits of nets on the group T.

References

[A] R. Azencott, *Espace de Poisson des groupes localement compact*, Lecture Notes in Math., vol. 148, Springer – Verlag, Berlin, 1970.

[D] D. Dokken, *μ – harmonic functions on locally compact groups*, J. d'Analyse. Math. **52** (1989).

[DE] D. Dokken and R. Ellis, *The Poisson flow associated with a measure*, Pacific J. Math. **141** (1990), no. 1.

[E] R. Ellis, *Lectures on topological dynamics*, W. A. Benjamin, New York, 1969.

[F1] H. Furstenberg, *Boundary theory and stochastic processes in homogeneous spaces*, Harmonic Analysis on Homogeneous Spaces, Symposia in Pure Mathematics, Williamstown, Mass., 1972.

[F2] H. Furstenberg, *A Poisson formula for semi-simple Lie groups*, Ann. of Math. **77**, 335–383.

[F3] H. Furstenberg, *Random walks and discrete subgroups of Lie groups*, Advances in Probability **I**, 2–63.

[G] S. Glasner, *Proximal flows*, Lecture Notes in Math. 517, Springer-Verlag, Berlin.

[Hel] S. Helgason, *Differential Geometry, Lie Groups, and Symmetric Spaces*, Academic Press, New York, San Francisco, London, 1978.

[Hey] H. Heyer, *Probability measures on locally compact groups*, Springer-Verlag, Berlin – Heidelberg – New York, 1976.

University of Saint Thomas, Department of Mathematics, Saint Paul, MN 55105
E-mail address: dpdokken@stthomas.edu

Contemporary Mathematics
Volume **215**, 1998

Another Proof of Moore's Ergodicity Theorem for $SL(2,\mathbb{R})$

S. Adams

We present here a proof due to R. Ellis and M. Nerurkar ([**EN1**], [**EN2**]) of the decay matrix coefficients for $SL(2,\mathbb{R})$. It is hoped the argument given in this case will illuminate the proof given by Ellis and Nerurkar to those not familiar with the Ellis-Nerurkar enveloping semigroup.

We have incorporated several suggestions of J. Feldman and D. Witte.

We do not specifically define the Ellis-Nerurkar enveloping semigroup ([**EN1**], [**EN2**]), but the idea is implicit in Claim 4 below.

THEOREM 1. *(The Decay of Matrix Coefficient) Let $\mathcal{H}$ be a separable Hilbert space and let $\pi : SL(2,\mathbb{R}) \to U(\mathcal{H})$ be a unitary representation with no non-zero fixed vectors in $\mathcal{H}$. Then $\langle \pi(g)v, w\rangle \to 0$ as $g \to \infty$, for all v, $w \in \mathcal{H}$.*

Let $\hat{G} := SL(2,\mathbb{R})$. Let $G := \pi(\hat{G})$. Let $\hat{e}$ denote the identity of $\hat{G}$ and let $e = \pi(\hat{e})$ be the identity operator on $\mathcal{H}$.

We may assume that $\mathcal{H} \neq \{0\}$; the result is otherwise obvious. Since $\hat{G}/\{\pm\hat{e}\}$ is a simple group, the kernel of π must be one of: $\{\hat{e}\}$, $\{\pm\hat{e}\}$, or $\hat{G}$. The case $ker(\pi) = \hat{G}$ is impossible, as $\mathcal{H}\backslash\{0\}$ does not admit fixed vectors; consequently, $ker(\pi) \subseteq \{\pm\hat{e}\}$.

Throughout we think of the algebra $B(\mathcal{H})$ of bounded operators on $\mathcal{H}$ as having the weak operator topology. So, for example, $\bar{G}$ denotes the weak operator closure of G in $B(\mathcal{H})$.

Let $Ball(\mathcal{H})$ denote the closed unit ball of $B(\mathcal{H})$. Then $Ball(\mathcal{H})$ is (weak operator) compact and metrizable. Since $G \subseteq U(\mathcal{H}) \subseteq Ball(\mathcal{H})$ and since $Ball(\mathcal{H})$ is closed, we see that $\bar{G} \subseteq Ball(\mathcal{H})$, so that $\bar{G}$ is metrizable. (Consequently, any element of $\bar{G}$ is a limit of a sequence of elements of G.) Further, $G \subseteq U(\mathcal{H})$, so $g^* = g^{-1}$, for all $g \in G$.

The adjoint map $R \mapsto R^*: \mathcal{B}(\mathcal{H}) \to \mathcal{B}(\mathcal{H})$ is (weak operator) continuous. On the other hand, multiplication $(R,S) \mapsto RS: B(\mathcal{H}) \times B(\mathcal{H}) \to B(\mathcal{H})$ is not jointly continuous, although it is continuous in each variable separately. Our replacement for joint continuity of multiplication is Lemma 3, below.

Let $\hat{K}, \hat{A}, \hat{U}$ *and* $\hat{V}$ be, respectively, the following subroups of $SL(2,\mathbb{R})$: the special orthogonal matrices, the diagonals of determinant one with positive entries, the upper triangular unipotents and the lower triangular unipotents. Let $K := \pi(\hat{K})$, $A := \pi(\hat{A})$, $U := \pi(\hat{U})$, $V := \pi(\hat{V})$.

1991 *Mathematics Subject Classification.* Primary 54H20, 28D15; Secondary 22E46.

It is a consequence of the finite-dimensional spectral theorem that $\hat{K}\hat{A}\hat{K} = \hat{G}$. This is called the *Cartan decomposition.* Elementary row and column operations will reduce any element of $\hat{G}$ to the identity matrix; equivalently, $\hat{U}$ and $\hat{V}$ generate $\hat{G}$. This implies that U and V generate G.

We proceed with several lemmas.

LEMMA 1. *Let S_i be a sequence of unitary operators. Assume that $S_i \to S$. If S is unitary, then $S_i v \to Sv$, for all $v \in \mathcal{H}$.*

PROOF. Fix $v \in \mathcal{H}$. We wish to show that $\|S_i v - Sv\|^2 \to 0$. Expanding $\langle S_i v - Sv, S_i v - Sv\rangle$ gives

$$\|S_i v - Sv\|^2 = \|S_i v\|^2 + \|Sv\|^2 - 2\ Re\langle S_i v, Sv\rangle.$$

Since S_i and S are unitary, $\|S_i v\| = \|Sv\| = \|v\|$. By definition of weak operator convergence, $\langle S_i v, Sv\rangle \to \langle Sv, Sv\rangle = \|Sv\|^2 = \|v\|^2$. Therefore,

$$\|S_i v - Sv\|^2 \to \|v\|^2 + \|v\|^2 - 2\ Re\|v\|^2 = 0.$$

□

LEMMA 2. *Let R_i be a sequence of unitary operators. Let S_i be a sequence of bounded operators such that $S_i v \to 0$, for all $v \in \mathcal{H}$. Then $R_i S_i \to 0$.*

PROOF. Fix $v, w \in \mathcal{H}$. We wish to show that $\langle R_i S_i v, w\rangle \to 0$. By hypothesis, $S_i v \to 0$. By Cauchy-Schwartz,

$$|\langle R_i S_i v, w\rangle| \leq \|R_i S_i v\| \cdot \|w\|.$$

Since the R_i 's are unitary, $\|R_i S_i v\| = \|S_i v\| \to 0$. The result follows. □

LEMMA 3. *Let R_i and S_i be two sequences of unitary operators. Assume that $R_i \to R$ and $S_i \to S$. If either R or S is unitary, then $R_i S_i \to RS$.*

PROOF. We will assume that S is unitary. The case where R is unitary follows from this case by taking adjoints.

By Lemma 1, $(S_i - S)v \to 0$, for all $v \in \mathcal{H}$. Then, by Lemma 2, $R_i(S_i - S) \to 0$. By continuity of multiplication in the first variable, $R_i S \to RS$. Adding these last two limits gives the desired $R_i S_i \to RS$. □

CLAIM 1. The set $\bar{A}$ is closed under multiplication, i.e., $\bar{A}\bar{A} \subseteq \bar{A}$.

PROOF. We begin by showing the simpler $A\bar{A} \subseteq \bar{A}$. So let $a \in A$. Since the map $u \mapsto au: B(\mathcal{H}) \to B(\mathcal{H})$ is (weak operator) continuous, it follows that the preimage of $\bar{A}$ under this map is closed. This preimage clearly contains A, as $aA \subseteq A \subseteq \bar{A}$. Combinining these two statements, we see that the preimage includes $\bar{A}$. In other words, $a\bar{A} \subseteq \bar{A}$, for all $a \in A$. That is, $A\bar{A} \subseteq \bar{A}$.

Now we show that $\bar{A}\bar{A} \subseteq \bar{A}$. Let $a \in \bar{A}$. By the preceding paragraph, $Aa \subseteq \bar{A}$, so A is contained in the preimage of $\bar{A}$ under the (weak operator) continuous map $b \mapsto ba: B(\mathcal{H}) \to B(\mathcal{H})$. Consequently $\bar{A}$ is contained in this preimage, i.e., $\bar{A}a \subseteq \bar{A}$, for all $a \in \bar{A}$. That is, $\bar{A}\bar{A} \subseteq \bar{A}$, as desired. □

Let the upper left and lower right entries of $\hat{a} \in \hat{A}$ be denoted respectively by $UL(\hat{a})$ and $LR(\hat{a})$. Note that the determinant one condition implies that $UL(\hat{a}) \cdot LR(\hat{a}) = 1$, for all $\hat{a} \in \hat{A}$.

CLAIM 2. If $\hat{a}_1, \hat{a}_2, \ldots$ is a sequence in $\hat{A}$ and $UL(\hat{a}_i) \to 0$, then $\hat{a}_i \hat{U} \hat{a}_i^{-1} \to \hat{e}$ and $\hat{a}_i^{-1} \hat{V} \hat{a}_i \to \hat{e}$, for all $\hat{u} \in \hat{U}, \hat{v} \in \hat{V}$.

PROOF. We will only prove the first limit, the other being similar. Let $s_i := UL(\hat{a}_i)$, so that $1/s_i = LR(\hat{a}_i)$. Let $\hat{u} \in \hat{U}$ be arbitrary. Then, for some $t \in \mathbb{R}$,

$$\hat{u} = \begin{bmatrix} 1 & t \\ 0 & 1 \end{bmatrix}.$$

We are assuming that $s_i = UL(\hat{a}_i) \to 0$, so

$$\hat{a}_i \hat{u} \hat{a}_i^{-1} = \begin{bmatrix} s_i & 0 \\ 0 & s_i^{-1} \end{bmatrix} \begin{bmatrix} 1 & t \\ 0 & 1 \end{bmatrix} \begin{bmatrix} s_i & 0 \\ 0 & s_i^{-1} \end{bmatrix}^{-1} = \begin{bmatrix} 1 & s_i^2 t \\ 0 & 1 \end{bmatrix} \to \begin{bmatrix} 1 & 0 \\ 0 & 1 \end{bmatrix} = \hat{e}.$$

□

CLAIM 3. If $\hat{a}_1, \hat{a}_2, \ldots$ is a sequence in $\hat{A}$ and $LR(\hat{a}_i) \to 0$, then $\hat{a}_i^{-1} \hat{u} \hat{a}_i \to \hat{e}$ and $\hat{a}_i \hat{v} \hat{a}_i^{-1} \to \hat{e}$, for all $\hat{u} \in \hat{U}, \hat{v} \in \hat{V}$.

PROOF. This can be proved similarly; it also follows from Claim 2, by taking transposes of the two limit formulas. □

In what follows, the statement "$UyV = y$" means "$uyv = y$, for all $u \in U, v \in V$", while "$Vz = z$" means "$vz = z$, for all $v \in V$". Similar statements have similarly obvious meanings.

CLAIM 4. If $\hat{a}_i \to \infty$ in $\hat{A}$ and $a_i := \pi(\hat{a}_i) \to y$ in $\bar{A}$, then either $UyV = y$ or $VyU = y$.

PROOF. First, $UL(\hat{a}_1), UL(\hat{a}_2), \ldots$ is a sequence of real numbers with no *nonzero* limit points. (Otherwise, passing to that subsequence, we could assume that $UL(\hat{a}_i) \to s$, for some $s \in \mathbb{R} \backslash 0$, whereupon $LR(\hat{a}_i) \to 1/s$. Then $\hat{a}_i$ would converge to the diagonal matrix with entries $s, 1/s$, contradicting $\hat{a}_i \to \infty$.) Hence we may pass to a subsequence and assume either that $UL(\hat{a}_i) \to 0$ or that $LR(\hat{a}_i) = 1/UL(\hat{a}_i) \to 0$.

Case 1: $UL(\hat{a}_i) \to 0$. In this case, we will show that $VyU = y$. Let $u \in U$ and $v \in V$. We wish to show that $yu = y$ and $vy = y$. Choose $\hat{u} \in \hat{U}$ and $\hat{v} \in \hat{V}$ so that $\pi(\hat{u}) = u$ and $\pi(\hat{v}) = v$. By Claim 2, $\hat{a}_i \hat{u} \hat{a}_i^{-1} \to \hat{e}$ and $\hat{a}_i^{-1} \hat{v} \hat{a}_i \to \hat{e}$. By continuity of multiplication in the first variable and by Lemma 3,

$$yu = \lim a_i u = \lim (a_i u a_i^{-1}) a_i = ey = y.$$

Similarly,

$$vy = \lim v a_i = \lim a_i (a_i^{-1} v a_i) = ye = y.$$

Case 2: $LR(\hat{a}_i) \to 0$. In an analogous way, using Claim 3, we show that $uy = y$ and $yv = y$, for all $u \in U$, $v \in V$. It then follows that $UyV = y$ in this case. □

CLAIM 5. If $z \in \bar{A} \backslash A$ and $z = z^*$, then $z = 0$.

PROOF. Choose a sequence $\hat{a}_i \in \hat{A}$ such that $a_i := \pi(\hat{a}_i) \to z$. Then $\hat{a}_i \to \infty$ in $\hat{A}$; otherwise, passing to a subsequence, we would have $\hat{a}_i \to \hat{a}$ in $\hat{A}$, so

$$z = \lim \pi(\hat{a}_i) = \pi(\lim \hat{a}_i) = \pi(\hat{a}) \in \pi(\hat{A}) = A.$$

By Claim 4, we conclude either that $UzV = z$ or that $VzU = z$. However, since $z = z^*$, since $U^* = U^{-1} = U$ and since $V^* = V^{-1} = V$, it follows that:

$$UzV = z \iff (UzV)^* = z^* \iff V^*z^*U^* = z^* \iff VzU = z.$$

Thus $UzV = z$ *and* $VzU = z$. These, respectively, imply that $Uz = z$ and $Vz = z$. Since G is generated by U and V, we conclude that $Gz = z$. Then, for all $v \in \mathcal{H}$, $[\pi(\hat{G})][zv] = G(zv) = (Gz)v = zv$, so zv is a π-invariant vector. One of the hypotheses of the theorem is that π has no non-zero invariant vectors, hence $zv = 0$, for all $v \in \mathcal{H}$. That is, $z = 0$. □

CLAIM 6. If $\hat{a}_i \to \infty$ in $\hat{A}$ and if $\pi(\hat{a}_i) \to y$ in $\bar{A}$, then $y = 0$.

PROOF. By Claim 4, either $UyV = y$ or $VyU = y$. Therefore, either $yV = y$ or $yU = y$.

As $\ker(\pi) \subseteq \{\pm\hat{e}\}$, we see that $\hat{U} \not\subseteq \ker(\pi)$ and $\hat{V} \not\subseteq \ker(\pi)$. That is, $U \neq \{e\}$ and $V \neq \{e\}$.

It then follows that y has no left inverse in $B(\mathcal{H})$: if there were $y' \in B(\mathcal{H})$ such that $y'y = e$, then either $V = (y'y)V = y'(yV) = y'y = e$ or $U = (y'y)U = y'(yU) = y'y = e$.

It now follows that $z := y^*y$ has no left inverse in $B(\mathcal{H})$: if there were $z' \in B(\mathcal{H})$ such that $z'z = e$, then $(z'y^*)y = z'(y^*y) = z'z = e$, so $z'y^*$ would be a left inverse for y.

By the weak operator continuity of adjoint and of π, we see that

$$y^* = \lim \pi(\hat{a}_i)^* = \lim \pi(\hat{a}_i)^{-1} = \lim \pi(\hat{a}_i^{-1}) \in \bar{A}.$$

By Claim 1, $z = y^*y \in \bar{A}\bar{A} \subseteq \bar{A}$. On the other hand, A is a group, so every element of A has a two-sided inverse in A. We therefore conclude that $z \notin A$.

By Claim 5, $y^*y = z = 0$. Then, for all $v \in \mathcal{H}$,

$$\langle yv, yv \rangle = \langle y^*yv, v \rangle = 0,$$

so $yv = 0$, for all $v \in \mathcal{H}$. That is, $y = 0$. □

PROOF OF THEOREM 1. The following claim is equivalent to the conclusion of the theorem.

CLAIM 7. If $\hat{g}_1, \hat{g}_2, \ldots \to \infty$ in $\hat{G}$, then $\pi(\hat{g}_1), \pi(\hat{g}_2), \ldots \to 0$ in $Ball(\mathcal{H})$.

PROOF. Assume for a contradiction that $g_i := \pi(\hat{g}_i)$ does not tend toward 0. Passing to a subsequence, we may assume that all the g_i live outside some fixed (weak operator) open neighborhood of 0.

By the Cartan decomposition, we may choose $\hat{k}_i, \hat{k}'_i \in \hat{K}$ and $\hat{a}_i \in \hat{A}$ such that $\hat{g}_i = \hat{k}_i\hat{a}_i\hat{k}'_i$, for all i. Passing to a subsequence, we may assume that $\hat{k}_i \to \hat{k}$ and $\hat{k}'_i \to \hat{k}'$ in the compact set $\hat{K}$.

Let $k_i := \pi(\hat{k}_i)$, $a_i := \pi(\hat{a}_i)$, $k'_i := \pi(\hat{k}'_i)$, for all i. Let $k := \pi(\hat{k})$, $k' := \pi(\hat{k}')$. Since $\bar{A}$ is compact, we may pass to a subsequence and assume that $a_i \to y$ for some $y \in \bar{A}$.

Since $\hat{g}_i \to \infty$ and since $\hat{K}$ is compact, it follows that $\hat{a}_i = \hat{k}_i^{-1}\hat{g}_i(\hat{k}_i')^{-1} \to \infty$ in $\hat{A}$. Then, by Claim 6, $y = \lim a_i = 0$. Applying Lemma 3, we have $k_i a_i \to ky = 0$. Another application of Lemma 3 gives $g_i = (k_i a_i)k_i' \to 0k' = 0$. On the other hand, every g_i lives outside a fixed neighborhood of 0, a contradiction.

This concludes the proof of Theorem 1. □

References

[EN1] R. Ellis and M. Nerurkar, *An enveloping semigroup in Ergodic theory*, Lecture notes in Mathematics **1342** (1988), Springer Verlag.

[EN2] R. Ellis and M. Nerurkar, *Weakly almost periodic flows*, Transactions of the American Mathematical Society **313** (1989), no. 1.

School of Mathematics, University of Minnesota, Minneapolis, MN 55455
E-mail address: adams@math.umn.edu

Contemporary Mathematics
Volume **215**, 1998

Multiple Recurrence and Doubly Minimal Systems

B. Weiss

ABSTRACT. A system (X, T) is *doubly minimal* if the orbit closure of any pair (x, y) under $T \times T$ is either all of $X \times X$ or the graph of a power of T. It is shown that certain variants of this imply zero entropy and that any zero entropy measure theoretic system has a uniquely ergodic model which is doubly minimal.

1. Introduction

Minimal systems have been a lifelong object of study for Bob Ellis who has made fundamental contributions to our understanding of them. I shall explore in this note, dedicated to him, some byroads that are related to this central concept. The starting point is the notion of a *weakly rigid system* introduced by E. Glasner and D. Maon [GM]. Recall that a *rigid system* (X, T) is one for which there is a sequence $n_i \to \infty$ such that for all $x \in X$,

$$\lim_{i \to \infty} T^{n_i} x = x.$$

This implies in particular, that any point in $X \times X \times \cdots \times X$ a finite product of X is recurrent under $T \times T \times \cdots \times T$. A system is said to be weakly rigid if this consequence holds, to repeat:

DEFINITION 1. A system (X, T) is *weakly rigid* if for all k, every point of $X^{(k)}$ is recurrent under $T^{(k)}$ where X is the k-fold product of X and $T^{(k)}$ represents the diagonal action of $\mathbb{Z}$ on $X^{(k)}$, i.e.

$$T^{(k)}(x_1, x_2, \ldots, x_k) = (Tx_1, Tx_2, \ldots, Tx_k).$$

It is easy to see that rigid systems must have zero topological entropy. Glasner and Maon asked does weak rigidity suffice for zero entropy. It appears to be quite likely that even the assumption that every point of $X \times X$ is recurrent under $T \times T$ (two-fold recurrence) would suffice to imply zero entropy but this remains open.

What is true is the following:

THEOREM 2. *If every point of $X \times X$ is positively recurrent under $T \times T$ then $h_{\text{top}}(X, T) = 0$.*

By positive recurrence of a point $x \in X$ we mean that for some $n_i \to +\infty$ $\lim_{i \to \infty} T^{n_i} x = x$. This theorem has as a corollary:

1991 *Mathematics Subject Classifications*: Primary 28D05; Secondary 54H20.

COROLLARY 3. *If (X,T) is four-fold recurrent, and in particular if it is weakly rigid, then $h_{\text{top}}(X,T)=0$.*

Two-fold recurrence of (X,T) implies that either every point of X is positively recurrent or every point of X is negatively recurrent. Indeed, if x_1 is not positively recurrent and x_2 is not negatively recurrent then (x_1,x_2) is not recurrent. If (X,T) is four-fold recurrent the previous analysis applies to $(X\times X, T\times T)$. Thus either every point of $X\times X$ is positively recurrent under $T\times T$ or under $T^{-1}\times T^{-1}$ and in either case the corollary follows. Unfortunately I do not know how to prove Theorem 2 without using ergodic theoretic ideas which do appear to be out of place here.

Having established Theorem 2 let us turn to the question of how prevalent are such systems. This turns out to be difficult since infinite symbolic systems cannot be doubly positively recurrent. However, say that a system (X,T) is doubly minimal if every (x_1,x_2) such that x_2 is not on the orbit of x_1 has a dense orbit under $T\times T$. Formally:

DEFINITION 4. A system (X,T) is *doubly minimal* if for all $x\in X$, $y\notin \{T^n x\}_{-\infty}^{\infty}$, $\{(T^j x, T^j y)\}$ is dense in $X\times X$.

These systems were called topologically minimal self-joining by A. del Junco [dJ]. But I think that doubly minimal is an apter name. The first example of such a system was given by J. King [K] who also proved that systems with this type of property for *four-fold* products cannot exist. One can show by topological methods that doubly minimal systems have zero topological entropy. These systems, which are of course doubly recurrent, are very prevalent as the following theorem shows:

THEOREM 5. *Any ergodic system with zero entropy $(Y,\mathcal{C},\nu,S)$ has a uniquely ergodic model (X,T) that is doubly minimal.*

In conclusion I would like to thank Eli Glasner for several very useful discussions of the material presented here.

2. Positively Recurrent Systems

Before proving Theorem 2, I would like to point out an analogy with *distality.* To say that a system is a union of minimal sets says very little about finer properties of the system since in particular every minimal system has this property in a trivial way. However, R. Ellis showed long ago that to say that a product $X\times X$ is a union of minimal sets under $T\times T$ is one of the characterizations of *distality* of (X,T). This implies severe restrictions on the dynamical behavior of (X,T) and in particular zero entropy. In an analogous fashion to say that every point of X positively recurs under T says little about (X,T) in the sense that all minimal systems have this property. For some results on non minimal systems with this property see [KW]. However, Theorem 2 asserts that the same property for $X\times X$ puts a severe entropy restriction on the original system (X,T).

What we shall actually establish is that if (X,T) has an invariant measure μ such that the metric entropy of (X,T,μ) is positive then there are pairs of points (x_1,x_2) in $X\times X$ that do not recur positively. Thus if (X,T) is doubly positively recurrent then for all invariant measures μ we have

$$h_\mu(X,T)=0$$

and then the variational principle implies that the topological entropy of (X, T) is zero.

Let us fix a metric d on X. Observe that for any fixed $x_0 \in X$ there can be only countably many values of r such that

$$\mu(\partial(B(x_0, r))) > 0,$$

where $B(x_0, r)$ is the d-ball of radius r centered at x_0. It follows easily from this observation that for any $\varepsilon > 0$ we can find a finite covering of X by closed sets $\{P_i\}_{i=1}^I$ satisfying:

(1) $\mathrm{diam}(P_i) < \varepsilon$, $1 \leq i \leq I$,
(2) for $i \neq j$, $\mathrm{int}P_i \cap \mathrm{int}P_j = \phi$,
(3) $\mu(\partial P_i) = 0$ for all $1 \leq i \leq I$.

Clearly as $\varepsilon \to 0$ such partitions generate the Borel σ-algebra and thus the positivity of the entropy implies that for some small $\varepsilon > 0$, the partition P described above will have positive entropy under T. The form we shall take for this is:

$$H(P | \bigvee_1^\infty T^{-n} P) > 0.$$

Now this clearly means that there are fibers of the σ-algebra $\bigvee_1^\infty T^{-n}P$ which intersect the interiors of at least two atoms of P since in fact that must happen for a set of such fibers with positive measure. Let x_1, x_2 be points that belong to such a fiber ξ with $x_1 \in \mathrm{int}P_i$, $x_2 \in \mathrm{int}P_j$, $i \neq j$. I claim that there is no sequence of positive integers n_k such that simultaneously $T^{n_k}x_1 \to x_1$ and $T^{n_k}x_2 \to x_2$. This is because $x_1, x_2 \in \xi$ implies that for all $n \geq 1$, $T^n x_1$ and $T^n x_2$ belong to the same atom of P. This completes the proof of Theorem 2. □

Doubly minimal systems are not covered by this theorem, however it is not hard to show that they must have zero topological entropy. One way to see this is through the notion of p-points of J. Auslander and K. Berg, see their paper [AB].

3. Doubly Minimal Systems

In this section we shall prove the main result of this note namely that any zero entropy system has a uniquely ergodic model that is *doubly minimal.* By a well known strengthening of the Jewett-Krieger theorem we may assume at the outset that we have a zero entropy uniquely ergodic subshift $Y \subset \{a, b\}^{\mathbb{Z}}$ with unique invariant measure μ. We will denote by $\{P_a, P_b\} = P$ the partition of Y defined by the value of 0-th coordinate. We will construct a sequence of generating partitions $\{P_n\}_{n \geq 0}$ with $P_0 = P$, and for $n \geq 1$ the sets of P_n will be labeled by $\{a, b\} \times \mathbb{N}$, and our model will be given by a limit of these partitions.

We need some preliminary results about towers. Since Y is minimal, any clopen set C in Y is the base of a bounded *Kakutani-Rohlin tower* (KR tower). By this we mean the partition of the space Y determined as follows:

Let the return time to C, r_c be defined for $y \in C$ by

$$r_C(y) = \min\{i \geq 1 : T^i y \in C\},$$

where we denote the shift by T. The minimality of T implies that r_C is bounded and thus the partition of C into sets of constancy of r_c is finite and thus clopen, denote by

$$C_i = \{y \in C : r_c(y) = i\}.$$

Then the columns of the KR tower over C are the sets

$$\hat{C}_i = \bigcup_{0 \le j < i} T^j C_i$$

and $Y = \bigcup_i \hat{C}_i$, the union being finite. We shall need a simple lemma concerning such sets:

LEMMA 1. *Given a clopen set C with $\max r_C(y) = L$, for all M sufficiently large we can find a clopen set $D \subset C$ such that*

$$-L + M \le r_D(y) \le M + L + 1,$$

for all $y \in D$.

PROOF. Since Y is infinite there is a clopen set $E \subset C$ such that $\min r_E(y) \ge M^2$. Purify the columns of the KR tower over E by C. That means that we partition E into sets E_u so that r_E is constant that on E_u, say with the value i, and for all $0 \le j < i$, if

$$T^j E_u \cap C \ne \emptyset$$

then $T^j E_u \subset C$. That is to say each level is either entirely in C or entirely in its complement. Now for each E_u, mark off levels separated by M or $M + 1$ so that the whole column is divided into blocks with length either M or $M + 1$. This is possible since the height is greater than M^2. If the level we marked happens to lie in C, leave it in place - otherwise move the mark up to the first level belonging to C that lies above it. By our assumption no marker moves more than L steps and so the new blocks have lengths in the range $(-L + M, M + L + 1)$. These marked levels are put together to form the desired set D. □

Next we describe the combinatorial fact that underlies the construction. Think of points of $\{0,1\}^N = \Omega_N$ as subsets of $N = \{0, 1, \dots, N-1\}$, and call a set $A_N \subset \Omega_N$ *approximately random* if:

(1) for any $\alpha, \beta \in A_N$, $\alpha \ne \beta$ and any $L \le \frac{9}{10}N$,

$$\alpha \cap (L + \beta) \ne \emptyset;$$

(2) for any $\alpha \in A_N$, and any $\frac{1}{10}N \le L \le \frac{9}{10}N$,

$$\alpha \cap (L + \alpha) \ne \emptyset.$$

LEMMA 2. *Given M, there is some $e = e(M)$, such that for all N sufficiently large, there is a set $A_N \subset \Omega_N$ satisfying*

(1) *A_N is approximately random,*
(2) *for each $\alpha \in A_N$, the gaps between successive members of α are at least M,*
(3) $|A_N| \ge 2^{eN}$.

We postpone the proof of this lemma to Section 4 and go on to show how it enters into the main construction. To begin we can take for C_1 any proper clopen subset of Y, and define P_1 by refining P_a and P_b by C_1 and $Y \setminus C_1$ labeling $Y \setminus C_1$ by 0 and C_1 by 1, explicitly if $C_0 = Y \setminus C_1$ then

$$P_{(a,i)} = P_a \cap C_i, \quad P_{(b,i)} = P_b \cap C_i, \quad i = 0, 1,$$

are the atoms of P_1. Next we look at the KR tower over C_1 and purify the columns with respect to P_0. A slight technical problem arises if the heights of these pure C_1 columns are not relatively prime. We will assume for the moment that this is not the case - i.e. at each stage we will assume that the column heights are relatively prime. Later on in Section 4 we will explain what to do if this is not the case.

Form now an abstract word w_1 built from the P_1-names of the pure columns for the tower over C_1 that contains all pairs of the form uv where u, v are such P_1-names. Label the first element of w_1 with a 3. More explicitly, a typical u looks like $(a1, b0, b0, a0, a0, b0)$ which corresponds to points y in C_1 that return for the first time to C_1 after 6 iterations of T, and such that

$$y \in P_a,\ Ty \in P_b,\ T^2y \in P_b,\ T^3y \in P_a,\ T^4y \in P_a,\ T^5y \in P_b.$$

Take some M_1 that is very large compared with the length of w_1 and apply Lemma 2 to find some constant e_1, as guaranteed by the lemma. Now since Y has zero topological entropy, we can find an N_1 large enough for both Lemma 2 to apply and so that the total number of P_1-words in Y of length N_1 is less than $2^{\frac{1}{2}e_1 N_1}$.

Apply now Lemma 1 to find a subset $C_2 \subset C_1$ with return times in the range $[N_1, N_1 + 2 \cdot \max r_{c_1}]$. We are now in a position to describe the basic step in the construction - the formation of P_2. Purify the KR tower over C_2 by the partition P_1. For all points in the base C_2 change the 1 to 2 in the labeling. Assign to each pure column a distinct element from the approximately random set A_{N_1} given to us by Lemma 2. Now the P_1-name of each columns will be modified by relabeling levels so that w_1 will occur at the places indicated by the element of A_{N_1} assigned to the column. Since we are assuming that column lengths of the tower over C_1 are relatively prime this can be done in such a way that the name along the C_2-column consists only of the P_1-names encountered in the tower over C_1. This requires a change in a region around each location which has size at most

$$|w_1| + 2 \cdot (\max r_{c_1})^2,$$

which is small compared to M_1 - thus we have made a small change in P_1.

The extra label 3 that we added to the first element of w_1 guarantees that even though we have changed the P_0-names, our new partition P_2 is still a generator, since the location of these w_1's is given by an assignment which was a bijection between the pure C_1-columns and the approximately random strings.

Our new partition P_2 is still a partition into clopen sets and hence is still uniform. It follows easily from the approximate randomness of A_{N_1}, that if we form any two infinite sequences x, x' by concatenating P_2-names of columns of the tower over C_2 then unless one is a shift of the other, any block of the form (u, u'), where u and u' are blocks of length less than $\max r_{c_1}$ occurring in P_1-names of points in Y, will appear somewhere in (x, x'). This property, which will be maintained during the later stages of the construction is what will lead to the double-minimality of the system that will be constructed as a limit of the systems defined by these partitions P_i. Since P_2 is uniform there is some $\hat{N}_2$ which will exhibit this uniformity for all blocks of lengths up to $2 \cdot \max r_{c_2}$ to within say 10^{-2}. We will make further changes in such a way that the regions of change will occur at a frequency much lower than $1/\hat{N}_2$. This will mean that the uniformity will be retained at the next level. Changes in P_2 will be made only via names of blocks that occur in the tower over C_2, so that the "level one" uniformities are indeed going to be main maintained.

We are now in a position to apply Lemma 2 again with M large compared to $\hat{N}_2$ (see below), to find an e such that there are at least 2^{eN} approximately random elements with gaps $\geq \hat{N}_2^2$. Since the entropy is zero we can choose N_2 so that the number of P_2 names of length N_2 is less than $2^{\frac{e}{2}N_2}$, and apply Lemma 1 to find a set $C_3 \subset C_2$ that satisfies the requirements of Lemma 1. Just as before we will purify the tower over C_3 by P_2 and label the base by the new symbol 4 (overwriting the 2 that was there from C_2). Letting w_2 denote a word that contains all pairs of P_2-names of columns over C_2 with a 5 added at the beginning to mark it, we now modify each pure column over C_3 by inserting w_2 at locations specified by the ones in a unique element of A_{N_2}-the approximately random set given by Lemma 2 - that we assign to the column. Once again, for simplicity, we assume that the column heights of the KR tower over C_2 are relatively prime so that these insertions can be made at the specified locations by changing the P_2-name locally and preserving the property that the new name of the column is built up out of entire C_2 column names. The new partition that is defined by this relabeling of some levels is called P_3. It differs from P_2 in a set of very small measure depending on M.

The key property is that any two infinite concatenations of these P_3-names of C_3-columns will exhibit all pairs of names of length N_2-unless one is a shift of the other. This is an immediate consequence of the approximate randomness of the set A_{N_2} which dictated where the w_2 was inserted. It should be clear by now how to continue this inductive procedure. The limiting partition P_∞ is no longer a partition into clopen sets - and in fact it is only on a set of full measure that P_∞ is well defined - since we are making infinitely many changes. The Borel-Cantelli lemma guarantees that indeed infinitely many changes *at a given point* occur only on null set. However, the uniformity of the partition is guaranteed since it gets better and better on the P_n-column names of the towers over C_{n+1} and these are preserved in the limit. Thus the P_∞-partition defines a uniquely ergodic system which by our previous discussion is both a generator and is manifestly doubly minimal. This completes the proof of Theorem 5 - except for the proof of Lemma 2 which we will give in the next section together with a discussion of what to do if column heights are not relatively prime. □

4. Technical Complements

A. PROOF OF LEMMA 2. Fix M, and define a process by concatenating independently, with equal probabilities, blocks 10^M, 10^{M+1}. This can be thought of as a markov chain x_i with states $\{0, 1, \ldots, M, M+1\}$ and transitions given by:

(i) for $0 \leq i \leq M-1$ move deterministically from i to $i+1$,
(ii) for $i = M$ move to 0 or $M+1$ with probability $1/2$,
(iii) for $i = M+1$ move deterministically to 0,

where we collapse states as follows: $0 \to 1$; $\{1, \ldots, M+1\} \to 0$. Let α be a fixed element in $\{0,1\}^n$ with ones separated by either M or $M+1$ zeros as in the above process. For $n \geq M^4$ I claim that P_α, the probability that the process x_i avoids the ones in α up to time n is uniformly bounded away from 1. More precisely, let

$$P_\alpha = \max_{\{\xi_i:\ i<0\}} \text{Prob}\{\alpha(j)x_j = 0, \text{ for all } 0 \leq j < n | \ x_i = \xi_i, \ i < 0\}.$$

For the chain $\{x_j\}$, conditioned on the values of x_i for $i < 0$, x_n is spread out like the sum of $n/(M + 1/2)$ independent random variables uniformly distributed on $\{0, 1\}$. If $n \geq M^4$ the central limit theorem easily gives that P_α is at most $\left(1 - \frac{1}{2M}\right)$ for any α as above.

Consider now N very large compared to M^4. We will carry out the construction of A_N in a random fashion using the measure on Ω_N given by the process that we have just described. For this type of construction the main thing to calculate is the size of the set of elements of Ω_N that are eliminated from consideration by the addition of a new element $\hat{\alpha}$ to the set A_N that we are building up. For this calculation, divide N into blocks of size M^4. The estimate of the previous paragraph gives that for a fixed overlap of size $\geq \frac{1}{10}N$, the set of elements of Ω_N that won't intersect $\hat{\alpha}$ is at most

$$(*) \qquad \left(1 - \frac{1}{2M}\right)^{N/10M^4}.$$

To begin with, we can use this estimate to eliminate from Ω_N all α's that don't satisfy condition (2) of the definition of approximately random sets. This set is exponentially small in N, by $(*)$. Having done this we choose in succession elements that will also satisfy condition (1). Each choice eliminates a set whose measure is at most N times the expression in (*) and we can therefore continue the process to obtain a set A_N that is approximately random and has cardinality at least 2^{eN} where e is a constant that depends only on M and thus we have established Lemma 2. □

B. In the proof of our main theorem we assumed at one point that the column heights of our KR tower are relatively prime. If there are roots of unity in the spectrum of the initial transformation this assumption may not hold. In this case one way to proceed is to give up on our strict requirement that the $(k+1)$-blocks shall be built only of previous k-blocks and allow one new block which is obtained from one of the odd k-blocks by adding a single symbol. Using this new block allows one to achieve the precise alignment that we need to get the double minimality from the approximately random sequences. If this new block is used sparingly then it won't interfere with the uniformity that was obtained from the odd k-blocks.

One point to observe is that when one does the same thing at the next stage one should take care that the new symbols don't accumulate. This can be easily done since only a single new block is created at each stage.

A different way to proceed is to begin with a topologically mixing model that is uniquely ergodic. Such models were constructed by E. Lehrer [L] for arbitrary measure preserving systems. In his general construction no attempt was made to obtain a symbolic representation with only two symbols for zero entropy systems. Such a modification of his construction is possible and thus it is possible to guarantee an initial model in which the simplifying assumption that we made in Section 3 is actually satisfied.

References

[AB] J. Auslander and K. Berg, *A condition for zero entropy*, Israel J. of Math. **69** (1990), 59–64.

[GM] E. Glasner and D. Maon, *Rigidity in topological dynamics*, Ergodic Theory and Dynamical Systems **9** (1989), 309–320.

[dJ] A. del Junco, *On minimal self-joinings in topological dynamics*, ibid **7** (1987), 211-227.

[K] J. King, *A map with topological minimal self-joinings in the sense of del Junco*, ibid **10** (1990), 745–761.

[KW] Y. Katznelson and B. Weiss, *When are all points recurrent/generic, in Ergodic Theory and Dynamical Systems - I*, Proceedings Special Year, Maryland 1979–80, Birkhauser, 1981.

[L] E. Lehrer, *Topological mixing and uniquely ergodic systems*, Israel J. of Math. **57** (1987), 239–255.

Institute of Mathematics, Hebrew University of Jerusalem, Jerusalem, Israel
E-mail address: weiss@math.huji.ac.il

Contemporary Mathematics
Volume **215**, 1998

Subset Dynamics and van der Waerden's Theorem

H. Furstenberg and E. Glasner

1. Introduction

The following is a slightly disguised version of van der Waerden's theorem on arithmetic progressions. Let $A \subset \mathbb{Z}$ be a "big" set in the following sense: for some $m \in \mathbb{N}$ and for every $M \in \mathbb{N}$, there exit intervals $(b, c) \subset \mathbb{Z}$ with $c - b \geq M$ such that every subinterval of (b, c) of length m meets A. The assertion is that under these conditions A contains arbitrarily long arithmetic progressions. This statement follows from the conventional formulation of van der Waerden's theorem which states that for a given length l, if a sufficiently long interval is partitioned into m subsets, one of these contains an arithmetic progression(a.p.) of length l. If we define $f_A(n) = \inf\{j \geq 0 : n + j \in A\}$ then f_A determines such a partition of (b, c), and if

$$f_A(a) = f_A(a + d) + \cdots = f_A(a + (l - 1)d)) = j,$$

then $a + j, a + j + d, \cdots, a + j + (l - 1)d$ is an a.p. of length l inside A. Conversely suppose we are given the reformulation above. Partition $\mathbb{Z}$ into finitely many sets. It is easy to see that in any partition of a big set into two sets (and therefore any finite number of sets) one of the subsets is big. (This will appear as a corollary in §3.) Thus one of the sets in the partition of $\mathbb{Z}$ is big and possesses arbitrarily long a.p.'s.

Now bigness can be defined for subsets of any abelian group Γ. First we recall the notion of a "replete" set. A set $A \subset \Gamma$ is called *replete* if any finite set $F \subset \Gamma$ has a translate $F + x$ contained in A. We now say that a set B is *big* if for some finite set $\{y_1, \ldots, y_r\}$, $\cup_{i=1}^{r}(B + y_i)$ is replete. Alternatively define a *syndetic* subset of Γ as a set finitely many translates of which cover Γ. Then a big set is the intersection of a syndetic set and a replete set. The name "big" was given to sets with this property by Ellis and Keynes in [EK], and this notion plays a prominent role in [GW].

For given l one can form the subgroup $AP_l(\mathbb{Z}) \subset \mathbb{Z}^l$ of all arithmetic progressions of length l, $(x_1, \ldots, x_l) \in AP_l(\mathbb{Z}) \Leftrightarrow x_l - x_{l-1} = \cdots = x_2 - x_1$. In what follows we shall strengthen van der Waerden's theorem by showing that a big subset of $\mathbb{Z}$ contains a "big" set of arithmetic progressions. More precisely set $AP_l(B) = AP_l(\mathbb{Z}) \cap B^l \subset \mathbb{Z}^l$. We will show

THEOREM 1.1. *If B is a big set in $\mathbb{Z}$, then $AP_l(B)$ is a big set in $AP_l(\mathbb{Z})$.*

1991 *Mathematics Subject Classifications*: Primary 54H20.

In more concrete terms what this says is that if B is big, there exists $m \in \mathbb{N}$, so that for every $M \in \mathbb{N}$, there exists two intervals (b_1, c_1), (b_2, c_2) of length at least M, so that for any $x_1 \in (b_1, c_1)$, $x_2 \in (b_2, c_2)$ there is an a.p. $x'_1, \dots, x'_l$ inside B with $|x_1 - x'_1| < m$ and $|x_2 - x'_2| < m$.

Our proof is closely related to the algebraic proof of van der Waerden's theorem given in [BFHK] which was based on the theory of compact semigroups applied to the Stone-Čech compactification of $\mathbb{Z}$, an idea inspired by the work of R. Ellis (see [FW] and [FK] for related results). What we show here is that the algebraic result on compact semigroups from which van der Waerden's theorem follows (as does also Hales-Jewett theorem) yields the stronger result contained in theorem 1.1. Unlike van der Waerden's theorem we do not know a strictly combinatorial proof of theorem 1.1.

We might remark that if "big" is replaced by "syndetic" then the parallel result is no longer valid. If $S = \{n \in \mathbb{Z} : n \equiv [\sqrt{|n|}] \ (mod\ 2)\}$ then S is a syndetic set but does not contain a syndetic set of 3-term arithmetic progression. On the other hand if "big" is replaced by "positive upper density" then the parallel result is valid and follows without much difficulty from the ergodic theoretic formulation of Szemeredi's theorem. But although Szemeredi's theorem implies the existence of arithmetic progressions in big sets, it doesn't guarantee the existence of a big set of these.

2. The Algebra of Compact Semigroups

Let E be a compact Hausdorff space with a multiplication so that E is a semigroup, and suppose that multiplication is continuous in the following sense: for fixed $y \in E$, $x \mapsto xy$ is a continuous map of E to itself. A subset I of E is a *left ideal* if $EI \subset I$, a *2-sided ideal* if $EI \cup IE \subset I$. An element $e \in E$ is an *idempotent* if $e^2 = e$.

LEMMA 2.1. *A compact semigroup (in the foregoing sense) contains some idempotent element.*

PROOF. We let

$$\mathcal{M} = \{M \subset E : M \text{ is closed and } MM \subset M\},$$

and order it by inclusion. Let M_0 be a minimal element of $\mathcal{M}$. Let $e \in M_0$; since $M_0eM_0e \subset M_0e$, and M_0 is compact and therefore closed, $M_0e \in \mathcal{M}$. By minimality of M_0, $M_0e = M_0$. So there exists $v \in M_0$ with $ve = e$. Let $M_1 = \{v \in M_0 : ve = e\}$. M_1 is closed and is a subsemigroup of M_0. But then $M_1 = M_0$ and in particular $e^2 = e$. □

Another use of Zorn's lemma implies the existence of minimal closed left ideals. Let I be such an ideal, then for $p \in I$, Ep is a closed left ideal contained in I. Hence $I = Ep$. We see incidentally that minimal left ideals are automatically closed. Such an ideal is again a semigroup and so possesses an idempotent e.

LEMMA 2.2. *If e is an idempotent of a minimal left ideal I then for every $x \in I$, $xe = x$.*

PROOF. $I = Ee$. But for $x = ye, xe = x$. □

We now consider product semigroups $E^l = E \times E \times \cdots \times E$. Let $\Delta_l(E) = \{(p, p, \dots, p) : p \in E\}$.

THEOREM 2.3. *Let Σ be a closed subsemigroup of E^l containing $\Delta_I(E)$. Then any closed 2-sided ideal of Σ meets $\Delta_l(E)$.*

PROOF. For $p \in E$, set $\tilde{p} = (p, p, \ldots, p) \in \Delta_l(E)$. Let Λ be a closed 2-sided ideal in Σ. $\Lambda\tilde{p}$ is a closed semigroup and so it contains an idempotent θ. Write $\theta = (q_1, q_2, \ldots, q_i)$. Then $q_i \in Ep = I$, and since $\theta^2 = \theta$, each q_i is an idempotent of I. It follows from the foregoing lemma that $p = pq_i$ for each i. But this means that $\tilde{p} = \tilde{p}\theta$ and since Λ is a left ideal, $\tilde{p} \in \Lambda$. Since $\tilde{p} \in \Delta_l(E)$ we see that $\Lambda \cap \Delta_l(E) \neq \emptyset$. □

3. Dynamics of Subsets of a Group

In this section we represent the Stone-Čech compactification of an abelian group Γ as the enveloping semigroup of the Γ dynamical system obtained by letting Γ act on its collection of subsets. We rederive the properties of the Stone-Čech compactification that will be needed in the proof of theorem 1.1 in the form that will be most convenient (see also [E]). Let Γ be an abelian group. We denote by 2^Γ the space of all subsets of Γ. For F a finite set in Γ and $A \in 2^\Gamma$ we define a neighborhood of A:

$$\mathcal{U}_F(A) = \{A' \in 2^\Gamma : A' \cap F = A \cap F\}.$$

With these sets as a basis 2^Γ becomes a compact Hausdorff space. Convergence in this topology is the usual one for sets: $A = \lim A_n$ if

$$\liminf A_n \equiv \cup_n \cap_{k \geq n} A_k = \cap_n \cup_{k \geq n} A_k \equiv \limsup A_n = A.$$

We shall write Ω for 2^Γ. We define an "action" of Γ on Ω: for $\gamma \in \Gamma$ set $T(\gamma)(A) = A - \gamma$, $A \in \Omega$. Now consider the semigroup of all maps of Ω to itself, continuous or not. This semigroup can be denoted Ω^Ω and it is compact in the usual product topology and has the one-sided continuity property that played a role in the foregoing section. For any $S \subset \Gamma$ let $T_\Gamma(S) = \{T(\gamma) : \gamma \in S\} \subset \Omega^\Omega$ and define $\beta\Gamma = \overline{T(\Gamma)}$ inside Ω^Ω, where we write $T(\Gamma)$ instead of $T_\Gamma(\Gamma)$. This is one version of the Stone-Čech compactification of Γ (see e.g. [G] lemma 4.1). $\beta\Gamma$ is a semigroup. This is based on the fact that for each γ, the composition map $f \mapsto T(\gamma)f$ is a continuous map of Ω^Ω to itself. So $T(\gamma)\overline{T(\Gamma)} \subset \overline{T(\gamma)T(\Gamma)} = \overline{T(\Gamma)}$. Hence $T(\Gamma)\beta(\Gamma) \subset \beta(\Gamma)$. Since for every $g \in \Omega^\Omega$, $f \mapsto f \circ g$ is continuous, $\overline{T(\Gamma)}g \subset \overline{T(\Gamma)g}$, and so $\beta(\Gamma)\beta(\Gamma) \subset \overline{T(\Gamma)\beta(\Gamma)} \subset \beta(\Gamma)$. In this form $\beta(\Gamma)$ is the *enveloping semigroup* of the dynamical system $(2^\Gamma, \Gamma)$.

For $\tau \in \beta(\Gamma)$ and $A \subset \Gamma$ we denote by τA the image of $A \in \Omega$ under the transformation τ. It is easy to see that τ preserves unions, intersections and complements since each $T(\gamma)$ does so. If $\mathcal{A} \subset \beta(\Gamma)$ and $\tau \in \beta(\Gamma)$ we use the notation

$$\mathcal{A}\tau^{-1} = \{\sigma \in \beta(\Gamma) : \sigma\tau \in \mathcal{A}\}.$$

By the one sided continuity condition for multiplication in Ω^Ω it follows that $\mathcal{A} \mapsto \mathcal{A}\tau^{-1}$ takes open sets to open sets and closed sets to closed sets.

LEMMA 3.1. *For any $S \subset \Gamma$ and $\tau \in \beta(\Gamma)$, $\tau \in \overline{T(S)} \Leftrightarrow 0 \in \tau S$.*

PROOF. If $\tau \in \overline{T(S)}$ we will have for some $s \in S$, $\tau S \cap \{0\} = T(s)S \cap \{0\} = \{0\}$, i.e., $0 \in \tau S$. Conversely, suppose $0 \in \tau S$, and let $\{s_\alpha\}$ be a net with $T(s_\alpha) \to \tau$. Eventually $0 \in T(s_\alpha)S = S - s_\alpha$. But then it follows that $\tau \in \overline{T(S)}$. □

LEMMA 3.2. *If $S \subset \Gamma$, then $\overline{T(S)}$ is both open and closed.*

PROOF. $\overline{T(S)}$ is closed by definition. Let $S' = \Gamma \setminus S$. Clearly $\beta(\Gamma) = \overline{T(S)} \cup \overline{T(S')}$. Let us show that $\beta(\Gamma) = \overline{T(S)} \cap \overline{T(S')} = \emptyset$ so that $\overline{T(S)}$ is the complement of a closed set. Suppose $\tau \in \overline{T(S)}$; then $0 \in \tau S$ by lemma 3.1, therefore $0 \notin \tau S'$ and then $\tau \notin \overline{T(S')}$. This completes the proof. □

LEMMA 3.3. *If $\mathcal{R} \subset \beta(\Gamma)$ and $\mathcal{R}$ is both open and closed, then $\mathcal{R} = \overline{\mathcal{R} \cap T(\Gamma)}$.*

PROOF. Let $\mathcal{R}' = \beta(\Gamma)\backslash\mathcal{R}$ so that $\beta(\Gamma) = \mathcal{R}\cup\mathcal{R}'$ and $\mathcal{R}\cap\mathcal{R}' = \emptyset$. $T(\Gamma) \subset \beta(\Gamma)$ so that $T(\Gamma) = (T(\Gamma) \cap \mathcal{R}) \cup (T(\Gamma) \cap \mathcal{R}')$ and so $\beta(\Gamma) = \overline{(T(\Gamma) \cap \mathcal{R})} \cup \overline{(T(\Gamma) \cap \mathcal{R}')}$. Now $\mathcal{R} \cap (T(\Gamma) \cap \mathcal{R}') = \emptyset$ and since $\mathcal{R}$ is open, $\mathcal{R} \cap \overline{T(\Gamma) \cap \mathcal{R}'} = \emptyset$. It follows that $\mathcal{R} \subset \overline{T(\Gamma) \cap \mathcal{R}} \subset \mathcal{R}$ and we have equality. □

LEMMA 3.4. *If $\tau \in \beta(\Gamma)$ and $S \subset \Gamma$,* $\overline{T(\tau S)} = \overline{T(S)}\tau^{-1}$.

PROOF. $\sigma \in \overline{T(\tau S)} \Leftrightarrow 0 \in \sigma\tau S \Leftrightarrow \sigma\tau \in \overline{T(S)} \Leftrightarrow \sigma \in \overline{T(S)}\tau^{-1}$. □

LEMMA 3.5. *For $S \subset \Gamma$, $\overline{T(S)} \cap \Gamma = S$.*

PROOF. Let $S' = \Gamma \setminus S$. Then $\overline{T(S')} \cap \overline{T(S)} = \emptyset$ and so $\overline{T(S)} \cap \Gamma \cap S' = \emptyset$. Thus $\overline{T(S)} \cap \Gamma \subset S$. But clearly $\overline{T(S)} \cap \Gamma \supset S$ so we have equality. □

The next theorem characterizes bigness of a set in terms of $\beta(\Gamma)$.

THEOREM 3.6. *The following conditions are equivalent for a subset $S \subset \Gamma$.*

1. *S is big*
2. *For some $\tau_1, \dots, \tau_r \in \beta(\Gamma)$, $\cup_{i=1}^r \tau_i S = \Gamma$.*
3. *For some $\tau_1, \dots, \tau_r \in \beta(\Gamma)$, $\cup_{i=1}^r \overline{T(S)}\tau_i^{-1} = \beta(\Gamma)$.*
4. *For some τ in a minimal left ideal of $\beta(\Gamma)$, $\tau S \neq \emptyset$.*

PROOF. (1) $\Rightarrow$ (2) Suppose first that A is a replete set. We have for any finite $F \subset \Gamma$, $T(\gamma)A \supset F$ for an appropriate $\gamma \in \Gamma$. Passing to a limit we find $\tau A = \Gamma$, for some $\tau \in \beta(\Gamma)$. Now suppose $\cup_{i=1}^k (S + \gamma_i)$ is replete. Then for an appropriate $\tau \in \beta(\Gamma)$, $\cup \tau T(-\gamma_i)S = \Gamma$.

(2) $\Rightarrow$ (3) $\cup \tau_i S = \Gamma$ implies that $\cup\overline{T(\tau_i S)} = \beta(\Gamma)$ and by lemma 3.4, $\overline{T(\tau_i S)} = \overline{T(S)}\tau_i^{-1}$.

(3) $\Rightarrow$ (4) Let p belong to a minimal left ideal $I \subset \beta(\Gamma)$. Assume $\beta(\Gamma) = \cup\overline{T(S)}\tau_i^{-1}$. p is in one of these sets; so $p\tau_i \in \overline{T(S)}$, or by lemma 3.1, $0 \in p\tau_i S$. Now the set $I\tau_i$ is again a left ideal and it is not hard to see that it is minimal. Set $\tau = p\tau_i$ and we have $\tau S \neq \emptyset$ for a τ in a minimal left ideal.

(4) $\Rightarrow$ (1) Let I be a minimal left ideal; let $\tau \in I$ and assume $\tau S \neq \emptyset$. We claim that τS is syndetic. If not, then for each finite F, $\tau S - F \neq \Gamma$ and there exists $\gamma_F \in \Gamma$ with $\gamma_F \notin \tau S - F$, or $F \cap T(\gamma_F)\tau S = \emptyset$. Passing to a limit of $T(\gamma_F)$ as $F \to \Gamma$ we find $\sigma \in \beta(\Gamma)$ with $\sigma\tau S = \emptyset$. But $\beta(\Gamma)\sigma\tau$ is a closed subideal of I; hence $\beta(\Gamma)\sigma\tau = I$ and there exists $\rho \in \beta(\Gamma)$ with $\rho\sigma\tau = \tau$. But then $\tau S = \rho\sigma\tau S = \emptyset$ which is a contradiction. We must therefore have that τS is syndetic. For appropriate $\gamma_1, \dots, \gamma_r \in \Gamma$ this give $\cup T(\gamma_i)\tau S = \Gamma$ or $\tau(\cup T(\gamma_i)S) = \Gamma$. Set $A = \cup T(\gamma_i)S$; we claim that A is replete. We have $\tau A = \Gamma$, so for any finite $F \subset \Gamma$, there exists $\gamma \in \Gamma$ with $T(\gamma)A \supset F$, i.e. $F + \gamma \subset A$. This is the condition for repleteness and this prove that S is big. □

As a corollary we have the assertion of §1:

COROLLARY 3.7. *If S is big and $S = S' \cup S''$, then one of S', S'' is big.*

PROOF. Let τ belong to a minimal left ideal and assume $\tau S \neq \emptyset$. $\tau S = \tau S' \cup \tau S''$ so either $\tau S' \neq \emptyset$ or $\tau S'' \neq \emptyset$. □

Finally we consider homomorphisms of groups, $\pi : \Gamma \to \Gamma'$, and we show that π induces a homomorphism $\hat{\pi} : \beta(\Gamma) \to \beta(\Gamma')$. $\hat{\pi}$ is uniquely determined by the condition of continuity and the condition $\hat{\pi}(T_\Gamma(\gamma)) = T_{\Gamma'}(\pi(\gamma))$. When π is onto it is easy to define $\hat{\pi}$, namely $\hat{\pi}(\tau)A = \pi(\tau\pi^{-1}(A))$. When π is 1-1 and Γ can be regarded as a subgroup of Γ' we define $\hat{\pi}(\tau)$ for $\tau \in \beta(\Gamma)$ coset by coset:

$$\hat{\pi}(\tau)A \cap (\Gamma + u) = \tau[(A - u) \cap \Gamma] + u$$

for $A \subset \Gamma$. Combining these two cases we obtain the induced homomorphism $\hat{\pi}$ for an arbitrary homomorphism π. In particular we see that for $\Gamma \subset \Gamma'$, $\beta(\Gamma) \subset \beta(\Gamma')$ and more specifically $\beta(\Gamma)$ can be identified by

$$\beta(\Gamma) \cong \overline{T_{\Gamma'}(\Gamma)} \subset \beta(\Gamma'). \tag{3.1}$$

4. Big Sets of Arithmetic Progressions

We apply the foregoing with $\Gamma = \mathbb{Z}, \mathbb{Z}^l$ and $AP_l(\mathbb{Z}) \subset \mathbb{Z}^l$. By the remarks at the end of the previous section we can identify $\beta(AP_l(\mathbb{Z}))$ with

$$\overline{T_{\mathbb{Z}^l}(AP_l(\mathbb{Z}))} \subset \beta(\mathbb{Z}^l)$$

and we will use this in discussing big sets in $AP_l(\mathbb{Z})$. We wish to show that if B is big in $\mathbb{Z}$, then $AP_l(B) \equiv AP_l(\mathbb{Z}) \cap B^l$ is big in $AP_l(\mathbb{Z})$.

There are l component projections $\pi_i : \mathbb{Z}^l \to \mathbb{Z}$, and these induce homomorphisms $\hat{\pi}_i : \beta(\mathbb{Z}^l) \to \beta(\mathbb{Z})$. Form

$$\tilde{\pi} = \hat{\pi}_1 \times \hat{\pi}_2 \cdots \times \hat{\pi}_l : \beta(\mathbb{Z}^l) \to \beta(\mathbb{Z})^l.$$

Denote the set $\overline{T_{\mathbb{Z}^l}(AP_l(\mathbb{Z}))} \subset \beta(\mathbb{Z}^l)$ by $\tilde{\Sigma}$, and set $\Sigma = \tilde{\pi}(\tilde{\Sigma}) \subset \beta(\mathbb{Z})^l$. Let $\Delta = \Delta_l(\beta(\mathbb{Z}))$ be the diagonal in the product space $\beta(\mathbb{Z})^l$. $\tilde{\Sigma} \cong \beta(AP_l(\mathbb{Z}))$ is a compact semigroup and so is Σ, its homomorphic image.

We are given that B is a big set in $\mathbb{Z}$. By theorem 3.6, there exist $\tau_i \in \beta(\mathbb{Z})$ with

$$\bigcup \overline{T_{\mathbb{Z}}(B)}\tau_i^{-1} = \beta(\mathbb{Z}). \tag{4.1}$$

$\overline{T_{\mathbb{Z}}(B)}$ is an open set and $\mathcal{V} = \overline{T_{\mathbb{Z}}(B)} \times \cdots \times \overline{T_{\mathbb{Z}}(B)}$, ($l$ copies) is open in $\beta(\mathbb{Z})^l$. From (4.1) we have

$$\begin{aligned}\Delta &\subset \bigcup_i \overline{T_{\mathbb{Z}}(B)}\tau_i^{-1} \times \cdots \times \overline{T_{\mathbb{Z}}(B)}\tau_i^{-1} \\ &= \bigcup_i \mathcal{V}(\tau_i^{-1}, \ldots, \tau_i^{-1}) = \mathcal{U} \subset \beta(\mathbb{Z})^l\end{aligned} \tag{4.2}$$

with $\mathcal{U}$ open.

PROPOSITION 4.1. *With $\mathcal{U}$ defined by (4.2),*

$$\bigcup_{\sigma \in \Sigma} \mathcal{U}\sigma^{-1} \supset \Sigma. \tag{4.3}$$

PROOF. Let $\Lambda = \Sigma \setminus \bigcup_{\sigma\in\Sigma} \mathcal{U}\sigma^{-1}$. Assume $\Lambda \neq \emptyset$. Clearly $\lambda \in \Lambda$ iff $\lambda \in \Sigma$ and $\lambda\sigma \notin \mathcal{U}$ for any $\sigma \in \Sigma$. It follows that if $\lambda \in \Lambda$ then $\lambda\sigma_1 \in \Lambda$ for any $\sigma_1 \in \Sigma$, and hence Λ is a closed right ideal: $\Lambda\Sigma \subset \Lambda$. Now if $(a, a+d, \dots, a+(l-1)d) \in AP_l(\mathbb{Z})$ then $(T_{\mathbb{Z}}(a), T_{\mathbb{Z}}(a+d), \dots, T_{\mathbb{Z}}(a+(l-1)d)) = \tilde{\pi}(T_{\mathbb{Z}^l}(a, a+d, \dots, a+(l-1)d)) \in \tilde{\pi}(\tilde{\Sigma}) = \Sigma$. Thus Λ is closed under multiplication by $(T_{\mathbb{Z}}(a), T_{\mathbb{Z}}(a+d), \dots, T_{\mathbb{Z}}(a+(l-1)d))$ to the right. But these elements commute with $\beta(\mathbb{Z})^l$ and so Λ is closed under multiplication by them to the left. Finally $\tilde{\pi}(T_{\mathbb{Z}^l}(AP_l(\mathbb{Z})))$ is dense in Σ, so that $\Sigma\Lambda \subset \Lambda$ in addition and Λ is a 2-sided ideal of Σ. If in the foregoing we take $d = 0$, we see that Σ contains the diagonal of $T_{\mathbb{Z}}(\mathbb{Z}) \times \cdots \times T_{\mathbb{Z}}(\mathbb{Z})$ and since Σ is closed,

$$\Sigma \supset \Delta = \Delta_l(\beta(\mathbb{Z})).$$

We now turn to theorem 2.3. From this theorem applied to $E = \beta(\mathbb{Z})$ it follows that if $\Lambda \neq \emptyset$, then $\Lambda \cap \Delta \neq \emptyset$. But $\Delta \subset \mathcal{U} \subset \bigcup_{\sigma\in\Sigma} \mathcal{U}\sigma^{-1}$ so that $\Lambda \cap \Delta = \emptyset$. We must therefore have $\Lambda = \emptyset$ which proves the proposition. □

Now apply $\tilde{\pi}^{-1}$ to the inclusion in (4.3). This gives

$$\tilde{\Sigma} \subset \bigcup_{\sigma\in\Sigma} \tilde{\pi}^{-1}(\mathcal{U}\sigma^{-1}) = \bigcup_{\tilde{\sigma}\in\tilde{\Sigma}} \tilde{\pi}^{-1}(\mathcal{U})\tilde{\sigma}^{-1}. \tag{4.4}$$

We wish to expand this further using (4.2). We have $(\tau_i, \tau_i, \cdots, \tau_i) \in \Delta \subset \tilde{\pi}(\tilde{\Sigma}) = \Sigma$ and so we can find $\tilde{\rho}_i \in \tilde{\Sigma}$ with $\tilde{\pi}(\tilde{\rho}_i) = (\tau_i, \tau_i, \cdots, \tau_i)$. It is easy to see that (4.2) gives

$$\tilde{\pi}^{-1}(\mathcal{U}) = \bigcup \tilde{\pi}^{-1}(\mathcal{V})\tilde{\rho}_i^{-1}.$$

Combining this with (4.4),

$$\tilde{\Sigma} \subset \bigcup_{i,\tilde{\sigma}} \tilde{\pi}^{-1}(\mathcal{V})\tilde{\rho}_i^{-1}\tilde{\sigma}^{-1} = \bigcup_{i,\tilde{\sigma}} \tilde{\pi}^{-1}(\mathcal{V})(\tilde{\sigma}\tilde{\rho}_i)^{-1}.$$

Using compactness we find

$$\tilde{\Sigma} \subset \bigcup_k \tilde{\pi}^{-1}(\mathcal{V})\tilde{\omega}_k^{-1}$$

for a finite set of $\tilde{\omega}_k$ in $\tilde{\Sigma}$. This of course implies

$$\tilde{\Sigma} \subset \bigcup_k [\tilde{\pi}^{-1}(\mathcal{V}) \cap \tilde{\Sigma}]\tilde{\omega}_k^{-1}. \tag{4.5}$$

Now $\tilde{\pi}^{-1}(\mathcal{V}) \cap \tilde{\Sigma}$ is open and closed in $\tilde{\Sigma} \cong \beta(AP_l(\mathbb{Z}))$. By lemma 3.3 applied to $AP_l(\mathbb{Z}) \subset \mathbb{Z}^l$,

$$\tilde{\pi}^{-1}(\mathcal{V}) \cap \tilde{\Sigma} = \overline{\tilde{\pi}^{-1}(\mathcal{V}) \cap T_{\mathbb{Z}^l}(AP_l(\mathbb{Z}))}.$$

Consider $\xi \in \tilde{\pi}^{-1}(\mathcal{V}) \cap T_{\mathbb{Z}^l}(AP_l(\mathbb{Z}))$. Write $\xi = T_{\mathbb{Z}^l}(x_1, \dots, x_l)$ with $(x_1, \dots, x_l) \in AP_l(\mathbb{Z})$. Then

$$\tilde{\pi}(\xi) = \hat{\pi}_1 \times \cdots \times \hat{\pi}_l(T_{\mathbb{Z}^l}(x_1, \dots, x_l)) = (T_{\mathbb{Z}}(x_1), \dots, T_{\mathbb{Z}}(x_l))$$

belongs to $\mathcal{V} = \overline{T_{\mathbb{Z}}(B)} \times \cdots \times \overline{T_{\mathbb{Z}}(B)}$. But, since $\overline{T_{\mathbb{Z}}(B)} \cap T_{\mathbb{Z}}(\mathbb{Z}) = T_{\mathbb{Z}}(B)$, by lemma 3.5, $T_{\mathbb{Z}}(x_j) \in T_{\mathbb{Z}}(B)$ for $j = 1, \dots, l$, and so $x_j \in B$. We conclude that $\xi \in T_{\mathbb{Z}^l}(AP_l(B))$. We can now rewrite (4.5) as

$$\beta(AP_l(\mathbb{Z})) = \tilde{\Sigma} \subset \bigcup_k \overline{T_{\mathbb{Z}^l}(AP_l(B))}\tilde{\omega}_k^{-1},$$

and so by theorem 3.6, $AP_l(B)$ is big in $AP_l(\mathbb{Z})$. This completes the proof of theorem 1.1.

References

[E] R. Ellis, *Lectures in topological dynamics*, Benjamin, New York, 1969.

[EK] R. Ellis and H. Keynes, *Bohr compactifications and a result of Fölner*, Israel J. of Math. **12** (1972), 314–330.

[FK] H. Furstenberg and Y. Katznelson, *Idempotents in compact semigroups and Ramsey theory*, Isreal J. of Math. **68** (1989), 257–270.

[FW] H. Furstenberg and B. Weiss, *Topological dynamics and combinatorial number theory*, J. Analyse Math. **34** (1978), 61–85.

[G] S. Glasner, *Divisible properties and the Stone-Čech compactification*, Can. J. Math. **32** (1980), 993–1007.

[GW] S. Glasner and B. Weiss, *Interpolation sets for subalgebras of $l^\infty(\mathbb{Z})$*, Israel J. Math. **44** (1983), 345–360.

Institute of mathematics, Hebrew university of Jerusalem
E-mail address: `harry@math.huji.ac.il`

Department of Mathematics , Tel Aviv University
E-mail address: `glasner@math.tau.ac.il`

Contemporary Mathematics
Volume **215**, 1998

Recurrence for Semigroup Actions and a non-Commutative Schur Theorem

V. Bergelson and R. McCutcheon

1. Introduction

This note is concerned with the topological, measure theoretical and combinatorial aspects of *recurrence* and their interplay. The classical Poincaré recurrence theorem asserts that for any finite measure preserving system $(X, \mathcal{B}, \mu, T)$ and any set $A \in \mathcal{B}$ with $\mu(A) > 0$, μ-almost every point $x \in A$ returns to A under a power of the transformation T, that is $\mu(\{x \in A : \text{ there exists } n \in \mathbf{N} \text{ with } T^n x \in A\}) = \mu(A)$. This result, which deals with recurrence of points, can easily be obtained from a result dealing with recurrence of sets: if $\mu(A) > 0$ then for some $n \in \mathbf{N}$ one has $\mu(A \cap T^{-n}A) > 0$. It is the recurrence of sets in measure preserving, topological and combinatorial set-ups which will be the focus of our attention in this paper. (A standing exercise, which we offer to the reader, is to detect the recurrence of points lying in the background.)

A far-reaching refinement of the Poincaré recurrence theorem was obtained by Furstenberg in [**F1**]: for any finite measure preserving system $(X, \mathcal{B}, \mu, T)$, any $A \in \mathcal{B}$ with $\mu(A) > 0$, and any $k \in \mathbf{N}$ there exists $n \in \mathbf{N}$ such that $\mu(A \cap T^{-n}A \cap \cdots \cap T^{-kn}A) > 0$. In the same paper Furstenberg showed that this *multiple recurrence* theorem implies Szemerédi's celebrated theorem on arithmetic progressions ([**Sz**]), thereby establishing a fruitful and mutually perpetuating link between ergodic theory and *density* Ramsey theory.

In [**FW**] Furstenberg and Weiss established a similar link between topological dynamics and *partition* Ramsey theory, which deals with theorems like van der Waerden's, Hindman's, etc. (see [**GRS**] for background; see also [**F2**] and [**B2**]). For example, one of the results in [**FW**] asserts that for any continuous self-mapping T of a compact metric space X, any $k \in \mathbf{N}$, and any $\epsilon > 0$, there exists $x \in X$ and $n \in \mathbf{N}$ such that the diameter of the set $\{x, T^n x, \cdots, T^{kn}x\}$ is smaller than ϵ. This result implies van der Waerden's theorem on arithmetic progressions, which is the natural counterpart in partition Ramsey theory to Szemerédi's theorem. Again, this result about recurrence of points follows from a corresponding result about recurrence of sets: if (X, T) is a minimal dynamical system, then for any $k \in \mathbf{N}$ and any non-empty open set U one has $U \cap T^{-n}U \cap \cdots \cap T^{-kn}U \neq \emptyset$ for some $n \in \mathbf{N}$. Alternatively: if (X, T) is a (not necessarily minimal) dynamical system and $\{U_1, \cdots, U_t\}$ is an open cover of X then for some i, $1 \leq i \leq t$, and some $n \in \mathbf{N}$ one has $U_i \cap T^{-n}U_i \cap \cdots \cap T^{-kn}U_i \neq \emptyset$.

1991 *Mathematics Subject Classifications*: Primary 28D05.

While the measurable and topological multiple recurrence results alluded to provide one with strong Ramsey-theoretical results, already there are some interesting questions and results concerning the relationship between these two modes of recurrence for the single recurrence case. A good illustration is given by a result due to Kriz ([**K**]), which we now describe. In an attempt to better understand the connection between results of density Ramsey theory and their counterparts in partition Ramsey theory, the first author asked whether it is true that if $\mathcal{R}$ is a shift-invariant family of finite sets in $\mathbf{N}$ having the property that some member of $\mathcal{R}$ can always be found in one cell of an arbitrary finite partition of $\mathbf{N}$, then any set of positive upper density in $\mathbf{N}$ contains a member of $\mathcal{R}$. Kriz demonstrated the answer to be no.

To more fully elucidate what he proved, we will introduce two notions. First, let a subset R of $\mathbf{N}$ be called *density intersective* provided that for every $A \subset \mathbf{N}$ with $\bar{d}(A) > 0$ there exists $n \in R$ such that $A \cap (A - n) \neq \emptyset$. Set R will be called *chromatically intersective* if for every partition of $\mathbf{N}$ into r cells, one of the cells, call it C, satisfies $C \cap (C - n) \neq \emptyset$ for some $n \in R$. What Kriz did was construct a chromatically intersective set R which is not density intersective. Letting then $\mathcal{R}$ be the set of configurations $\{a, a+n\}$, where $n \in R$ and $a \in \mathbf{N}$, $\mathcal{R}$ is easily seen to indeed be a shift-invariant class having the property that for every finite partition of $\mathbf{N}$, some cell contains a member of $\mathcal{R}$, yet there exists a set A with $\bar{d}(A) > 0$ which does not contain a member of $\mathcal{R}$.

Due to the aforementioned connection between ergodic theory and density combinatorics (see also Theorem 2.1), one may show that a set $R \subset \mathbf{N}$ is density intersective if and only if for every measure preserving transformation T of a probability space $(X, \mathcal{B}, \mu)$ and every $A \in \mathcal{B}$ with $\mu(A) > 0$ there exists $n \in R$ such that $\mu(A \cap T^{-n}A) > 0$. For this reason, density intersective sets are also called *sets of measurable recurrence.* Similarly, the connection between Ramsey theory and topological dynamics provides a dynamically formulated condition equivalent to that of being a set of chromatic intersectivity. There are several equivalent forms, however the one most commonly used ([**Fo**], [**M**]) has been the following: R is chromatically intersective if and only if for every invertible, minimal dynamical system (X, T), where X is a compact metric space, and every non-empty open set U, there exists $n \in R$ such that $U \cap T^{-n}U \neq \emptyset$. For this reason, chromatically intersective sets in $\mathbf{N}$ are also called *sets of topological recurrence.* Thus, the set constructed by Kriz is a set of topological recurrence which is not a set of measurable recurrence.

One of our purposes is to extend these notions, defined here in $\mathbf{N}$, to arbitrary countable semigroups S, exploring their interrelationships both with each other and with combinatorics. We remark that most of the definitions and results of this paper make sense for actions of uncountable discrete semigroups S (in other words, actions with no continuity restrictions), however we do not pursue this notion, in part because we have no combinatorial results for uncountable semigroups which are not already obtainable as corollaries of results for countable semigroups.

Moreover, usually when one considers actions of uncountable semigroups, these actions are taken to be continuous with respect to some given topology, and for this reason there may be some cause for confusion in what should be considered a set of recurrence in a group such as $\mathbf{R}$. Indeed, by [**BBB**], p.43] for any $\alpha > 0$, any continuous measure preserving flow $\{T_t : t \in \mathbf{R}\}$ on a probability space $(X, \mathcal{B}, \mu)$ and any $A \in \mathcal{B}$ with $\mu(A) > 0$ there exists $n \in \mathbf{N}$ such that $\mu(A \cap T_{n^\alpha}A) > 0$. On the other hand, for all but countably many α, $\{n^\alpha : n \in \mathbf{N}\}$ is linearly independent over $\mathbf{Q}$ (see [**BBB**], Lemma 2.9 and Remark on p. 38]). For such an $\alpha > 1$, [**BBB**], Theorem D] allows one to find a measurable set $E \subset \mathbf{R}$ such that

$$d(E) = \lim_{t\to\infty} \frac{m(E \cap [0,t])}{t} = \frac{1}{2}$$

and $E \cap (E - n^\alpha) = \emptyset$ for all $n \in \mathbf{N}$. Using the methods of Section 2 (namely the proof of Theorem 2.1, which works for uncountable *discrete* semigroups) one may use this example to construct a (non-continuous!) measure preserving $\mathbf{R}$-action $\{T_g\}_{g\in\mathbf{R}}$ on a probability space $(X, \mathcal{B}, \mu)$ such that for some $A \in \mathcal{B}$ with $\mu(A) > 0$ we have $\mu(A \cap T_{n^\alpha} A) = 0$ for all $n \in \mathbf{N}$. Therefore, whether or not one should call $\{n^\alpha : n \in \mathbf{N}\}$ a set of measurable recurrence appears to be a matter of perspective. This is another reason we choose not to pursue this issue here.

DEFINITION 1.1. Let S be a countable semigroup and suppose $R \subset S$.

(a) R is called a *set of measurable recurrence* if for every measure preserving S-action $\{T_g\}_{g\in S}$ on a probability space $(X, \mathcal{B}, \mu)$ and every $A \in \mathcal{B}$ satisfying $\mu(A) > 0$ we have $\mu(A \cap T_g^{-1} A) > 0$ for some $g \in R$.

(b) R is called a *set of topological recurrence* if for every minimal S-action $\{T_g\}_{g\in S}$ by continuous self-mappings of a compact metric space X and every open set U we have $U \cap T_g^{-1} U \neq \emptyset$ for some $g \in R$.

Unfortunately, for non-abelian semigroups S the property of Definition 1.1 (b) is not what is needed for the Ramsey theoretic connection. The correct notion, which for abelian semigroups (but not in general) is easily shown (see Theorem 2.6 (b)) to be equivalent, is the following.

DEFINITION 1.2. A subset R of a countable semigroup S is called a *set of chromatic recurrence* if for every S-action $\{T_g\}_{g\in S}$ by continuous self-mappings of a compact metric space X and every open covering $\{U_1, \cdots, U_r\}$ of X there exists i with $1 \le i \le r$ such that $U_i \cap T_g^{-1} U_i \neq \emptyset$ for some $g \in R$.

REMARK. Our definitions of sets of recurrence are such that if S contains an identity e then $\{e\}$ is automatically a set of recurrence. This is a departure from the definition of sets of recurrence for $S = \mathbf{Z}$ given in [**F2**], in which R is a set of recurrence if and only if $R \setminus \{0\}$ is a set of recurrence. We are changing things a little for the sake of convenience. The difference, of course, is more philosophical than actual.

Equivalent combinatorial formulations of each of these three types of sets of recurrence are given in Section 2. For sets of topological recurrence and sets of chromatic recurrence, these equivalences hold for all countable semigroups S, however for sets of measurable recurrence the equivalence is only valid in the event that S is countable and left amenable. We then show that sets of topological recurrence are always sets of chromatic recurrence, and that for countable left amenable semigroups sets of measurable recurrence are always sets of topological recurrence. Meanwhile, since for abelian semigroups chromatic and topological recurrence are equivalent, [**K**] provides us with a set of chromatic recurrence which is not a set of measurable recurrence.

The reason sets of measurable recurrence in countable left amenable semigroups must also be sets of topological recurrence is that for any action of such a semigroup S by continuous self-mappings of a compact metric space X, there exists an invariant measure. For non-amenable (semi)groups, no such measure is guaranteed, removing any a priori reason why a set of measurable recurrence should be a set of topological recurrence. Indeed, one may almost effortlessly find sets of measurable recurrence in free groups which are not sets of topological recurrence (see Theorem 2.6 (c)). A more interesting question is whether one may find sets of measurable recurrence which are not sets of *chromatic* recurrence. There is some evidence (Theorem 4.5) which suggests that there may be such examples, but as of now we haven't found any. Section 4 is a short section devoted to expounding upon this and various other open questions, a few of which pertain to *multiple* recurrence.

Section 3 contains a non-commutative version of Schur's theorem. Recall that given a finite partition of a semigroup S, $S = \bigcup_{i=1}^{r} C_i$, one of the cells C_i contains a configuration of the form $\{x, y, xy\}$. This result, which in the case $S = (\mathbf{N}, +)$ is due to Schur (see [**S**], [**GRS**]), has interesting consequences in number theory (cf. [**Si**], [**B1**]).

Schur's theorem is extended considerably by the following marvelous theorem of Hindman ([**H**]): for any finite partition of a semigroup S there is a sequence $\{x_i\}_{i\in\mathbf{N}}$ with the property that all its elements together with all finite products of the form $x_{i_1}x_{i_2}\cdots x_{i_k}$, where $k \in \mathbf{N}$ is arbitrary and $i_1 < \cdots < i_k$, belong to the same cell of the partition.

We use a result about *double recurrence* recently obtained in [**BMZ**] to advance Schur's theorem in a non-commutative direction. Namely, we address the following question inspired by Hindman's theorem and stemming from the general philosophy of Ramsey theory: Is it true that for any finite partition of any "sufficiently non-commutative" group and any natural number k, one of the cells of the partition contains pairwise non-commuting elements $x_1, \cdots, x_k$ together with all possible products $x_{i_1}x_{i_2}\cdots x_{i_m}$ (where $i_r \neq i_t$ for $r \neq t$)? The vague formulation ("sufficiently non-commutative") we have given is intentionally open-ended, however at this stage its meaning for us is that, letting A be the subgroup (to see that A is in fact a subgroup, see the beginning of the proof of Theorem 3.4 in Section 3) of G consisting of all elements x such that $[G : C(x)] < \infty$, where $C(x)$ is the centralizer of x, we have $[G : A] = \infty$. In other words, there are neglectably few elements which commute with many elements of G. For amenable groups at least, this seems to be appropriate. Plenty of countable amenable groups satisfy this condition, for example the Heisenberg group and the group S_∞ of finite permutations of $\mathbf{N}$. Indeed, we know of no example of a countable amenable group G having no abelian subgroups of finite index which fails the condition. At any rate, our modest contribution to the question introduced above is the following theorem.

THEOREM 3.4. *Suppose that G is a countable amenable group and that $G = \bigcup_{i=1}^{r} C_i$ is a finite partition. Let $A = \{g \in G : [G : C(g)] < \infty\}$. If $[G : A] = \infty$, then there exist $x, y \in G$ and i, $1 \leq i \leq r$, with $xy \neq yx$ and such that $\{x, y, xy, yx\} \subset C_i$.*

Although on the face of it, this theorem belongs strictly to partition Ramsey theory, it belongs to a sub-class of results in this field for which we have at present no proof which avoids ergodic theory (see [**BMZ**], Corollary 7.2] for another). In Section 5 we offer a very short philosophical discussion which hints at some of the reasons for this phenomenon (at least in non-abelian groups) by indicating where the available topological methods fail.

2. Combinatorial Formulations of and Relationships Between the Various Types

In this section we will formulate equivalent combinatorial conditions for the three kinds of sets of recurrence defined in Section 1. Then, we will point out a few of the inclusive and exclusive relationships these concepts have with each other. First we handle sets of measurable recurrence. To this end, we give a general form of a correspondence principle due to Furstenberg which is valid for countable left amenable semigroups. Recall that a semigroup S is left amenable if there exists a *left invariant mean* m on $l^\infty(S)$. Namely, m is a member of $l^\infty(S)^*$ with $m(\mathbf{1}) = 1$, $m(g) \geq 0$ if $g(s) \geq 0$ for all $s \in S$, and $m({}_sg) = m(g)$ for all $s \in S$ (where ${}_sg(t) = g(st)$). Given a left invariant mean $m \in l^\infty(S)^*$ and $E \subset S$, we will follow convention and write $m(E)$ for $m(1_E)$. Written this way, m becomes a finitely additive, shift invariant (i.e. $m(s^{-1}E) = m(E)$ for all $s \in S$ and $E \subset S$, where $s^{-1}E = \{g \in S : sg \in E\}$) probability measure on $\mathcal{P}(S)$, the power set of S.

THEOREM 2.1. *Let S be a countable left amenable semigroup, let m be a left invariant mean and suppose $A \subset S$. There exists a probability space $(X, \mathcal{B}, \mu)$ (X is in fact a compact metric space and $\mathcal{B}$ is the σ-algebra of Borel sets), a set $U \in \mathcal{B}$ (which is open and closed), and a measure preserving S-action $\{T_g\}_{g \in S}$ on X (the transformations are continuous) such that for every $h_1, \cdots, h_k \in S$ we have*

$$\mu(T_{h_1}^{-1}U \cap T_{h_2}^{-1}U \cap \cdots \cap T_{h_k}^{-1}U) = m(h_1^{-1}A \cap h_2^{-1}A \cap \cdots \cap h_k^{-1}A). \tag{2.1}$$

PROOF. By adding it if necessary we will assume that S contains an identity e. Let $X = \{0,1\}^S$. With the product topology, X is compact and metric. Let $\mathcal{C}$ be the collection of *cylinder sets*, that is sets of the form

$$C = \{\gamma \in X : \gamma(h_1) = \epsilon_1, \gamma(h_2) = \epsilon_2, \cdots, \gamma(h_k) = \epsilon_k\}, \tag{2.2}$$

where $k \in \mathbf{N}$, $h_i \in S$ are distinct and $\epsilon_i \in \{0,1\}$, $1 \leq i \leq k$. Note that $\mathcal{C}$ is closed under finite intersections. Let $\mathcal{A}$ be the algebra generated by $\mathcal{C}$, that is, the set of finite unions of members of $\mathcal{C}$. For $C \in \mathcal{C}$ given by (2.2), let $\lambda(C) = m(h_1^{-1}A_1 \cap h_2^{-1}A_2 \cap \cdots \cap h_k^{-1}A_k)$, where $A_i = A$ if $\epsilon_i = 1$ and $A_i = A^c$ if $\epsilon_i = 0$, $1 \leq i \leq k$. One easily checks (due to additivity of m) that if C_1 and C_2 are disjoint members of $\mathcal{C}$ and $(C_1 \cup C_2) \in \mathcal{C}$ then $\lambda(C_1 \cup C_2) = \lambda(C_1) + \lambda(C_2)$. It follows that λ may be extended to a finitely additive set function on $\mathcal{A}$. Furthermore, every member of $\mathcal{A}$ is both open and closed in X. Therefore if $(B_i)_{i=1}^\infty$ is a sequence of pairwise disjoint members of $\mathcal{A}$ whose union lies in $\mathcal{A}$ then by compactness all but finitely many of the sets B_i must be empty. In particular, $\lambda(\bigcup B_i) = \sum \lambda(B_i)$. It follows that λ is a *premeasure* on $\mathcal{A}$ and hence extends to a measure μ on the σ-algebra $\mathcal{B}$ of Borel sets. One easily checks that, due to invariance of m, μ is invariant under the shift action $\{T_g\}_{g \in S}$, where $T_g\gamma(h) = \gamma(hg)$, $\gamma \in X$, $h, g \in S$ (it suffices to check this on $\mathcal{A}$). Finally we let $U = \{\gamma \in X : \gamma(e) = 1\}$. Then

$$\begin{aligned}\mu(T_{h_1}^{-1}U \cap T_{h_2}^{-1}U \cap \cdots \cap T_{h_k}^{-1}U) &= \mu(\{\gamma \in X : \gamma(h_1) = 1, \gamma(h_2) = 1, \cdots, \gamma(h_k) = 1\}) \\ &= m(h_1^{-1}A \cap h_2^{-1}A \cap \cdots \cap h_k^{-1}A).\end{aligned}$$

□

This in hand, we are able to give combinatorial necessary and sufficient conditions for a set R to be a set of measurable recurrence. These conditions involve the return of a "large" set $A \subset S$ to itself, in the sense that $A \cap g^{-1}A \neq \emptyset$ for some $g \in R$. (See also Theorem 3.2 and the discussion following Definition 4.1.)

THEOREM 2.2. *Suppose that S is a countable left amenable semigroup. Then $R \subset S$ is a set of measurable recurrence if and only if for every left invariant mean m and every $A \subset S$ with $m(A) > 0$ we have $A \cap g^{-1}A \neq \emptyset$ for some $g \in R$.*

PROOF. ($\Rightarrow$) Suppose m is a left invariant mean and $A \subset S$ with $m(A) > 0$. Let $(X, \mathcal{B}, \mu)$ and $U \in \mathcal{B}$ be as guaranteed by Theorem 2.1. In particular, (2.1) holds. Since R is a set of measurable recurrence there exists $g \in R$ such that $\mu(U \cap T_g^{-1}U) > 0$. Therefore by (2.1) $m(A \cap g^{-1}A) > 0$.

($\Leftarrow$) Let $\{T_g\}_{g \in S}$ be a measure preserving S-action on a probability space $(X, \mathcal{B}, \mu)$ and suppose that $U \in \mathcal{B}$ with $\mu(U) > 0$. Delete from X every set of the form $U \cap T_g^{-1}U$ which is of measure 0, $g \in S$. The reason we are doing this is that in the space which remains, $U \cap T_g^{-1}U \neq \emptyset$ implies $\mu(U \cap T_g^{-1}U) > 0$. For every $x \in X$, let $E_x = \{g \in S : x \in T_g^{-1}U\}$. We claim that $m(E_x) > 0$ for some $x \in X$ and some left invariant mean m. Supposing the claim to be valid, by hypothesis there exists $g \in R$ such that $E_x \cap g^{-1}E_x \neq \emptyset$. Letting $k \in E_x \cap g^{-1}E_x$, we have $T_k x \in U$ and $T_{gk}x \in U$, so that $T_k x \in U \cap T_g^{-1}U$. According to our earlier stipulation, $\mu(U \cap T_g^{-1}U) > 0$. Therefore, all we must do is establish the claim.

We take $l^1(S)$ to consist of functions $f : S \to \mathbf{R}$ satisfying

$$||f|| = \sum_{s \in S} |f(s)| < \infty.$$

Members f of $l^1(S)$ induce elements $\tilde{f}$ of $l^\infty(S)^*$ by the formula

$$\tilde{f}(g) = \sum_{s \in S} f(s)g(s), \quad g \in l^\infty(S).$$

Let $\Phi \subset l^1(S)$ consist of all functions $f : S \to [0, 1]$ such that $\{s : f(s) > 0\}$ is finite and $\sum_{s \in S} f(s) = 1$. For all $f \in \Phi$ one has $\tilde{f}(\mathbf{1}) = 1$ and $\tilde{f}(g) \geq 0$ if $g \geq 0$, hence members of Φ are called *finite means*. For $s \in S$ and $f \in \Phi$, define $s * f \in \Phi$ by $s * f(t) = \sum_{\{h : sh = t\}} f(h)$. A theorem of Day ([**D**]) states that left amenability of S is equivalent to existence of a net $(f_\gamma) \subset \Phi$ satisfying $\lim_\gamma ||s * f_\gamma - f_\gamma|| = 0$. Using the fact that S is countable, one can assume this net to be a *sequence* $(f_i)_{i=1}^\infty$.

For $E \subset S$, let $\overline{m}(E) = \limsup_{n \to \infty} \tilde{f}_n(E)$ (by $\tilde{f}_n(E)$ we mean $\tilde{f}_n(1_E) = \sum_{s \in E} f_n(s)$). Suppose for some $x \in X$ we were to have $\overline{m}(E_x) > 0$. Then for some increasing sequence $(n_k) \subset \mathbf{N}$, $\lim_k \tilde{f}_{n_k}(E) > 0$ and letting m be any weak-* limit point of $\{\tilde{f}_{n_k}\}_{k=1}^\infty$ in $l^\infty(S)^*$, m will be a left invariant mean satisfying $m(E_x) > 0$, completing the proof. We have therefore reduced the problem to finding x with $\overline{m}(E_x) > 0$.

Suppose then that $\overline{m}(E_x) = 0$ for all $x \in X$ (we will arrive at a contradiction). For every x there exists some least number $n_x \in \mathbf{N}$ having the property that for all $n \geq n_x$ we have $\tilde{f}_n(E_x) < \frac{\mu(U)}{2}$. That n_x is a measurable function of x follows from the fact that $x \to \tilde{f}_n(E_x) = \sum_{x \in T_s^{-1}U} f_n(s)$ is measurable for each n. Let $n_0 \in \mathbf{N}$ have the property that $\mu(\{x : n_x > n_0\}) < \frac{\mu(U)}{2}$. For every $n \geq n_0$, we have

$$\int \tilde{f}_n(E_x) \, d\mu(x) < \frac{\mu(U)}{2} + \frac{\mu(U)}{2} = \mu(U).$$

On the other hand,

$$\begin{aligned}\int \tilde{f}_n(E_x)\,d\mu(x) &= \int \sum_{s\in E_x} f_n(s)\,d\mu(x)\\ &= \int \sum_{s\in S} f_n(s)1_{E_x}(s)\,d\mu(x)\\ &= \int \sum_{s\in S} f_n(s)1_{T_s^{-1}U}(x)\,d\mu(x)\\ &= \sum_{s\in S} f_n(s)\mu(T_s^{-1}U) = \mu(U).\end{aligned}$$

This contradiction completes the proof. □

Thus we see that sets of positive measure return to themselves under R if and only if sets of positive density return to themselves in the sense that $A\cap g^{-1}A \neq \emptyset$ for some $g \in R$. (Of course for **Z**-actions this is already well known and Theorem 2.2 is a routine extension of this fact.) For topological recurrence we need a different notion of largeness for subsets of S, namely that of *syndeticity*. A subset of E of **N** is said to be syndetic if it has bounded gaps. Equivalently, $E \subset \mathbf{N}$ is syndetic if **N** is the union of finitely many shifts of E. For non-abelian semigroups, there are two versions of syndeticity, namely left and right, depending on which direction one shifts from. An idea related to syndeticity is that of *thickness*. In an abelian semigroup, a set is thick if it contains a shifted copy of every finite configuration (hence in **Z** a set is thick if and only if it contains arbitrarily long intervals). For non-abelian semigroups there are again two versions. Since we deal with left semigroup actions, left syndeticity and left thickness are the right notions for us (irony unintended).

DEFINITION 2.3. Let S be a countable semigroup and suppose $E \subset S$;

(a) E is said to be *left* (respectively *right*) *syndetic* if there exists a finite set $H \subset S$ such that $S = \bigcup_{t\in H} t^{-1}E$ (respectively $S = \bigcup_{t\in H} Et^{-1}$);

(b) E is said to be *left* (respectively *right*) *thick* if for every finite set $H \subset S$ there exists $x \in S$ such that $Hx \subset E$ (respectively $xH \subset E$).

It is easy to show that E is left (respectively right) syndetic if and only if E^c fails to be left (respectively right) thick. This observation is used in our next proof, as is the well-known fact that if $\{T_g\}_{g\in S}$ is an S-action by continuous self-mappings of a compact space Y then there exists a closed set $X \subset Y$ which is invariant (i.e. $T_gX \subset X$ for all $g \in S$) and with respect to which $\{T_g\}_{g\in S}$ is minimal. That is, if $W \subset X$ is a closed invariant set then $W = X$ or $W = \emptyset$.

THEOREM 2.4. *In a countable semigroup S, a set R is a set of topological recurrence if and only if for every left syndetic set $A \subset S$ there exists $g \in R$ such that $A \cap g^{-1}A \neq \emptyset$.*

PROOF. ($\Rightarrow$) Let $A \subset S$ be left syndetic. By adjoining one if necessary, we will assume that S has an identity e. Put $Z = \{0,1\}^S$. Endowed with the product topology, Z is a compact metric space. A left S-action $\{T_g\}_{g\in S}$ by continuous self-mappings of Z may be given by $T_g\gamma(h) = \gamma(hg)$, $\gamma \in Z$. Let $Y = \overline{\{T_g1_A : g \in S\}}$. One easily checks that Y is S-invariant. Choose a closed invariant set $X \subset Y$ with respect to which the action $\{T_g\}_{g\in S}$ is minimal. Let $U = \{\gamma \in X : \gamma(e) = 1\}$, U is clearly open in X. We claim it is also non-empty. Indeed, for any $\gamma \in X$, $T_g\gamma \in U$ for some $g \in S$. Equivalently, $\gamma(g) = 1$ for some $g \in S$, that is, the function which sends every element of S to 0 does not lie in X. Otherwise, for every finite set $H \subset S$ there would exist $k \in G$ such that $T_k 1_A(s) = 0$ for all $s \in H$, in other words $Hk \subset A^c$, which since H is arbitrary is to say that A^c is left thick, a contradiction. This establishes the claim. Since U is non-empty, and R is a set of topological recurrence, there exists $g \in R$ such that $U \cap T_g^{-1}U \neq \emptyset$. In particular, there exists $\gamma \in Y$ and $g \in R$ such that $\gamma(e) = \gamma(g) = 1$, which implies that for some $h \in S$ we have $T_h 1_A(e) = T_h 1_A(g) = 1$, meaning that h and gh lie in A. That is, $h \in A \cap g^{-1}A$.

($\Leftarrow$) Let $\{T_g\}_{g\in S}$ be a minimal S-action on a compact metric space X and let $U \subset X$ be open. There exists a finite set $H \subset S$ such that $X = \bigcup_{h\in H} T_h^{-1}U$. Let $x \in X$ and put $E = \{g \in S : T_g x \in U\}$. Notice that for every $g \in S$ we have $T_g x \in T_h^{-1}U$ for some $h \in H$, that is $T_{hg}x \in U$, which is to say $g \in h^{-1}E$. Hence $S = \bigcup_{h\in H} h^{-1}E$, so that in particular E is left syndetic and by hypothesis there exists $g \in R$ such that $E \cap T_g^{-1}E \neq \emptyset$. Let $a \in E \cap T_g^{-1}E$ and put $z = T_a x$. Then $z \in U$ and $T_g z \in U$. In other words, $z \in U \cap T_g^{-1}U$. □

Finally, here is the equivalence result for sets of chromatic recurrence.

THEOREM 2.5. *In a semigroup S, a set R is a set of chromatic recurrence if and only if for every partition of S into finitely many cells $A_1, \cdots, A_r$ there exists i, $1 \leq i \leq r$ such that $A_i \cap g^{-1}A_i \neq \emptyset$ for some $g \in R$.*

PROOF. ($\Rightarrow$) Suppose a partition $S = A_1 \cup \cdots \cup A_r$ is given. If S contains an identity, call it e and let $A_{r+1} = \emptyset$. Otherwise, adjoin an identity e to S and put $A_{r+1} = \{e\}$. Let $Z = \{1, 2, \cdots, r+1\}^S$ and define $\gamma \in Z$ by $\gamma(g) = i$ if and only if $g \in A_i$. Let $X = \overline{\{T_g\gamma : g \in S\}}$, where $T_g\alpha(h) = \alpha(hg)$, $g \in S$, $\alpha \in Z$. Then X is S-invariant. Let $U_i = \{\alpha \in X : \alpha(e) = i\}$, $1 \leq i \leq r+1$. Then $\{U_1, \cdots, U_{r+1}\}$ is an open covering of X. Since R is a set of chromatic recurrence, there exists i, $1 \leq i \leq r+1$, and $g \in R$ such that $U_i \cap T_g^{-1}U_i \neq \emptyset$. This implies that $T_k\gamma \in U_i \cap T_g^{-1}U_i$ for some $k \in S$. Hence $k \in A_i$ and $gk \in A_i$, that is, $k \in A_i \cap g^{-1}A_i$. If $i \neq r+1$ we are done, but if $i = r+1$ then $k = g = e$, a contradiction since $g \in R \subset \bigcup_{i=1}^r A_i$.

($\Leftarrow$) Let $\{T_g\}_{g\in S}$ be an S-action by continuous self-mappings of a compact metric space X and suppose that $\{U_1, \cdots, U_r\}$ is an open covering. Let $x \in X$ and for every $g \in S$ let i_g be the minimal i such that $T_g x \in U_i$. Let $A_i = \{g \in S : i_g = i\}$. Then $S = A_1 \cup \cdots \cup A_r$ is a finite partition. By hypothesis, there exists i, $1 \leq i \leq r$, and $g \in R$ such that $A_i \cap g^{-1}A_i \neq \emptyset$. Let $k \in A_i \cap g^{-1}A_i$. Then $T_k x \in U_i$ and $T_{gk}x \in U_i$. It follows that $T_k x \in U_i \cap T_g^{-1}U_i$. □

The following theorem indicates in part the relative strength of the three notions of sets of recurrence we have defined.

THEOREM 2.6. *Suppose that S is a countable semigroup and $R \subset S$.*

(a) *If R is a set of topological recurrence then R is a set of chromatic recurrence.*
(b) *If R is a set of chromatic recurrence and $R \subset Z(S) = \{s \in S : st = ts \text{ for all } t \in S\}$ (in particular, if S is abelian) then R is a set of topological recurrence.*

(c) *There exists an example where S is a group and R is both a set of measurable recurrence and a set of chromatic recurrence, but R is not a set of topological recurrence.*
(d) *If R is a set of measurable recurrence and S is left amenable then R is both a set of topological recurrence and a set of chromatic recurrence.*
(e) *There exists an example in which $S = \mathbf{N}$, R is both a set of topological recurrence and a set of chromatic recurrence, but R is not a set of measurable recurrence.*

PROOF.
(a) Suppose that $\{T_g\}_{g\in S}$ is an S-action by continuous self-mappings of a compact metric space X and suppose that $\{U_1, \cdots, U_r\}$ is an open covering of X. Let $Y \subset X$ be a minimal invariant set. For some i we have $U = Y \cap U_i \neq \emptyset$. Since R is a set of topological recurrence and U is a non-empty open subset of Y, $U \cap T_g^{-1}U \neq \emptyset$ for some $g \in R$. Clearly we have $U_i \cap T_g^{-1}U_i \neq \emptyset$ as well. (The reader is urged to supply an alternative proof of (a) using the equivalent combinatorial characterizations of Theorems 2.4 and 2.5.)
(b) Suppose that R is a set of chromatic recurrence contained in $Z(S)$ and that $L \subset S$ is left syndetic. Then there exists a finite set $H \subset S$ such that $S = \bigcup_{h\in H} h^{-1}L$. Since R is a set of chromatic recurrence there exist $h \in H$ and $g \in R$ such that $(h^{-1}L)\cap g^{-1}(h^{-1}L) \neq \emptyset$. Let $k \in (h^{-1}L)\cap g^{-1}(h^{-1}L)$. Then $hk \in L$ and $hgk \in L$. But g commutes with h, so $hk \in L \cap g^{-1}L$. By Theorem 2.4, R is a set of topological recurrence.
(c) Let S be the free group on the letters a and b and let R be the set $\{a^k : k \in \mathbf{N}\}$. Let $S = A_1 \cup \cdots \cup A_r$ be any partition of S and choose $m < n$ such that a^m and a^n lie in the same cell A_i. Then $a^m \in A_i \cap (a^{n-m})^{-1}A_i$. This shows that R is a set of chromatic recurrence. Let $\{T_g\}_{g\in S}$ be a measure preserving S-action on a probability space $(X, \mathcal{B}, \mu)$ and suppose $U \in \mathcal{B}$ with $\mu(U) > 0$. Choose $m < n$ with $\mu(T_{a^m}^{-1}U \cap T_{a^n}^{-1}U) > 0$. Then $\mu(U \cap T_{a^{n-m}}^{-1}U) > 0$. This shows that R is a set of measurable recurrence. To see that R is not a set of topological recurrence, consider the left syndetic set B consisting of all words that begin with either b or b^{-1}. Clearly $B \cap (a^k)^{-1}B = \emptyset$ for all $k \in \mathbf{N}$. By Theorem 2.4, R cannot be a set of topological recurrence.
(d) Suppose that R is a set of measurable recurrence and S is left amenable. Let m be a left invariant mean and suppose $L \subset S$ is left syndetic. For some finite set $\{h_1, \cdots, h_n\} \subset S$, $S = \bigcup_{i=1}^n h_i^{-1}L$. We must have $m(L) \geq \frac{1}{n}$. Since R is a set of measurable recurrence, by Theorem 2.2 there exists $g \in R$ such that $L \cap g^{-1}L \neq \emptyset$. Therefore by Theorem 2.4, R is a set of topological recurrence and by (a) a set of chromatic recurrence as well.
(e) This non-trivial result, mentioned in the introduction, is due to Kriz ([**K**], see also [**Fo**] and [**M**]). □

We now see that, while sets of measurable recurrence and sets of chromatic recurrence are partition regular, sets of topological recurrence are not.

THEOREM 2.7. *Let S be a countable semigroup. Suppose $R_1, R_2, \cdots, R_k \subset S$ and let $R = R_1 \cup R_2 \cup \cdots \cup R_k$.*

(a) *If R is a set of measurable recurrence then R_i is a set of measurable recurrence for some i, $1 \leq i \leq k$.*

(b) *If R is a set of chromatic recurrence then R_i is a set of chromatic recurrence for some i, $1 \le i \le k$.*
(c) *There exists an example in which S is a group, R is a set of topological recurrence and no R_i is a set of topological recurrence, $1 \le i \le k$.*

PROOF.
(a) Suppose not. Then there exist probability spaces $(X_i, \mathcal{B}_i, \mu_i)$, positive measure sets $A_i \in \mathcal{B}_i$, and measure preserving S-actions $\{T_g^{(i)}\}_{g\in S}$ such that

$$\mu_i(A_i \cap (T_g^{(i)})^{-1}A_i) = 0, \quad g \in R_i, \ 1 \le i \le k.$$

Let $X = \prod_{i=1}^k X_i$, $\mu = \prod_{i=1}^k \mu_i$, $T_g = \prod_{i=1}^k T_g^{(i)}$ and put $A = \prod_{i=1}^k A_i$. Then $\mu(A) > 0$ and yet $\mu(A \cap T_g^{-1}A) = 0$ for all $g \in R$, a contradiction.
(b) Suppose not. Then there exist S-actions $\{T_g^{(i)}\}_{g\in S}$ by continuous self-mappings of compact metric spaces X_i, and open covers $\{U_1^{(i)}, \cdots, U_{t_i}^{(i)}\}$ of X_i such that

$$U_j^{(i)} \cap (T_g^{(i)})^{-1}U_j^{(i)} = \emptyset, \quad 1 \le i \le k, \ 1 \le j \le t_i, \ g \in R_i.$$

Let $X = \prod_{i=1}^k X_i$, $T_g = \prod_{i=1}^k T_g^{(i)}$ and let $\mathcal{U}$ be the open cover of X consisting of sets of the form $\prod_{i=1}^k U_{j_i}^{(i)}$, where $1 \le j_i \le t_i$, $1 \le i \le k$. Then for every $U \in \mathcal{U}$ and every $g \in R$ we have $U \cap T_g^{-1}U = \emptyset$, a contradiction.
(c) Let S be the free group on the letters $\{g, h, k\}$. $R = S \setminus \{e\}$ is left thick and hence a set of topological recurrence (see Example 4.4 (b)). If $w_1, w_2 \in R$, let $w_1 \sim w_2$ if the first letter of w_1 in its reduced form is the same as or the inverse of the first letter of w_2 in its reduced form, and if the last letter of w_1 in its reduced form is the same as or the inverse of the last letter of w_2 in its reduced form. Let $R_1, \cdots, R_9$ be the set of equivalence classes under $\sim$. Given i, $1 \le i \le 9$, one easily finds a left syndetic set L such that $L \cap g^{-1}L = \emptyset$ for all $g \in R_i$. For example, suppose that R_i is the set of words beginning with g or g^{-1} and ending with h or h^{-1}. Let L be the set of words beginning with k or k^{-1}. Then L is left syndetic, but $L \cap g^{-1}L = \emptyset$ for $g \in R_i$. By Theorem 2.5 none of the sets R_i is a set of topological recurrence. □

REMARK. At this stage the alert reader may wonder why we even care about sets of "topological recurrence". Insofar as they fail to provide a connection to partition Ramsey theory, and insofar as they fail to possess many of the good properties of their counterparts, sets of measurable recurrence and sets of chromatic recurrence, they may appear in many ways to be a useless and inappropriate idea. Nevertheless, we have our reasons, which will be very briefly expounded upon in Section 5.

We now move to discussion which will serve as a preparation for the non-commutative Schur theorem we aim to prove in the next section. Schur's theorem, which states that for any finite coloring of a semigroup S a configuration of the type $\{x, y, xy\}$ can be found in one color, has been given many different proofs. Schur's original method ([**S**], see also [**Si**], [**H**]) employed counting techniques. Another way is to use *Ramsey's theorem.* A proof along these lines may be found in [**GRS**].

A third idea is to derive Schur's theorem as a consequence of Hindman's theorem or its proof, for example by using the existence of an idempotent ultrafilter in βS.

We do not see any way to generalize any of these methods to get $\{x, y, xy, yx\}$ in one color with $xy \neq yx$. Consider, however, the following proof of Schur's theorem which uses partition regularity of sets of recurrence and is valid for countable amenable semigroups (this proof is an adaptation of the proof of a *density Schur theorem* in [**B1**]): Suppose $S = \bigcup_{i=1}^{r} C_i$ is a finite partition of a countable, left amenable semigroup S and let m be a left invariant mean. Renumbering if necessary we may assume that there exists r_0 such that $m(C_i) > 0$ when $1 \leq i \leq r_0$, and $m(C_i) = 0$ when $r_0 < i \leq r$. Then $m(\bigcup_{i=1}^{r} C_i) = 1$, so that in particular $\bigcup_{i=1}^{r} C_i$ is left thick ([**P**], Proposition 1.21]) and hence a set of measurable recurrence (Example 4.4 (a) gives something even stronger). By Theorem 2.7 (a), therefore, there exists i with $1 \leq i \leq r_0$ such that C_i is a set of measurable recurrence. In particular, by Theorem 2.2 we have $C_i \cap x^{-1}C_i \neq \emptyset$ for some $x \in C_i$. Letting $y \in C_i \cap x^{-1}C_i$, one obtains $\{x, y, xy\} \subset C_i$.

The preceding proof uses a recurrence fact, namely the fact that left thick sets are sets of measurable recurrence. The proof in the next section substitutes for this assertion the following stronger result from [**BMZ**]: in amenable groups, left thick sets are sets of measurable *double recurrence.* What exactly this means will be explained presently.

3. Double Recurrence and a Non-Commutative Schur Theorem

A special case of Szemerédi's theorem due to K. Roth ([**R**]) asserts that if $E \subset \mathbf{N}$ has positive upper density then E contains an arithmetic progression of length 3. Roth's theorem is easily shown via Furstenberg correspondence to be a consequence of the following *double recurrence* strengthening of the Poincaré recurrence theorem: for any finite measure preserving system $(X, \mathcal{B}, \mu, T)$ and any $A \in \mathcal{B}$ with $\mu(A) > 0$ there exists $n \in \mathbf{N}$ such that $\mu(A \cap T^{-n}A \cap T^{-2n}A) > 0$. This of course is a special case of Furstenberg's multiple recurrence theorem, however it may also be obtained (see [**F1**], [**F2**]) as a consequence of what Furstenberg calls the *ergodic Roth theorem*, which asserts that $\frac{1}{N}\sum_{n=1}^{N} T^n f T^{2n} g$ converges in $L^2(X, \mathcal{B}, \mu)$ for all $f, g \in L^\infty(X, \mathcal{B}, \mu)$ and identifies the limit.

The phenomenon of double recurrence prompts the question of which sets it occurs along. In the following definition and elsewhere, we take two S-actions $\{T_g\}_{g\in S}$ and $\{S_g\}_{g\in S}$ to be commuting if $T_g S_h = S_h T_g$ for all $g, h \in S$.

DEFINITION 3.1. Let R be a subset of a countable semigroup S. Then R is called a *set of measurable 2-recurrence* if for any pair of commuting measure preserving S-actions $\{T_g\}_{g\in S}$ and $\{S_g\}_{g\in S}$ on a probability space $(X, \mathcal{B}, \mu)$ and every $A \in \mathcal{B}$ with $\mu(A) > 0$ there exists $g \in R$ such that

$$\mu(A \cap T_g^{-1}A \cap (T_g S_g)^{-1}A) > 0. \tag{3.1}$$

Like sets of single measurable recurrence, this notion has an equivalent, combinatorial formulation which may be proved by methods similar to those used for Theorem 2.2.

THEOREM 3.2. *Let $R \subset S$, where S is a countable left amenable semigroup. Then R is a set of measurable 2-recurrence if and only if for every left invariant mean m on $l^\infty(S \times S)$ and every $A \subset S \times S$ with $m(A) > 0$, A contains a configuration of the form $\{(a,b),(ga,b),(ga,gb)\}$, where $g \in R$.*

We would now like to say a few words about inequality (3.1). If S is abelian and (3.1) were to be replaced by the alternate expression $\mu(A \cap T_g^{-1}A \cap S_g^{-1}A) > 0$, the resulting definition for sets of 2-recurrence would be equivalent to the given one. However, if S is non-abelian, the corresponding alternative definition is neither thought to be equivalent to Definition 3.1, nor has this alternate version been conducive to study. Indeed, it is unknown whether or not for every countable amenable group S, given commuting measure preserving S-actions $\{T_g\}_{g\in S}$ and $\{S_g\}_{g\in S}$ on a probability space $(X,\mathcal{B},\mu)$, and $A \in \mathcal{B}$ with $\mu(A) > 0$ there exists $g \neq e$ with $\mu(A \cap T_g^{-1}A \cap S_g^{-1}A) > 0$. On the other hand, one has the following.

THEOREM 3.3. *([**BMZ**]) If G is a countable amenable group, $\{T_g\}_{g\in G}$ and $\{S_g\}_{g\in G}$ are commuting measure preserving G-actions on a probability space $(X,\mathcal{B},\mu)$, and $A \in \mathcal{B}$ with $\mu(A) > 0$ then the set $\{g \in G : \mu(A \cap T_g^{-1}A \cap (T_gS_g)^{-1}A) > 0\}$ is both left and right syndetic. In particular, any left or right thick subset of G is a set of measurable 2-recurrence.*

As an application of Theorem 3.3 we present the main original result of this paper, namely a non-commutative Schur theorem for amenable groups. Recall that a sequence of finite subsets $\{\Phi_n\}_{n=1}^\infty$ in a countable amenable group G is called a *left (respectively right) Følner sequence* if for every $s \in G$ we have $\frac{|\Phi_n \cap s\Phi_n|}{|\Phi_n|} \to 1$ (respectively $\frac{|\Phi_n \cap \Phi_n s|}{|\Phi_n|} \to 1$) as $n \to \infty$. A sequence $\{\Phi_n\}_{n=1}^\infty$ will be called a *two-sided Følner sequence* if it is both a left and a right Følner sequence. It is well known that every countable amenable group admits a two-sided Følner sequence. If $\{\Phi_n\}_{n=1}^\infty$ is a (left, right or two-sided) Følner sequence and $E \subset G$, the *upper density* of E with respect to $\{\Phi_n\}_{n=1}^\infty$ is defined to be the number

$$\overline{d}(E) = \limsup_{n\to\infty} \frac{|E \cap \Phi_n|}{|\Phi_n|}.$$

One easily checks that if $\{\Phi_n\}_{n=1}^\infty$ is a left (respectively right) Følner sequence then $\overline{d}(sE) = \overline{d}(E)$ (respectively $\overline{d}(Es) = \overline{d}(E)$) for all $s \in G$ and $E \subset G$. If $\lim_{n\to\infty} \frac{|E\cap\Phi_n|}{|\Phi_n|}$ exists, we call the limit $d(E)$ and say E has *density* $d(E)$. Working with d has an advantage over working with $\overline{d}$ in that d is additive in the following sense: if $E \cap F = \emptyset$ and $d(E)$ and $d(F)$ both exist then $d(E \cup F) = d(E) + d(F)$.

The following well known facts will be used in the proof of Theorem 3.4:

(a) Suppose $E_i \subset G$, $i \in \mathbf{N}$, and $\{\Phi_n\}_{n=1}^\infty$ is a left Følner sequence for G. There exists an increasing sequence $(n_k)_{k\in\mathbf{N}} \subset \mathbf{N}$ such that for all $i \in \mathbf{N}$, $d(E_i)$ exists with respect to the left Følner sequence $\{\Phi_{n_k}\}_{k\in\mathbf{N}}$.
(b) If $\{\Phi_n\}_{n=1}^\infty$ is a left Følner sequence then there exists a left invariant mean m having the property that for every $E \subset G$ for which $d(E)$ (taken with respect to $\{\Phi_n\}_{n=1}^\infty$) exists, one has $m(E) = d(E)$.
(c) If G is a countable amenable group then $G \times G$ is as well. Furthermore, if $\{\Phi_n\}_{n=1}^\infty$ and $\{\Psi_n\}_{n\in\mathbf{N}}$ are any two left Følner sequences for G then $\{\Phi_n \times \Psi_n\}_{n\in\mathbf{N}}$ is a left Følner sequence for $G \times G$.

(d) If G is a countable amenable group, $A \subset G$ is a subgroup, and $\{\Phi_n\}_{n=1}^{\infty}$ is any left Følner sequence for G then $d(A)$ exists and is equal to $\frac{1}{[G:A]}$ if $[G : A] < \infty$ and 0 if $[G : A] = \infty$. (This is a consequence of shift invariance of $\bar{d}$ and the fact that G is the disjoint union of the left cosets of A.)

(e) If G is a group having two subgroups of finite index A and B then $A \cap B$ is of finite index as well. (If $[G : A] = m$ and $[G : B] = n$, consider the homomorphism $g \rightarrow (\phi_1(g), \phi_2(g))$ of G into $S_m \times S_n$ (S_m being the group of permutations of the left cosets of A and S_n being the group of permutations of the left cosets of B), where $\phi_1(g)(hA) = (gh)A$ and $\phi_2(g)(hB) = (gh)B$. The kernel lies in $A \cap B$.)

THEOREM 3.4. *Suppose that G is a countable amenable group and that $G = \bigcup_{i=1}^{r} C_i$ is a finite partition. Let $A = \{g \in G : [G : C(g)] < \infty\}$. If $[G : A] = \infty$ then there exist $x, y \in G$ and i, $1 \leq i \leq r$, with $xy \neq yx$ and such that $\{x, y, xy, yx\} \subset C_i$.*

PROOF. That A is a subgroup of G is a consequence of (e), since $C(g) \cap C(h)$ is contained in $C(gh)$. Let $\{\Phi_n\}_{n=1}^{\infty}$ be a *two-sided* Følner sequence for G. Passing if necessary to a sub-sequence of $\{\Phi_n\}_{n=1}^{\infty}$ we may by (a) assume that $d(C_i)$ exists, $1 \leq i \leq r$. Renumbering if necessary, we may also assume that $d(C_i) > 0$, $1 \leq i \leq r_0$, and $d(C_i) = 0$, $r_0 < i \leq r$. Then $d(\bigcup_{i=1}^{r_0} C_i) = 1$. Let $E = G \setminus A$ and replace C_i by $C_i \cap E$, $1 \leq i \leq r_0$. By (d) we have $d(E) = 1$, so we still have $d(\bigcup_{i=1}^{r_0} C_i) = 1$, and moreover $\bigcup_{i=1}^{r_0} C_i \subset E$. By (b) there exists some left (or right, since $\{\Phi_n\}_{n=1}^{\infty}$ is two-sided) invariant mean l such that $l(\bigcup_{i=1}^{r_0} C_i) = 1$. In particular by [**P**], Theorem 1.21] $\bigcup_{i=1}^{r_0} C_i$ is both left and right thick. Since a subset F of G is left (respectively right) thick if and only if F^{-1} is right (respectively left) thick, it follows that $\bigcup_{i=1}^{r_0} C_i^{-1}$ is also both left and right thick.

Let $(k_n)_{n\in\mathbf{N}}$ be a sequence in $\mathbf{N}$ having the property that

$$\frac{|\Phi_{k_n} \cap \Phi_{k_n} g^{-1}|}{|\Phi_{k_n}|} > 1 - \frac{1}{n} \tag{3.2}$$

for all $n \in \mathbf{N}$ and all $g \in \Phi_n$ (this is possible since $\{\Phi_n\}_{n=1}^{\infty}$ is a right Følner sequence). By (c), $\{\Phi_{k_n} \times \Phi_n\}_{n\in\mathbf{N}}$ is a left Følner seqence for $G \times G$. For $C \subset G$ let $\tilde{C} = \{(a, b) \in G \times G : ab^{-1} \in C\}$. We claim that if $d(C)$ exists then $d(\tilde{C})$ (measured with respect to $\{\Phi_{k_n} \times \Phi_n\}_{n\in\mathbf{N}}$) exists and equals $d(C)$. To see this, consider that

$$\begin{aligned}
\lim_{n\to\infty} \frac{|\tilde{C} \cap (\Phi_{k_n} \times \Phi_n)|}{|\Phi_{k_n} \times \Phi_n|} &= \lim_{n\to\infty} \frac{1}{|\Phi_{k_n}||\Phi_n|} \sum_{g\in\Phi_n} |\tilde{C} \cap (\Phi_{k_n} \times \{g\})| \\
&= \lim_{n\to\infty} \frac{1}{|\Phi_{k_n}||\Phi_n|} \sum_{g\in\Phi_n} |\Phi_{k_n} \cap Cg| \\
&= \lim_{n\to\infty} \frac{1}{|\Phi_n|} \sum_{g\in\Phi_n} \frac{|\Phi_{k_n} g^{-1} \cap C|}{|\Phi_{k_n}|} \\
&= \lim_{n\to\infty} \frac{1}{|\Phi_n|} \sum_{g\in\Phi_n} \frac{|\Phi_{k_n} \cap C|}{|\Phi_{k_n}|} = d(C).
\end{aligned}$$

(We have used (3.2) in the final equality.) In particular for every i, $1 \leq i \leq r_0$, $d(\tilde{C}_i)$ exists and is positive. By (a), we may by passing to a sub-sequence of the

Følner sequence $\{\Phi_{k_n} \times \Phi_n\}_{n\in\mathbf{N}}$ assume that for all $(g_1,h_1),(g_2,h_2) \in G\times G$,

$$d(\tilde{C}_i \cap (g_1,h_1)^{-1}\tilde{C}_i \cap (g_2,h_2)^{-1}\tilde{C}_i)$$

exists, $1 \le i \le r_0$. Using (b), let m be a left-invariant mean on $G\times G$ having the property that $m(E) = d(E)$ for all $E \subset G\times G$ for which $d(E)$ exists. By Theorem 2.1, for $1 \le i \le r_0$ there exist probability spaces $(X_i,\mathcal{B}_i,\mu_i)$, sets $A_i \in \mathcal{B}_i$ with $\mu_i(A_i) > 0$, and measure preserving $G\times G$-actions $\{T^{(i)}_{(g,h)}\}_{(g,h)\in G\times G}$ such that for any (g_1,h_1) and (g_2,h_2) in $G\times G$ we have

$$\begin{aligned} d(\tilde{C}_i \cap (g_1,h_1)^{-1}\tilde{C}_i \cap (g_2,h_2)^{-1}\tilde{C}_i) &= m(\tilde{C}_i \cap (g_1,h_1)^{-1}\tilde{C}_i \cap (g_2,h_2)^{-1}\tilde{C}_i) \\ &= \mu(A_i \cap (T^{(i)}_{(g_1,h_1)})^{-1}A_i \cap (T^{(i)}_{(g_2,h_2)})^{-1}A_i), \end{aligned}$$

$1 \le i \le r_0$. Let $X = \prod_{i=1}^{r_0} X_i$, $\mu = \prod_{i=1}^{r_0}\mu_i$, $T_{(g,h)} = \prod_{i=1}^{r_0} T^{(i)}_{(g,h)}$, and $A = \prod_{i=1}^{r_0} A_i$. Then $\mu(A) > 0$, so by Theorem 3.3 the set

$$R = \{g \in G : \mu(A \cap T^{-1}_{(g,e)}A \cap T^{-1}_{(g,g)}A) > 0\}$$

is both left and right syndetic in G. We have

$$d(\tilde{C}_i \cap (g^{-1},e)\tilde{C}_i \cap (g^{-1},g^{-1})\tilde{C}_i) > 0,\ 1 \le i \le r_0,\ g \in R.$$

Since R is left and right syndetic and $\bigcup_{i=1}^{r_0} C_i^{-1}$ is left and right thick, there exists $g \in R \cap \bigcup_{i=1}^{r_0} C_i^{-1}$. Hence for some i, $1 \le i \le r_0$, we have $g \in R \cap C_i^{-1}$. Letting $C_0 = G \setminus C(g)$, we have by (d) that $d(C_0) = 1$ with respect to $\{\Phi_n\}_{n=1}^{\infty}$ (since $g^{-1} \in \bigcup_{i=1}^{r_0} C_i \subset E$), so that $d(\tilde{C}_0) = 1$ with respect to $\{\Phi_{k_n}\times\Phi_n\}_{n\in\mathbf{N}}$. Since

$$d(\tilde{C}_i \cap (g^{-1},e)\tilde{C}_i \cap (g^{-1},g^{-1})\tilde{C}_i) > 0,$$

there exists

$$(a,b) \in (\tilde{C}_0 \cap \tilde{C}_i \cap (g^{-1},e)\tilde{C}_i \cap (g^{-1},g^{-1})\tilde{C}_i).$$

Then ab^{-1} doesn't commute with g and $\{g^{-1}, ab^{-1}, gab^{-1}, gab^{-1}g^{-1}\} \subset C_i$. Letting $x = g^{-1}$ and $y = gab^{-1}$ we have $xy \ne yx$ and $\{x,y,xy,yx\} \subset C_i$. □

4. k-Recurrence, Questions, and Examples

As an extension of the ideas brought forth in the previous two sections, we now discuss those classes of sets along which one has higher orders of recurrence. The reader will notice that the case $k = 1$ in what is to follow gives the notions of Section 2, while the case $k = 2$ in part (a) corresponds to Definition 3.1.

DEFINITION 4.1. Let $k \in \mathbf{N}$ and suppose that R is a subset of a countable semigroup S:

(a) R is called a *set of measurable k-recurrence* if for every k commuting measure preserving S-actions $\{T^{(i)}_g\}_{g\in S}$, $1 \le i \le k$, on a probability space $(X,\mathcal{B},\mu)$ and every $A \in \mathcal{B}$ with $\mu(A) > 0$ we have

$$\mu(A \cap (T^{(1)}_g)^{-1}A \cap (T^{(1)}_g T^{(2)}_g)^{-1}A \cap \cdots \cap (T^{(1)}_g \cdots T^{(k)}_g)^{-1}A) > 0$$

for some $g \in R$;

(b) R is called a *set of topological k-recurrence* if for every k commuting S-actions $\{T^{(i)}_g\}_{g\in S}$, $1 \le i \le k$, by continuous self-mappings of a compact metric space X, which is minimal with respect to the resulting S^k-action, and every open set $U \subset X$ we have

$$U \cap (T_g^{(1)})^{-1}U \cap (T_g^{(1)}T_g^{(2)})^{-1}U \cap \cdots \cap (T_g^{(1)} \cdots T_g^{(k)})^{-1}U \neq \emptyset$$

for some $g \in R$;

(c) R is called a *set of chromatic k-recurrence* if for every k commuting S-actions $\{T_g^{(i)}\}_{g \in S}$, $1 \leq i \leq k$, by continuous self-mappings of a compact metric space X, and every open covering $\{U_1, \cdots, U_r\}$ of X there exists i with $1 \leq i \leq r$ such that

$$U_i \cap (T_g^{(1)})^{-1}U_i \cap (T_g^{(1)}T_g^{(2)})^{-1}U_i \cap \cdots \cap (T_g^{(1)} \cdots T_g^{(k)})^{-1}U_i \neq \emptyset$$

for some $g \in R$.

If a set R is a set of measurable (respectively topological, chromatic) k-recurrence for every $k \in \mathbf{N}$, then R is said to be a set of measurable (respectively topological, chromatic) *multiple* recurrence. The following remarks are given as motivation for the questions to follow.

REMARKS 4.2.

A. In $S = \mathbf{N}$ there exist sets of measurable 1-recurrence which are known to fail 2-recurrence ([**F2**], p.177]), or for which it is unknown whether or not they are sets of 2-recurrence, eg. $\{p-1 : p \text{ prime}\}$ (for a proof that this set is density intersective and hence a set of measurable recurrence see [**Sa**]). However, every set $R \subset \mathbf{N}$ which is known to be a set of 2-recurrence is also known to be a set of multiple recurrence.

B. In $S = \mathbf{N}$ there exist sets, e.g. the example of Kriz, which are sets of chromatic 1-recurrence but fail to be sets of measurable 1-recurrence.

C. According to Theorem 2.6 (d) every set of measurable recurrence in a countable left amenable semigroup S is a set of chromatic recurrence.

D. According to Theorem 2.6 (b), for countable abelian groups any set of chromatic recurrence is a set of topological recurrence, but there exist sets of chromatic recurrence in free groups which are not sets of topological recurrence.

QUESTIONS 4.3.

A. Does there exist a set $R \subset \mathbf{N}$ which is a set of measurable 2-recurrence but not a set of measurable multiple recurrence?

B. Does there exist a set $R \subset \mathbf{N}$ which is a set of chromatic 2-recurrence but not a set of measurable 1-recurrence?

C. Does there exist an example of a countable semigroup S and a set $R \subset S$ which is a set of measurable recurrence but is not a set of chromatic recurrence?

D. For which classes of countable semigroups, if any, intermediate to abelian and free, are all sets of chromatic recurrence sets of topological recurrence?

REMARK. Our guess is that the answers to Questions 4.3 A, B, and C should all be *yes*.

For the purposes of describing the current state of knowledge, we offer a few examples of types of sets that are known to be sets of k-recurrence for various k. The two most commonly encountered, non-trivial classes of sets known to have strong recurrence properties in semigroups are left thick sets (see Definition 2.3 (b)), and *IP sets*. Recall that in a semigroup S, the (left) *IP set* generated by a sequence $\langle y_n\rangle_{n\in\mathbf{N}} \subset S$ is the set of products of finite subsets of the sequence taken in order of increasing indices, namely the set $FP(\langle y_n\rangle_{n\in\mathbf{N}}) = \{y_{i_1}\cdots y_{i_k} : i_1 < \cdots < i_k\}$. Every left thick set L contains an IP set. Indeed, let $y_1 \in L$, and having chosen $y_1, y_2, \cdots, y_n \in S$ such that $FP(\langle y_k\rangle_{1\le k\le n}) \subset L$, choose y_{n+1} with $FP(\langle y_k\rangle_{1\le k\le n})y_{n+1} \subset L$. It follows that in the examples to follow any assertion made about IP sets is true for left thick sets as well.

EXAMPLE 4.4.

(a) Every IP set in a countable semigroup is both a set of measurable recurrence and a set of chromatic recurrence.
(b) Every left thick set in a countable semigroup is a set of topological recurrence.
(c) Every IP set in a countable group is a set of chromatic 2-recurrence ([**BH**], Corollary 2.7]). IP sets in countable groups need not, however, be sets of chromatic 3-recurrence ([**BH**], Corollary 4.6]), nor are IP sets in semigroups necessarily sets of chromatic 2-recurrence (see Theorem 4.5 below).
(d) Every IP set in a countable abelian group is a set of measurable multiple recurrence ([**FK**]) and a set of topological (or chromatic, which in the abelian case is the same) multiple recurrence ([**FW**]).
(e) ([**BMZ**]) Every left thick set in a countable amenable group is a set of measurable 2-recurrence, a set of topological 2-recurrence, and a set of chromatic 3-recurrence.

Although it does not provide an answer to Question 4.3 C, consider that if R is a set of topological recurrence in a countable semigroup S, $(X, \mathcal{B}, \mu)$ is a finite probability space, $A \in \mathcal{A}$ with $\mu(A) > 0$ and $\{R_g\}_{g\in S}$, $\{T_g\}_{g\in S}$ are commuting measure preserving S-actions on X, for some $g \in R$ we have $\mu(R_g^{-1}A \cap (R_gT_g)^{-1}A) = \mu(A \cap T_g^{-1}A) > 0$. The analogous "chromatic" statement is false:

THEOREM 4.5. *There exist a countable semigroup S, a set $R \subset S$ which is a set of chromatic (and measurable) recurrence, a compact space X, two commuting S-actions by continuous self-mappings of X $\{T_g\}_{g\in S}$ and $\{R_g\}_{g\in S}$, and an open cover $\{U_1, \cdots, U_r\}$ of X such that $R_g^{-1}U_i \cap (R_gT_g)^{-1}U_i = \emptyset$ for all $g \in R$ and $1 \le i \le r$.*

PROOF. Let S be the free semigroup on the countable family of letters $\langle y_n\rangle_{n\in\mathbf{N}}$ with identity e, and let R be the IP set $FP(\langle y_n\rangle_{n\in\mathbf{N}})$. Then R is a set of chromatic and measurable recurrence. According to [**BH**], Theorem 4.3], there exists a finite partition $\{A_1, \cdots, A_r\}$ of $S \times S$ such that no cell contains a configuration of the form $\{(ga, b), (ga, gb)\}$ for any $g \in FP(\langle y_n\rangle_{n\in\mathbf{N}})$ (any reader who looks up this reference is advised that we are performing a left-right switch in the semigroup multiplication due to the fact that IP sets are taken with multiplication in *decreasing* order of indices there).

Let $Z = \{1, \cdots, r\}^{S\times S}$. Let ξ be the element of Z defined by $\xi(s,t) = i$ if and only if $(s,t) \in A_i$, $1 \le i \le r$, $s, t \in S$. Define commuting S-actions $\{R_g\}_{g\in S}$ and $\{T_g\}_{g\in S}$ on X by $R_g\gamma(h,k) = \gamma(hg,k)$ and $T_g\gamma(h,k) = (h, kg)$. Put $X = \overline{\{T_gR_h\xi : (g,h) \in S \times S\}}$. Then X is invariant under both actions $\{R_g\}_{g\in S}$ amd $\{T_g\}_{g\in S}$. Let $U_i = \{\gamma \in X : \gamma(e,e) = i\}$, $1 \le i \le r$. Then $\{U_1, \cdots, U_r\}$ is an open cover of X.

Suppose now for some i, $1 \leq i \leq r$, and some $g \in R$, we have $R_g^{-1}U_i \cap (R_gT_g)^{-1}U_i \neq \emptyset$. Then for some $(a,b) \in S \times S$ we have $R_aT_b\xi \in R_g^{-1}U_i \cap (R_gT_g)^{-1}U_i$. Then $R_{ga}T_b\xi$ and $R_{ga}T_{gb}\xi$ both lie in U_i, that is $\{(ga,b),(ga,gb)\} \subset A_i$, a contradiction. □

5. Topological Recurrence Vs. Chromatic Recurrence

We would like to conclude with a few words concerning the relationship of topological k-recurrence to chromatic k-recurrence. For abelian groups these notions are the same, and a close inspection reveals that it is this equivalence which lies at the heart of the inductive proofs of multiple recurrence results in topological dynamics, e.g. the topological proof of van der Waerden's theorem. One is able to show that if a class of sets (such as thick sets or IP sets) having certain translation properties (for example, that any set in the class may be shifted by many of its own members and yield another set in the class) are sets of topological k-recurrence then they are also sets of topological $k+1$-recurrence. In the non-abelian case, this induction breaks down, but in an interesting way. One is still able to show that if members of the class are sets of topological k-recurrence then they are sets of chromatic $k+1$-recurrence. An argument of this nature was carried out in [**BMZ**], Theorem 7.1], where it was shown that left thick sets are sets of chromatic 3-recurrence for countable amenable groups. (This example, incidentally, marks one of our reasons for even bothering about "topological recurrence", which owing to results such as 2.6 (c) and 2.7 (c) seems otherwise to be a rather ineffectual notion.)

The difficulty is that one may not automatically conclude (as one does in the abelian case) that members of the class in question are sets of *topological* $k+1$-recurrence, which seems necessary in order for the induction to proceed. Indeed, the only way we have at present to circumvent this snag, and this method is valid only in the presence of amenability, is to establish *measurable* $k+1$-recurrence. For example, a consequence of Theorem 3.3 is that any left or right thick subset of a countable amenable group is a set of topological 2-recurrence. Oddly, we have no proof of this topological fact which doesn't use (measurable) ergodic theory. It would be interesting to see whether there are legitimate reasons for this being so or whether there exists a more elementary method, perhaps one which would yield arbitrarily high orders of chromatic recurrence, without any appeal to measurable methodology.

References

[B1] V. Bergelson, *A density statement generalizing Schur's theorem*, J. Combinatorial Theory (A) **43** (1986), 338-343.

[B2] V. Bergelson, *Ergodic Ramsey theory–an update Ergodic Theory of $\mathbf{Z}^d$-actions*, vol. 228, M. Pollicott and K. Schmidt London Math. Soc. Lecture Note Series, Cambridge University Press, 1996, pp. 1-61.

[BBB] V. Bergelson, M. Boshernitzan and J. Bourgain, *Some results on non-linear recurrence*, J. d'Analyse Math. **62** (1994), 29-46.

[BH] V. Bergelson and N. Hindman, *Some topological semicommutative van der Waerden type theorems and their combinatorial consequences*, J. London Math. Soc.(2) **45** (1992), 385-403.

[BMZ] V. Bergelson, R. McCutcheon, and Q. Zhang, *A Roth theorem for amenable groups*, American Journal of Mathematics, to appear.

[D] M. Day, *Amenable semigroups*, Illinois J. Math **1** (1957), 509-544.

[Fo] A. Forrest, *Ph.D. thesis, The Ohio State University* (1990).

[F1] H. Furstenberg, *Ergodic behavior of diagonal measures and a theorem of Szemerédi on arithmetic progressions*, J. d'Analyse Math. **31** (1977), 204-25.

[F2] H. Furstenberg, *Recurrence in Ergodic Theory and Combinatorial Number Theory*, Princeton Univ. Press, Princeton, New Jersey,, 1981.

[FK] H. Furstenberg and Y. Katznelson, *An ergodic Szemerédi theorem for IP-systems and combinatorial theory*, J. d'Analyse Math. **45** (1985), 117-168.

[FW] H. Furstenberg and B. Weiss, *Topological dynamics and combinatorial number theory*, J. d'Analyse Math. **34** (1978), 61-85.

[GRS] R. Graham, B. Rothschild and J. Spenser, *Ramsey Theory*, Wiley, New York.

[H] N. Hindman, *Ultrafilters and combinatorial number theory*, vol. 751, Number heory Carbondale M. Nathanson, ed. Lecture Notes in Math., 1979, pp. 119-184.

[K] I. Kriz, Large independent sets in shift-invariant graphs Solution of Bergelson's problem, Graphs and Combinatorics **3** (1987), 145-158.

[M] R. McCutcheon, *Three results in recurrence, Proceedings of Alexandria Conference on Harmonic Analysis and its Applications to Ergodic Theory*, vol. 205, K. Petersen I. Salama, London Math. Soc. Lecture Note Series, 1995, pp. 349-358.

[P] A. Paterson, *Amenability*, American Mathematical Society, Providence, 1988.

[R] K. Roth, *Sur quelques ensembles d'entiers*, t C.R. Acad. Sci. Paris **234** (1952), 388-390.

[Sa] A. Sárközy, *On difference sets of sequences of numbers III,*, Acta. Acad. Sci. Hung. **31** (1978), 355-386.

[S] I. Schur, *Uber die Kongruenz* $x^m + y^m = z^m$ *(mod p)*, Jber. Deutsch. Math.-Verein. Math.-Verein. **25** (1916), 114-116.

[Si] W. Sierpiński, *Elementary Theory of Numbers, Warszawa*, 1959.

[Sz] E. Szemerédi, *On sets of integers containing no k elements in arithmetic progression*, Acta. Arith. **27** (1975), 199-245.

THE OHIO STATE UNIVERSITY, COLUMBUS, OH 43210
E-mail address: `vitaly@math.mps.ohio-state.edu`

CASE WESTERN RESERVE UNIVERSITY, CLEVELAND, OH 44106
E-mail address: `vitaly@math.mps.ohio-state.edu`

Contemporary Mathematics
Volume **215**, 1998

A Note on Livšic's Periodic Point Theorem

Z. Coelho, W. Parry and R. Williams

For Robert Ellis

ABSTRACT. We consider two examples of diffeomorphisms of the connected sum of two anchor rings. The first is a hyperbolic diffeomorphism which induces an automorphism on the first cohomology group (of the attractor) having 1 as an eigenvalue. The second fails to be hyperbolic on two transverse intersecting discs and is non-uniformly hyperbolic on the rest of the attractor. The Livšic periodic point theorem is not true for the latter example.

Introduction

Let M be a compact manifold (possibly with boundary) and let $S\colon M \to S(M) \subseteq \mathrm{int}(M)$ be a diffeomorphism. A closed S-invariant subset $\Omega \subseteq M$ is called a *hyperbolic attractor* if Ω is an attractor for S endowed with a hyperbolic structure. By an *attractor* we mean a maximal and isolated attractor, i.e. there exists an open neighbourhood $\mathcal{U}$ of Ω such that $S(\mathcal{U}) \subseteq \mathcal{U}$; $\bigcap_{n\geq 0} S^n\mathcal{U} = \Omega$; and the non-wandering set of S restricted to $\mathcal{U}$ is Ω. By a *uniform hyperbolic structure* on Ω we mean there exists a continuous splitting of the tangent bundle restricted to Ω as a direct sum of two subbundles $T_\Omega M = E^s \oplus E^u$, and there are constants $0 < \rho < 1$ and $C > 0$ such that

(a) $DS_x(E^s_x) = E^s_{Sx}$, $DS_x(E^u_x) = E^u_{Sx}$; and
(b) $||DS^n_x(v)|| \leq C\,\rho^n\,||v||$, $||DS^{-n}_x(w)|| \leq C\,\rho^n\,||w||$,

for all $n \geq 0$, $v \in E^s_x$, $w \in E^u_x$, and all $x \in \Omega$. We say the hyperbolic structure on Ω is *non-uniform* if we allow $\rho = \rho_x$ to be dependent on $x \in \Omega$. When M itself has a (uniform) hyperbolic structure then we say that S is an *Anosov* diffeomorphism. We will also use, loosely, the term *pseudo-hyperbolic* to mean an attractor Ω for which a (non-uniform) hyperbolic structure exists on all but a 'small' set of exceptional points and on these exceptional points the Jacobian of S is an isometry.

Bowen ([B2], p.21) mentions a problem first raised by Hirsch [H]: Is there an Anosov map which induces an automorphism on the first homology group having 1 as an eigenvalue? When M is a nilmanifold the answer is negative since Manning

1991 *Mathematics Subject Classification.* Primary 58F11, 58F15; Secondary 28D05, 58F03, 58F20.

[M] shows that eigenvalues of modulus 1 cannot occur. Bowen ([B2], p.27) also remarks that this problem is equivalent to the existence of a continuous function $f: M \to \mathbb{R}$ such that $f^n(x) = f(x) + \cdots + f(S^{n-1}x) \in \mathbb{Z}$ whenever $S^n x = x$, but there is no constant $N \in \mathbb{Z}$ such that $f^n(x) = nN$ whenever $S^n x = x$.

In fact this is also the case for a connected hyperbolic attractor as we will show. (Here as elsewhere in this paper, in place of the first homology group of the manifold, we consider the first Čech cohomology group $H^1(\Omega, \mathbb{Z})$ interpreted as the Bruschlinsky group of continuous maps to the circle modulo null homotopic maps.) In connection with this problem we give an example of a hyperbolic diffeomorphism of the connected sum of two anchor rings which induces an automorphism on $H^1(\Omega, \mathbb{Z})$ having 1 as an eigenvalue.

The proof of the equivalence of the two conditions mentioned above depends on the celebrated

Livšic's periodic point theorem. *Let Ω be a hyperbolic attractor for S, and let $f: \Omega \to \mathbb{R}$ ($\Omega \to K$, the circle) be Hölder continuous. Then the following are equivalent:*

(i) *There exists a continuous map $h: \Omega \to \mathbb{R}$ ($\Omega \to K$) such that $f = h \circ S - h$ ($f = h \circ S / h$).*

(ii) *$f^n(x) = f(x) + \cdots + f(S^{n-1}x) = 0$ ($f^n(x) = f(x) \cdots f(S^{n-1}x) = 1$) for all $x \in \Omega$ such that $S^n x = x$.*

The example above does provide us with a real function which satisfies the Bowen condition and this will be made explicit.

Our second example is also of a diffeomorphism T of the connected sum of two anchor rings but having a pseudo-hyperbolic attractor Ω for which the additive version of Livšic's theorem fails.

Both examples are modelled on maps of the 1-dimensional branched (figure eight) manifold – the wedge of two circles. Both maps expand the circles and wrap each circle around itself twice and once around the other circle, leaving the branch point fixed. The difference between these maps lies in the order of the 'wrappings'. The diffeomorphisms we shall present have for their attractors the 1-dimensional complexes which (up to topological conjugacy) are the inverse limits of the figure eight under these one dimensional maps.

Our presentation of the diffeomorphisms will be geometric and largely descriptive with details of the analysis only indicated. Following these descriptions we establish (rigourously) the claims made above.

1. The Examples

Consider the (oriented) 1-dimensional branched manifold Λ as illustrated in:

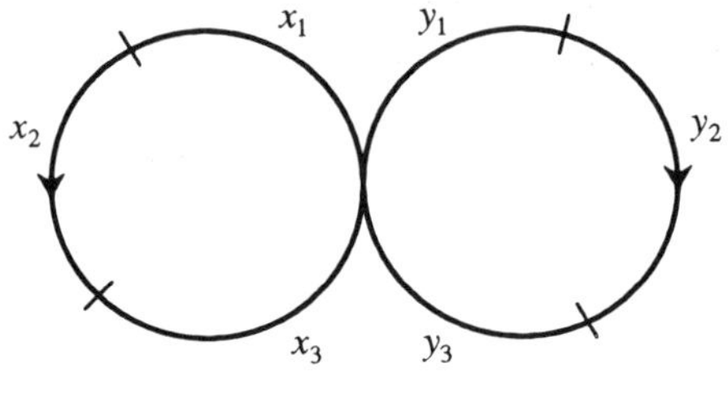

and the map σ which expands the first circle (by a factor of 3) and wraps each of x_1, x_2 successively around x ($x = x_1x_2x_3$) and x_3 around y ($y = y_1y_2y_3$) preserving orientation. In short $\sigma: x \mapsto x^2y$. Similarly σ maps y accordingly to $\sigma: y \mapsto xy^2$. We can mimic this behaviour by constructing a diffeomorphism S of the solid of 'genus' 2:

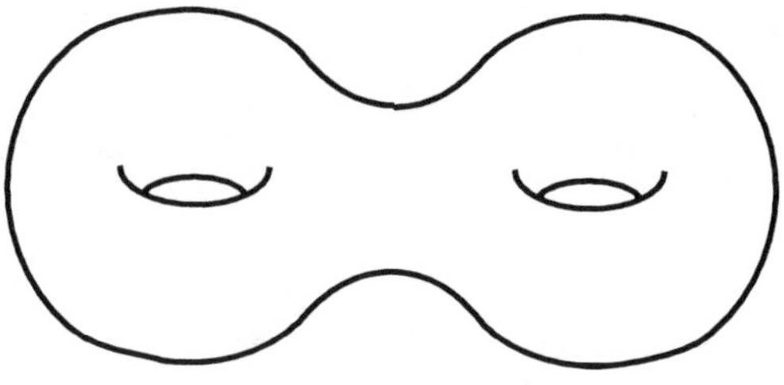

obtained by thickening the one dimensional branched manifold. The diffeomorphism S expands (like σ) in one direction and contracts the transverse discs. Mimicking σ we have the following picture of the manifold and its image under S:

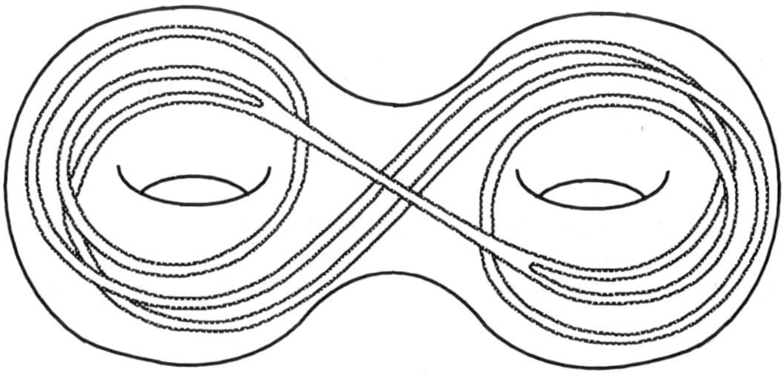

Well known arguments (cf. [W1],[W2]) will show that the attractor $\Omega = \cap S^n M$ is topologically conjugate to the shift $\hat{\sigma}: \hat{\Lambda} \to \hat{\Lambda}$ on the inverse limit solenoid defined by $\sigma: \Lambda \to \Lambda$ which is locally the direct product of a Cantor set and an arc. The branch point presents no problem here since the image under σ of a (branched) neighbourhood is an arc and S is hyperbolic (cf. [W2]). Clearly there is a projection (contracting transverse discs) which maps M onto $\hat{\Lambda}$ which in turn projects to Λ. Hence any Hölder continuous function defined on Λ can be lifted to $\hat{\Lambda}$ or M.

The second example T is again a diffeomorphism of M onto its image and is also modelled on a map $\tau: \Lambda \to \Lambda$. This time τ is symbolically represented by $x \mapsto xyx$, $y \mapsto yxy$. The map τ has a convenient representation as $\tau x = 3x \pmod 1$ defined on the unit interval with 0, 1/2 and 1 identified. One may picture T as follows:

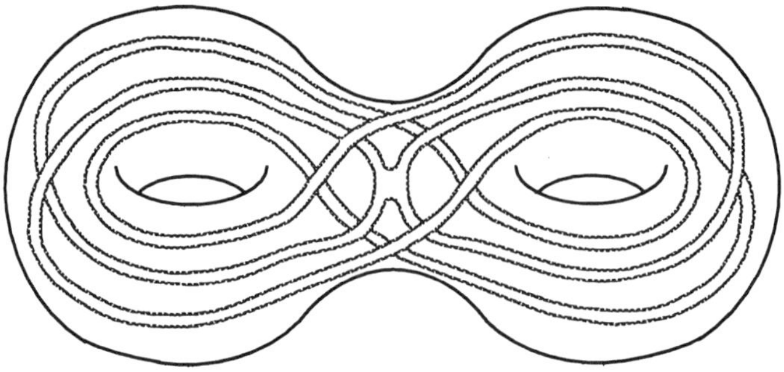

The idea here is again to expand in one direction and contract along the transverse discs. However, one must take care defining T on the 'bridge' joining the 2 anchor rings. The transverse discs bend toward perpendicular discs. Although the manifold expands (in one direction) the expansion is not uniform and in fact the coefficient of expansion decreases to 1 as one approaches the exceptional discs above. One can show that the non-uniformity of expansion is essential for such an example since the branch in a neighbourhood of the fixed point cannot be abolished. In other words, the inverse limit $\hat{\Lambda}$ contains a branched submanifold. Again a Hölder continuous function defined on Λ can be lifted to such a function on $\hat{\Lambda}$ and on M.

2. The Bruschlinsky Group $H^1(\Omega, \mathbb{Z})$

Since Ω is homeomorphic to $\hat{\Lambda}$ (in each case) we have $H^1(\Omega, \mathbb{Z})$ is the direct limit of

$$H^1(\Lambda, \mathbb{Z}) \longrightarrow H^1(\Lambda, \mathbb{Z}) \longrightarrow \cdots$$

with respect to the endomorphism induced by the appropriate map of the wedge of 2 circles. However, $H^1(\Lambda, \mathbb{Z})$ is $\mathbb{Z}^2$ generated by (maps of) the two circles. Moreover in each case it is easy to see that the induced endomorphism is given by the matrix $A = \begin{pmatrix} 2 & 1 \\ 1 & 2 \end{pmatrix}$.

So in each case there is a cohomology class fixed by A, namely $(1, -1)$. A circle valued function representing $(1, -1)$ is the tent function F:

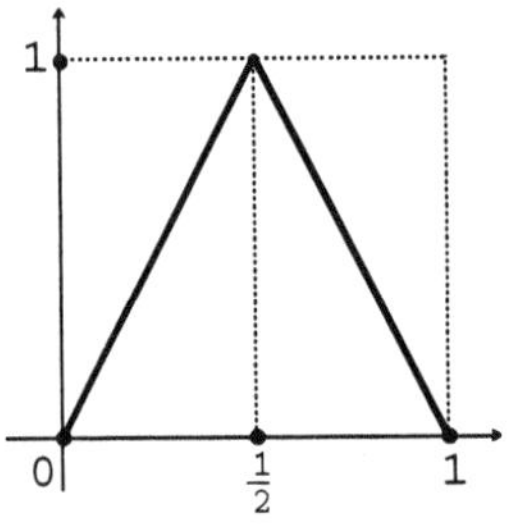

Here we have marked the identification points $\{0, 1/2, 1\}$ for the wedge domain and the identification points $\{0, 1\}$ for the circle image. For our hyperbolic example, $F \circ S/F$ has the symmetric graph

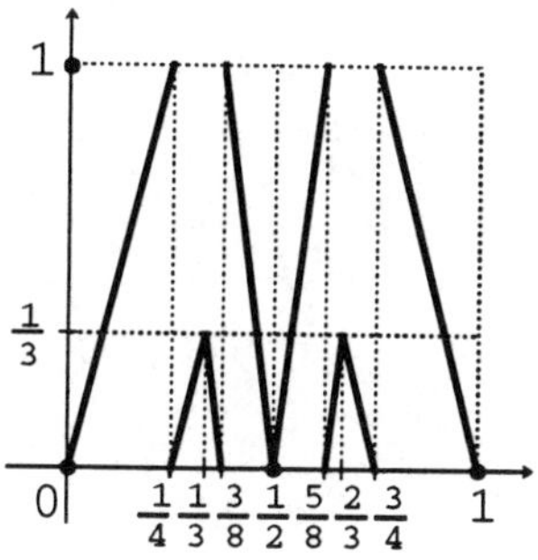

and therefore $F(Sx)/F(x) = e^{2\pi i r(x)}$ where r is the continuous real valued function with graph

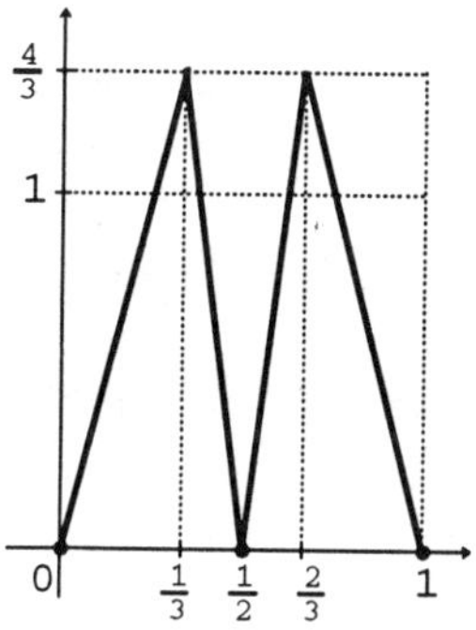

in conformity with the fact that A has a non-trivial fixed vector. Notice that we cannot write $r(x) = h(Sx) - h(x) + c$ (a.e.) for some constant c, where h is integrable with respect to Lebesgue measure (which is preserved by S). This is because integrating this equation we would have c as the integral of r, which is not an integer. Hence such an equation would lead to

$$\frac{F \circ S}{F} = \frac{e^{2\pi i h \circ S}}{e^{2\pi i h}} \cdot e^{2\pi i c}$$

and this would violate the fact that S is mixing. Alternatively, one would conclude that F is an inessential function, which is not the case. Moreover we see that $r^n(x) \in \mathbb{Z}$ when $S^n x = x$ and yet there is no integer N for which $r^n(x) = nN$ when $S^n x = x$, for otherwise we would have $r^1(0) = r(0) = 0$ so that $N = 0$ and $r^1(1/4) = r(1/4) = 0$ which is not the case.

If we now pass to our pseudo-hyperbolic example T we notice that the function F can be written $F(x) = e^{2\pi i u(x)}$, where u is a real valued function defined everywhere except at $\{0, 1/2, 1\}$. We can also write $F(Tx)/F(x) = e^{2\pi i s(x)}$, where s is a continuous real valued function on Ω with graph

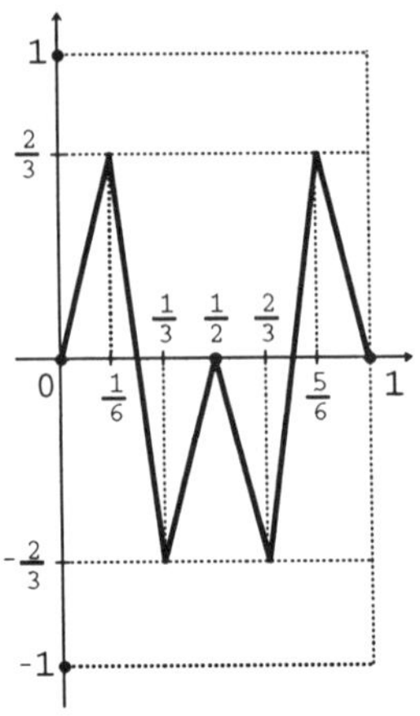

Let N be the set $\{x \in \Omega\colon\ T^n x = 0, 1/2, 1,$ for some $n\}$ then we have $s(x) = u(Tx) - u(x)$ for all $x \in \Omega \setminus N$. Notice also that $s^n(x) = s(x) + \cdots + s(T^{n-1}x) = 0$ whenever $T^n x = x$ since for such points $x \notin N$ and $s(x) = u(Tx) - u(x)$, unless $x = 0, 1/2, 1$ where s vanishes.

We show next that Livšic's (additive) periodic point theorem does not hold for T. If Livšic's theorem were true in this contex, we would conclude that $s = h \circ T - h$ for some continuous function h (defined everywhere on Ω). This implies that

$$\frac{F(Tx)}{F(x)} = \frac{e^{2\pi i h(Tx)}}{e^{2\pi i h(x)}}$$

and therefore $F(x)\, e^{-2\pi i h(x)}$ would be constant, contradicting the fact that F is *not* null-homotopic.

Now for completeness we include the proof of Bowen's observation. In what follows let Ω be a connected hyperbolic attractor for S-a diffeomorphism onto its image. Let $f\colon \Omega \to \mathbb{R}$ be a Hölder continuous function and let A be the endomorphism of $H^1(\Omega, \mathbb{Z})$ induced by S.

THEOREM (BOWEN). *The following are equivalent:*
(i) $\det(A - I) \neq 0$;
(ii) $f^n(x) \in \mathbb{Z}$ *whenever* $S^n x = x$ *implies that* $f^n(x) = nN$ *for some* $N \in \mathbb{Z}$ *whenever* $S^n x = x$.

PROOF. Assume (i) and let $f^n(x) \in \mathbb{Z}$ whenever $S^n x = x$, then

$$e^{2\pi i f(x)} \cdots e^{2\pi i f(S^{n-1}x)} = 1$$

whenever $S^n x = x$. By Livšic's theorem, we have $e^{2\pi i f(x)} = h(Sx)/h(x)$ for some continuous map $h\colon \Omega \to K$. Therefore $A[h] = [h]$ and by assumption $[h]$ is trivial i.e. $h(x) = e^{2\pi i g(x)}$ with g continuous. Thus

$$e^{2\pi i f(x)} = e^{2\pi i (g(Sx) - g(x))} ,$$

and by connectedness

$$f(x) = g(Sx) - g(x) + N$$

for some constant $N \in \mathbb{Z}$. Hence $f^n(x) = nN$ whenever $S^n x = x$.

Now assume *not* (i), then there exists a non-trivial cohomology class $[h]$ such that $A[h] = [h]$, i.e. $h(Sx)/h(x) = e^{2\pi i g(x)}$ for some continuous function g. Therefore $e^{2\pi i g^n(x)} = 1$ whenever $S^n x = x$, i.e. $g^n(x) \in \mathbb{Z}$ whenever $S^n x = x$. In order to complete the proof we need only show that nevertheless it is *not* true that $g^n(x) = nN$ for some $N \in \mathbb{Z}$. If on the contrary $g^n(x) = nN$ whenever $S^n x = x$ then $(g-N)^n(x) = 0$ whenever $S^n x = x$ and so by Livšic's theorem $g - N = k{\circ}S - k$ for some Hölder continuous k. Thus $h{\circ}S/h = e^{2\pi i k \circ S}/e^{2\pi i k}$ which implies $h\, e^{-2\pi i k}$ is a constant. Therefore $[h]$ is trivial – a contradiction. □

REMARK 1. Although Livšic's theorem was needed to prove the Bowen's result it is still true that the Bowen's statements hold for T, even though Livšic's theorem fails in this case. One can see this by lifting the function f to the hyperbolic system consisting of $x \mapsto 3x$ (mod 1) on the circle (where Livšic's is now true). The lift $\tilde{f}$ now satisfies (ii) and it follows from this that f satisfies (ii).

REMARK 2. The diffeomorphism T above also serve as an example where Livšic's regularization theorem does not hold. This follows from the fact that the Hölder continuous function $s(x)$ satisfies the equation

$$s(x) \;=\; u(Tx) - u(x)$$

everywhere, except for a countable number of points, and the integrable function u cannot be replaced by a continuous function satisfying this property.

REMARK 3. Examples such as S and T can also be realized as hyperbolic (and respectively, pseudo-hyperbolic) diffeomorphisms of the 3-sphere onto itself. This is because the 3-sphere can be viewed as the union of two solids of genus 2 with disjoint interiors. By a standard argument one can extend S say, to the complementary solid $\widetilde{M}$ by making $S|_{\widetilde{M}} \simeq S^{-1}|_{S(M)}$ (cf. [G]). The same procedure applies to T producing a pseudo-hyperbolic diffeomorphism of the 3-sphere.

References

[B1] R. Bowen, *Entropy versus homology for certain diffeomorphisms*, Topology **13** (1974), 61-67.

[B2] R. Bowen,, *On Axiom A diffeomorphisms, Regional Conference series*, Amer. Math. Soc., Providence, R.I. **35** (1978).

[BF] R. Bowen and J. Franks, *Homology for zero dimensional basic sets*, Ann. of Math. **106** (1977), 73–92.

[F1] J. Franks, *Anosov diffeomorphisms, Proc.Sympos. Pure Math.*, Amer. Math. Soc., Providence, R.I. **14** (1970), 61–94.

[F2] J. Franks, *Homology and dynamical systems, Regional Conference series*, Amer. Math. Soc., Providence, R.I. **49** (1982).

[G] J.C. Gibbons, *One dimensional basic sets in the three-Sphere*, Transactions of the AMS **114** (1972), 163–178.

[H] M. Hirsch, *Anosov maps, polycyclic groups and homology,*, Topology **10** (1971), 177–183.

[L1] A.N. Livšic, *Homology properties of Y-systems*, Mat. Zametki **10** (1971), 758–763.

[L2] A.N. Livšic, *Cohomology of dynamical systems, Math. USSR Izv.* **6** (1972), 1278–1301.

[M] A. Manning, *There are no new Anosov diffeomorphisms on tori*, Amer.J.Math. (1974), 422–429.

[PP] W. Parry and M. Pollicott, *Stability of mixing for toral extensions of hyperbolic systems, Warwick, preprint , 23/1996.*

[W1] R. Williams, *One dimensional non-wandering sets*, Topology **6** (1967), 473–487.
[W2] R. Williams, *Expanding attractors jour IHES Publ.Math.* **43** (1974), 169–203.

Departamento de Matemática, Universidade de Aveiro, 3810 Aveiro, PORTUGAL
E-mail address: zaqueu@mat.ua.pt

Mathematics Institute, University of Warwick, Coventry CV4 7AL, UK
E-mail address: etds@maths.warwick.ac.uk

Department of Mathematics, University of Texas at Austin, Texas 78712, USA
E-mail address: bob@ma.utexas.edu

Contemporary Mathematics
Volume **215**, 1998

A zero-one law for dynamical properties

Eli Glasner & Jonathan L. King

ABSTRACT. For any countable group Γ satisfying the "weak Rohlin property", and for each dynamical property, the set of Γ-actions with that property is either residual or meager. The class of groups with the weak Rohlin property includes each lattice $\mathbb{Z}^{\times d}$; indeed, all countable discrete amenable groups.

For Γ an arbitrary countable group, let $\mathbb{A}$ be the set of Γ-actions on the unit circle Y. We establish an Equivalence theorem by showing that a dynamical property is Baire/meager/residual in $\mathbb{A}$ if and only if it is Baire/meager/residual in the set of shift-invariant measures on the product space $Y^{\times\Gamma}$.

§1 Introduction

Halmos's book *Ergodic Theory* introduced many of us to the study of determining which dynamical properties are ***generic*** (i.e, topologically residual) in the so-called "coarse topology" on transformations. For instance, "weak-mixing" is generic, whereas "mixing" is not, [**Hal**, pp. 77,78]. The exploration of this notion of genericity became an active research area; see [**CP**] and [**CN**] for results and extensive bibliographies.

It has often been the case, when a property has failed to be generic, that further investigation has shown its negation to be generic. This suggests that there is a type of "zero-one" law operating for dynamical properties –each is either ***meager*** or ***residual***; that is, either a first-category set (a countable union of nowhere-dense sets) or the complement of such.

In 1993, we found a demonstration of this for certain acting groups Γ, for dynamical properties which are Baire-measurable. We first present the brief proof in the abstract framework of a group Φ acting as homeomorphisms of a topological space $\mathbb{A}$, where $\mathbb{A}$ fulfills the Baire Category Theorem in that each residual set is dense. We call such a space a ***BaireCat space***, and say that a subset of a topological space is ***BaireCat*** if it is a BaireCat space in its induced topology.†

We then show how this abstract framework applies when $\mathbb{A}$ is the space of measure-preserving Γ-actions with Φ playing the role of its group of isomorphisms.

1991 *Mathematics Subject Classification.* 28D15. Secondary: 54E52 60F20 54H05.

Key words and phrases. Ergodic theory, Baire Category, Genericity.

J.L. King partially supported by NSF grant DMS-9112595.

†A "Baire space" is the traditional name for what we call a BaireCat space. The traditional meanings of "Baire set" and "Baire space" are not related by the induced topology. In view of this, we have revised terminology for this article.

In this instance, $\mathbb{A}$ will be a ***Polish space***, that is, homeomorphic to a complete separable metric space.

Miscellany. In this article, a "field" is a sigma-algebra and "measure" means a probability measure on the Borel-field of a topological space. All sets and mappings are Borel-measurable unless specified otherwise. Given a measure μ, we call $\mu(A)$ the ***μ-mass*** of A. Let ***mp*** abbreviate "measure preserving".

For a set or measure or transformation A, let $A^{\times 3}$ denote the cartesian power $A \times A \times A$. For a set G, let $A^{\times G}$ means the cartesian power $\bigotimes_{\gamma \in G} A_\gamma$, where each $A_\gamma = A$.

What is a Dynamical Property? Leaving the definition of coarse topology until later, consider the case of a measure-preserving transformation (***mpt***) T of a probability space $\mathbb{Y} = (Y, \mathcal{Y}, \mu)$. Those transformations isomorphic to T are of the form $\varphi T \varphi^{-1}$, where φ is a member of the group, Φ, of bi-mpts. [We write "$\varphi(T)$" for $\varphi\ T\varphi^{-1}$, to notationally emphasize that Φ *acts* on the set of transformations.] Consequently, a dynamical property, such as the set of mixing transformations, is necessarily a union of isomorphism classes, that is, a union of Φ-orbits.

But Φ-invariance is not enough. For an invariant set, B, of transformations to be considered a "dynamical property", it is reasonable to require that it satisfy some kind of measurability condition. We can prove the Zero-One Law when B is Borel-measurable or even (co-)analytic, because our proof applies when B is a ***Baire set*** –that is, when B can be written as the symmetric-difference $V \triangle M$ of an open set with a meager set M. Call $V \triangle M$ a "picture" of B; a Baire set may have many pictures. The Baire sets♮ form a field which includes the Borel sets—indeed, even the analytic sets; see [**Kur**, pp. 482, 92]. As evidence that dynamical properties might transcend the Borel-field, the Prime Example, below, considers the set of actions without factors.

Before showing its application to transformation-groups, we now state and prove the result in an abstract setting. The abstract lemma, we later found out, appears as "folklore" (private communication) in [**Kec**, 1995], theorem 8.46. The earliest reference to the result that we have found is in [**Oxt2**, 1937], where it is stated in the case that Φ is the group of integers.

ZERO-ONE LEMMA. *Suppose Φ is a group of homeomorphisms of a BaireCat space $\mathbb{A}$. If there is some member $\mathbf{T} \in \mathbb{A}$ whose Φ-orbit is dense‡ in $\mathbb{A}$, then: Each Baire-measurable Φ-invariant set B is either meager or residual.*

The Baire necessities. (Proofs of the following facts can be found in [**Oxt**, Chap. 4].) A subset of a topological space is ***regular-open*** if it equals the interior of its closure. The fact we need from elementary topology is that a Baire set B has a "regular-picture" $B = U \triangle M$, where U is regular-open and M is meager.

♮We use the term 'Baire set' in place of the more conventional, but ponderous, phrase "a set with the Property of Baire". Our usage differs from the way 'Baire set' is used in topological measure theory.

The Baire-field is likely to include most dynamical properties of interest to Ergodic theorists, as the following two models, due to Solovay 1970, show. There is a model of ZFC, set theory with Choice, in which all the "projective sets" (any finite number of projections and complements of Borel sets) are Baire-measurable. Indeed, if our Ergodic theorist is willing to forego Choice, there is a model of ZF in which *all* subsets of Polish spaces are Baire sets.

‡Density implies (and –when $\mathbb{A}$ is Polish– is equivalent to) topological transitivity of the action of Φ on $\mathbb{A}$. This latter is all that the proof requires; however, in applications one usually establishes a dense orbit.

The following four properties of a topological space $\mathbb{A}$ are easily equivalent: *(a) Each Baire set $B \subset \mathbb{A}$ has at most one (hence exactly one) regular-picture $B = U \triangle M$. (b/c) The only open/regular-open meager set is empty. (d) $\mathbb{A}$ is BaireCat.* (By the way, the Baire Category Theorem is not reversible, in that there are BaireCat spaces which are neither locally-compact Hausdorff, nor complete-metric; [**FK**, 1978].)

PROOF OF ZERO-ONE LAW. Applying a homeomorphism $\varphi \in \Phi$ to the regular-picture yields

$$B = \varphi(B) = \varphi(U) \triangle \varphi(M).$$

Since φ is a homeomorphism, $\varphi(U)$ is regular-open and $\varphi(M)$ is meager. Then uniqueness of the regular-picture implies that both U and M are Φ-invariant.

If U is void, then $B = M$ and is consequently meager. Otherwise, U intersects the hypothesized dense-orbit and so, by Φ-invariance, U is dense. Thus U is residual and consequently so is B. □

Application to Group Actions

In classical Ergodic Theory, the role of "time" was usually played by $\mathbb{Z}$. Later, actions of more general topological groups Γ (usually amenable) were studied. In this article, we will restrict ourselves to *countable discrete groups* Γ, although both the definitions and the results apply more generally. Although Γ may be non-abelian, we write Γ additively, with identity $0 \in \Gamma$. As names for "times", we use $\alpha, \beta, \gamma \in \Gamma$.

On our probability space $\mathbb{Y}$, an ***action*** of Γ is a collection $\mathbf{T} = \{T^\gamma \mid \gamma \in \Gamma\}$ of μ-mpts so that $T^{\beta+\alpha} = T^\beta \circ T^\alpha$ and the mapping $(\gamma, y) \mapsto T^\gamma(y)$ is a measurable map

$$\big(\Gamma \times Y, \text{Borel} \times \mathcal{Y}\big) \longrightarrow \big(Y, \mathcal{Y}\big).$$

We call T^γ "the transformation at time γ".

Each bi-mp map $\varphi\colon (Y, \mu) \to (Y', \mu')$ defines a conjugate action $\mathbf{R} := \varphi \mathbf{T} \varphi^{-1}$ on (Y', μ') by $R^\gamma := \varphi T^\gamma \varphi^{-1}$. Two actions $(\mathbf{T} : Y, \mu)$ and $(\mathbf{R} : Y', \mu')$ are ***isomorphic*** if they are conjugate. If the measure spaces are understood, we write $\mathbf{T} \cong \mathbf{R}$.

The coarse or weak-operator topology. On the set of measure-preserving transformations of $\mathbb{Y}$, the ***coarse topology*** is defined by net-convergence,

$$T_i \underset{i}{\rightarrowtail} R \quad \textit{iff} \quad \mu\big(T_i^{-1}(E) \triangle R^{-1}(E)\big) \underset{i}{\rightarrowtail} 0,$$
$$\textit{for all measurable subsets } E \subset Y.$$

A countable algebra $(E_k)_{k=1}^\infty$ of sets, which separates points of Y, gives rise to a metric which induces the coarse topology:

$$\operatorname{dist}(T, R) := \sum_{k=1}^{\infty} \frac{1}{2^k} \mu\big(T^{-1}(E_k) \triangle R^{-1}(E_k)\big). \tag{1a}$$

With this metric, the set of mpts becomes a Polish space.

On $\mathbb{Y}$, let $\mathbb{A} = \mathbb{A}_\Gamma$ denote the set of Γ-actions. It also has a ***coarse topology*** from net-convergence: $\mathbf{T}_i \underset{i}{\rightarrowtail} \mathbf{R}$ *iff* $[\forall \gamma : T_i^\gamma \underset{i}{\rightarrowtail} R^\gamma]$. When Γ is countable, the coarse topology makes the action space $\mathbb{A}$ a Polish space, via the metric

$$d(\mathbf{T}, \mathbf{R}) := \sum_{j=1}^{\infty} \frac{1}{2^j} \operatorname{dist}(T^{\gamma_j}, R^{\gamma_j}), \tag{1b}$$

where $(\gamma_j)_{j=1}^{\infty}$ is a fixed enumeration of Γ.

Weak Rohlin Property. Zero-One applies to Γ whenever Φ acts topologically transitively on $\mathbb{A}_\Gamma$. In the case $\Gamma = \mathbb{Z}$, the Rohlin Lemma implies this, since any two aperiodic transformations, T and R, have arbitrarily large congruent Rohlin stacks. Thus some isomorphic copy $\varphi(T)$ is close to R, and so the set of copies, $\Phi(T)$, is dense among the aperiodics. The argument is finished by showing the aperiodics to be dense in $\mathbb{A}_\mathbb{Z}$.

With this as motivation, say that Γ has the ***weak Rohlin Property*** (WRP) if $\Phi(\mathbf{T})$ is dense in $\mathbb{A}_\Gamma$, for some Γ-action $\mathbf{T}$. The analog of aperiodicity is "freeness"; an action $\mathbf{T}$ is ***free*** if, after deleting a Y-nullset: *For each point y, the map $\gamma \mapsto T^\gamma(y)$ is injective.* The Appendix has a brief argument for the following:

$$\text{When } \Gamma \text{ is countable, the set of free } \Gamma\text{-actions is dense –indeed, residual– in } \mathbb{A}_\Gamma. \tag{2}$$

Since there is a Rohlin lemma for $\Gamma = \mathbb{Z}^{\times d}$ (first shown in [**KW**] and later in [**Con**, thm 3.1]), the integer lattice has the WRP. More generally, the seminal paper by Ornstein & Weiss, [**OW**, p. 59], proves a multiple-tower Rohlin lemma. Weiss and Rudolph have indicated to us that this can be used to establish the WRP for all countable discrete amenable groups, thereby establishing that the Zero-One Law applies to these groups Γ.

Question. How large is the class of WRP groups; in particular, does it include the free group on two generators?

Prime Example, 3. A Γ-action $(\mathbf{T}\colon Y, \mathcal{Y}, \mu)$ is ***prime*** if it has no proper non-trivial factors. *Is primeness generic?* Even when $\Gamma = \mathbb{Z}$, this question is open, [**dJR**, P. 556]. What we do know, for weak-Rohlin groups Γ, is that Zero-One is applicable, since –as shown next– primeness is co-analytic.

Let $\mathcal{M}$ be the "measure-algebra" of $\mathcal{Y}$ (i.e, Borel sets equal a.e are identified) but without the null and co-null sets. Since $\mathcal{M}$ is Polish, $\mathcal{M} \times \mathcal{M}$ is Polish. Thus, to show that the set of non-prime actions, $\mathbf{N} \subset \mathbb{A}_\Gamma$, is analytic, we find a Borel condition (i.e, subset) $\mathbf{B}$ in $\mathcal{M} \times \mathcal{M} \times \mathbb{A}$ so that: *The projection of $\mathbf{B}$ on the $\mathbb{A}$-component, equals $\mathbf{N}$.*

And $\mathbf{N}$ *is* a projection, since it comprises those $\mathbf{T}$ for which: $\exists I, D \in \mathcal{M}$ such that *"D is not in the $\mathbf{T}$,$\mathcal{P}$-factor"*, where here –and henceforth– $\mathcal{P}$ abbreviates the two-set partition $(I, Y \setminus I)$. Restating the quoted condition,

$$\exists \varepsilon\colon\ \forall\, \mathbf{T},\mathcal{P}\text{-cylinder-sets } C\colon \quad \mu(C \triangle D) \geq \varepsilon\,. \tag{$*$}$$

So $\mathbf{B}$ is the set of triples $(I, D, \mathbf{T})$ fulfilling $(*)$. It remains to show that $\mathbf{B}$ is Borel in $\mathcal{M} \times \mathcal{M} \times \mathbb{A}$.

Cylinders. Each finite subset $\mathcal{F} \subset \Gamma$ together with a tuple $\vec{v} = (v_1, \ldots, v_L)$ of maps $\mathcal{F} \to \{0, 1\}$ specifies a **"T,$\mathcal{P}$-cylinder set"**. It is $C_{\vec{v}}[I, \mathbf{T}] := \bigcup_{\ell=1}^{L} C_{v_\ell}$, where $I_0 := I$ and $I_1 := Y \setminus I$, and C_v abbreviates $\bigcap_{\gamma \in \mathcal{F}} T^{-\gamma}(I_{v(\gamma)})$. Observe that the mass on the lefthand side of

$$\mu\big(C_{\vec{v}}[I, \mathbf{T}] \triangle D\big) \geq \varepsilon \tag{$**$}$$

varies continuously as a function of I and $\mathbf{T}$ and D. So the set of triples $(I, D, \mathbf{T})$ which fulfill $(**)$, is a closed set. Consequently, first intersecting over $\mathcal{F}$ and $\vec{v}$, then unioning over countably many $\varepsilon \searrow 0$, shows that $\mathbf{B}$ is an $\mathcal{F}_\sigma$-set. □

Question. Is primeness Borel measurable? We do not know. Allowing Γ to range over continuous as well as discrete groups, the answer in principle could depend on Γ. For the related property of *simplicity*, [**dJR**], both its Borel and meager/residual status (the 0-1 law applies) are unknown.

§2 Dynamically generically-equivalent settings

Two of the classical settings for studing Γ-actions were: *The action space,* $\mathbb{A}$, *where the measure was fixed and the transformation varied*, and *The set of shift-invariant measures, where the transformation was fixed, and the measure varied.* Our goal, the Equivalence Theorem♯ below, is to show that these two settings give identical genericity results for Baire-measurable properties.

To discuss this generally, a ***setting*** is a topological space Ω and mapping $\omega \mapsto (\mathbf{R}_\omega\colon X_\omega, \mu_\omega)$ from Ω to Γ-actions. A subset $B \subset \Omega$ is ***saturated*** if $\mathbf{R}_\omega \cong \mathbf{R}_\eta$ implies that both ω and η, or neither, are in B. In principle, a saturated property might be Baire-measurable in one setting and not in another. Say that settings Ω and Ω' are ***dynamically generically-equivalent*** if saturated properties are Baire/meager/residual in Ω iff they are Baire/meager/residual in Ω'.

The space of Shift-Invariant Measures

We now specialize our measure space to be $\mathbb{Y} = (Y, \mathcal{Y}, \mathbf{m})$, where $Y := [0,1)$ is topologized to be the unit circle, and $\mathbf{m}$ is Lebesgue measure (restricted to $\mathcal{Y}$, the Borel-field of the circle).

For our new setting, $^\bullet\mathbb{SIM}$, we need the definitions which follow. A superscript "$\bullet$" on a set of measures is meant to suggest that these measures might have atoms.

- $^\bullet\mathbb{M}$ is the set of (Borel probability) measures on Y.
- $\mathbb{M}$ comprises those non-atomic $\mu \in {}^\bullet\mathbb{M}$ which have "full-support", that is, each non-void open set $V \subset Y$ has positive μ-mass. We'll call these the ***good*** measures.
- $\widehat{Y}$ denotes the product space $Y^{\times\Gamma}$, which is the infinite torus equipped with the product topology. Use $\widehat{y}, \widehat{z} \in \widehat{Y}$. Let $\widehat{y}[\gamma]$ denote the γth-coordinate of $\widehat{y}$.
- $\mathbf{S}$ is the ***shift*** on $\widehat{Y}$. It is the Γ-action $\{S^\alpha \mid \alpha \in \Gamma\}$ defined by

$$S^\alpha(\widehat{y}) := \widehat{z}, \qquad \text{where} \quad \widehat{z}[\gamma] := \widehat{y}[\gamma + \alpha].$$

  Note that $S^\beta \circ S^\alpha = S^{\beta+\alpha}$.
- $^\bullet\mathbb{SIM}$ is the set of shift-invariant measures on $\widehat{Y}$, that is, invariant under each S^α, for $\alpha \in \Gamma$. We'll call such a measure a ***sim***. For a sim $\widehat{\mu}$, let $\mathrm{Marg}[\widehat{\mu}]$ be the measure in $^\bullet\mathbb{M}$ which is the (1-dimensional) marginal of $\widehat{\mu}$; this marginal is independent of which copy of Y we condition on, since $\widehat{\mu}$ is shift-invariant.
- $\mathbb{SIM}$ is the set of ***good sims***, that is, those $\widehat{\mu} \in {}^\bullet\mathbb{SIM}$ such that $\mathrm{Marg}[\widehat{\mu}]$ is in $\mathbb{M}$.
- $\mathbb{SIM}_\nu$ comprises those good sims whose marginal is ν, where ν is a good measure. Thus $\mathbb{SIM}$ is the disjoint union, over all $\nu \in \mathbb{M}$, of "fibers" $\mathbb{SIM}_\nu$.

♯In the case that $\Gamma = \mathbb{Z}$, essentially this same result was independently obtained by Dan Rudolph in his work [**Rud**] on orbit-equivalence.

Weak-$*$ preliminaries. Both $^{\bullet}\mathbb{M}$ and $^{\bullet}\mathbb{SIM}$ become compact metric spaces under the weak-$*$ topology: $\mu_i \to \nu$ iff $\left[\int t\,d\mu_i \longrightarrow \int t\,d\nu\right]$, for all continuous real-valued "test" functions t. For U open, the indicator $\mathbf{1}_U$ can be arbitrarily well approximated by functions $t \leq \mathbf{1}_U$. Consequently, whenever $\mu_i \to \nu$,

$$\nu(\overline{B}) \geq \limsup_i \mu_i(\overline{B}) \geq \liminf_i \mu_i(B^\circ) \geq \nu(B^\circ) \tag{4}$$

for all sets B. In particular: *$\mu \mapsto \mu(B)$ is continuous on the collection of measures for which the boundary, ∂B, is a nullset.*

A ***partition*** $\mathcal{P}$ divides Y (disjointly) into finitely many pieces; each ***piece*** $I \in \mathcal{P}$ is an interval of positive length. Fix a finite set $\mathcal{F} \subset \Gamma$ of times. An "$\mathcal{F}$,$\mathcal{P}$-***cylinder set***", $C \subset \widehat{Y}$, is of the form $\bigotimes_{\gamma\in\mathcal{F}} I_\gamma$, where each I_γ is in $\mathcal{P}$. This cylinder C comprises those $\widehat{y}$ such that $\left[\forall\gamma \in \mathcal{F}\colon\ \widehat{y}[\gamma] \in I_\gamma\right]$. Observe that ∂C is a nullset, for good sims $\widehat{\mu}$, since $\mathrm{Marg}[\widehat{\mu}]$ is non-atomic. Thus

$$\text{On } \mathbb{SIM}: \qquad \widehat{\mu}_i \to \widehat{\nu} \iff \forall\text{cylinders } C\colon \widehat{\mu}_i(C) \to \widehat{\nu}(C) \tag{5}$$

Implication ($\Leftarrow$) holds since, if sim $\widehat{\nu}$ differs from $\widehat{\eta}$, then $\widehat{\nu}(C) \neq \widehat{\eta}(C)$ for some $\mathcal{G}_\delta$-set C, hence for some open C, hence for some cylinder.

Finally, we leave to the Appendix a proof that

$$\mathbb{SIM} \text{ is a residual subset of } {}^{\bullet}\mathbb{SIM}. \tag{6}$$

Courtesy of this, *henceforth all measures and sims discussed are good*. Certainly $\mathbf{m}$ is good. We use names $\mu, \nu \in {}^{\bullet}\mathbb{M}$ for other good measures, and use $\widehat{\mu}, \widehat{\nu}, \widehat{\eta}$ to name good sims.

EQUIVALENCE THEOREM, 7. *Settings $\mathbb{A}$ and $^{\bullet}\mathbb{SIM}$ are dynamically generically-equivalent.*

Overview of the proof. Produce a BaireCat space $\mathbb{H}$ and an embedding E, which is a homeomorphism from $\mathbb{H} \times \mathbb{A}$ into $\mathbb{SIM}$, so that

(8a) *For each $h \in \mathbb{H}$ and $\mathbf{T} \in \mathbb{A}$: $E(h, \mathbf{T})$ is isomorphic to $\mathbf{T}$.*

(8b) *The set,* Range(E), *is residual in* $\mathbb{SIM}$.

The map E makes $\mathbb{H} \times \mathbb{A}$ into a setting. That $\mathbb{A}$ and $\mathbb{H} \times \mathbb{A}$ are dynamically generically-equivalent follows from (8a) and the plausible fact, [**Oxt**, P. 57] theorems 15.2 and 15.3, that for arbitrary topological spaces with $\mathbb{H}$ a BaireCat space and $\mathbb{A}$ countably-generated,

A subset $C \subset \mathbb{A}$ is Baire/meager/residual in $\mathbb{A}$ iff the product $\mathbb{H} \times C$ is Baire/meager/residual in $\mathbb{H} \times \mathbb{A}$.

Lastly, (8b) and (6) allow Proposition 9, below, to show that Range(E) and $\mathbb{SIM}$ and $^{\bullet}\mathbb{SIM}$ are dynamically generically-equivalent.

Relative Baire necessities. Given subsets B and G of a topological space Ω, say that "B is G-open" (closed, nowhere-dense, meager, residual, Baire) iff $B \cap G$ is open (closed, etc.) relative to the induced topology on G.

PROPOSITION 9. *Suppose M, B and R are subsets of a topological space Ω, with R residual.*

(i) If M is R-meager then M is meager. Also: B is R-Baire $\Longrightarrow$ B is Baire.

(ii) Suppose Ω is a BaireCat space. Then:

M is meager $\Longrightarrow$ M is R-meager;
B is Baire $\Longrightarrow$ B is R-Baire.

Consequently, when Ω is BaireCat, a subset is Baire/meager/residual iff it is R-Baire/meager/residual.

PROOF OF (9*i*). Since the part of M protruding from R is meager, being inside of $\Omega \smallsetminus R$, we may assume that M is included in R. Further, we may assume that M is R-closed and has no R-interior.

Note that $\overline{R} \smallsetminus R$ has no interior. Our goal is to establish that $V := \text{Interior}(\overline{M})$ is empty. But $V \subset \overline{M} \subset \overline{R}$. So it suffices to show that

$$V \text{ does not intersect } R,$$

for then $V \subset \overline{R} \smallsetminus R$ and is consequently void.

For the sake of contradiction, fix a point $\omega \in V \cap R$. Each neighborhood $W \ni \omega$ hits M and so each R-neighborhood $W \cap R$ of ω also hits M, since $M \subset R$. Recall that $\omega \in R$ and M is R-closed; thus $\omega \in M$.

This shows that $V \cap R \subset M$. But M has no R-interior and so $V \cap R$ is empty, as desired.

Finally, for our R-Baire set B, we may assume that $B \subset R$, since $B \smallsetminus R$ is meager, hence Baire. So we can write $B = (V \cap R) \triangle M$, with V open and with M an R-meager set. From the foregoing, M is meager, hence Baire. Since the Baire sets form an algebra, and V, R, M are all Baire, B must be a Baire set. □

PROOF OF (9*ii*). Our meager M can be assumed to be nowhere-dense and closed; hence M is R-closed. To show that M is R–nowhere-dense, take an open V so that $V \cap R$ is the R-interior of M. Now write $R = \Omega \smallsetminus M'$, with M' meager. Thus $V \subset M \cup M'$. Since this union is meager, the openness of V forces V to be void. Consequently, $M \cap R$ is R-meager.

Finally, write Baire set B as $U \triangle M$ with U open and M meager. Then

$$B \cap R \;=\; (U \cap R) \triangle (M \cap R) \;=\; R\text{-open} \triangle R\text{-meager}$$

and is consequently a Baire set relative to R. □

Pushing measures forward. Given a measurable map $f\colon (X, \mathcal{X}) \to (X', \mathcal{X}')$, each measure μ on X pushes forward to a measure $f\langle\mu\rangle$ on $\mathcal{X}'$ defined by

$$f\langle\mu\rangle(B') \;:=\; \mu\big(f^{-1}(B')\big).$$

A (good) measure ν gives rise to a homeomorphism, h, of the circle Y, by

$$h(z) := y, \text{ where } z = \nu\big([0, y)\big)$$

for $y \in Y$. Consequently, $\nu\big([0, y)\big) = \mathbf{m}\big([0, z)\big) = \mathbf{m}\big(h^{-1}\big([0, y)\big)\big)$. Thus h "adapts", so-to-speak, Lebesgue measure into ν,

$$\nu = h\langle\mathbf{m}\rangle\,.$$

This h is an order-preserving homeomorphism of Y which fixes 0. We will call such a homeomorphism an ***adaptation***, and let $\mathbb{H}$ denote the set of adaptations. Easily, $\mathbb{H}$ is a $\mathcal{G}_\delta$-subset of the Polish space $C(Y)$ of continuous functions—hence $\mathbb{H}$

is Polish [**Oxt**, pp. 47–50] and thus is BaireCat, as needed by **Overview**. Finally, note that

$$\textit{Each fiber } \mathrm{SIM}_{\nu} \textit{ is the homeomorphic image of } \mathrm{SIM}_{\mathbf{m}}\,, \quad \textit{via the map } \widehat{\mu} \mapsto \widehat{h}\langle\widehat{\mu}\rangle\,; \tag{10a}$$

where here –and henceforth– for each adaptation h, we let $\widehat{h}$ denote the product $h^{\times\Gamma}$, which is a homeomorphism of $\widehat{Y}$.

Pushing actions to sims. Given an action $\mathbf{T} \in \mathbb{A}$, let $\widehat{\mathbf{T}}$ denote the map $y \mapsto \mathbf{T}(\cdot, y)$. This is an injection from $Y \hookrightarrow \widehat{Y}$ and gives a commutative diagram.

$$\begin{array}{ccc} Y & \xrightarrow{T^{\gamma}} & Y \\ \big\downarrow \widehat{\mathbf{T}} & & \big\downarrow \widehat{\mathbf{T}} \\ \widehat{Y} & \xrightarrow{S^{\gamma}} & \widehat{Y} \end{array}$$

FIGURE 11 For each time $\gamma \in \Gamma$, the mapping $\widehat{\mathbf{T}}$ yields a commutative diagram. Lebesgue measure $\mathbf{m}$, being T^{γ}-invariant, pushes forward to an S^{γ}-invariant measure $\widehat{\mu} := \widehat{\mathbf{T}}\langle\mathbf{m}\rangle$. So $(\mathbf{S}\colon \widehat{Y}, \widehat{\mu})$, the action downstairs, is isomorphic to $(\mathbf{T}\colon Y, \mathbf{m})$.

This mapping $\mathbf{T} \mapsto \widehat{\mathbf{T}}\langle\mathbf{m}\rangle$ is an embedding $\mathbb{A} \hookrightarrow \mathrm{SIM}_{\mathbf{m}}$. Its range is the set of "graph-sims in $\mathrm{SIM}_{\mathbf{m}}$". As a first step towards residuality, we show that these are dense.

LEMMA 10*b*. *The set of "graph-sims in* $\mathrm{SIM}_{\mathbf{m}}$*" is dense in* $\mathrm{SIM}_{\mathbf{m}}$.

PROOF. Fix a sim $\widehat{\nu}$ whose 1-marginal is $\mathbf{m}$. Given ε and an $\mathcal{F}$,$\mathcal{P}$-cylinder set, we wish to find an action $\mathbf{T} \in \mathbb{A}$ which gives ε-equality,

$$\mathbf{m}\Bigl(\bigcap_{\gamma\in\mathcal{F}} [T^{-\gamma}](I_{\gamma})\Bigr) \;\overset{\varepsilon}{\approx}\; \widehat{\nu}\Bigl(\bigotimes_{\gamma\in\mathcal{F}} I_{\gamma}\Bigr)\,. \tag{$*$}$$

Fixing $\mathcal{P}$, it turns out that there is a $\mathbf{T}$ giving actual equality above, for all subsets $\mathcal{F} \subset \Gamma$ and all $\mathcal{F}$,$\mathcal{P}$-cylinders.

For each piece $I \in \mathcal{P}$, let $\widehat{I}$ be corresponding time-0 set; the collection of $\widehat{y}$ for which $\widehat{y}[0] \in I$. Tautologically, $\widehat{\nu}(\widehat{I}) = \mathbf{m}(I)$ and the intersection $\bigcap_{\gamma\in\mathcal{F}} [S^{-\gamma}](\widehat{I}_{\gamma})$ equals $\bigotimes_{\gamma\in\mathcal{F}} I_{\gamma}$.

Each $\widehat{I}$ is Borel-equivalent to an interval. Consequently there is a bi-mp Borel isomorphism φ carrying probability space $(\widehat{Y}, \widehat{\nu})$ to $(Y, \mathbf{m})$, with $\varphi^{-1}(I) = \widehat{I}$ for each piece. Thus $\mathbf{T} := \varphi\mathbf{S}\varphi^{-1}$ gives equality in $(*)$. □

Measures which live on a graph. Temporarily, let M denote the set of measures on $Y \times Y$ whose two marginals are equal and good. Consider an $\eta \in M$ satisfying, for all $B \in \mathcal{Y}$, that

$$\exists A \in \mathcal{Y}: \quad \eta(A \times Y) = \eta(A \times B) = \eta(Y \times B)\,. \tag{12}$$

Letting μ denote the marginal of η, this is equivalent to asserting that $\eta(A{\times}B)$ equals $\mu\bigl(A \cap R^{-1}(B)\bigr)$ for some μ-mpt R. The graph of R has full η-mass, and so η is called a "graph" joining.

Let $\mathcal{C} \subset \mathcal{Y}$ be the countable algebra of sets formed by finite unions of rational-endpoint intervals in Y. It is straightforward to check that a measure η is a graph joining iff (12) holds for each $B \in \mathcal{C}$. Equivalently, for ε and all such B,

$$\exists A \in \mathcal{C}: \quad \mathrm{Diameter}\bigl\{\eta(A \times Y), \eta(A \times B), \eta(Y \times B)\bigr\} \;<\; \varepsilon\,.$$

Let U_B^ε comprise such η. Since η varies over measures with non-atomic marginals, the diameter above varies continuously with η. Thus U_B^ε is open and, in consequence, the set $\bigcap_{B\in\mathcal{C}}\bigcap_{\varepsilon\searrow 0} U_B^\varepsilon$ –the graph joinings– is $\mathcal{G}_\delta$ in M.

General graph-sims. If each 2-dimensional marginal of sim $\widehat{\nu}$ is a graph joining, then $\widehat{\nu}$ is called a ***graph-sim***. Letting ν be $\mathrm{Marg}[\widehat{\nu}]$, the shift-invariance of $\widehat{\nu}$ implies that it is of the form $\widehat{\nu} = \widehat{\mathbf{R}}\langle\nu\rangle$, for a (unique) action $(\mathbf{R}\colon Y,\nu)$.

Fix any two distinct times $\alpha,\beta\in\Gamma$ and consider the set of sims whose 2-marginal on $Y_\alpha\times Y_\beta$ is a graph. By the paragraphs above, this is a $\mathcal{G}_\delta$-set of sims. Since there are but countably many pairs in Γ, the set $\mathbb{SIM}^{\mathrm{Graph}}$ of graph-sims is $\mathcal{G}_\delta$ in $\mathbb{SIM}$. Consequently,

$$\mathbb{SIM}^{\mathrm{Graph}} \textit{ is a residual subset of } \mathbb{SIM} \tag{10c}$$

since, courtesy of (10a,*b*), the graph-sims are dense in each fiber SIM_ν and are thereby dense in $\mathbb{SIM}$.

The embedding E

The preparation done, we can finally define the embedding $E\colon \mathbb{H}\times\mathbb{A}\hookrightarrow\mathbb{SIM}$ which will fulfill the conditions of **Overview**. Let

$$E(h,\mathbf{T}) := \widehat{h}\langle\widehat{\mathbf{T}}\langle\mathbf{m}\rangle\rangle\,, \tag{13}$$

where, as usual, $\widehat{h}$ means $h^{\times\Gamma}$. Evidently, the range of E is simply $\mathbb{SIM}^{\mathrm{Graph}}$. In the next several paragraphs, we establish that E is a homeomorphism by writing it as a composition of two homeomorphisms. The first one we consider is

$$D(h,\widehat{\mu}) := \widehat{h}\langle\widehat{\mu}\rangle\,,$$

which is a mapping from $\mathbb{H}\times\mathrm{SIM}_{\mathbf{m}}$ onto $\mathbb{SIM}$.

Is D a homeomorphism? Evidently D is a bijection. In order to simplify the discussion of continuity properties of D, observe that for each adaptation $f\in\mathbb{H}$:

- *The composition map $h\mapsto fh$ is a homeomorphism of $\mathbb{H}$. Thus $\psi(h,\widehat{\mu}) := (fh,\widehat{\mu})$ is a homeomorphism of $\mathbb{H}\times\mathrm{SIM}_{\mathbf{m}}$.*
- *D carries ψ to the mapping $\widehat{\eta}\mapsto\widehat{f}\langle\widehat{\eta}\rangle$, which is a homeomorphism of $\mathbb{SIM}$.* (To see this, let $\widehat{\eta}$ be $\widehat{h}\langle\widehat{\mu}\rangle$.)

Consequently, when arguing that D is continuous at a point $(f,\widehat{\nu})$, or that D^{-1} is continuous at $\widehat{f}\langle\widehat{\nu}\rangle$, we may assume that f is the identity map *Id* on Y. We now make these two arguments, having fixed: *A partition $\mathcal{P}$, a $\mathcal{P}$-cylinder set $C = \bigotimes_{\gamma\in\mathcal{F}} I_\gamma$, and a small $\varepsilon>0$. Also, let δ denote $\varepsilon/|\mathcal{F}|$.*

D is continuous at $(Id,\widehat{\nu})$. Fix $\widehat{\nu}$ and consider any $\widehat{\mu}$ sufficiently near $\widehat{\nu}$ that $\widehat{\mu}(C)$ is ε-close to $\widehat{\nu}(C)$. Take any adaptation h for which

$$\left\|h - Id\right\|_{\sup} \le \delta\,. \tag{$*$}$$

Then $\mathbf{m}\big(h^{-1}(I)\,\triangle\, I\big)\le 2\delta$, since each piece I is an interval. Consequently,

$$\begin{aligned}\left|\widehat{\mu}\Big(\bigotimes_{\gamma\in\mathcal{F}} h^{-1}(I_\gamma)\Big) - \widehat{\mu}(C)\right| &\le \sum_{\gamma\in\mathcal{F}} \mathbf{m}\big(h^{-1}(I_\gamma)\,\triangle\, I_\gamma\big)\\ &\le |\mathcal{F}|\cdot 2\delta = 2\varepsilon\,.\end{aligned} \tag{$**$}$$

Applied to C, then, $\widehat{h}\langle\widehat{\mu}\rangle$ is 2ε-close to $\widehat{\mu}$ and thus is 3ε-close to $\widehat{\nu} = \widehat{Id}\langle\widehat{\nu}\rangle$. □

The inverse, D^{-1}, is continuous at $\widehat{\nu} = \widehat{Id}\langle\widehat{\nu}\rangle$. If $\widehat{h}\langle\widehat{\mu}\rangle$ is sufficiently near $\widehat{\nu}$, then their 1-marginals are close enough, choosing an n with $\frac{1}{n} < \delta/2$, that

$$\mathbf{m}\Big(h^{-1}\big([0,y)\big)\Big) \overset{\delta/2}{\approx} \mathbf{m}\Big([0,y)\Big), \quad \text{for } y = \tfrac{1}{n}, \tfrac{2}{n}, \ldots, \tfrac{n}{n}.$$

Since h preserves order on Y and fixes zero, $h^{-1}(y)$ is $(\delta/2)$-close to y, for these points y. Thus $(*)$ holds. Therefore on C, as $(**)$ argued, sims $\widehat{h}\langle\widehat{\mu}\rangle$ and $\widehat{\mu}$ are 2ε-close. So if $\widehat{h}\langle\widehat{\mu}\rangle$ is near enough to $\widehat{\nu}$ that they are ε-close on C, then $\widehat{\mu}$ is 3ε-close (on C) to $\widehat{\nu}$.

Together with $(*)$, this establishes that the pair $(h, \widehat{\mu})$ is close to $(Id, \widehat{\nu})$. □

REMARK. The two preceding arguments show that D is a homeomorphism. By restricting D, we have the following:

The map $(h, \widehat{\mu}) \mapsto \widehat{h}\langle\widehat{\mu}\rangle$ is a homeomorphism from $\mathbb{H} \times \mathbb{SIM}_{\mathbf{m}}^{\mathrm{Graph}}$ onto $\mathbb{SIM}^{\mathrm{Graph}}$, (14)

where $\mathbb{SIM}_{\mathbf{m}}^{\mathrm{Graph}}$ is the set of graph-sims with marginal $\mathbf{m}$.

A homeomorphism $\mathbb{A} \longrightarrow \mathbb{SIM}_{\mathbf{m}}^{\mathrm{Graph}}$. Recall, from Figure 11, the injection $\mathbf{T} \mapsto \widehat{\mathbf{T}}\langle\mathbf{m}\rangle$. An argument similar to the continuity of D shows that $\mathbf{T} \mapsto \widehat{\mathbf{T}}\langle\mathbf{m}\rangle$ is continuous. We now show that its inverse is continuous at an arbitrary point $\widehat{\mathbf{R}}\langle\mathbf{m}\rangle$, where $\mathbf{R} \in \mathbb{A}$.

Fix ε and a time γ and a subinterval $I \subset Y$. Let J denote the subset $R^{-\gamma}(I)$. Regard $J \times I$ as a product set, on times 0 and γ, and compute:

$$\begin{aligned}
\widehat{\mathbf{R}}\langle\mathbf{m}\rangle(J \times I) &= \mathbf{m}(J \cap R^{-\gamma}I) = \mathbf{m}(J)\,; \\
\widehat{\mathbf{T}}\langle\mathbf{m}\rangle(J \times I) &= \mathbf{m}(J \cap T^{-\gamma}I)\,.
\end{aligned}$$

If sim $\widehat{\mathbf{T}}\langle\mathbf{m}\rangle$ is near enough to $\widehat{\mathbf{R}}\langle\mathbf{m}\rangle$, then the lefthand terms are ε-close.† So the righthand terms are ε-close. Hence $2\varepsilon > \mathbf{m}\big(T^{-\gamma}(I) \triangle J\big)$, which equals $\mathbf{m}\big(T^{-\gamma}(I) \triangle R^{-\gamma}(I)\big)$

This argument applies for all pieces I in a partition, and for all times γ in any specified finite set $\mathcal{F}$, so actions $\mathbf{T}$ and $\mathbf{R}$ have been shown to be close. □

Final step. The foregoing shows that the mapping $\big(h, \mathbf{T}\big) \mapsto \big(h, \widehat{\mathbf{T}}\langle\mathbf{m}\rangle\big)$ is a homeomorphism of $\mathbb{H} \times \mathbb{A}$ onto $\mathbb{H} \times \mathbb{SIM}_{\mathbf{m}}^{\mathrm{Graph}}$. Composing this with (14) shows that E is a homeomorphism onto $\mathbb{SIM}^{\mathrm{Graph}}$, thus completing the proof of the Equivalence Theorem.

Acknowledgments

We thank our friends Karl Petersen –for the **[Oxt2]** reference– and Andrés del Junco, Dan Rudolph, and Benjy Weiss, for illuminating conversations on genericity issues. For $\mathbb{Z}$-actions, Dan put us wise to Lemma 10b, which he proved by a different argument. The second author gratefully thanks the University of Toronto for its hospitality while this paper was being completed.

†This, by approximating J with a finite union J' of intervals, so that $\mathbf{m}(J' \triangle J) \le \varepsilon/2$.

§A Appendix

PROOF OF (6), FIRST STEP: *$\mathbb{M}$ is a $\mathcal{G}_\delta$-subset of ${}^{\bullet}\mathbb{M}$.* Write $\mathbb{M} = \mathbf{F} \cap \mathbf{N}$, where these denote the sets, respectively, of full-support and non-atomic measures in ${}^{\bullet}\mathbb{M}$. For two measures μ and ν, their convolution is the measure

$$[\nu * \mu](A) := \int_Y \mu(A \ominus y)\, d\nu(y)\,, \tag{15}$$

where $A \ominus y$ is the group translation $\{z \ominus y \mid z \in A\}$ in Y.

From (4), if $U \subset Y$ is open then $V_U^\varepsilon := \{\mu \in {}^{\bullet}\mathbb{M} \mid \mu(U) > \varepsilon\}$ is open. Letting $\mathcal{B}$ be a countable base for the topology of Y, then,

$$\mathbf{F} = \bigcap_{U \in \mathcal{B}} \bigcup_{\varepsilon \searrow 0} V_U^\varepsilon$$

is a $\mathcal{G}_\delta$-set.

For $C \subset Y$ closed, the set $\{\mu \in {}^{\bullet}\mathbb{M} \mid \mu(C) < \varepsilon\}$ is open, by (4). So V_ε^K is open, where it comprises those μ such that, for each $n = 1, \ldots, K$: $\mu\Big(\big[\frac{n-1}{K}, \frac{n}{K}\big]\Big) < \varepsilon$. Consequently,

$$\mathbf{N} = \bigcap_{\varepsilon \searrow 0} \bigcup_{K=1}^{\infty} V_\varepsilon^K$$

and so is $\mathcal{G}_\delta$ in ${}^{\bullet}\mathbb{M}$.

In order to show that $\mathbf{N}$ is dense in ${}^{\bullet}\mathbb{M}$, take δ a small positive number and let $\mathbf{m}_\delta := \frac{1}{\delta} \cdot \mathbf{m}|_{[0,\delta)}$ denote normalized Lebesgue measure on $[0, \delta)$. For each $\nu \in {}^{\bullet}\mathbb{M}$, then, the convolution $\nu * \mathbf{m}_\delta$ tends to ν, as $\delta \searrow 0$. And $\nu * \mathbf{m}_\delta$ is non-atomic, since $\mathbf{m}_\delta$ is non-atomic. □

SECOND STEP: *$\mathbb{SIM}$ is residual in ${}^{\bullet}\mathbb{SIM}$.* Let $\mathbf{F}$ and $\mathbf{N}$ now denote the set of sims $\widehat{\mu} \in {}^{\bullet}\mathbb{SIM}$ whose time-0 marginal is a full-support or non-atomic measure. The foregoing shows that these are $\mathcal{G}_\delta$ sets in ${}^{\bullet}\mathbb{SIM}$. And since all 1-marginals of a sim are equal, $\mathbf{F} \cap \mathbf{N} = \mathbb{SIM}$.

To see that $\mathbf{F}$ is dense, take a particular sim $\widehat{\mu} \in \mathbf{F}$ (e.g, product measure $\mathbf{m}^{\times\Gamma}$). For each sim $\widehat{\nu}$, the average $(1-\delta)\widehat{\nu} + \delta\widehat{\mu}$ is in $\mathbf{F}$, and tends to $\widehat{\nu}$ as $\delta \searrow 0$.

To show $\mathbf{N}$ dense in ${}^{\bullet}\mathbb{SIM}$, fix $\widehat{\nu}$ and $\delta > 0$. Again let $\mathbf{m}_\delta$ be normalized Lebesgue measure on $[0, \delta)$, and let $\widehat{\mathbf{m}_\delta}$ be the product measure $(\mathbf{m}_\delta)^{\times\Gamma}$. Taken over the torus $\widehat{Y}$, the convolution $\widehat{\nu} * \widehat{\mathbf{m}_\delta}$ is shift-invariant and has marginal $\mathrm{Marg}[\widehat{\nu}] * \mathbf{m}_\delta$, which is non-atomic. As $\delta \searrow 0$, furthermore, $\widehat{\mathbf{m}_\delta}$ tends to atomic measure on the identity $\widehat{0} \in \widehat{Y}$. Thus $\widehat{\nu} * \widehat{\mathbf{m}_\delta} \to \widehat{\nu}$. □

PROOF OF (2): *Free actions are dense in $\mathbb{A}$.* Let $\mathbf{Fr}$ be the set of "free sims" $\widehat{\mu}$, i.e, $(\mathbf{S}\colon \widehat{Y}, \widehat{\mu})$ is a free action. By showing $\mathbf{Fr}$ residual in ${}^{\bullet}\mathbb{SIM}$, the Equivalence Theorem will tell us that the free actions are residual in $\mathbb{A}$.

Fix a time β and let $C_\beta := \{\widehat{y} \mid S^\beta(\widehat{y}) = \widehat{y}\}$. Since S^β is continuous, C_β is closed and consequently the collection $\{\widehat{\mu} \in {}^{\bullet}\mathbb{SIM} \mid \widehat{\mu}(C_\beta) < \varepsilon\}$ is open. Intersecting over ε and all times β shows that $\mathbf{Fr}$ is $\mathcal{G}_\delta$. Certainly, $\mathbf{Fr}$ is dense in ${}^{\bullet}\mathbb{SIM}$ since, from above, $\widehat{\nu} * \widehat{\mathbf{m}_\delta} \to \widehat{\nu}$. And $\widehat{\nu} * \widehat{\mathbf{m}_\delta}$ is free. □

References

Library call numbers, where available, follow each reference.

[Con] J. P. Conze, *Entropie d'un groupe abélien des transformations*, Z. Wahrscheinlichkeitstheorie Verw. Geb. **26** (1973), 11–30, cn: *QA273.Z4.*

[CP] J.R. Choksi & V.S. Prasad, *Approximation and Baire category theorems in ergodic theory*, Measure Theory and its Applications, (Proc. of a conf. held at Sherbrooke, Canada, June 7-18, 1982, ed. by J.M. Belley, J. Dubois and P. Morales), vol. Springer Lecture notes in Mathematics 1033, Springer-Verlag, New York, 1983, pp. 94–113, cn: *QA3.L28.*

[CN] J.R. Choksi & M.G. Nadkarni, *Baire Category in Spaces of Measures, Unitary Operators, and Transformations*, Invariant subspaces and allied topics, (Proc. of conference held at Delhi in December 1986), Narosa Publishing House, New Delhi, 1990, pp. 147–163, cn: *QA322.I58.*

[dJR] A. del Junco, D.J. Rudolph, *On ergodic actions whose self-joinings are graphs*, Ergodic Theory and Dynamical Systems **7** (1987), 531–557, cn: *QA614.8.E736.*

[FK] W.G. Fleissner & K. Kunen, *Barely Baire spaces*, Fundamenta Mathematicae **101** #3 (1978), 229–240, cn: *QA1.F8.*

[Hal] P. Halmos, *Ergodic Theory*, Chelsea Publishing Co., New York, 1956, cn: *QA614.H3.*

[KW] Y. Katznelson & B. Weiss, *Commuting measure preserving transformations*, Israel J. Math. **12** (1972), 16–173, cn: *QA1.I92.*

[Kec] A.S. Kechris, *Classical Descriptive Set Theory*, Springer-Verlag, New York, 1995, cn: *QA248. K387.*

[Kur] K. Kuratowski, *Topology*, vol. 1, Academic Press, New York, 1966, cn: *QA611 K8413.*

[OW] D.S. Ornstein & B. Weiss, *Entropy and isomorphism theorems for actions of amenable groups*, J. d'Analyse Math. **48** (1987), 1–141, cn: *QA1.J85.*

[Oxt] J.C. Oxtoby, *Measure and Category*, Springer-Verlag, New York, 1971, cn: *QA312.O9.*

[Oxt2] J.C. Oxtoby, *Note on transitive transformations*, Proc. Nat. Acad. Sci. USA **23** (1937), 443–446, cn: *Q11.N26.*

[Rud] D.J. Rudolph, *Residuality and Orbit Equivalence*, Contemporary Mathematics **215** (1998), 245-256.

[Sol] R.M. Solovay, *A model of set theory in which every set of reals is Lebesgue measurable*, Annals of Math. **92** (1970), 1–56, cn: *QA1.A6.*

Contemporary Mathematics
Volume **215**, 1998

Residuality and Orbit Equivalence

D. Rudolph

This work is offered in celebration of the mathematics of Robert Ellis upon his retirement.

ABSTRACT. Classically the study of residual behavior for finite measure preserving actions takes place in one of two contexts: the weak* topology on finite invariant measures for a shift map on $X^{\mathbb{Z}}$ where X is a compact metric space without isolated points, or the strong operator topology on L^1 for measure preserving transformations of a fixed nonatomic standard probability space. We will introduce a third context, the space of all actions orbit equivalent to some fixed action, topologized naturally as a Polish space. What we will see is that all three of these give precisely the same notions of residuality of conjugacy invariant properties. To say this more precisely, in each of these three one can consider those sets of actions which are conjugacy invariant. These form Boolean algebras. There is a canonical Boolean isomorphism between them. Any conjugacy invariant set which is residual in one of these algebras (contains a dense G_δ) is in all the others as well.

1. Introduction

The notion of residuality of various dynamic properties in the space of ergodic and measure preserving actions, or joinings of two or more, or of cocycles over some fixed action, provides an interesting and often useful perspective on what is possible in these spaces. Often one discovers that what is difficult to find as a unique object is easier to find as a residual set of such objects. The basic step to begin such work is to identify an appropriate Polish space whose elements can be identified with the objects. I say this as opposed to saying "one must topologize the objects of study as a Polish space" as quite regularly each object of study (action or cocycle) will have many representatives in the Polish space. Certainly that is the case for the three situations we will look at.

What we will examine here are three Polish spaces on which one can naturally identify points with measure preserving and ergodic actions. The first is the weak* topology on the space of shift invariant probability measures on some $X^{\mathbb{Z}}$ where X is a compact metric space with no isolated points. The second is the strong operator

1991 *Mathematics Subject Classification.* Primary 28D15, 28D20.
The author gratefully acknowledges the support of NSF grant DMS9401538.

topology on the measure preserving transformations of a fixed standard probability space $(\Omega, \mathcal{F}, \mu)$ as maps on $L^1(\mu)$. These two are the classical situations. We will introduce a third, and this is our main interest, by topologizing a complete set of orbit equivalent actions on a fixed ergodic and measure preserving orbit relation on a standard probability space.

In each of these cases any particular measure preserving and ergodic action will have many representatives, in fact a dense family all of which are measurably conjugate to it. In the first two cases this is a well-known and not so difficult application of the Rokhlin lemma. In the third it will become apparent as we continue. We will only be interested in subsets of these Polish spaces that are conjugacy invariant, that is to say contain full conjugacy classes of elements. These sets form Boolean algebras. Notice that in this case any such set in one of these Polish spaces gives rise to a unique such set in the others consisting of all elements conjugate to some element of the original set. This identification is of course a Boolean isomorphism. We will often speak as if such a set is its collection of conjugacy classes and hence identify the corresponding sets in the three distinct spaces as if they were in fact the same set.

Our interest is in those conjugacy invariant sets that are residual, that is to say contain a dense G_δ. Our conclusion will be that if such a set is residual in any of these examples, its version in the others is also residual. For the first two this is (I believe) folklore. Glasner and King [**GK**] have given a particularly elegant proof of this. We will show a bit more. In the second and third spaces conjugacy invariant and dense G_δ sets will also be identical, and this will almost be true between the first and the other two in that it will be true of a fixed dense G_δ subset of the first. This last fact is easily evident also in the argument of Glasner and King [**GK**]. Glasner and King [**GK**] have also shown that under very mild assumptions any conjugacy invariant property will be either residual or meager, there is no middle ground.

We will first lay out the three topological spaces that interest us. We will then show how to identify each of these spaces as subsets of the previous. That is to say, we will identify the space of all maps preserving some fixed measure as a G_δ subset of the space of all shift invariant measures on a product space. We will then identify the set of all actions orbit equivalent to some given action as a G_δ subset of this space. In neither case will the subset be dense of course. In fact each space will be fibered by copies of the next. (This is not quite accurate, but the deficit will be explained later.) It is this fibering of the first two by copies of the third that we will exploit to prove our main result.

2. The Three Topologies and Conjugacy Invariant Sets

We will say very little here about the first two topological spaces other than to give them names. Fix a compact metric space (X, d) with no isolated points. Let $\mathbb{T}_1$ be the set of all shift invariant and ergodic Borel probability measures on $X^{\mathbb{N}}$. It is classical (see exercise 7.1 of [**R**] for example) that this is a dense G_δ subset of the space of all shift invariant probability measures, which is a compact metric space in the weak* topology. For our second Polish space let $(\Omega, \mathcal{F}, \mu)$ be a nonatomic standard probability space and let $\mathbb{T}_2$ be the space of all μ preserving free and ergodic actions of $\mathbb{Z}$ on Ω in the strong operator topology (this is also called the weak or coarse topology on transformations). That is to say $T_i \to T$ in

$\mathbb{T}_2$ iff $f \circ T_i \to f \circ T$ in $L^1(\mu)$ for all $f \in L^1(\mu)$. It is not difficult to see that this is a G_δ subset of the space of all μ preserving actions, which form a Polish space. Thus $\mathbb{T}_2$ is itself Polish.

For the third example we give more detail. Let $(\Omega, \mathcal{F}, \mu)$ again be a fixed nonatomic standard probability space, and now let T_0 be a fixed measure preserving, free, and ergodic $\mathbb{Z}$ action on Ω. Let $\mathbb{T}_3$ be the space of all $\mathbb{Z}$ actions on Ω that are orbit equivalent to T_0. We need to topologize $\mathbb{T}_3$. We do this using notions from [**KR**]. There the fundamental objects are *arrangements* rather than orbit equivalent actions. These two notions are simply different perspectives on the same objects. For the benefit of the reader we will avoid introducing arrangements and instead we will stick to the vocabulary of actions.

Define a metric as follows. For any two elements T_1 and T_2 in $\mathbb{T}_3$ and $\omega \in \Omega$ consider the bijection of $\mathbb{Z}$ given by

$$h_\omega^{T_1,T_2}(i) = j \text{ if } T_1^i(\omega) = T_2^j(\omega).$$

Notice that $h_\omega^{T_1,T_2}$ is not only a bijection but also fixes 0. Let $\mathcal{I}$ be the semi-group of all injections of $\mathbb{Z}$ which fix 0 and $\mathcal{H} \subseteq \mathcal{I}$ be the subgroup of all bijections of $\mathbb{Z}$ fixing 0. We can topologize $\mathcal{I}$ as a Polish semi-group as follows. For any two elements $h_1, h_2 \in \mathcal{I}$ set $J(h_1, h_2)$ to be the largest value J so that h_1 and h_2 agree on the interval $[-J, J]$. Now set

$$D(h_1, h_2) = (J(h_1, h_2) + 1)^{-1}.$$

It is easy to check that a D-Cauchy sequence of injections must converge to an injection so that for this metric $\mathcal{I}$ is complete and separable. Moreover, $D(hh_1, hh_2) = D(h_1, h_2)$. The space $\mathcal{H}$ is not complete in D but the map $h \to h^{-1}$ is continuous on $\mathcal{H}$ and hence the metric

$$D_1(h_1, h_2) = D(h_1, h_2) + D(h_1^{-1}, h_2^{-1})$$

gives the same topology on $\mathcal{H}$ as D does. With respect to D_1, the group $\mathcal{H}$ is complete, we have though lost the left-invariance of the metric, which we need to construct a metric on $\mathbb{T}_3$. We do though conclude that $\mathcal{H}$ is a G_δ-subset of $\mathcal{I}$.

Using the left-invariance of D we can now put a metric on $\mathbb{T}_3$. For T_1 and T_2 in $\mathbb{T}_3$ set

$$D_3(T_1, T_2) = \int D(h_\omega^{T_1,T_2}, \mathrm{id})\, d\mu(\omega).$$

LEMMA 2.1. *The metric D on $\mathbb{T}_3$ generates a Polish topology.*

PROOF. Let $\hat{\mathbb{T}}_3$ be the completion of $\mathbb{T}_3$ in the metric D_3. If T_i are Cauchy in D_3 by dropping to a subsequence we can assume $D(T_i, T_{i+1}) < 2^{-i}$. It now follows that for μ-a.e. ω the sequence of bijections $h_\omega^{T_1,T_i}$ is Cauchy in D and hence the family of sequences $\{h_\omega^{T_1,T_i}\}$ converges for a.e. ω to a measurable family of injections $\{h_\omega\}$ of $\mathbb{Z}$. If two Cauchy sequences in $\mathbb{T}_3$ represent the same element in $\hat{\mathbb{T}}_3$ then they give a.s. the same family of injections $\{h_\omega\}$. Hence the families $\{h_\omega\}$ that are reachable by Cauchy sequences in D_3 faithfully represent $\hat{\mathbb{T}}_3$, and D_3 extends to these families as the metric

$$D_3(\{h_\omega\}, \{h'_\omega\}) = \int D(h_\omega, h'_\omega)\, d\omega.$$

We identify $\hat{\mathbb{T}}_3$ with these measurable families of injections.

In this form $\mathbb{T}_3$ consists of those $\{h_\omega\} \in \hat{\mathbb{T}}_3$ where h_ω is a.s. a bijection. Notice that for any $j \in \mathbb{N}$ the sets

$$\mathcal{O}_J = \{\{h_\omega\} : \mu(\{\omega : [-J, J] \subseteq \mathit{Range}(h_\omega)\}) > 1 - 2^{-J})\}$$

are D_3 open. As each contains $\mathbb{T}_3$, they are also dense. But now

$$\mathbb{T}_3 = \cap_J \mathcal{O}_J$$

and so is a G_δ subset of its completion.

To see separability, fix a block $[-j, j]$ and for each T partition Ω according to the countable list of values $h_\omega^{T_0,T}|_{[-j,j]}$. If for two T's these partitions are close in the partition metric, then the two actions will be close, and vice versa, if two actions are close, then these partitions will be close. As the partition metric is separable, this forces our metric D_3 to generate a seperable topology. □

For the first two topological spaces $\mathbb{T}_1$ and $\mathbb{T}_2$ it is well known and easy to show that any particular conjugacy class occurs densely. For $\mathbb{T}_3$ this may not be so well known.

LEMMA 2.2. *For any given measure preserving and ergodic transformation of a nonatomic space, in the space $\mathbb{T}_3$ there are actions conjugate to this given action. Furthermore, any given conjugacy class in $\mathbb{T}_3$ is dense.*

PROOF. What is in fact true is that the full group Γ of the orbit relation of T_0 acts minimally on $\mathbb{T}_3$ where the action is $T \to \phi T \phi^{-1}$. The reason for this can be found in **[R]** on page 14, where orbit equivalence as an example of restricted orbit equivalence is discussed. It is shown there that if T_1 and T_2 are two orbit equivalent actions then by successively "collapsing" the T_2 arrangement of orbits back onto the T_1 arrangement one obtains a sequence of full group elements ϕ_i with $\phi_i T_1 \phi_i^{-1}$ converging to T_2. □

In each of our three examples we can consider the Boolean algebras of conjugacy invariant sets. There is a canonical Boolean isomorphism among these three. This is simply because each is Boolean isomorphic to the Boolean algebra of subsets of conjugacy classes of actions. Rather than give these sets and conjugacies names, we will simply refer to them as conjugacy invariant sets. We will identify such a set in any of the spaces with its collection of conjugacy classes.

3. Embedding $\mathbb{T}_3$ in $\mathbb{T}_2$ in $\mathbb{T}_1$

Returning to $\mathbb{T}_1$, the space of ergodic and shift invariant Borel probability measures on $X^{\mathbb{Z}}$, our first step will be to restrict to a nice residual subset of $\mathbb{T}_1$. Earlier we said that we were essentially going to fiber $\mathbb{T}_1$ by copies of $\mathbb{T}_2$. What we meant by that is that we will in fact fiber the dense G_δ subset that we now describe.

LEMMA 3.1. *The free orbits for σ, the shift map on $X^{\mathbb{Z}}$, are a dense G_δ and hence those elements of $\mathbb{T}_1$ supported on these free orbits also form a dense G_δ.*

PROOF. The free orbits of σ are easily seen a dense G_δ by noting $\{\vec{x} : \sigma^n(\vec{x}) = \vec{x}\}$ is closed and nowhere dense for each value n (remember X has no isolated points.) The measures giving full support to an open and dense subset of a compact metric space are easily seen to be a dense G_δ completing the result. □

Much more is known of course. The residual behavior of a measure in $\mathbb{T}_1$ is weakly mixing, rigid, and hence of zero entropy.

Notice that when one models a dynamical system $(\Omega, \mathcal{F}, \mu, T)$ as a σ-invariant measure on $X^{\mathbb{Z}}$ what one does is to first assume that Ω is X and μ is some Borel measure and T some Borel map. One then maps a point x to its infinite name $\{T^j(x)\}_{j\in\mathbb{Z}}$. Notice that in doing this one obtains a very special sort of measure in $\mathbb{T}_1$ in that the "time zero" σ-algebra is almost-surely the full algebra. That is to say knowing the time-zero value x will determine a.s. all the rest of the values $T^j(x)$. Although this may seem a very special situation it is in fact residual. We cite Glasner and King [**GK**] for this fact. They prove it true for actions of general countable discrete groups.

DEFINITION 3.2. In $\mathbb{T}_1$ let $\mathbb{T}_1'$ consist of those measures μ supported on the free orbits of σ and for which the time-zero σ-algebra is μ-a.s. the full algebra of measurable sets.

THEOREM 3.3. [Glasner and King] *The subset $\mathbb{T}_1'$ is a dense G_δ subset of $\mathbb{T}_1$.*

For our next step we will fiber $\mathbb{T}_1'$ according to the projection of the measure μ on its time-zero coordinate.

DEFINITION 3.4. For any Borel probability measure μ_0 on X let $\mathbb{T}_{1,\mu_0}$ consist of those measures $\mu \in \mathbb{T}_1'$ whose restriction to the zero'th coordinate is μ_0.

LEMMA 3.5. *For any nonatomic Borel μ_0 on X, the set of measures $\mathbb{T}_{1,\mu_0}$ is a nonempty closed subset of $\mathbb{T}_1'$ containing elements of all conjugacy classes of ergodic and free actions.*

PROOF. As all nonatomic standard probability spaces are Borel equivalent, for any ergodic free action there will be μ_0-preserving Borel actions on X conjugate to any given action. Mapping a point x to its name under this action will lift μ_0 to some $\mu \in \mathbb{T}_{1,\mu_0}$. □

To identify $\mathbb{T}_2$ as a subset of $\mathbb{T}_1$, first use the above observation about nonatomic standard probability spaces to identify $(\Omega, \mathcal{F}, \mu)$ as $(X, \mathcal{B}, \mu_0)$. We have just seen how any element of $\mathbb{T}_2$ can be lifted to an element of $\mathbb{T}_{1,\mu_0}$, and it is equally clear that conversely any element of $\mathbb{T}_{1,\mu_0}$ gives a μ_0-preserving map of X. Let

$$\Phi : \mathbb{T}_2 \to \mathbb{T}_{1,\mu_0}$$

be this bijection identifying actions with measures.

COROLLARY 3.6. *For any fixed nonatomic Borel measure μ_0 on X, the map $\Phi : \mathbb{T}_2 \to \mathbb{T}_{1,\mu_0}$ is a homeomorphism.*

PROOF. We already know the map Φ is a bijection. To see it is a homeomorphism notice that any $\mu \in \mathbb{T}_{1,\mu_0}$ can be regarded as a coupling of countably many copies of μ_0, one on each coordinate. Hence to say $\mu_i \to \mu$ in $\mathbb{T}_{1,\mu_0}$ is to say, taking any list of bounded Borel functions $f_j : X \to \mathbb{R}$, with only finitely many not identically 1,

$$\int \Pi_j f_j(x_j)\, d\mu_i(\vec{x}) \underset{i}{\to} \int \Pi_j f_j(x_j)\, d\mu(\vec{x}).$$

This is identical to saying

$$\int \Pi_j f_j(T^j_{\mu_i}(x_0))\, d\mu_0(x_0) \underset{i}{\to} \int \Pi_j f_j(T^j_{\mu}(x_0))\, d\mu_0(x_0),$$

where by T_{μ_i} we mean the map on X whose names give rise to μ_i.

This certainly implies that for all bounded Borel f,

$$\int f(T_{\mu_i}(x_0))\, d\mu_0 \underset{i}{\to} \int f(T_{\mu}(x_0))\, d\mu_0.$$

This implies that the maps T_{μ_i} converge in the strong operator topology to T_μ.

For the other direction, let $\| \ \|_1^\mu$ be the L^1 norm relative to the measure μ. If for all bounded Borel f we have

$$\|f \circ T_{\mu_i} - f \circ T_\mu\|_1^{\mu_0} \underset{i}{\to} 0,$$

then equally well for all j we have

$$\|f \circ T^j_{\mu_i} - f \circ T^j_\mu\|_1^{\mu_0} \underset{i}{\to} 0.$$

Hence taking any collection of bounded Borel functions f_i, only finitely many of which are not identically 1, we have

$$\|\Pi_j f_j \circ T^j_{\mu_i} - \Pi_j f_j \circ T^j_\mu\|_1^{\mu_0} \underset{i}{\to} 0$$

and hence $\mu_i \to \mu$ weak*. □

Notice that we can now immediately say that any conjugacy invariant and G_δ subset of $\mathbb{T}_1$ is a conjugacy invariant and G_δ subset of $\mathbb{T}_2$, as the restriction of the set to $\mathbb{T}_2$ must be a G_δ and dense, as it is nonempty. This does not so immediately show that any conjugacy invariant and *residual* subset of $\mathbb{T}_1$ is residual in $\mathbb{T}_2$. We postpone that until we have built up more machinery.

We now move to the problem of embedding $\mathbb{T}_3$ in $\mathbb{T}_2$ and in fact fibering $\mathbb{T}_2$ by copies of $\mathbb{T}_3$. This is significantly easier as again regarding the measure space as fixed, and selecting some element T_0 in $\mathbb{T}_2$ to give us an orbit relation, the version of $\mathbb{T}_3$ we obtain is directly a subset of $\mathbb{T}_2$. To be more precise, let $\mathbb{T}_{2,T_0}$ be the subset of actions in $\mathbb{T}_2$ that are orbit equivalent to T_0. We can lift this back to $\mathbb{T}_1$ by setting $\mathbb{T}_{1,\mu_0,T_\mu}$ to be the subset of measures in $\mathbb{T}_{1,\mu_0}$ whose corresponding actions are orbit equivalent to the given action $T_\mu = \Phi^{-1}(\mu)$.

THEOREM 3.7. *Each set $\mathbb{T}_{2,T_0}$ is a G_δ subset of $\mathbb{T}_2$ and contains representatives of all conjugacy classes.*

PROOF. We assume that Ω is X, a compact metric space. For $T \in \mathbb{T}_2$ and $x \in X$ let $r_T(x) = \{T^j(x)\}_{j\in\mathbb{Z}}$ be the indexed T-orbit of X. Define a metric D on $X^{\mathbb{Z}}$ by

$$D(\vec{x}, \vec{x}') = \min\{(j+1)^{-1} : d(x_i, x_i') < 1/j \text{ for all } i \in [-j, j]\}.$$

Now define a metric D_2 giving the topology on $\mathbb{T}_2$ by

$$D_2(T, T') = \int D(r_T(\vec{x}), r_{T'}(\vec{x}))\, d\mu_0.$$

Notice that for T and T' in $\mathbb{T}_3$ that $D_2(T, T') \leq D_3(T, T')$ as for D_2 to be $< (j+1)^{-1}$, the T and T' orbits must stay within $1/j$ on $[-j, j]$, but for D_3 to be $< 1/j$ they must be identical.

For $\varepsilon > 0$ and $x \in X$ consider $\{j : d(T(x), T_0^j(x)) < \varepsilon\}$. For μ_0-a.e. x this set will be nonempty as T_0 is ergodic. Consider the two values

$$\begin{aligned} j^+ &= \min\{j > 0 : d(T(x), T_0^j(x)) < \varepsilon\} \text{ and} \\ j^- &= \max\{j < 0 : d(T(x), T_0^j(x)) < \varepsilon\}. \end{aligned}$$

Define

$$J(\varepsilon, x) = \begin{cases} j^+ & \text{if } |j^+| \leq |j^-| \text{ and} \\ j^- & \text{otherwise, i.e. if } |j^-| < |j^+|. \end{cases}$$

For any $B > 0$ define a function

$$F_{\varepsilon,B}(T) = \mu_0(\{x : |J(\varepsilon, x)| \leq B\}).$$

Notice that if $T_i \to T$ in $\mathbb{T}_2$ then for μ_0 a.e. x,

$$\liminf_{i\to\infty} F_{\varepsilon,B}(T_i) \geq F_{\varepsilon,B}(T).$$

Setting $O(B, C, \varepsilon)$ to be $\{T : F_{\varepsilon,B}(T) > C\}$, these are open subsets of $\mathbb{T}_2$. Thus

$$O'(B, C, \varepsilon) = \bigcap_{C' \nearrow C} O(B, C', \varepsilon) = \{T : F_{\varepsilon,B}(T) \geq C\}$$

is a G_δ. Hence

$$\begin{aligned} O(B, C) &= \bigcap_{\varepsilon \searrow 0} O'(B, C, \varepsilon) \\ &= \{T : \mu_0\{x : T(x) = T_0^j(x) \text{ for some } j \in [-B, B]\} > C\} \end{aligned}$$

is again a G_δ. Now

$$O(C) = \bigcap_{B \nearrow \infty} O(B, C) = \{T : \mu_0\{x : T(x) \text{ is on the } T_0 \text{ orbit of } x\} > C\}$$

is again a G_δ. Finally

$$\mathbb{T}_{2,T_0} = \bigcap_{C \nearrow 1} O(C)$$

is a G_δ

As each $\mathbb{T}_{2,T_0}$ is a copy of $\mathbb{T}_3$, it must intersect all conjugacy classes by Dye's Theorem [**D**]. □

Sadly the enherited topology on $\mathbb{T}_{2,T_0}$ is not the topology of $\mathbb{T}_3$. We have already seen why as the metric D_3 on $\mathbb{T}_3$ is simply a much stronger metric than D_2.

COROLLARY 3.8. *Fixing T_0 in $\mathbb{T}_2$, the two spaces $\mathbb{T}_3$ and $\mathbb{T}_{2,T_0}$ are identical as sets. The topology of the first is stronger than that of the second. That is to say if $T_i \to T$ in $\mathbb{T}_3$ then $T_i \to T$ in $\mathbb{T}_{2,T_0}$.*

We now can also say that any G_δ and dense conjugacy invariant subset of either $\mathbb{T}_1$ or $\mathbb{T}_2$ will be a dense G_δ subset of $\mathbb{T}_3$. It will obviously be G_δ by the previous theorem, and will be dense as once it is nonempty it must be. We now set out to show that residuality of a conjugacy invariant set in each of the $\mathbb{T}_i$ is the same notion. As indicated earlier we will prove also a related fact, that in each of $\mathbb{T}_1'$, $\mathbb{T}_2$, and $\mathbb{T}_3$ to be a conjugacy invariant G_δ set is the same notion.

4. Residuality in The Three Spaces

We now consider $\mathbb{T}_1'$ to be our largest space. Certainly the residual sets in it and in $\mathbb{T}_1$ are the same. Hence we can now assume we are working on just a single copy of the compact metric space (X, d). The space $\mathbb{T}_1'$ consists of a choice for a nonatomic Borel probability measure μ_0 and a μ_0 preserving action T_0. The topology is the weak* topology on the lift of μ_0 to the space of T_0-names in $X^{\mathbb{Z}}$. Let D_1 be a metric giving this topology on the space of pairs (μ_0, T_0). The space $\mathbb{T}_1'$ is fibered by choices for the measure μ_0 into closed subsets $\mathbb{T}_{1,\mu_0}$ all of which are homeomorphic (in the induced topology) to $\mathbb{T}_2$. This topology is given explicitely by the metric D_2, which we call D_{2,μ_0} on this fiber.

These fibers are in turn fibered as copies of $\mathbb{T}_3$ by selecting some member T_0 and considering only those actions which are orbit equivalent to T_0. We write this fiber as $\mathbb{T}_{1,\mu_0,T_0}$. This has its enherited topology but also has the stronger topology obtained from $\mathbb{T}_3$, given by the metric D_3. We write this metric, when placed on this fiber, as D_{3,μ_0,T_0}.

We now make a simple but quite crucial observation. Suppose $\phi : X \to X$ is a Borel bijection. Then for any of the fibers $\mathbb{T}_{1,\mu_0,T_0}$,

$$\phi : \mathbb{T}_{1,\mu_0,T_0} \to \mathbb{T}_{1,\phi^*(\mu_0),\phi T_0 \phi^{-1}}$$

is an *isometry* from the metric D_{3,μ_0,T_0} to the metric $D_{3,\phi^*(\mu_0),\phi T_0\phi^{-1}}$ as conjugation leaves fixed the way one orbit is *rearranged* to form another. Further, as any conjugacy between conjugate elements of $\mathbb{T}_1'$ can be assumed Borel, it will automatically extend to an isometry on all the $\mathbb{T}_3$ fibers of the two actions.

The key fact we now demonstrate is that if two elements are close in $\mathbb{T}_1'$ then we can find such an isometry that is in fact small. The reader may find the proof we give of this unsatisfying as we refer to results in [**R**] and [**KR**]. We do this for efficiency. A direct proof is certainly possible modelled on the proof of Dye's theorem.

THEOREM 4.1. *Given any element (μ_0, T_0) in $\mathbb{T}_1'$ and $\varepsilon > 0$ there is a $\delta > 0$ so that if $(\mu_1, T_1) \in \mathbb{T}_1'$ satisfy*

$$D_1((\mu_0, T_0), (\mu_1, T_1)) < \delta$$

then there is a Borel map ϕ *with* $\phi^*(\mu_0) = \mu_1$, $\phi T_0 \phi^{-1} \in D_{1,\mu_1,T_1}$, *(i.e. is orbit equivalent to* T_1*) and moreover both*

$$\int d(x, \phi(x))\, d\mu_0 < \varepsilon \text{ and}$$

$$D_{3,\mu_1,T_1}(\phi T_0 \phi^{-1}, T_1) < \varepsilon.$$

That is to say if (μ_1, T_1) *is close enough to* (μ_0, T_0) *in* $\mathbb{T}'_1$ *then we can construct an isometry of the* $\mathbb{T}_3$ *fibers through them that is small pointwise relative to* μ_0 *and moves* T_0 *close to* T_1 *in the* D_3 *metric.*

PROOF. This argument is an application of the restricted orbit equivalence machinery (see [**KR**] and [**R**]) to the special case of orbit equivalence. We sketch what we need of it here. For two actions $(X_0, \mathcal{B}_0, \mu_0, T_0)$ and $(X_1, \mathcal{B}_1, \mu_1, T_1)$ an *orbit joining* is any third space $(Y, \mathcal{G}, \hat{\mu})$ on which we have two free, ergodic, and orbit equivalent actions $\hat{T}_0$ and $\hat{T}_1$ and a pair of factor maps ϕ_0 taking the action of $\hat{T}_0$ to that of T_0 and ϕ_1 taking the action of $\hat{T}_1$ to that of T_1. In keeping with rather common usage we do not bother to point out that these maps preserve the associated measures. Suppose we also have two finite partitions P_0 of X_0 and P_1 of X_1, both using the same finite labeling set Σ. As is usually done we indicate the processes they generate as (T_0, P_0) and (T_1, P_1). We can now define a distance between processes labeled by Σ:

$$\overline{m}_0(T_0, P_0; T_1, P_1) = \inf_{\text{orbit joinings}} (\hat{\mu}(\phi_0^{-1}(P_0) \Delta \phi_1^{-1}(P_1)) + D_3(\hat{T}_0, \hat{T}_1)).$$

A process (T_0, P_0) is called m_0*-finitely determined* (in fact m_0 is just D_3 in our notation) if for any $\varepsilon > 0$ there is a δ and an N so that:

(1) If the probability vector of μ_0 masses of the names in $\vee_{j=-N}^{N} T_0^{-j}(P_0)$ is within ℓ_1 distance δ of the vector of μ_1 masses of the corresponding names in $\vee_{j=-N}^{N} T_1^{-j}(P_1)$,
(2) then $\overline{m}_0(T_0, P_0; T_1, P_1) < \varepsilon$.

This particular notion of finitely determined, associated with the size m_0, is the weakest of all sizes in the restricted orbit equivalence development. In particular from lemma 2.8 of [**R**] any two orbit equivalent actions are m_0-equivalent and all ergodic processes are m_0-finitely determined (see page 146 of [**R**]).

Using this we now can say more about the orbit joinings of any two processes and in particular the joinings that bring two processes close in $\overline{m}_0$ in (2) above. The space of m-joinings of any two ergodic actions T_0 and T_1 can itself be topologized as a Polish space. Suppose one of the two actions, say T_0, is in fact m-finitely determined, and the other has at least as much m-entropy i.e. $h_m(T_1) \geq h_m(T_0)$. (If you are unfamiliar with m-entropy don't worry, in our case h_{m_0} is always identically zero.) One concludes that those m-joinings where $\phi_0^{-1}(\mathcal{B}_0) \subseteq \phi_1^{-1}(\mathcal{B}_1)$ and even more, the choice of maps $h_y^{\hat{T}_1, \hat{T}_0}$ are $\phi_1^{-1}(\mathcal{B}_1)$ measurable form a residual subset of the m-joinings. Said another way, consider those m-joinings for which

(i) $Y = X_1$,
(ii) ϕ_1 is the identity, and
(iii) the action $\hat{T}_1$ is just T_1.

The set of m-joinings satisfying (i), (ii), and (iii) form a residual set in the space of all m-joinings. This fact is the core conclusion of the restricted orbit equivalence theorem (Section 7.3 of [**KR**]).

In our case (as m_0-entropy is always zero and all ergodic actions are m_0-finitely determined), those joinings which are in fact orbit equivalences, that is to say $\hat{T}_1 = T_1$ and ϕ_0 is a measure preserving bijection from X_1 to X_0, form a residual set. In particular we can assume that the orbit joining giving (2) in the definition of m_0-finitely determined is an orbit equivalence.

Now to apply this to prove our result. Let P_0 be a finite partition of X into sets of diameter less than $\varepsilon/2$, all of which have boundaries of μ_0 measure zero. We conclude from this that for any N and set $A \in \vee_{j=-N}^{N} T_0^{-j}(P_0)$ that if $(\mu_i, T_i) \to (\mu_0, T_0)$ in D_1 then we must have $\mu_i(A) \to \mu_0(A)$.

But then, as (T_0, P_0) is m_0-f.d., there must be a δ associated with $\varepsilon/2$ so that if

$$D_1((\mu_0, T_0), (\mu_1, T_1)) < \delta$$

then

$$\overline{m}_0(T_0, P_0; T_1, P_0) < \varepsilon/2,$$

and we can assume that this $\overline{m}_0$ distance is made small by an orbit equivalence of T_0 to an action T_1' that is conjugate to T_1. That is to say there is a map $\phi : X \to X$ with $\phi^*(\mu_0) = \mu_1$ conjugating T_1' to T_1 and we have

$$D_{3,\mu_0,T_0}(T_0, T_1') < \varepsilon/2$$

and

$$\mu_0(\phi^{-1}(P_0)\triangle P_0) < \varepsilon/2.$$

Notice that whenever x and $\phi(x)$ belong to the same element of P_0 then $d(x, \phi(x)) < \varepsilon/2$. Assuming that diam $(X) \leq 1$ we conclude that

$$\int d(x, \phi(x))\, d\mu_0 < \varepsilon$$

completing the proof. □

We can reinterpret this result in a rather simple fashion. Fix a choice for $(\mathbb{T}_{1,\mu_0,T_0}, D_{3,\mu_0,T_0})$. Let $(\mathbb{F}, D)$ be the Polish space of all Borel bijections of X with the metric

$$D(\phi_1, \phi_2) = \int d(\phi_1(x), \phi_2(x))\, d\mu_0 = \int d(x, \phi_2\phi_1^{-1}(x))\, d\phi_1^*(\mu_0).$$

On the product space $(\mathbb{T}_{1,\mu_0,T_0} \times \mathbb{F})$ we have canonical maps

$$\Phi_1 : \mathbb{T}_{1,\mu_0,T_0} \to \mathbb{T}_1' \text{ and}$$
$$\Phi_3 : \mathbb{T}_{1,\mu_0,T_0} \to \mathbb{T}_{1,\mu_0,T_0}$$

given by

$$\Phi_1(T, \phi) = (\phi^*(\mu_0), \phi T \phi^{-1}) \text{ and}$$
$$\Phi_3(T, \phi) = T.$$

Notice we can also restrict $\mathbb{F}$ to those ϕ for which $\phi^*(\mu_0) = \mu_0$. This is a closed subset of $\mathbb{F}$ we call $\mathbb{F}_1$. Restricting Φ_1 to $\mathbb{F}_1$ we get a map

$$\Phi_2 : \mathbb{T}_{1,\mu_0,T_0} \times \mathbb{F}_1 \to \mathbb{T}_{1,\mu_0}.$$

COROLLARY 4.2. *All three maps Φ_1, Φ_2, and Φ_3 are continuous, onto, and open. Moreover if $A_1 \subseteq \mathbb{T}_1$ and $A_2 \in \mathbb{T}_{1,\mu_0}$ and $A_3 \in \mathbb{T}_{1,\mu_0,T_0}$ are three conjugacy invariant sets, all corresponding to the same collection of conjugacy classes, then*

$$\Phi_1^{-1}(A_1) = \Phi_2^{-1}(A_2) = \Phi_3^{-1}(A_3) = A_3 \times \mathbb{F}.$$

We now quote a basic result of Sierpinski on Polish spaces (see [**K**] Theorem 8.19).

THEOREM 4.3. *Suppose A is Polish, and $\phi : A \to B$ is continuous, open, and onto where B is known to be separable and metrizable. Then B is in fact Polish.*

COROLLARY 4.4. *Suppose A is Polish, B is separable and metrizable, and $\phi : A \to B$ is continuous, open, and onto. Then if $B' \subseteq B$ is a dense G_δ so is $\phi^{-1}(B') \subseteq A$ (this is trivial). On the other hand, if $A' \subseteq A$ is a dense G_δ, then so is $\phi(A') \subseteq B$ (this follows from Sierpinski's result above).*

We can now apply these to prove our main theorem.

THEOREM 4.5. *The residual and conjugacy invariant subsets of each of the three spaces $\mathbb{T}_1$, $\mathbb{T}_2$, and $\mathbb{T}_3$ are the same. One can also say that the conjugacy invariant and dense G_δ subsets of each of the three spaces $\mathbb{T}_1'$, $\mathbb{T}_2$, and $\mathbb{T}_3$ are the same.*

PROOF. The only point not directly obvious is that the maps Φ_1 and Φ_2 are open. But this is just a rereading of the Theorem 4.1. □

We end with a few comments. The core of this work was Theorem 4.1, the simplest case of the restricted orbit equivalence theorem [**KR**]. This offers two directions for generalization. First, one can ask if this result is true for more general group actions. The machinery of [**OW**] and [**KR**] make the answer yes for actions of discrete and amenable groups. One can ask if it is possible to push beyond this to not necessarily discrete but still amenable groups. The techniques described here still have some hope of working via the machinery of [**OW**], but to push beyond amenable groups would require significantly new ideas. As a second comment, if one now is willing to accept the orbit equivalence class as a reasonable place to discuss issues of residuality then one can ask, in the context of general restricted orbit equivalences, what is the residual behavior of an action. The point here is that the restricted orbit equivalence machinery is built around the idea of topologizing an equivalence class as a Polish space. It was this realization that led the author to the investigation given here.

Glasner and King in [**GK**] show that $\mathbb{T}_1'$ is in fact homeomorphic to a direct product $\mathbb{T}_{1,\mu_0} \times \mathbb{X}$ for $\mathbb{X}$ an extremely natural Polish space. This avoids the need for Sierpinski's theorem in proving the commonality of residual properties between these two spaces. Is the same true for T_{1,μ_0,T_0}? That is to say, is there some Polish space $\mathbb{Y}$ so that $\mathbb{T}_1'$ is homeomorphic to $\mathbb{T}_{1,\mu_0,T_0} \times \mathbb{Y}$?

References

[D] H.A. Dye, *On groups of measure preserving transformations*, Amer. J. Math. **81** (1959), 119–159.

[GK] S. Glasner and J. King, *A zero-one law on the space of dynamical properties, (published in this volume)..*

[K] A.S. Kechris, *Classical descriptive set theory*, Springer Verlag, New York, 1995.

[KR] J. Kammeyer and D.J. Rudolph, *Restricted orbit equivalence for actions of discrete amenable groups, Preprint.*.

[OW] D. Ornstein and B. Weiss, *Entropy and isomorphism theorems for actions of amenable groups*, J. d'Analyse Math **48** (1987), 1–141.

[R] D. Rudolph, *Restricted orbit equivalence*, Memoirs of the AMS **54 No. 323** (1985).

Department of Mathematics, University of Maryland, College Park, MD 20742
E-mail address: djr@math.umd.edu

Contemporary Mathematics
Volume **215**, 1998

Uncountably Many Vershik-Inequivalent Group Actions of Equal Entropy

J. Feldman

ABSTRACT. Let G be an increasing union of finite abelian groups $\{0\} = G_0 \subset G_1 \subset G_2 \subset \cdots$, where $G_n = H_1 + \cdots + H_n$, and H_n has cardinality > 1 for infinitely many n. Write G^* for G together with the sequence of subgroups $(G_n)_{n=0}^{\infty}$. Two free actions α and β of G on (X, μ) and (Y, ν) respectively will be called *G^*-equivalent* if there is a measure-preserving bijection $\phi : (X, \mu) \mapsto (Y, \nu)$ which carries *a.e.* orbit of $\alpha|G_n$ onto an orbit of $\beta|G_n$. When every H_n is of the form Z/p_nZ this has also been called *Vershik equivalence.* Our main theorem is:

MAIN THEOREM *For every r in $[0, \infty]$ there exist uncountably many free ergodic actions of G of entropy r no two of which are G^*-equivalent.*

1. Decreasing Sequences of Sigma-Fields

The G^*-equivalence classes of actions of the groups described above are in natural one-one correspondence with certain (isomorphism classes of) decreasing sequences of sigma-fields on probability spaces. Such sequences were investigated by A.M. Vershik in a sequence of papers [V1],[V2],[V3],[V4]. These sequences were used in a fundamental way in [DFST], where they were called "reverse filtrations"; we shall use this name here. The present section and the following one will be entirely about reverse filtrations. Throughout, except where otherwise specified, all measure spaces will be Lebesgue spaces.

A *reverse filtration* $(X, \mathbf{F}, P)$ is a probability space $(X, \mathcal{F}_0, P)$ equipped with a decreasing sequence $\mathbf{F} = (\mathcal{F}_n)_{n=0}^{\infty}$ of sigma-fields: $\mathcal{F}_0 \supset \mathcal{F}_1 \supset \mathcal{F}_2 \supset \ldots$ (all sigma-fields are taken to be complete for P) such that $\bigcap_n \mathcal{F}_n$ is the trivial sigma-field consisting of the sets of measure zero and one. For brevity we will usually just write *r.f.*

There is a natural notion of isomorphism between two such objects:

DEFINITION 1.1. An r.f. $(X, \mathbf{F}, P)$ is *isomorphic* to $(X', \mathbf{F}', P')$ if there exists an almost one-one map $\phi : X \to X'$ sending P to P', and such that $\phi^{-1}\mathcal{F}'_n = \mathcal{F}_n$.

Any sequence of random variables $\xi_0, \xi_1, \ldots$ on $(X, \mathcal{F}, P)$ determines a decreasing sequence of sigma-fields: $\mathcal{F}_n$ is the sigma-field generated by the variables

1991 *Mathematics Subject Classifications*: Primary 28D15; Secondary 28D20.
Supported in part by NSF Grant # DMS 9113642.

$\xi_n, \xi_{n+1}, \dots$. If $\mathcal{F}_0 = \mathcal{F}$ and the intersection of the $\mathcal{F}_n$ is trivial, then we call $(X, (\mathcal{F}_n)_{n=0}^\infty, P)$ the r.f. *generated* by $(\xi_n)_{n=1}^\infty$.

DEFINITION 1.2. An *extension* of an r.f. $(X, \mathbf{F}, P)$ consists of an r.f. $(\tilde{X}, \tilde{\mathbf{F}}, \tilde{P})$ and a measure-preserving map π from $\tilde{X}$ to X with $\pi^{-1}(\mathcal{F}_n) \subset \tilde{\mathcal{F}}_n$, $n = 0, 1, \dots,$ such that the equality

$$\tilde{\mathbf{E}}(f \circ \pi | \tilde{\mathcal{F}}_n) = \big(\mathbf{E}(f|\mathcal{F}_n)\big) \circ \pi$$

holds almost everywhere for any bounded real-valued measurable function f.

Vershik calls one of his decreasing sequences of sigma-fields *standard* if it is generated by a sequence of independent random variables, each variable being either non-atomic, or purely atomic with all its atoms having the same probability. Clearly any r.f. isomorphic to a standard one is also standard.

Let $\mathbf{p}$ be a sequence of positive integers $p_1, p_2, \cdots$, infinitely many of which are greater than one. An r.f. will be called $\mathbf{p}$-*adic* if for each positive integer n the restriction of the measure P to $\mathcal{F}_{n-1}$, when conditioned by $\mathcal{F}_n$, is *a.e.* atomic, with exactly p_n atoms of equal mass. If $\mathbf{p}$ is $2, 2, 2, \cdots$ then the r.f. is called *dyadic.* Note that if some $p_n = 1$ then $\mathcal{F}_{n-1} = \mathcal{F}_n$. While this possibility was not envisioned in previous work, it provides no extra difficulties; all arguments we need go through as well in this case as when *all* $p_n > 1$.

For any two r.f.'s $(X', \mathbf{F}', P')$ and $(X'', \mathbf{F}'', P'')$, their product $(X' \times X''$, $\mathbf{F}' \times \mathbf{F}''$, $P' \times P'')$ is an r.f.; here $\mathbf{F}' \times \mathbf{F}'' = (\mathcal{F}'_n \times \mathcal{F}''_n)_{n=0}^\infty$. If the r.f.'s are respectively $\mathbf{p}'$-adic and $\mathbf{p}''$-adic then their product is $\mathbf{p}$-adic, where $\mathbf{p} = p'_1 p''_1,\ p'_2 p''_2, \cdots$. The product $(X' \times X''$, $\mathbf{F}' \times \mathbf{F}''$, $P' \times P'')$ together with the map $\pi : X' \times X'' \to X'$, $\pi(x', x'') = x'$ is a special case of an extension.

A useful concept is that of having a standard extension. Since an extension of an extension is itself an extension, it follows that if an r.f. has an extension which has a standard extension, then the original r.f. itself has a standard extension (see [V4]). A standard r.f. obviously has at least *one* standard extension, namely itself. Conversely, it is a (deep) fact that any $\mathbf{p}$-adic r.f. which has a standard extension is itself standard. This is stated in [DFST], but only one direction is proved there; for the other, more difficult direction see [V4].

Do nonstandard r.f.'s exist? The answer is yes, but it is far from evident. Vershik was the first to construct examples of such r.f.'s. Initially he made nonstandard *dyadic* examples, in [V2]; later he also indicated briefly a method to construct, for *arbitrary* $\mathbf{p}$, nonstandard $\mathbf{p}$-adic r.f.'s. This is in his discussion of Examples 5 and 6 in Chap.3, Sec.5 of [V4]. We will make essential use of the existence of *one* nonstandard $\mathbf{p}$-adic r.f. for every $\mathbf{p}$.

2. Connection Between Decreasing Sequences of Sigma-Fields and Actions of Certain Groups

Let G etc. be as in the abstract. The cardinality of G_n is then $q_n = p_1 \cdots p_n$ (note that p_0 is undefined, but $q_0 = 1$). Any free ergodic action α of G on a probability space $(X, \mathcal{F}_0, P)$ produces (using G^*) a decreasing sequence of σ-fields as follows: $\mathcal{F}_n = \{A \in \mathcal{F}_0 : \alpha|G_n$ leaves A *a.e.* invariant $\}$. The tail-triviality of $\mathbf{F} = (\mathcal{F}_n)_{n=0}^\infty$ corresponds to ergodicity of α, and the fact that the r.f. is $\mathbf{p}$-adic comes from the definition of p_n and freeness of the action. Conversely, any $\mathbf{p}$-adic r.f. comes from some such action, and presumably from many nonisomorphic ones (see B in Section 7). This is discussed, for example, in [S].

PROPOSITION 2.1. *Necessary and sufficient that two actions of G, α on $(X, \mathcal{F}_0, P)$ and β on $(Y, \mathcal{G}_0, Q)$, produce isomorphic r.f.'s by using G^* is that they be G^*-equivalent in the sense of the abstract, i.e. that there is a measure-preserving bijection from X to Y carrying a.e. $\alpha|G_n$ orbit onto a $\beta|G_n$ orbit. The bijection implements the r.f. isomorphism.*

The proof is easy and completely routine, and is left to the reader.

EXAMPLE 1. Because of the fact that a finite abelian group is isomorphic with its own dual group, G may be identified with a dense subgroup of its dual group, a compact abelian group which is here called X. If we equip X with its Haar measure, then translation by G gives an action which we call the *standard* action; up to isomorphism this action does not depend on on the choices of the isomorphisms of the finite subgroups with their duals. The **p**-adic r.f. produced by using the sequence G^* of subgroups is indeed the standard one, as suggested by our terminology.

EXAMPLE 2. Let $(S, \mathcal{E}, m)$ be a probability space (for example, a finite one) and $X = S^G$ equipped with the infinite product measure P obtained from m on the completion $\mathcal{F}_0$ of the σ-algebra generated by the coordinates. The corresponding *Bernouilli* action of G is the action obtained from G acting by translation on the coordinates. The entropy of this action is precisely the entropy of the probability measure m, and from [OW] it is known that two such actions are isomorphic if and only if they have the same entropy.

3. Producing a Nonstandard p-adic R.F. Via a Group Action of Entropy Zero

Given G as $H_1 + H_2 + \cdots$, with H_n of cardinality p_n, where $p_n > 1$ for infinitely many n, we divide the positive integers into subsets N' and N'' , each containing infinitely many integers n for which $p_n > 1$. We set $H'_n = H_n$ if $n \in N'$ and $H'_n = \{0\}$ if $n \in N''$; similarly, we set $H''_n = H_n$ if $n \in N''$ and $H''_n = \{0\}$ if $n \in N'$. Then $H'_n \times H''_n$ is isomorphic to H_n for all positive integers n. If we set $G' = H'_1 + H'_2 + \cdots$ and $G'' = H''_1 + H''_2 + \cdots$ then G is isomorphic to $G' \times G''$ in a way that is consistent with these isomorphisms, and which therefore carries the the corresponding G_n to $G'_n \times G''_n$.

Let $(X', \mathbf{F}', P')$ be a nonstandard $\mathbf{p}'$-adic r.f., and let α' be an action of G' producing it via G'^*. Let α'' be a free ergodic action of G'', producing (via G''^*) a $\mathbf{p}''$-adic r.f. $(X'', \mathbf{F}'', P'')$. Let $(X, \mathbf{F}, P)$ be the product of the two r.f.'s; Then it is produced by the action α of $G' \times G''$ via the sequence of subgroups $G'_n \times G''_n$, where $\alpha(g', g'')(x', x'') = (\alpha'(g')x', \alpha''(g'')x'')$. (This is what is sometimes called the *group product* of the two actions. Note that the two groups involved, G' and G'', are not necessarily isomorphic.) Since $(X', \mathbf{F}', P')$ is nonstandard, its extension $(X, \mathbf{F}, P)$ is again nonstandard. We will show that it is produced (via G^*) by an action of G which has entropy zero. Since G is isomorphic to $G' \times G''$ in a way which takes the system of subgroups G^* to the corresponding system for $G' \times G''$, we need only to show that the action α has entropy zero. This is universally true of the group product of two actions, but in this special case it is especially easy to calculate the entropies.

PROPOSITION 3.1. *The action α of the group $G' \times G''$ described above has entropy zero.*

PROOF. Let$\mathcal{P}'$ and $\mathcal{P}''$ be finite measurable partitions of X' and X'' respectively, with cardinalities respectively k' and k''.. Then $\alpha|(G' \times G'')_n(\mathcal{P}' \times \mathcal{P}'')$ is just $\alpha'|G'_n(\mathcal{P}') \times \alpha''|G''_n(\mathcal{P}'')$, whose entropy is exactly

$$H(\alpha'|G'_n(\mathcal{P}') + H(\alpha''|G''_n(\mathcal{P}'') \leq (q'_n)logk' + (q''_n)logk'',$$

where $H(\mathcal{P})$ denotes the usual entropy of a partition $\mathcal{P}$ in a probability space. Dividing by the cardinality of $G'_n \times G''_n$ gives $(logk')/q'_n + (logk'')/q''_n$, which goes to zero as n increases to infinity. Since this holds for all such partitions, the entropy of α is zero.

4. Making the Entropy Positive

Another sort of product, of actions of the *same* group, will now be used; it is sometimes called their *Cartesian product.* Given actions α and β of G on $(X, \mathcal{F}_0, P)$ and $(Y, \mathcal{F}_0, Q)$ respectively, the Cartesian product action α' of G on the product space is defined by $\alpha'(g)(x, y) = (\alpha(g)x, \beta(g)y)$. This is the product action commonly used in ergodic theory, and it is well known that entropy adds: $h(\alpha') = h(\alpha) + h(\beta)$. Suppose α and β are free and ergodic, α' is obviously free; we would like it also to be ergodic. This may be achieved by choosing β to be weak mixing, as will be the case if β is a Bernoulli action. Thus, if β is a Bernoulli action of entropy r, where $0 < r \leq \infty$ and α is a free ergodic action of entropy zero then α' is a free ergodic action of entropy r. Thinking of G as coming from a fixed sequence G^* of finite subgroups, we may use α, β, and α' to produce $\mathbf{p}$-adic r.f.'s.

PROPOSITION 4.1. *The r.f. $(X \times Y, \mathbf{F}', P \times Q)$ produced by α' above is an extension of the r.f. $(X, \mathbf{F}, P)$ produced by α, via the map $\pi : (x, y) \to x$.*

PROOF. First we observe that $\pi^{-1}(\mathcal{F}_n) = \pi^{-1}(\mathcal{F}_0) \cap \mathcal{F}'_n$. This may be done by checking that the atoms of the two σ-algebras are the same. The atoms of $(\mathcal{F}_n)$ are the sets of the form $\alpha(G_n)x$, so those of $\pi^{-1}(\mathcal{F}_n)$ are the sets of the form $\pi^{-1}(\alpha(G_n)x = (\alpha(G_n)x) \times Y = \alpha'(G_n)(x \times Y)$. These are invariant under $\alpha'|G_n$, so they are in $\mathcal{F}'_n$; and they are clearly minimal with respect to this property together with the property of being of the form $A \times Y$, the latter being the condition for being in $\pi^{-1}(\mathcal{F}_0)$.

Now: if f is in $\mathcal{L}^\infty(P)$ then

$$\begin{aligned}\mathbf{E}[f|\mathcal{F}_n] \circ \pi &= \mathbf{E}[f \circ \pi|\pi^{-1}\mathcal{F}_n] = \mathbf{E}[f \circ \pi|\pi^{-1}(\mathcal{F}_0) \cap \mathcal{F}'_n] \\ &= \mathbf{E}[\mathbf{E}[f \circ \pi^{-1}(\mathcal{F}_0)]|\mathcal{F}'_n] = \mathbf{E}[\mathbf{E}[f|\mathcal{F}_0] \circ \pi|\mathcal{F}'_n] = \mathbf{E}[f|F'_n],\end{aligned}$$

so π is indeed an extension map.

From the previous section, α could have been so chosen that the corresponding r.f. is nonstandard. Then the r.f. produced by α' will also be nonstandard.

5. Cutting Down to a Subsequence

Let $\mathbf{s}$ be an increasing sequence of integers $0 = s(0) < s(1) < \cdots$, and let $(X, \mathbf{F}, P)$ be an r.f., $\mathbf{F}|\mathbf{s}$ will denote the decreasing sequence $\mathcal{F}_{s(0)} = \mathcal{F}_0$, $\mathcal{F}_{s(1)}, \cdots$ of σ-algebras, so $(X, \mathbf{F}|\mathbf{s}, P)$ is an r.f. If the original r.f. is $\mathbf{p}$-adic then the new one is $\mathbf{p}|\mathbf{s}$-adic, where $\mathbf{p}|$ $\mathbf{s}$ is the sequence $(p_1 p_2 \cdots p_{s(1)})$, $(p_{s(1)+1} p_{s(1)+2} \cdots p_{s(2)}), \cdots$. Writing $\mathbf{q}$ for the sequence $q_0 = 1, q_1 = p_1, q_2 = p_1 p_2, \cdots$, the corresponding products of finite initial subsequences of $\mathbf{p}|\mathbf{s}$ turn out to be simply $q_{s(0)} = 1, q_{s(1)}, \cdots$. If

$(X, \mathbf{F}, P)$ is produced by an action of G via the system $G^* = (G_0, G_1, \cdots)$ of subgroups, then $(X, \mathbf{F}|\mathbf{s}, P)$ is produced via the system $G^*|\mathbf{s}= (G_0 = G_{s(0)}, G_{s(1)}, \cdots)$.

The following was proved by Vershik in [V1]:

PROPOSITION 5.1. *For any r.f.* $(X, \mathbf{F}, P)$ *there is some* $\mathbf{s}$ *so that* $(X, \mathbf{F}|\mathbf{s}, P)$ *is standard.*

The set of sequences $\mathbf{s}$ for which $(X, \mathbf{F}|\mathbf{s}, P)$ is standard is obviously an isomorphism invariant which is called in [V1] its *fundamental invariant.*

THEOREM 5.2. *For* G *made from the system* G^* *of subgroups, any* $r \in [0, \infty]$, *and any* $\mathbf{s}$, *there is some free ergodic action* α *of* G *of entropy* r *so that if* $(X, \mathbf{F}, P)$ *is the r.f. produced by* α *via* $G*$ *then* $(X, \mathbf{F}|\mathbf{s}, P)$ *is nonstandard.*

PROOF. Apply the results of Section 4 to G with the system $G^*|\mathbf{s}$ of subgroups to obtain an action α of G of entropy r which produces, via $G^*|\mathbf{s}$, a nonstandard $\mathbf{p}|\mathbf{s}$-adic r.f. $(X, \mathbf{F}', P)$. This same α can be used to produce a $\mathbf{p}$-*adic* r.f. $(X, \mathbf{F}, P)$ via G^*; from the definition of the σ-algebras $\mathcal{F}_n$ and $\mathcal{F}'_n$ we see that $\mathcal{F}'_n$ is precisely $\mathcal{F}_{s(n)}$, i.e. $\mathbf{F}'$ is precisely $\mathbf{F}|\mathbf{s}$, so we are done.

6. Uncountably Many G^*-Inequivalent Actions of Equal Entropy

THEOREM 6.1. *For any* r *in* $[0, \infty]$ *and any* G *with system of subgroups* G^*, *there exist uncountably many free ergodic actions of* G *with entropy* r, *which are pairwise* G^**-inequivalent.*

PROOF. Let $\alpha^1, \alpha^2, \cdots$ be any sequence of such actions; we will produce another such action α which is not G^*-equivalent to any of them.

Let $(X^j, \mathbf{F}^j, P^j)$ be the r.f. produced via G^* by α^j. There exist sequences $\mathbf{s}^1$, $\mathbf{s}^2, \cdots$ as in Section 5 so that $(X^j, \mathbf{F}^j|\mathbf{s}^j, P^j)$ is standard. By proceeding inductively, applying Proposition 5.1 successively to each new $(X^j, \mathbf{F}^j|\mathbf{s}^j, P^j)$ we can arrange that each $\mathbf{s}^{j+1}$ is a subsequence of $\mathbf{s}^j$. Let $\mathbf{s}$ be the sequence defined by $s(j) = s^j(j)$. Then for each j, $\mathbf{s}$ is a subsequence of $\mathbf{s}^j$, except for its first j terms. Now we make two observations.

1. If $(Y, \mathbf{G}', Q)$ is standard and $\mathbf{p}'$-adic, and $\mathbf{G}$ is obtained by omitting some of the σ-algebras of $\mathbf{G}'$, then $(Y, \mathbf{G}, Q)$ is also standard.
2. If $(Y, \mathbf{G}', Q)$ is $\mathbf{p}'$-adic, and $\mathbf{G}$ is obtained by omitting *finitely many* of the σ-algebras of $\mathbf{G}'$, and $(Y, \mathbf{G}, Q)$ is standard, then so is $(Y, \mathbf{G}', Q)$.

These are both shown in [V1] and are not difficult to see directly from the definition of standardness. The second involves realizing that if $\mathcal{G}$ and $\mathcal{G}'$ are σ-fields in a probability space with $\mathcal{G} \subset \mathcal{G}'$, and the conditional probabilities on $\mathcal{G}'$ given $\mathcal{G}$ all consist of a fixed number p of atoms of equal mass, then there is a partition $\mathcal{P}$ into p sets of equal measure which is an independent complement of $\mathcal{G}$ in $\mathcal{G}'$.

Thus each $(X^j, \mathbf{F}^j|\mathbf{s}, P^j)$ is standard. But by Theorem 5.2 we can construct a free ergodic action α of G of entropy r so that if $(X, \mathbf{F}, P)$ is the r.f. produced by it via G^*, then $(X, \mathbf{F}|\mathbf{s}, P)$ is *not* standard; therefore $(X, \mathbf{F}, P)$ is not isomorphic to any of the $(X^j, \mathbf{F}^j, P^j)$, and α is not G^*-equivalent to any of the α^j.

7. Remarks and Questions

(A) It should be noted that the entire construction depends on starting with a single given strictly nonstandard $\mathbf{p}$-adic r.f.

(B) It ought to be the case that each G^*-equivalence class contains uncountably many nonisomorphic actions of G; but we have not attempted to show this, and the question is open.

(C) G^*-equivalence may be studied in the context of "restricted orbit equivalence" (r.o.e.), as developed for actions of discrete amenable groups in [KR]. For the case where H_n is the integers mod p_n this has been done by D.Heicklin in [H]; what we have called G^*-equivalence is there called *Vershik equivalence* or *Vershik* (p_n)*-equivalence* (for this particular G^*). If $\sum_{n=0}^{\infty} p_n/q_{n-1}$ is finite then Vershik [V3] and Stepin [S] showed that equivalence of two actions of G in this sense implies equality of entropy. This is called in r.o.e. theory the *entropy-determined* case. The existence of Bernouilli actions of all positive entropies thus already shows that in this case there are uncountably many nonisomorphic r.f.'s. **p**-adic r.f .'s. However if the above sum is infinite then Heicklin shows in [H], using r.o.e. theory, that we are in the *entropy-free* case: that in fact every equivalence class contains actions of zero entropy. However our results in the present paper have shown in particular that even for such **p** there are uncountably many nonisomorphic r.f.'s.

(D) For the G^* discussed in the previous paragraph the following question remains open: does every equivalence class in the entropy-free case contain actions of *arbitrary* entropy, or even a single one of positive entropy? This question seems difficult. We mention that in the dyadic case, i.e. each $H_n = Z/2Z$, it has been shown by J.Hasfura [Ha] that every free ergodic action is equivalent to one which is strongly mixing; and in joint work with A.Fieldsteel [FH] that every free ergodic action of positive entropy is equivalent to one of *completely* positive entropy. It is to be expected that these will continue to hold for arbitrary **p**.

(E) The use of more general H_n rather than just Z/p_nZ was motivated not by any desire for generality, but by the fact that our proof, even in the case $G = Z/p_1Z + Z/p_2Z + \cdots$, required shifting to other groups at certain points, specifically in Sections 3 and 5. In Section 3 this could have been avoided by use of r.o.e. theory; but Section 5 is another matter.

References

[DFST] L.Dubins, J.Feldman, M.Smorodinsky, B.Tsirelson, *Decreasing sequences of sigma-fields and a measure change for Brownian motion to appear*, Annals of Probability (1996).

[FH] A.Fieldsteel and J.R.Hasfura-Buenaga, *Dyadic equivalence to completely positive entropy, preprint* (1996).

[H] D.Heicklen, *Entropy and Vershik equivalence, preprint* (1996).

[Ha] J.R.Hasfura-Buenaga, *Mixing for dyadic equivalence*, Act. Math. Univ. Comenienae **LXIV** (1970), 141-152.

[KR] J.W.Kammeyer and D.J.Rudolph, *Restricted orbit equivalence for actions of discrete amenable groups, preprint* (1996).

[OW] D.S.Ornstein and B.Weiss, *Entropy and isomorphism theorems for actions of amenable groups*, J.Analyse Math. **48** (1987), 1-41.

[S] A.M.Stepin, *On an entropy invariant of decreasing sequences of measurable partitions*, Funct. Anal. and Appl. **5** (1971), 80-84.

[V1] A.M.Vershik, *On lacunary isomorphism of monotone sequences of partitions*, Funct. Anal. and Appl. **2** (1968, #3).

[V2] A.M.Vershik, *Decreasing sequences of measurable partitions and their applications*, Soviet Math. Dokl. **7** (1970).

[V3] A.M.Vershik, *A continuous family of pairwise nonisomorphic dyadic sequences*, Funct. Anal. and Appl. **5** (1971 #3).

[V4] A.M.Vershik, *Theory of decreasing sequences of measurable partitions*, St.Petersburg Math. Jour. **4** (1994).

DEPARTMENT OF MATHEMATICS, UNIVÉRSITY OF CALIFORNIA, BERKELEY 94729 CA, U.S.A.
E-mail address: `feldman@math.berkeley.edu`

Part III.
Applications to Differential Equations

Contemporary Mathematics
Volume **215**, 1998

Positive exponents for a dense set of continuous $SL(2, \mathbf{R})$ valued cocycles which arise as solutions to strongly accessible linear differential systems

M. Nerurkar

1. Introduction

There has been a great deal of interest and activity in the study of Lyapunov exponents in the past two decades. Lyapunov exponents play a central role in smooth ergodic theory as well as in the spectral theory of random, one-dimensional Schrödinger operators. Positivity of Lyapunov exponents yield important information about the underlying system. In smooth ergodic theory, absence of zero exponents is a sign of existence of a Bernoulli component, whereas the same condition on exponents is related to "localizability" in the spectral theory of random Schrödinger operators. Numerical investigations of various specific examples (in particular specific $SL(2, \mathbf{R})$ valued cocycles) suggests that positive Lyapunov exponents are quite frequent. In this paper we shall precisely formulate and prove this observation. The fundamental technique developed in this note is modelled over a method introduced by R. Ellis [**E1**] to produce minimal, compact group extensions. This paper was written in honour of Prof. Robert Ellis, on the occasion of his retirement and I wish to dedicate it to him with love, respect and gratitude.

First, we shall discuss what is known about the density (in appropriate classes) of cocycles with non-zero exponent. Basically there are two known results. The first is due to O. Knill [**Kn**] and the second due to R. Johnson [**J2**]. Both results deal with $SL(2, \mathbf{R})$ valued cocycles. O. Knill proves his result in the class of all L^∞ cocycles based on aperiodic discrete dynamical system whereas R. Johnson's result is in the class of all continuous cocycles based on flows. O. Knill's method relies heavily on estimates for the exponents obtained by M. Herman. On the other hand R. Johnson employs work of S. Kotani followed by a perturbation argument.

In this paper we set out to prove a density result about continuous cocycles with non-zero exponents, in the classes of cocycles which are "extremly thin" within the class of all cocycles. Our cocycles will be generated as the fundamental matrix solutions to certain classes of non-autonomous linear differential equations. These classes of linear differential equations are required to satisfy a certain Lie algebraic condition - the condition of strong accessibility. In our proof, we follow R. Johnson's approach of employing S. Kotani's result to reduce the problem to producing a dense set of "unbounded cocycles". In solving this later problem we differ from R.

1991 *Mathematics Subject Classifications*: Primary 34C35; Secondary 34D05.

Johnson. We shall employ a "non-compact" modification of a technique developed in [**NS**].

This technique will allow us to prove the result in "extremly thin" classes of cocycles. The perturbation technique that R. Johnson uses is not suitable when one is dealing with the class of cocycles arising from differential equations of specific form. The specific form of the equations impose additional (and usually very severe) constraints and one has to construct a perturbation which will "respond" to these constraints. In some special situations, may be this can be done, for example, R. Johnson has constructed perturbation that will respond to the constraints coming from the one-dimensional Schrödinger equation. However these specific techniques become heavily dependent on the specific nature of the constraints. On the other hand, the condition of strong accessibility that we impose on the underlying class of cocycles is easily verifiable in concrete examples and modulo this condition, our method is independent of the specific form of the underlying differential equations. In particular, we shall see that the classes of cocycles arising from one-dimensional Schrödinger equation or from various other specific systems will satisfy these conditions and subsequently contain a dense set of cocycles with non-zero Lyapunov exponents. Our technique is based on blending methods of Dynamical Systems and Control Theory.

We shall keep our discussion of Lyapunov exponents in the context of linear differential equations, i.e. in the context of cocycles based on flows rather than those based on a discrete dynamical system. The first reason for doing so is, of course, our technique is best suited for flows. Secondly, our motivation comes from spectral problems in Mathematical Physics, in particular the spectra of one dimensional Schrödinger operator with recurrent coefficients.

Lyapunov exponents were introduced to study the logarithmic growth rate of a solution of a given differential equation. For linear equations with constant coefficients,

$$x' = Ax, \quad x \in \mathbf{R}^n,$$

where $A \in M(n, \mathbf{R})$ - the set of $n \times n$ real matrices, these growth rates are exactly the real parts of eigenvalues of A. When A is time dependent, i.e. $A \equiv A(t)$, the eigenvalues of individual $A(t)$'s do not have much significance. This is well demonstrated by a well known example of Markus and Yamabe, where the eigenvalues of each $A(t)$ have negetive real part but there is still an unbounded solution. To capture the appropriate notion of growth rate, rather than studying the given single equation it becomes necessary to study collectively the entire family of equations

$$x' = A_\tau(t)x, \quad x \in \mathbf{R}^n,$$

where $A_\tau(t) \equiv A(t+\tau)$, $(t \in \mathbf{R})$ is the translate of the function A. The time dependence of $A(t)$ will be "recurrent", in the sense that we will think of the matrix function $t \to A(t)$ as arising from some given continuous matrix-valued function A, defined on a space Ω where some recurrent flow $T \equiv \{T_t\}_{t\in\mathbf{R}}$ is given, by evaluating A along trajectories of T. In other words given a single continuous map $\mathbf{R} \to M(n, \mathbf{R}) : t \to A(t)$, one can take Ω to be its "hull" (in a suitable topology) and extend the map to all of Ω. Thus the following basic set up is fairly general.

The formal description of our basic set up involves a *flow* $(\Omega, \{T_t\}_{t\in\mathbf{R}})$, i.e. a compact metric space Ω and a one parameter group $\{T_t\}_{t\in\mathbf{R}}$ of homeomorphisms of

Ω such that the action $(\omega, t) :\longrightarrow T_t(\omega) \in \Omega$ is jointly continuous. Let $L \subseteq M(n, \mathbf{R})$ be a Lie algebra of a connected Lie group $G \subset GL(n, \mathbf{R})$ and let $A : \Omega :\longrightarrow L$ be a given continuous function. Consider the family of linear differential equations parametrized by points of Ω:

$$(1.1.\omega) \qquad x' = A(T_t\omega)x, \qquad x \in \mathbf{R}^n,\ \omega \in \Omega.$$

For each $\omega \in \Omega$, let $t \to X_A(\omega, t)$ denote the fundamental matrix solution to $(1.1.\omega)$, i.e. the solution $x(\cdot)$ of $(1.1.\omega)$ such that $x(0) = I$-the $n \times n$ identity matrix. Then the map $X_A : \Omega \times \mathbf{R} \to G$ is continuous and satisfies the following cocycle identity:

$$(1.2) \qquad X_A(\omega, t+s) = X_A(T_t(\omega), s)X_A(\omega, t) \qquad \text{for all } \omega \in \Omega,\ t, s \in \mathbf{R}.$$

The classes of cocycles we shall be considering are described via a constraint on the values of the map $A : \Omega \to L$. The constraint is described by a closed convex subset S of L and our class of cocycles consists of those which arise as a fundamental matrix solution to $(1.1.\omega)$ from S valued continuous maps A. Typically S will be "thin" in L, in the sense that the number $\dim(S)$ of the free parameters that we are allowed to vary may be considerabily smaller than the dimension of the Lie group G. This is demonstrated by the following important example.

EXAMPLE. Let $(\Omega, \{T_t\}_{t \in \mathbf{R}})$ be a flow and $Q : \Omega \to F \subseteq \mathbf{R}$ be a continuous function, where F is a given closed, convex subset of $\mathbf{R}$. For each $\omega \in \Omega$ consider the family L_ω, $(\omega \in \Omega)$ of one-dimensional Schrödinger operators acting on $y \in L^2(\mathbf{R})$ by

$$L_\omega(y) = \Big(-\frac{d^2}{dt^2} + Q_\omega \Big) y,$$

where $Q_\omega(t) = Q(T_t(\omega))$, $t \in \mathbf{R}$. For a fixed $\nu \in \mathbf{R}$, the spectral problem $L_\omega y = \nu y$ can be written as a 2×2 family of linear system

$$\begin{pmatrix} y \\ y' \end{pmatrix}' = \begin{pmatrix} 0 & 1 \\ Q(\omega) - \nu & 0 \end{pmatrix} \begin{pmatrix} y \\ y' \end{pmatrix}, \qquad \omega \in \Omega.$$

Notice that the map A generating this family of linear system takes values in the Lie algebra $sl(2, \mathbf{R})$ and is given by

$$A(\omega) = \begin{pmatrix} 0 & 1 \\ Q(\omega) - \nu & 0 \end{pmatrix}.$$

Thus we describe this type of system in terms of $S \equiv S_{F,\nu}$ valued maps, where the closed, convex constaining set $S_{F,\nu}$ is given by

$$S_{F,\nu} = \{ B \in sl(2, \mathbf{R}) \mid B = \begin{pmatrix} 0 & 1 \\ r & 0 \end{pmatrix} \ r \in F - \nu \}.$$

Notice that the cocycle generated by this system takes values in $SL(2, \mathbf{R})$ - a three dimensional Lie group, however $\dim(S) = 1$.

Now we describe the Lie algebraic condition imposed on the constraining set S. Let $L(S)$ be the Lie subalgebra of L generated by S, and let $L_0(S)$ be the ideal of $L(S)$ generated by the difference set $S - S = \{x - y \mid x \in S,\ y \in S\}$. The following notion of strong accessibility is of fundamental importance in Control Theory.

DEFINITION 1.1. *A subset S of the Lie algebra L has the strong accessibility property if $L_0(S) = L$.*

EXAMPLE 1. (The Schrödinger-system): Consider the set $S_{F,\nu} \subset sl(2,\mathbf{R})$ described in the example before. If the closed, convex subset F is non-singleton, then $S \equiv S_{F,\nu}$ has the strong accessibility property. To show this, first notice that $\begin{pmatrix} 0 & 0 \\ r & 0 \end{pmatrix} \in S - S$, for all $r \in F - F$ and hence $\begin{pmatrix} 0 & 0 \\ r & 0 \end{pmatrix} \in L_0(S)$ for all $r \in \mathbf{R}$. Furthermore,

$$\left[\begin{pmatrix} 0 & 0 \\ r & 0 \end{pmatrix}, \begin{pmatrix} 0 & 1 \\ s-\nu & 0 \end{pmatrix}\right] = \begin{pmatrix} -r & 0 \\ 0 & r \end{pmatrix} \in L_0(S), \text{ for all } r \in \mathbf{R}, s \in F, \text{ and}$$

$$\left[\begin{pmatrix} -r & 0 \\ 0 & r \end{pmatrix}, \begin{pmatrix} 0 & 1 \\ s-\nu & 0 \end{pmatrix}\right] = \begin{pmatrix} 0 & -2r \\ 2rs - 2r\nu & 0 \end{pmatrix} \in L_0(S), \text{ for all } r \in \mathbf{R}, s \in F.$$

Since $L_0(S)$ is an ideal (and hence a subspace), the above computation implies that $\begin{pmatrix} 0 & -2r \\ 0 & 0 \end{pmatrix} \in L_0(S)$ for all $r \in \mathbf{R}$. In particular,

$$\begin{pmatrix} r & 0 \\ 0 & -r \end{pmatrix}, \begin{pmatrix} 0 & 0 \\ r & 0 \end{pmatrix}, \begin{pmatrix} 0 & r \\ 0 & 0 \end{pmatrix} \in L_0(S), \text{ for all } r \in \mathbf{R}.$$

Hence $L_0(S) = sl(2,\mathbf{R})$.

EXAMPLE 2. (The Vinograd system, [**V**]) : Given a flow $(\Omega, \{T_t\}_{t\in\mathbf{R}})$ and a continuous map $\varphi : \Omega \to F$ into a closed convex subset $F \subseteq \mathbf{R}$, consider the linear system

$$x' = \begin{pmatrix} 0 & 1+\varphi(T_t(\omega)) \\ 1-\varphi(T_t(\omega)) & 0 \end{pmatrix} x, \quad x \in \mathbf{R}^2.$$

Here the set S is given by $S \equiv S_F = \{\begin{pmatrix} 0 & 1+r \\ 1-r & 0 \end{pmatrix} \mid r \in F\}$. Again, if F is non-singleton, then S has the strong accessibility property. To see this, notice that $\begin{pmatrix} 0 & p \\ -p & 0 \end{pmatrix} \in S - S$, for all $p \in F - F$, furthermore

$$\left[\begin{pmatrix} 0 & p \\ -p & 0 \end{pmatrix}, \begin{pmatrix} 0 & 1+r \\ 1-r & 0 \end{pmatrix}\right] = \begin{pmatrix} 2p & 0 \\ 0 & -2p \end{pmatrix},$$

$$\left[\begin{pmatrix} 0 & p \\ -p & 0 \end{pmatrix}, \begin{pmatrix} q & 0 \\ 0 & -q \end{pmatrix}\right] = \begin{pmatrix} 0 & -2qp \\ -2qp & 0 \end{pmatrix}.$$

This shows that $L_0(S)$ contains matrices of the form $\begin{pmatrix} 0 & p \\ -p & 0 \end{pmatrix}, \begin{pmatrix} m & 0 \\ 0 & -m \end{pmatrix}$, $\begin{pmatrix} 0 & n \\ n & 0 \end{pmatrix}$ for non zero p, m, n. Thus $L_0(S) = sl(2,\mathbf{R})$ and S satisfies the strong accessibility condition.

Notice that if $F \cap [-1,1] = \emptyset$, then both eigenvalues of every $A \in S \equiv S_F$ are purely imaginary. However, we shall see that by Theorem 1.8, a generic Vinograd system corresponding to F has positive Lyapunov exponents. Once again this illustrates that the eigenvalues of the individual $A(\omega)$, $(A \in C(\Omega, S))$ do not have much significance.

Next, we define the Lyapunov exponents of a $GL(n, \mathbf{R})$ valued cocycle and introduce some notation and results we shall need.

DEFINITION 1.2. *Let $X_A : \Omega \times \mathbf{R} \to GL(n, \mathbf{R})$ be a cocycle based on an ergodic flow $(\Omega, \{T_t\}_{t\in\mathbf{R}}, \mu)$, where μ is a flow invariant, ergodic Borel probability measure on Ω. Let $(v, \omega) \in \mathbf{R}^n \times \Omega$. Define $\lambda(v, \omega)$ by setting*

$$\lambda(v, \omega) = \lim_{t\to\infty} \frac{1}{t} ln||X_A(\omega, t)v||.$$

The Oseledets's multiplicative ergodic theorem guarantees the existence of this limit a.e. ω with respect to μ for each $v \in \mathbf{R}^n$. Furthermore, there is a finite set $\Sigma_\mu(A)$ (depending on X_A and μ) such that for μ almost all ω we have

$$\Sigma_\mu(A) = \{\lambda(v, \omega) \mid v \in \mathbf{R}^n\}.$$

This set is called the Lyapunov spectrum of the cocycle X_A with respect to μ. We also introduce the largest (smallest) exponent with respect to μ by setting

$$\lambda_\mu^+ \equiv \lambda_\mu^+(A) = Max\ \{\lambda \mid \lambda \in \Sigma_\mu(A)\}\ and\ \lambda_\mu^- \equiv \lambda_\mu^-(A) = Min\ \{\lambda \mid \lambda \in \Sigma_\mu(A)\}.$$

Yet another way to capture the logarithmic growth rate of the solutions is through the notion of uniform hyperbolicity (or exponential dichotomy). This description of exponents is purely in terms of topological dynamics and does not involve invariant measures on the base flow.

DEFINITION 1.3. *A cocycle $X_A : \Omega \times \mathbf{R} \to GL(n, \mathbf{R})$ based on a flow $(\Omega, \{T_t\}_{t\in\mathbf{R}})$ is said to be uniformly hyperbolic if there exists continuous vector subbundles V^s, V^u of $\mathbf{R}^n \times \Omega$ with the following properties*

1. *$V^s \oplus V^u = \mathbf{R}^n \times \Omega$, (Whitney sum),*
2. *V^s and V^u are invariant under the skew product flow $(\mathbf{R}^n \times \Omega, \{T_t^A\}_{t\in\mathbf{R}})$ defined by*

$$T_t^A(v, \omega) = (X_A(\omega, t)v, T_t(\omega)), \qquad (v, \omega) \in \mathbf{R}^n \times \Omega. \tag{1.3}$$

3. *There exist constants $K > 0$ and $\alpha > 0$ such that*

$$||X_A(\omega, t)v|| \le Ke^{-\alpha t}||v|| \quad if\ (v, \omega) \in V^s\ and\ t \ge 0,$$
$$||X_A(\omega, t)v|| \le Ke^{\alpha t}||v|| \quad if\ (v, \omega) \in V^u\ and\ t \le 0.$$

The set $\Sigma(A)$ defined by

$$\Sigma(A) = \{\lambda \in \mathbf{R} \mid the\ cocycle\ e^{-\lambda t}X_A(\omega, t)\ is\ uniformly\ hyperbolic\},$$

is called the dichotomy spectrum of the cocycle X_A.

When the base flow is chain recurrent, the following spectral theorem (Theorem 1.6) describes the general nature of $\Sigma(A)$ and its relationship with $\Sigma_\mu(A)$, (see [**JPS**]). First we recall the following definition.

DEFINITION 1.4. *The flow* $(\Omega, \{T_t\}_{t\in\mathbf{R}})$ *is chain recurrent if given numbers* $\varepsilon > 0, T > 0$ *and points* $\omega, \overline{\omega} \in \Omega$, *there exists points* $\omega = \omega_0, \omega_1, \cdots, \omega_n = \overline{\omega}$ *and numbers* $t_1 > T, \cdots, t_n > T$ *such that*

$$d(T_{t_i}(\omega_i), \omega_{i+1}) < \varepsilon, \quad (1 \leq i \leq n).$$

Here d *is some metric on* Ω *generating the topology on* Ω.

Before stating the spectral theorem, we note the following sufficiency condition for chain recurrence.

LEMMA 1.5. *Let* $(\Omega, \{T_t\}_{t\in\mathbf{R}}, \mu)$ *be an ergodic flow such that* $Supp(\mu) = \Omega$. *Then the flow is chain recurrent.*

PROOF. It suffices to prove that Ω contains a point whose positive semi-orbit, i.e. the set $\{T_t(\omega_0) \mid t > 0\}$ is dense in Ω. Since the measure μ is supported on all of Ω, the existence of points with dense positive semi-orbit follows from the Birkhoff ergodic theorem. □

PROPOSITION 1.6. *Let* X_A *be a* $GL(n, \mathbf{R})$ *valued continuous cocycle based on a chain recurrent flow* $(\Omega, \{T_t\}_{t\in\mathbf{R}})$. *Then,*

1. *the dichotomy spectrum* $\Sigma(A)$ *is a disjoint union of* k $(k \leq n)$ *intervals,* $\Sigma(A) = \cup_{1\leq i\leq k}[a_i, b_i]$. *Furthermore, if* $k = n$, *all of these intervals cannot be non-degenerate (i.e.* a_i *and* b_i *cannot be distinct for all i's);*
2. $\Sigma_\mu(A) \subseteq \Sigma(A)$ *for all flow invariant, ergodic Borel probability measures* μ *on* Ω *and each endpoint of every spectral interval (i.e. each* a_i *and* b_i*) is contained in* $\Sigma_\mu(A)$ *for some invariant, ergodic* μ, *(*μ *depends on the endpoint).*

If the map A takes values in the Lie algebra of infinitesimally symplectic matrices, then it is fairly easy to check that the dichotomy spectrum of the cocycle X_A is symmetric about the origin. This observation and the above proposition implies that if A takes values in $sl(2, \mathbf{R})$ (or equivalently X_A takes values in $SL(2, \mathbf{R})$) then there are exactly three possibilities, namely, either (1) $\Sigma(A)$ consists of two (and hence necessarily degenerate) intervals, i.e. $\Sigma(A)$ is a two point set $\{-\beta, \beta\}$, $\beta > 0$ or $\Sigma(A)$ consist of a single interval. In the later case either (2) the spectral interval is non-degenarate, i.e. $\Sigma(A) = [-\beta, \beta]$, $\beta > 0$ or (3) the interval is degenarate and $\Sigma(A) = \{0\}$ is a singleton set. Furthermore, it follows that

$$\beta \equiv \beta(A) = \sup_{\mu\in M(\Omega)} \lambda_\mu^+(A)$$

where $M(\Omega)$ is the set of all flow invariant, ergodic Borel probability measures on Ω. Now we are ready to state our results.

THEOREM 1.7. *Let* $(\Omega, \{T_t\}_{t\in\mathbf{R}})$ *be a flow. Let* G *be a closed subgroup of* $GL(n, \mathbf{R})$ *with Lie algebra* L *and let* S *be a subset of* L. *Let* $C(\Omega, S)$ *denote the metric space of all continuous maps from* Ω *to* S *with the supremum metric. Assume that*

1. G *is connected, non-compact and unimodular,*
2. $\omega_0 \in \Omega$ *is a non-periodic point,*
3. S *is closed and convex,*
4. S *has the strong accessibility property.*

For each $A \in C(\Omega, S)$, consider the cocycle X_A generated by the differential system (1.1.ω). Let

$$C_{unb}(\Omega, S) = \{A \in C(\Omega, S) \mid \text{the set } \{X_A(\omega_0, t) \mid t \in \mathbf{R}\} \text{ is unbounded}\}.$$

Then $C_{unb}(\Omega, S)$ is a residual subset of $C(\Omega, S)$.

THEOREM 1.8. *Let $(\Omega, \{T_t\}_{t\in\mathbf{R}}, \mu)$ be a flow. Let $G = SL(2, \mathbf{R})$ and $S \subseteq L = sl(2, \mathbf{R})$ be a subset. Assume that*

1. *μ is an ergodic probability measure such that $Supp(\mu) = \Omega$ and μ is not supported on a single orbit,*
2. *S is closed and convex,*
3. *S has the strong accessibility property.*

Let

$$C^{\mu}_{pos}(\Omega, S) = \{A \in C(\Omega, S) \mid \lambda^+_\mu(A) > 0\}.$$

Then $C^{\mu}_{pos}(\Omega, S)$ is a dense subset of $C(\Omega, S)$.

In particular, if the flow $(\Omega, \{T_t\}_{t\in\mathbf{R}})$ is minimal, then the set

$$C_{pos}(\Omega, S) = \{A \in C(\Omega, S) \mid \Sigma(A) \neq \{0\}\}$$

is a dense subset of $C(\Omega, S)$.

2. Proofs

First we shall derive Theorem 1.8 from Theorem 1.7. This argument is due to R. Johnson and is based on a generalization of a theorem of S. Kotani. The fundamental technique in studying the spectral theory of systems such as (1.1.ω) is to introduce a complex "spectral parameter" $\eta \in \mathbf{C}$ into the family and then to examine the dependence of the Lyapunov exponents on the parameter η. A proper way to introduce this spectral parameter into system (1.1.ω) with $A(\omega) \in sl(2, \mathbf{R})$ is described by the following system:

$$x' = \big[A(T_t\omega) - \eta J\big]x, \qquad x \in \mathbf{R}^2,\ \omega \in \Omega, \tag{2.1}$$

where $J = \begin{pmatrix} 0 & 1 \\ -1 & 0 \end{pmatrix}$. Let $X_{A,\eta}$ denote the cocycle (i.e. the fundamental matrix solution) generated by (2.1). Given an invariant, ergodic probability measure μ on Ω, the largest Lyapunov exponent $\lambda^+_\mu(\eta) \equiv \lambda^+_\mu(A - \eta J)$ of the system (2.1) is a function of η. The following is a generalization (see [**J1**]) of a result of S. Kotani, [**Ko**].

THEOREM 2.1. *Consider the system (2.1) with $A \in C(\Omega, sl(2, \mathbf{R}))$. Let μ be an invariant, ergodic probability measure on Ω such that $Supp(\mu) = \Omega$. Suppose $I \subset \mathbf{R}$ is a non-empty open interval containing 0 such that $\lambda^+_\mu(\eta) = 0$ for almost all $\eta \in I$ with respect to the Lebesgue measure. Then there exists a compact set $K \subset SL(2, \mathbf{R})$ such that*

$$X_A(\omega, t) \in K, \quad \text{for all } t \in \mathbf{R}, \text{ and for all } \omega \in \Omega.$$

Proof of Theorem 1.8.

Fix a sequence $(K_n)_{n\in\mathbf{N}}$ of compact subsets of G such that $K_n \subseteq int(K_{n+1})$, for all $n \in \mathbf{N}$ and $G = \cup_{n\in\mathbf{N}} K_n$. For contradiction, suppose that there is an open set $V \subseteq C(\Omega, S)$ such that for each $B \in V$ all Lyapunov exponents (with respect to measure μ) of the cocycle X_B are zero. Fix $B_0 \in V$ and consider the family of equations

$$x' = [B_0 T_t(\omega) - \eta J]x, \qquad x \in \mathbf{R}^2. \tag{2.2}$$

Since B_0 is in the interior of V, there is an interval $I \subseteq \mathbf{R}$ containing zero such that

$$\lambda_\mu^+(\eta) \equiv \lambda_\mu^+(B_0 - \eta J) = 0, \quad \text{for all } \eta \in I.$$

Hence by Theorem 2.1 there exists a compact set K_m (for some $m \in \mathbf{N}$) such that

$$X_{B_0,\eta}(\omega, t) \in K_m, \quad \text{for all } (\omega, t) \in \Omega \times \mathbf{R}.$$

For $n \in \mathbf{N}$, set

$$W_n = \{B \in V \subseteq C(\Omega, S) \mid X_B(\omega, t) \in K_n, \text{ for all } (\omega, t) \in \Omega \times \mathbf{R}\}.$$

Since B_0 in the above argument was an arbitrary point in V, it follows that

$$V \subseteq \cup\{W_n \mid n \in \mathbf{N}\}.$$

Since each W_n is closed, for some $n_0 \in \mathbf{N}$, the set W_{n_0} contains a non-empty open subset of V. This contradicts the density of the set $C_{unb}(\Omega, S)$ of unbounded cocycles guaranteed by Theorem 1.7, (notice that our assumption guarantees the existence of a non-periodic point).

For the second assertion note that minimality of the flow on Ω implies that every invariant, ergodic measure on Ω is supported on all of Ω and for any such measure μ one has the following inclusion: $C^\mu_{pos}(\Omega, S) \subseteq C_{pos}(\Omega, S)$, (see [**JPS**]).□

Now we proceed to prove Theorem 1.7. This proof is a "non-compact" modification of a technique used to prove of Theorem 1.1 of reference [**NS**], and the reader may find it helpful to refer to it for additional details.

Sketch of the Proof of Theorem 1.7.

As before, fix a sequence $(K_n)_{n\in\mathbf{N}}$ of compact subsets such that $K_n \subseteq int(K_{n+1})$, $n \in \mathbf{N}$ and $G = \cup_{n\in\mathbf{N}} K_n$. Given $N \in \mathbf{N}$, set

$$E(N) = \{A \in C(\Omega, S) \mid X_A(\omega_0, \tau) \notin K_N \text{ for some } \tau > 0\}.$$

We shall show that $E(N)$ is open and dense in $C(\Omega, S)$. Then the desired conclusion follows from the Baire category theorem by considering $\cap_{N\in\mathbf{N}} E(N)$. Openness of $E(N)$ follows easily from the compactness of Ω and continuous dependence of solutions on the matrix defining the linear system. Now we proceed to prove the density.

Let $A_0 \in C(\Omega, S)$ and $\varepsilon > 0$ be given. From now on, we choose an inner product $\langle\cdot,\cdot\rangle$ on L and let $||\cdot||$ be the corresponding norm. We want to construct $A \in C(\Omega, S)$ that is ϵ-close (in the chosen norm) to A_0 and such that $X_A(\omega_0, \tau) \in G\backslash K_N$ for some $\tau > 0$. Now we sketch this construction.

The property $X_A(\omega_0, \tau) \in G\backslash K_N$ only depends on the values of A on the set $R(\tau) = \{T_t(\omega_0) \mid 0 \leq t \leq \tau\}$, which is homeomorphic to the interval $[0, \tau]$ via the map $t \to T_t(\omega_0)$, because ω_0 is a non-periodic point. So we only need to construct the desired perturbation A on $R(\tau)$, and then extend it to all of Ω using Tietze's extension theorem. (This extension can be kept S valued and ε-close to A_0 if these properties hold on $R(\tau)$.) We therefore have to show that if τ is large enough, and $\gamma_{0,\tau}$ denotes the restriction to $[0, \tau]$ of the curve $\gamma_0 : \mathbf{R} \to S$ given by $\gamma_0(t) = A_0(T_t(\omega_0))$, then $\gamma_{0,\tau}$ can be ε perturbed in the space $C([0, \tau], S)$ of continuous S valued functions on $[0, \tau]$ so that for the resulting curve γ, if we solve the Cauchy problem $g'(t) = \gamma(t)g(t)$, $g(0) = e_G$-the identity of G, then the solution will satisfy $g(\tau) \in G\backslash K_N$.

Equations such as

$$g'(t) = \gamma(t)g(t), \tag{2.3}$$

in which γ is an arbitrary curve with values in some subset S of the Lie algebra L of a Lie group G are known in Control Theory as right invariant control systems on a Lie group. The curves γ are the *controls*, and each control gives rise to a *trajectory* for each initial condition $g(0)$. If $\overline{g} \in G$, then the set of all points of the form $g(\tau)$, for all trajectories $g(\cdot)$-corresponding to all possible controls-that satisfy $g(0) = \overline{g}$, is known as the *time* τ *reachable set* from $\overline{g}$. In particular, our problem can now be stated as follows: given a connected, non-compact, unimodular Lie group G, a subset S of the Lie algebra L of G, an $\varepsilon > 0$, and a curve $\gamma_0 : \mathbf{R} \to S$, we want to show (under the strong accessibility hypothesis on S) that given a compact set $K \equiv K_N \subset G$, there exists some $\tau > 0$ such that the time τ reachable set from e_G for the control system (2.3) with time-varying constrints $\gamma(t) \in S$, $||\gamma(t) - \gamma_0(t)|| < \varepsilon$, intersects $G\backslash K$.

To achieve this, we shall show that there is a $p > 0$ such that the time p reachable set (from the identity e_G) by means of piecewise continuous controls has nonempty interior. This will be true for an arbitrary curve γ_0, so in particular we can apply it to all the translates $t \to \gamma_0(t + kp) = \gamma_0^k(t)$, for $k = 0, 1, 2, \cdots$. In each case, we get a nonempty open subset V_k of G such that every point g_k of V_k can be reached from e_G in time p by means of a trajectory of $g' = \gamma(t)g$ corresponding to a piecewise continuous, g_k dependent control $\gamma^k : [0, p] \to S$ that is ε close to γ_0^k on $[0, p]$. If we translate the γ^k's back to the intervals $[kp, (k+1)p]$, and concatenate them, we get, for each $m \in \mathbf{N}$, a piecewise continuous control $\hat{\gamma}^m : [0, mp] \to S$ that is ε close to γ_0 and gives rise to a trajectory going from e_G to $g_m g_{m-1} \cdots g_1$ in time mp. Now suppose that for some m we could guarantee that

$$V_m V_{m-1} \cdots V_1 \cap (G\backslash K) \neq \emptyset. \tag{2.4}$$

In that case choose $\gamma = \hat{\gamma}^m$ so that the solution of (2.3) that goes through e_G at time 0 will satisfy $g(mp) \in G\backslash K$. Then taking τ (in the previous discussion) to be mp, this γ will be our desired perturbation. However, γ is only piecewise continuous, and we need it to be continuous. This last point, however is easily taken care of by approximating γ by a continuous S valued maps γ^ν that are ε close to γ_0, in such a way that the corresponding trajectory g^ν converges uniformly to $g(\cdot)$. Naturally, this approximation is not possible in the uniform topology, but it is possible, for example in L^1 norm, and this suffices to get uniform convergence of the solutions. Then for sufficiently large ν, the function γ^ν will be our desired perturbation.

Now the question is whether the V_k's can be chosen so that (2.4) holds for some m. If we could guarantee that the V_k's do not get "too small" then this can indeed be done. We shall prove this in the following proposition. We remark that this proposition is a "non-compact" as well as a "measure theoretic" analogue of the statement that compact, connected Lie groups have "Property A", (see [**NS**]).

PROPOSITION 2.2. *Let*

1. *G be a non-compact, locally compact, Hausdorff, connected unimodular topological group.*
2. *Let $K \subseteq G$ be a compact subset.*
3. *Let $U_1, U_2, \cdots, U_n \cdots$ be a given sequence of open subsets of G such that for some $\delta > 0$,*

$$\mu(U_i) \geq \delta \quad \textit{for all } i \in \mathbf{N},$$

where μ is a left Haar measure on G.

Then there exists $m \in \mathbf{N}$ such that

$$U_1 U_2 \cdots U_m \cap (G \backslash K) \neq \emptyset.$$

PROOF. First we shall replace the given sequence $\{U_i \mid i \in \mathbf{N}\}$ by another sequence $\{B_i \mid i \in \mathbf{N}\}$ of open sets such that each B_i contains the identity element e_G and at the same time, for any $n \in \mathbf{N}$, the product of the first n sets in both sequences is a translate of each other. This is done by first selecting a sequence $a_n \in G$ as follows. Let $a_1 \in G$ be such that $e_G \in U_1 a_1$ and then assuming that $a_1, \cdots, a_{n-1}$ are selected, pick a_n inductively so that

$$e_G \in a_{n-1}^{-1} U_n a_n, \quad n \in \mathbf{N},$$

where $a_0 = e_G$. Set, $B_n = a_{n-1}^{-1} U_n a_n$, $n \in \mathbf{N}$. Then

(i) each B_n is an open set containing e_G,
(ii) $\mu(B_n) = \mu(U_n) \geq \delta$, $n \in \mathbf{N}$ (by the unimodularity),
(iii) $(U_1 \cdots U_n) a_n = B_1 \cdots B_n$, for all $n \in \mathbf{N}$, and
(iv) $a_n \in U_n^{-1} \cdots U_1^{-1} = (U_1 \cdots U_n)^{-1}$ for all $n \in \mathbf{N}$.

Now suppose that $U_1 U_2 \cdots U_m \cap (G \backslash K) = \emptyset$ for all $m \in \mathbf{N}$. Then by (iii) and (iv) it follows that

$$B_1 B_2 \cdots B_m = (U_1 \cdots U_m) a_m \subseteq K K^{-1}, \quad \text{for all } m \in \mathbf{N}. \tag{2.5}$$

Now we shall show that the inclusion (2.5) is false.

Let

$$\begin{aligned} S &= \{x \in G \mid x \in B_i \text{ for infinitely many indices } i\} \\ &= \cap_{k=1}^\infty \cup_{i=k}^\infty B_i. \end{aligned}$$

Now, by (i), for any $m \in \mathbf{N}$, $B_m \subseteq B_1 \cdots B_m \subset K K^{-1}$. Hence,

(a) $\mu(\cup_{m \in \mathbf{N}} B_m) \leq \mu(K K^{-1}) < \infty$. Furthermore, by (ii)
(b) $\mu(\cup_{i=k}^\infty B_i) \geq \delta$.

Since the sequence $k \to \cup_{i=k}^{\infty} B_i$, $k = 1, 2, \cdots$ of sets is monotonic decreasing, using (a) and (b) we conclude that

$$\mu(S) \geq \delta.$$

Let S^* be the semigroup generated by S. Then

$$S^* = \cup_{n=1}^{\infty} S^n \subseteq \cup_{m=1}^{\infty} B_1 B_2 \cdots B_m \subseteq K.K^{-1}$$

(here, once more we have used the fact $e_G \in B_j$ for all $j \in \mathbf{N}$). This shows that $S^* \subset G$ is a compact semigroup. But compact, Hausdorff semigroups in which cancellation law holds is a group (see [E2]), hence S^* is a compact subgroup of G. On the other hand, since $\mu(S) \geq \delta$, S^* has positive Haar measure. Hence $S^* = S^*(S^*)^{-1}$ must contain a neighbourhood of the identity e_G. Since G is connected, S^* must generate G which is a contradiction since G is non-compact.□

Going back to the discussion of our proof, notice that the sets V_k depend on the restriction $\overline{\gamma}_0^k$ of γ_0^k to $[0, p]$. If we could guarantee that every V_k contains a translate of some fixed neighbourhood W of identity e_G, then the uniform estimate in the hypothesis (3) above is valid and we can apply Proposition 2.2. Now the crucial point is that, although the neighbourhoods V_k depend on the curves $\overline{\gamma}_0^k \in C([0, p], S)$, the desired uniformity can be achieved because the set of curves $\overline{\gamma}_0^k$ that interest us is precompact. The precompactness follows easily from the fact that the curves $t \to A_0(T_t(\omega))$, $0 \leq t \leq p$, form a compact set as ω varies over Ω. The fact that this implies the existence of W is the content of our main accessibility result. This is already proved for compact groups G, (see Proposition 2.1 in [**NS**]) and we shall indicate the modification needed to extend it to the non-compact case (Proposition 2.3 below).

To state this result precisely, first we need to introduce some notation. For a closed interval $I \subseteq \mathbf{R}$ containing 0, let $C^{pc}(I, S)$ denote the set of all piecewise continuous maps (with finite set of jump discontinuities) from I to S. If $A \in C^{pc}(I, S)$, let Γ_A denote the unique G-valued absolutely continuous solution curve of the initial value problem

$$g' = A(t)g(t), \quad t \in I,$$
$$g(0) = e_G.$$

If $A \in C^{pc}(I, S)$ and $\delta > 0$, define the set $N_{I,S}(A, \delta)$ by setting

$$N_{I,S}(A, \delta) = \{B \in C^{pc}(I, S) \mid \sup_{t \in I} ||A(t) - B(t)|| < \delta\}.$$

Now we state our main "accessibility result".

PROPOSITION 2.3. *Let G be a Lie group with Lie algebra L. Let S be as in Theorem 1.7. Let $p^* > 0$, and $\mathcal{F} \subseteq C([0, p^*], S)$ be a compact subset. Then given $\delta > 0$ there exists a $p \in (0, p^*)$ and a neighbourhood W of identity e_G of G, depending on δ but independent of $A \in \mathcal{F}$, such that for each $A \in \mathcal{F}$ the set $\{\Gamma_B(p) \mid B \in N_{[0,p],S}(A, \delta)\}$ contains some right translate of W.*

Before proving Proposition 2.3, we show how it allows us to prove Theorem 1.7 by following the strategy outlined above.

Proof of Theorem 1.7.

Recall that $A_0 \in C(\Omega, S)$, $N \in \mathbf{N}$ and $\varepsilon > 0$ are given and we want to find $A \in C(\Omega, S)$ that is ε close to A_0 and such that $X_A(\omega_0, t) \in G\backslash K_N$ for some $t > 0$. Now we give the step by step constructions of A.

Step 1 : Let $\delta = \frac{\varepsilon}{2}$.

Step 2 : Apply Proposition 2.3 with $p^* = 1$ and

$$\mathcal{F} = \{(A_0)_\omega \mid \omega \in \Omega\} \subseteq C([0,1], S),$$

where $(A_0)_\omega(t) = A_0(T_t(\omega))$, $t \in [0,1]$. We then get a $p \in (0,1)$ and a neighbourhood W of e_G depending only on δ, such that for each $\omega \in \Omega$ there exists some $h(\omega) \in G$ satisfying

$$Wh(\omega) \subseteq \{\Gamma_B(p) \mid B \in N_{[0,p],S}((A_0)_\omega, \delta)\}. \tag{2.6}$$

Step 3 : Let $\xi_j = T_{pj}(\omega_0)$, $j \in \mathbf{N} \cup \{0\}$ and $h_j = h(\xi_{j-1})$, $j \in \mathbf{N}$. Define a sequence of open sets $\{U_j\}_{j \in \mathbf{N}}$ by setting

$$U_j = Wh_j, \quad j \in \mathbf{N}. \tag{2.7}$$

Notice that this sequence satisfies hypothesis (3) of Proposition 2.2. Pick $m \in \mathbf{N}$ as in the conclusion of Proposition 2.2. Thus

$$U_1 U_2 \cdots U_m \cap (G\backslash K_N) \neq \emptyset. \tag{2.8}$$

Step 4 : Consider $A_j \in \mathcal{F}$ defined by

$$A_j(t) = A_0(T_t(\xi_{j-1})), \quad t \in [0,p],\ 1 \leq j \leq m.$$

Thus, (2.6) and (2.7) yield

$$U_j \subset \{\Gamma_B(p) \mid B \in N_{[0,p],S}(A_j, \delta)\}. \tag{2.9}$$

Step 5 : Using (2.8), we can find $g \in G$ such that

$$g_m g_{m-1} \cdots g_1 \equiv g \notin K_N, \quad \text{where } g_j \in U_j \text{ for } j = 1, \cdots, m. \tag{2.10}$$

Step 6 : Using (2.9), select $\hat{A}_j \in N_{[0,p],S}(A_j, \delta)$ such that $g_j = \Gamma_{\hat{A}_j}(p)$.

Step 7 : Define

$$\tilde{A}(t) = \hat{A}_j(t - (j-1)p) \text{ if } (j-1)p \leq t \leq jp,\ 1 \leq j \leq m,$$

$$\tilde{A}(t) \equiv A_0(T_t(\omega_0)) \text{ if } t \in \mathbf{R}\backslash[0, mp).$$

Then $\tilde{A} \in C^{pc}(\mathbf{R}, S)$, $\Gamma_{\tilde{A}}(mp) = g \notin K_N$ and $||\tilde{A}(t) - A_0(T_t(\omega_0))|| < \delta$ for all $t \in \mathbf{R}$.

Step 8 : Let $\{A^\nu\}$ be a sequence of continuous S-valued functions on $\mathbf{R}$ that converges to $\tilde{A}$ in L^1 norm on each compact interval, and satisfies

$$||\tilde{A}^\nu(t) - A_0(T_t(\omega_0))|| < 2\delta \text{ for all } t \in \mathbf{R}.$$

(For an explicit construction of the sequence $\{A^\nu\}$, we refer to the proof of Theorem 1.1 of [**NS**].)
Step 9 : It follows from Gronwall's inequality that $\Gamma_{\tilde{A}^\nu}(mp) \to \Gamma_{\tilde{A}}(mp)$ as $\nu \to \infty$. Since $\Gamma_{\tilde{A}}(mp) \notin K_N$ and K_N is closed, we conclude that $\Gamma_{\tilde{A}^\nu}(mp) \notin K_N$ for some ν. Pick a ν for which this is true, and let $A^* : [0, mp] \to S$ be the restriction of $\tilde{A}^\nu$ to $[0, mp]$. Then A^* is continuous and S-valued on $[0, mp]$, satisfies $\Gamma_{A^*}(mp) \notin K_N$, and is ε close to $t \to A_0(T_t(\omega_0))$ in the supremum norm on $[0, mp]$.
Step 10 : Using the fact that ω_0 is non- periodic, we may identify the interval $[0, mp]$ with the compact subset $C = \{T_t(\omega_0) \mid t \in [0, mp]\}$ of Ω. So we may regard A^* as defined on C. Using a version of Tietze Extension Theorem (refer to [**NS**] for complete details), it is easy to see that A^* has a continuous extension $A : \Omega \to L$ that satisfies

$$A(\omega) \in S \quad \text{for all } \omega \in \Omega,$$

$$||A(\omega) - A_0(\omega)|| < \varepsilon, \quad \text{for all } \omega \in \Omega.$$

Notice that by construction

$$X_A(\omega_0, mp) = \Gamma_A(mp) \in G \backslash K_N,$$

thus the proof of Theorem 1.7 is complete. □

Finally we turn to the proof of Proposition 2.3. This is a slight alteration of Proposition 2.1 of [**NS**] which yields a stronger conclusion namely, p can be taken to be p^* itself, under an additional assumption of compactness of G. Unlike results in [**NS**], here we do not need this stronger conclusion. For the reader's convinience we shall essentially reproduce that proof without the compactness assumption on G. First we recall the following lemma (Lemma 2.3 of [**NS**]).

LEMMA 2.4. *Let G be an m dimensional Lie group with Lie algebra L, and let S be a subset of L that has the strong accessibility property. Then there exists an m-tuple $X = (X_1, \cdots, X_m) \in S^m$ such that for every $\eta > 0$ the set*

$$\{e^{s_1 X_1} \cdots e^{s_m X_m} \mid 0 < s_j, \sum_{j=1}^m s_j = \eta\}$$

has nonempty interior.

Proof of Proposition 2.3

Let $m = \dim G$. Recall that we are give a compact set $\mathcal{F} \subseteq C([0, p^*], S)$, $p^* > 0$, and a $\delta > 0$. Let $K = \{A(t) \mid t \in [0, p^*],\ A \in \mathcal{F}\}$. Then $K \subseteq S$ is compact. Let $\{U_j \mid 1 \leq j \leq q\}$ be a cover of K by relatively open subsets of S with diameter less that δ. Using a Lebesgue number for this cover and the fact that $\mathcal{F}$ is compact and hence equicontinuous, we can find p such that $0 < p < p^*$ with the property that for each $A \in \mathcal{F}$ there is a $j = j(A)$ such that

$$\{A(t) \mid 0 \leq t \leq p\} \subseteq U_j.$$

We now let Σ denote the set of all m-tuples $s = (s_1, \cdots, s_m) \in \mathbf{R}^m$ such that $s_k > 0$ for $k = 1, \cdots, m$ and $\sum_{k=1}^m s_k = p$. If $X = (X_1, \cdots, X_m) \in L^m$ and $s = (s_1, \cdots, s_m) \in \mathbf{R}^m$, write $e^{sX} = e^{s_m X_m} \cdots e^{s_1 X_1}$.

Since S is convex, it is easy to verify that every nonempty relatively open subset of S has the strong accessibility property. Then, applying Lemma 2.4 with U_j in the role of S, we get, for every $j \in \{1, \cdots, q\}$, an m-tuple

$$X^j = (X_1^j, \cdots, X_m^j) \in U_j \times \cdots \times U_j$$

such that the set $Q_j = \{e^{sX^j} \mid s \in \Sigma\}$ contains a translate $W_j h_j$ of some neighbourhood W_j of the identity. Let

$$W = \cap_{j=1}^q W_j.$$

Then W is a neighbourhood of e_G and clearly $Wh_j \subseteq Q_j$ for every j. Notice that W depends only on p, which in turn depends only on ε and not on individual $A \in \mathcal{F}$. Now given any $s \in \Sigma$, $1 \le j \le q$, define $H_{s,j} : [0,p] \to S$ to be the piecewise constant map whose value on the interval $[s_1 + \cdots + s_{k-1}, s_1 + \cdots + s_k)$ is X_k^j. Given $A \in \mathcal{F}$, $s \in \Sigma$, define $B_{s,A} : [0,p] \to S$ by setting $B_{s,A}(t) = H_{s,j(A)}(t)$ for $0 \le t \le p$.

Then, if $A \in \mathcal{F}$, it is clear that $B_{s,A} \in N_{[o,p],S}(A, \delta)$ for every $s \in \Sigma$. Moreover,

$$Wh_{j(A)} \subseteq Q_{j(A)} = \{\Gamma_{H_{s,j(A)}}(p) \mid s \in \Sigma\} = \{\Gamma_B(p) \mid B \in N_{[o,p],S}(A, \delta)\}.$$

This shows that W has the desired property. □

References

[E1] R. Ellis, *The construction of minimal discrete flows*, Amer. J. of Math. **LXXXVII no.3** (1965), 564–574..

[E2] R. Ellis, *Lectures on Topological Dynamics* (1967), Benjamin, New York.

[J1] R. Johnson, *Remarks on linear differential equations with measurable coefficients*, AMS Proceeding **100** (1987), 491-504.

[J2] R. Johnson, *Hopf bifurcation from non-periodic solutions of differential equations I- Linear theory*, Journal of Dynamics and Differential Equations **1, no.2** (1989), 179- 198..

[JPS] R. Johnson, K. Palmer and G. Sell, *Ergodic properties of linear dynamical systems*, SIAM Journal of Mathematical Analysis **18** (1987), 1-33.

[Kn] O. Knill, *Positive Lyapunov exponents for a dense set of bounded measurable $SL(2, \mathbf{R})$ cocycles*, Ergodic Theory and Dynamical Systems **12** (1992), 319-331.

[Ko] S. Kotani, *Lyapunov indices determine absolutely continuous spectrum of stationary random Schrödinger operators*, Proc. Taniguchi Symposium SA, Katata (1982), 225-247.

[NS] M. Nerurkar and H. Sussmann, *Construction of minimal cocycles arising from specific differential equations*, to appear in the Israel Journal of Mathematics.

[V] R. Vinograd, *A problem suggested by N. P. Erugin*, Differential Equations **11** (1975), 474-478.

[W] T. Wu, *Notes on compact topological groups (unpublished)*.

DEPARTMENT OF MATHEMATICS, RUTGERS UNIVERSITY, NJ 08102

E-mail address: `nmahesh@ crab.rutgers.edu`

Contemporary Mathematics
Volume **215**, 1998

Topological Dynamics and Differential Equations

G. Sell, W. Shen, and Y. Yi

ABSTRACT. By reviewing our previous works on lifting dynamics in skew-product semi-flows and also the work of Johnson on almost periodic Floquet theory, we show several significant applications of the abstract theory of topological dynamics to the qualitative study of non-autonomous differential equations. The paper also contains some detailed discussions on a conjecture of Johnson.

1. Introduction

We discuss connections between the abstract theory of topological dynamics, especially the algebraic theory of Ellis, and the qualitative study of non-autonomous differential equations. As remarked by Ellis in his book ([**E1**]), the abstract theory of topological dynamics usually plays less of a role in the qualitative study of autonomous differential equations because, not only is the differential structure ignored but the topological properties of the reals are not used in an essential manner. Nevertheless, since the introduction of the notion of a continuous *skew-product flow* by Miller ([**M**]) and Sell ([**S1**]) in 1960's, the theory of topological dynamics has found significant applications in many essential ways to the study of non-autonomous ordinary, partial and functional differential equations.

To show how topological dynamics comes into play, let us recall the construction of a skew-product flow from a non-autonomous ordinary differential equation:

$$x' = f(x,t), \quad x \in \mathbb{R}^n, \tag{1.1}$$

where f is C^2 *admissible*, that is, f is C^2 in x and Lipschitz in t, and moreover, for any compact set $K \subset \mathbb{R}^n$, f as well as all its partial derivatives are bounded and uniformly continuous on $K \times \mathbb{R}$. Due to the time dependence, (1.1) does not generate a flow on $\mathbb{R}^n$ itself. One alternative would be to add one dimension and make it autonomous since the system

$$\begin{cases} x' = f(x,t) \\ t' = 1 \end{cases} \tag{1.2}$$

1991 *Mathematics Subject Classification.* 34C27, 34C28, 54H20.
The second author is partially supported by NSF grant DMS-9402945.
The third author is partially supported by NSF grant DMS-9501412

clearly defines a flow $\tilde{\Pi}$ on $\mathbb{R}^n \times \mathbb{R}$. Unfortunately, in the dynamical system (1.2), all ω, α-limit sets will be empty due to the lack of compactness of solutions.

Therefore, to study the dynamics of (1.1), one has to take into account the effect of its coefficient structure. Such a structure should mainly capture the topological rather than differential natures of the coefficient space because, the differentiability of f with respect to t usually only contributes to the regularity of solutions but not to their dynamical behavior. To be more precise, let the coefficient space be $H(f)$, the *hull* of f, that is, $H(f) = cl\{f_\tau | \tau \in \mathbb{R}\}$ under compact open topology, where $f_\tau(x,t) \equiv f(x,t+\tau)$. The time translate $g \cdot t = g_t$ then defines a natural flow $(H(f), \mathbb{R})$, and moreover, this flow is minimal or almost periodic minimal if f is a minimal or an almost periodic function in t uniformly with respect to other variables. For each $g \in H(f)$, $x_0 \in \mathbb{R}^n$, let $x(x_0, g, t)$ denote the solution of

$$(1.3)_g \qquad x' = g(x,t)$$

with initial value x_0. By the standard local existence, uniqueness and continuity results for ordinary differential equations, equation (1.1) gives rise to a (local) *skew-product flow* $\Pi : \mathbb{R}^n \times H(f) \times \mathbb{R} \to \mathbb{R}^n \times H(f)$,

$$(1.4) \qquad \Pi(x_0, g, t) = (x(x_0, g, t), g \cdot t),$$

where $x(x_0, g, t)$ is C^1 in x_0.

Since the family $(1.3)_g$ $(g \in H(f))$ consists of only translated and limiting equations of (1.1), the (local) skew-product flow (1.4) precisely reflects the 'dynamics' of (1.2) especially when long time behavior of solutions are concerned. Such a topological setting allows one to apply general techniques developed in the abstract theory of topological dynamics since the natural projection $p : \mathbb{R}^n \times H(f) \to H(f)$ induces a flow homomorphism from a compact invariant set of (1.4) to $H(f)$.

Concerning with the qualitative study of a non-autonomous differential equation, there are essentially three types of problems which are closely related to the abstract theory of topological dynamics: a) Global structure of a system (see the almost periodic Floquet theory in Section 3 for example). b) Asymptotic behavior of bounded solutions. In this context, one tends to study the lifting property of an ω-limit set in the associated skew-product flow. c) The existence of certain recurrent or oscillatory solutions. For example, if f in (1.1) is almost periodic in t, then one is often interested in an almost periodic or an almost automorphic minimal lifting of (1.4) from the almost periodic base flow $(H(f), \mathbb{R})$. Clearly, these lifting properties will heavily depend on certain differential structures (e.g. stability, hyperbolicity and monotonicity) in a particular differential equation.

Motivated by applications, there has been a tremendous amount of studies on non-autonomous differential equations and (continuous) skew-product flows or semi-flows in the past twenty years or so. The aim of the current article is however not to give a survey to this broad and active area. Instead, by reviewing several existing works, we are trying to explore the fact that the application of the abstract theory of topological dynamics can be essential in the qualitative study of non-autonomous differential equations.

The paper is organized as follows. In Section 2, we summarize basic concepts and fundamental results from topological dynamics which are used in the current paper. For our particular applications, we shall deal with only real flows, that is, transformation groups with $\mathbb{R}$ as the acting group. In Section 3, we review some early results of Sacker and Sell ([**SS2**]) and recent results of Shen and Yi ([**SY4**]) on lifting properties in skew-product semi-flows. Section 4 is a brief review of the almost periodic Floquet theory due to Johnson ([**J1**]). We will also give some discussions on a conjecture of Johnson in [**J1**] and indicate several fundamental topological dynamics issues related to it. It should be pointed out that the current paper is by no means even a complete review of these works mentioned above. Also, an apology must be made in advance for failure to mention many (in fact most of) important works in the area of non-autonomous differential equations and skew-product (semi) flows.

We would like to thank R. A. Johnson for helpful discussions and references. We are indebted to the referee for comments and suggestions which lead to improvements of the current paper.

2. Algebraic Theory of Topological Dynamics

An algebraic way to study the nature of a compact flow was introduced by R. Ellis ([**E1**]). One basic idea of the algebraic theory is to associate a semigroup, the *Ellis semigroup*, or *enveloping semigroup*, to a compact flow.

Let $(X, \mathbb{R})$ be a compact flow. The space X^X of self maps of X, when furnished with the point open topology, is a compact T_2 space, and, composition of maps provides a natural semigroup structure on X^X. For each $t \in \mathbb{R}$, we note that $\Pi_t : X \to X$, $\Pi_t x =: x \cdot t$ defines a homeomorphism, hence an element of X^X.

DEFINITION 2.1. The set $E(X) = cl\{\Pi_t | t \in \mathbb{R}\} \subset X^X$ is called the *Ellis semigroup* associated to $(X, \mathbb{R})$.

Clearly, $E(X)$ is a sub-semigroup of X^X with identity $e = \Pi_0$, and the composition $\Pi_t \circ \gamma =: \gamma \cdot t, (\gamma \in E(X), t \in \mathbb{R})$ defines a compact pointed flow $(E(X), e, \mathbb{R})$. Throughout the paper, using the identification $\Pi_t x =: x \cdot t$ $(x \in X, t \in \mathbb{R})$, we shall choose the left action of $E(X)$ on a compact flow $(X, \mathbb{R})$ although the action of $\mathbb{R}$ on X has been assumed on the right.

THEOREM 2.1. (Ellis [**E1**]) *Let $p : (X, \mathbb{R}) \to (Y, \mathbb{R})$ be an epimorphism of compact flows. Then there exists a unique epimorphism $\tilde{p} : (E(X), \mathbb{R}) \to (E(Y), \mathbb{R})$ such that $p(\gamma x) = (\tilde{p}\gamma)(px)$ for all $\gamma \in E(X), x \in X$.*

In what follows, we will write $\tilde{p}$ as p if no confusion occurs.

DEFINITION 2.2. 1) A (left) *ideal* in $E(X)$ is a non-empty subset I in $E(X)$ with $E(X)I \subset I$. A (left) ideal I in $E(X)$ is said to be *minimal* if it contains no non-empty proper (left) sub-ideal in $E(X)$.

2) An *idempotent point* $u \in E(X)$ is such that $u^2 = u$.

It is observed in [**E1**] that I is an (left) (minimal) ideal in $E(X)$ if and only if I is an invariant (minimal) subset of the compact flow $(E(X), \mathbb{R})$. It follows that a minimal (left) ideal in $E(X)$ always exists. The structure of a minimal (left) ideal is as follows.

THEOREM 2.2. (Ellis [**E1**]) *Let I be a minimal (left) ideal in $E(X)$ and $J(I)$ be the set of idempotent points of $E(X)$ in I. Then the following holds:*

1) *$J(I) \neq \emptyset$;*
2) *For each $u \in J(I)$, uI is a group with identity u and the family $\{uI\}_{u \in J(I)}$ forms a partition of I.*

DEFINITION 2.3. 1) Points $x_1, x_2 \in X$ are said to be *distal* (*positively distal, negatively distal*) if there is a pseudo-metric d on X generating its topology and such that

$$\inf_{t\in\mathbb{R}(t\in\mathbb{R}^+, t\in\mathbb{R}^-)} d(x_1 \cdot t, x_2 \cdot t) > 0.$$

Points x_1, x_2 are said to be *proximal* (*positively proximal, negatively proximal*) if they are not distal (positively distal, negatively distal).

2) A point $x \in X$ is a *distal point* if it is only proximal to itself. A flow $(X, \mathbb{R})$ is a *point distal flow* if there is a distal point $x_0 \in X$ with dense orbit. A flow $(X, \mathbb{R})$ is a *distal flow* if every point in X is a distal point.

3) Let $p : (X, \mathbb{R}) \to (Y, \mathbb{R})$ be an epimorphism of compact flows. Then $(X, \mathbb{R})$ is a *distal* (*proximal*) *extension* of $(Y, \mathbb{R})$ if for all $y \in Y$, any two points x_1 and x_2, on the same fiber $p^{-1}(y)$, are distal (proximal).

Clearly, a distal extension of a distal flow is distal.

DEFINITION 2.4. The set $P(X) = \{(x_1, x_2) \in X \times X | x_1, x_2 \text{ are proximal}\}$ is referred to as the *proximal relation* of $(X, \mathbb{R})$.

Some consequences of distality and proximality are summarized below.

THEOREM 2.3. (Ellis [**E1**]) 1). *A flow $(X, \mathbb{R})$ is distal if and only if $E(X)$ is a group.*

2) *If $(X, \mathbb{R})$ is distal, then it laminates into minimal sub-flows, that is, X is a union of minimal sets.*
3) *Let $x_1, x_2 \in X$, then $(x_1, x_2) \in P(X)$ if and only if there exists a minimal (left) ideal I in $E(X)$ such that $\gamma x_1 = \gamma x_2$ $(\gamma \in I)$.*
4) *The relation $P(X)$ is an equivalence relation if and only if there is only one minimal (left) ideal in $E(X)$.*
5) *If $P(X)$ is closed, then it is an equivalence relation.*

THEOREM 2.4. *(See* [**A**], [**F**]) *Let $p : (X, \mathbb{R}) \to (Y, \mathbb{R})$ be a homomorphism of distal minimal flows. Then p is an open map.*

DEFINITION 2.5. 1) A flow $(X, \mathbb{R})$ is (uniformly) *almost periodic* if $\{\Pi_t | t \in \mathbb{R}\} \subset X^X$ forms an equicontinuous family.

2) A minimal flow $(X, \mathbb{R})$ is *almost automorphic* if there is $x_0 \in X$ such that whenever t_α is a net with $x_0 \cdot t_\alpha \to x_*$, then also $x_* \cdot (-t_\alpha) \to x_0$.

The notion of almost automorphy, as a generalization to almost periodicity, was first introduced by S. Bochner in 1955 in a work on differential geometry ([**B**]). Fundamental properties of almost automorphic functions on groups and abstract almost automorphic minimal flows were studied by W. Veech ([**V**]-[**V2**]) and others (see [**Fl**], [**FW**], [**R**], [**T**], [**T1**]). Recently, Shen and Yi ([**SY**]-[**SY6**], [**Y**]) gave a systematic study on almost automorphic phenomena in almost periodic differential equations (see also Section 3).

THEOREM 2.5. (Ellis [**E1**]) *A flow $(X, \mathbb{R})$ is almost periodic if and only if $E(X)$ is a group of continuous maps of X into X.*

If $(X, \mathbb{R})$ is almost periodic minimal, then it is isomorphic to $(E(X), \mathbb{R})$ (see [**E1**]). Using Theorem 2.5, it is easy to see that $(X, \mathbb{R})$ is almost periodic minimal if and only if $E(X)$ is a compact abelian topological group ([**Y**]).

THEOREM 2.6. ([**V**], 3.4) *A minimal flow* $(X, \mathbb{R})$ *is almost automorphic if and only if it is an* almost automorphic extension *of an almost periodic minimal flow* $(Y, \mathbb{R})$, *that is, there is a* x_0 *with* $p^{-1}p(x_0) = \{x_0\}$, *where* $p : (X, \mathbb{R}) \to (Y, \mathbb{R})$.

We note that if X is a metric space, then the almost automorphic extension in the above theorem becomes an almost 1 to 1 extension ([**V1**]).

3. Lifting Properties in Skew–Product Semi–Flows

In this section, we give a brief review of works by the current authors concerning lifting properties in skew-product semi-flows related to qualitative study of non-autonomous differential equations.

Throughout the section, we assume that $\sigma = (Y, \mathbb{R})$ is a minimal flow with compact metric phase space Y, X is a Banach space, $\Pi = (X \times Y, \mathbb{R}_+)$ is a *skew-product semi-flow*, namely, a semi-flow of the following form:

$$\Pi(x, y, t) = (\phi(x, y, t), y \cdot t), \quad (x, y) \in X \times Y, \quad t \in \mathbb{R}_+. \tag{3.1}$$

Let $p : X \times Y \to Y$ be the natural projection.

Skew–product semi–flows generated by PDE's and FDE's.

In the introduction, we have shown how to construct a skew product flow from a non-autonomous ordinary differential equation. We now give two examples of skew-product semi-flows generated by non-autonomous parabolic equations and delay differential equations.

EXAMPLE 1. Consider a scalar parabolic equation

$$\begin{cases} u_t = \Delta u + f(u, \nabla u, x, t), & t > 0, \quad x \in \Omega \\ u|_{\partial\Omega} = 0 \quad \text{or} \quad \dfrac{\partial u}{\partial n}|_{\partial\Omega} = 0, & t > 0, \end{cases} \tag{3.2}$$

where Ω is a bounded, connected, smooth domain in $\mathbb{R}^n$, $f : (\mathbb{R}^1 \times \mathbb{R}^n \times \mathbb{R}^n) \times \mathbb{R} \to \mathbb{R}^1$ is a C^2 admissible and (uniform) minimal function. By the standard theory of parabolic equations, for each $U_0 \in C^1(\bar{\Omega})$ which satisfies the boundary condition of (3.2) and for each $g \in H(f)$, the equation

$$\begin{cases} u_t = \Delta u + g(u, \nabla u, x, t), & t > 0, \quad x \in \Omega \\ u|_{\partial\Omega} = 0 \quad \text{or} \quad \dfrac{\partial u}{\partial n}|_{\partial\Omega} = 0, & t > 0, \end{cases} \tag{3.3$_g$}$$

has (locally) a unique classical solution $u(U_0, g, x, t)$ with initial value U_0.

We now define a (local) skew-product semi-flow over $H(f)$ similarly to the case of (1.1). To do so, the phase space X can be chosen as a suitable fractional power space ([**He**]) which is embedded in $C^1(\bar{\Omega})$. Having chosen such a X, one can show that if $U_0 \in X$, $g \in H(f)$, then $u(U_0, g, \cdot, t) \in X$ is C^2 in U_0 and is continuous in g, t within its (time) interval of existence. In other word, there is a well defined (local) skew-product semi-flow $\Pi : X \times H(f) \times \mathbb{R}^+ \to X \times H(f)$:

$$\Pi(U_0, g, t) = (u(U_0, g, \cdot, t), g \cdot t), \quad t > 0 \tag{3.4}$$

associated to (3.2), where $u(U_0, g, t)$ is C^2 in U_0.

By the standard *a priori* estimates of parabolic equations, if $u(U_0, g, \cdot, t)$ is a bounded solution of $(3.3)_g$ for t in its interval of existence, then $u(U_0, g, \cdot, t)$ exists for all $t > 0$. Furthermore, for $\delta > 0$, $\{u(U_0, g, \cdot, t) | t \geq \delta\}$ is relatively compact, hence its ω-limit set $\omega(U_0, g)$ is well defined and compact. Moreover, the restriction of Π to $\omega(U_0, g)$ is a (global) semi-flow which admits a flow extension $(\omega(U_0, g), \mathbb{R})$. A minimal set E of (3.4) can be defined in the same fashion, that is, $E = \omega(x_0, g)$ for some $(x_0, g) \in E$ and $(E, \mathbb{R})$ is minimal in the usual sense.

We remark that, with the introduction of a skew-product semi-flow (3.4), dynamics of (3.2) is relatively independent of the choice of a phase space X as long as the class of solutions under investigation possess enough regularity. In fact, by the standard *a priori* estimates of parabolic equations, for any $U_0 \in X$, if $u(U_0, g, \cdot, t)$ is X-bounded, then it is $H^{2,p}$ bounded and moreover $\omega(U_0, g)|_{X \times H(f)}$ coincides with $\omega(U_0, g)|_{H^{2,p}(\Omega) \times H(f)}$ $(p > n)$.

Finally, we note that by the comparison principle for scalar parabolic equations, one can easily find a natural condition which guarantees the existence of a C^2-bounded (hence X-bounded) solution for (3.2).

EXAMPLE 2. Consider a delay differential equation

$$x'(t) = f(x(t), x(t-1), t), \tag{3.5}$$

where $f : \mathbb{R}^n \times \mathbb{R}^n \times \mathbb{R} \to \mathbb{R}^n$ is a C^2 admissible and (uniform) minimal function. Let $X = C([-1, 0], \mathbb{R}^n)$. Then by the standard theory of delay differential equation ([**Ha**]), for each $\phi \in X$ and each $g \in H(f)$, the equation

$$x'(t) = g(x(t), x(t-1), t) \tag*{$(3.6)_g$}$$

has (locally) a unique solution $x(\phi, g, t)$ with initial value ϕ, that is, $x(\phi, g, t) = \phi(t)$ for $t \in [-1, 0]$. Let $x_t(\phi, g) \in X$ $(t > 0)$ be defined as follows: $x_t(\phi, g)(\theta) \equiv x(\phi, g, t + \theta)$ $(\theta \in [-1, 0])$. Then $x_t(\phi, g)$ is C^2 in $\phi \in X$ and Lipschitz in $g \in H(f)$ (see [**Ha**]). Therefore, there is a well defined (local) skew-product semi-flow $\Pi : X \times H(f) \times \mathbb{R}^+ \to X \times H(f)$,

$$\Pi(\phi, g, t) = (x_t(\phi, g), g \cdot t) \tag{3.7}$$

associated to (3.5).

It can be proved that if $x(\phi, g, t) \in \mathbb{R}$ is a bounded solution of $(3.6)_g$ for t in the existence interval, then $x_t(\phi, g)$ is defined for all $t > 0$ and $\{x_t(\phi, g) | t \geq 1 + \delta\}$ is relatively compact in X for any $\delta > 0$, hence $\omega(\phi, g)$ is well defined (see [**SY5**]). Moreover, under certain conditions (see [**SY5**]), Π restricted to $\omega(\phi, g)$ extends to a usual flow.

Lifting Dynamics

We now consider the skew-product semi-flow (3.1) and assume that $\sigma = (Y, \mathbb{R})$ is minimal and distal. Throughout rest of the section, we let $M \subset X \times Y$ be a compact invariant set of Π (that is, the semi-flow on M admits a flow extension and shall refer to it as "the flow $(M, \mathbb{R})$").

Motivated by the qualitative study of differential equations, one is often interested in the lifting properties of $p : (M, \mathbb{R}) \to (Y, \mathbb{R})$. In case that $(Y, \mathbb{R})$ is almost periodic, one would like to ask when the lifted flow $(M, \mathbb{R})$ is also almost periodic or almost automorphic of finite type (this is closely related to the classical study of the existence of a harmonic or sub-harmonic almost periodic solution in differential equations with almost periodic time dependence). It turns out that such a lifting is not generally possible without a stability condition.

DEFINITION 3.1. 1) The flow $(M, \mathbb{R})$ is *uniformly stable* if for any $\epsilon > 0$, there is a $\delta = \delta(\epsilon) > 0$ such that whenever $(\hat{x}, y) \in M$, $(x, y) \in X \times Y$ satisfy $\|\hat{x} - x\| < \delta(\epsilon)$, then

$$\|\phi(\hat{x}, y, t) - \phi(x, y, t)\| < \epsilon \quad \text{for all } t \geq 0.$$

2) The flow $(M, \mathbb{R})$ is *uniformly asymptotically stable* if it is uniformly stable and there is a $\delta_0 > 0$ such that if $(\hat{x}, y) \in M$, $(x, y) \in X \times Y$ and $\|\hat{x} - x\| \leq \delta_0$, then

$$\lim_{t \to \infty} \|\phi(\hat{x}, y, t) - \phi(x, y, t)\| = 0.$$

THEOREM 3.1. (Sacker-Sell [**SS1**][**SS2**]) *If* $(M, \mathbb{R})$ *is uniformly stable, then* $p : (M, \mathbb{R}) \to (Y, \mathbb{R})$ *is a distal extension.*

PROOF. First, we claim that $(M, \mathbb{R})$ is negatively distal, that is, for any (x_1, y), $(x_2, y) \in M$, $x_1 \neq x_2$,

$$\inf_{t \leq 0} \|\phi(x_1, y, t) - \phi(x_2, y, t)\| > 0. \tag{3.8}$$

Suppose (x_1, y), $(x_2, y) \in M$ are two points for which (3.8) fails. Let $\epsilon_0 = \frac{1}{2}\|x_1 - x_2\|$ and let $\delta(\epsilon_0)$ be as in Definition 3.1. Then there is a $\tau < 0$ such that $\|\phi(x_1, y, \tau) - \phi(x_2, y, \tau)\| < \delta(\epsilon_0)$. Thus, $\|\phi(x_1, y, t + \tau) - \phi(x_2, y, t + \tau)\| < \epsilon_0$ $(t \geq 0)$. In particular, for $t = -\tau$, one has $\|x_1 - x_2\| = 2\epsilon_0 < \epsilon_0$, a contradiction.

Let $E(M)$ denote the Ellis semigroup of $(M, \mathbb{R})$ and e be the identity of $E(M)$. Since the α-limit set $E_-(M) \equiv \alpha(e)$ of e is compact invariant, it contains a minimal set I, which is a minimal (left) ideal of $E(M)$. Let $u \in J(I)$ be an idempotent point of I. For any $(x, y) \in M$, since $u(u(x, y)) = u(x, y)$, (x, y) and $u(x, y)$ are negatively proximal. Hence $u(x, y) = (x, y)$. Since (x, y) is arbitrary, $u = e$, that is, $I = eI$ is a group (Theorem 2.2). Now $E(M) = E(M)e \subset E(M)I \subset I$. One has that $E(M) = I$ is a group, that is, $(M, \mathbb{R})$ is distal (Theorem 2.3, 1)). □

THEOREM 3.2. (Sacker-Sell [**SS2**]) *If* $(M, \mathbb{R})$ *is uniformly asymptotically stable, then there is a positive integer* N *such that* $p : (M, \mathbb{R}) \to (Y, \mathbb{R})$ *is an* N *to* 1 *extension, that is,* $\operatorname{card} M \cap p^{-1}(y) = N$ $(y \in Y)$. *Moreover,* M *is an* N*-fold covering of* Y *if* $(Y, \mathbb{R})$ *is distal.*

PROOF. Suppose that $\operatorname{card} M \cap p^{-1}(y_0) = \infty$ for some $y_0 \in Y$. Since $M \cap p^{-1}(y_0)$ is compact, it must contain an accumulation point, say (x_0, y_0). Let δ_0 be as in Definition 3.1, 2). Then there exists a $(x_*, y_0) \in M \cap p^{-1}(y_0) \setminus \{(x_0, y_0)\}$ such that $\|x_* - x_0\| \leq \delta_0$. It follows that $\|\phi(x_*, y_0, t) - \phi(x_0, y_0, t)\| \to 0$ as $t \to \infty$, that is, (x_*, y_0), (x_0, y_0) are proximal, a contradiction to Theorem 3.1. Thus $\operatorname{card} M \cap p^{-1}(y_0) < \infty$ for all $y \in Y$.

Let $y_1, y_2 \in Y$ be arbitrary and let $t_n \to \infty$ be a sequence such that $y_1 \cdot t_n \to y_2$. Since $p : (M, \mathbb{R}) \to (Y, \mathbb{R})$ is a distal extension according to Theorem 3.1, points on $M \cap p^{-1}(y_1)$ converge to distinct points on $M \cap p^{-1}(y_2)$ following a subsequence of $\{t_n\}$. This implies that $\operatorname{card} M \cap p^{-1}(y_2) \geq \operatorname{card} M \cap p^{-1}(y_1)$. Similarly, $\operatorname{card} M \cap p^{-1}(y_1) \geq \operatorname{card} M \cap p^{-1}(y_2)$. Therefore, there exists a positive integer N for which $\operatorname{card} M \cap p^{-1}(y) = N$ $(y \in Y)$.

In the case that $(Y, \mathbb{R})$ is distal, one can apply Theorem 2.4 to conclude that M is an N-fold covering of Y. □

REMARK 3.1. It is shown in [**SS2**] that if $p : (M, \mathbb{R}) \to (Y, \mathbb{R})$ is an N to 1 extension (e.g., $(M, \mathbb{R})$ satisfies the condition of Theorem 3.2) and if $(Y, \mathbb{R})$ is almost periodic, then $(M, \mathbb{R})$ is almost periodic. It is also shown in [**SS2**] that if M is a uniformly stable ω-limit set, then $p : (M, \mathbb{R}) \to (Y, \mathbb{R})$ is in fact a minimal extension.

We now assume that ϕ is C^2 in x in (3.1) and denote $\Phi(x, y, t) \equiv \phi_x(x, y, t)$. Then Φ defines a *semicocycle* on $X \times Y$. Due to the fact that dynamics of a conservative system (e.g., a Hamiltonian) is usually complicated, we pay particular attentions to lifting properties in skew-product semi-flows of strongly monotone natures.

DEFINITION 3.2. A flow Π is *strongly monotone* if the following conditions hold:

1) The Banach space X is strongly ordered, that is, there is a closed convex cone $X_+ \subset X$ with $IntX_+ \neq \emptyset$ and $X_+ \cap (-X_+) = \{0\}$;
2) if $v \in X_+$, then for any $(x, y) \in X \times Y$, $\Phi(x, y, t)v \in IntX_+$ $(t > 0)$.

DEFINITION 3.3. The flow $(M, \mathbb{R})$ is *linearly stable* if its upper Lyapunov exponent λ_M is non-positive, that is,

$$\lambda_M = \sup_{(x,y)\in M} \limsup_{t\to\infty} \frac{\ln \|\Phi(x, y, t)\|}{t} \leq 0.$$

Typical strongly monotone skew-product semi-flows are those generated from almost periodic co-operative systems of ordinary and delay differential equations (see [**SY5**]). Also, a skew-product semi-flow of type (3.4) generated by a parabolic equation is always strongly monotone (see [**SY5**]).

By [**SY4**], a strongly monotone skew-product flow Π is *strongly order preserving* in the following sense:

Define a strong partial ordering $\geq$ on $X \times Y$ as follows:

$$(x_1, y_1) \geq (x_2, y_2) \iff y_1 = y_2 \text{ and } x_1 - x_2 \in X_+;$$
$$(x_1, y_1) > (x_2, y_2) \iff (x_1, y_1) \geq (x_2, y_2) \text{ and } x_1 \neq x_2;$$
$$(x_1, y_1) \gg (x_2, y_2) \iff y_1 = y_2 \text{ and } x_1 - x_2 \in \operatorname{Int} X_+.$$

Then $u(x_1, y, t) \gg u(x_2, y, t)$ $(t > 0)$ if $(x_1, y) > (x_2, y)$.

THEOREM 3.3. (Shen-Yi [**SY4**]) (1) *If $(M, \mathbb{R})$ is linearly stable and minimal, then there exists a distal and N to 1 extension $p_* : (Y_*, \mathbb{R}) \to (Y, \mathbb{R})$ for some positive integer N such that $\tilde{p} : (M, \mathbb{R}) \to (Y_*, \mathbb{R})$ is almost* 1 *to* 1*, where $p = p_* \circ \tilde{p}$.*

(2) *If $(M, \mathbb{R})$ is also uniformly stable, then $p : (M, \mathbb{R}) \to (Y, \mathbb{R})$ is an N to* 1 *extension for some positive integer N.*

PROOF. (2) This is an easy consequence of 1) and Theorem 3.1.

(1) To prove (1), let $O(M) \subset M \times M$ denote the order relation on M, that is,

$$O(M) = \{(x_1, y_1), (x_2, y_2)) \in M \times M \mid (x_1, y_1) \geq (x_2, y_2) \text{ or } (x_2, y_2) \geq (x_1, y_1)\}.$$

Then $O(M)$ is clearly a closed and positively invariant relation.

Consider 2^M with the Hausdorff metric and consider the map $q : Y \to 2^M$, $y \mapsto M \cap p^{-1}(y)$. Since q is upper semi-continuous, the set Y_0 of continuity points of q is a residual subset of Y.

We now claim that the following *non-ordering principle* holds: for any $y \in Y_0$, there is no ordered pair in $M \cap p^{-1}(y)$. Suppose for contradiction that there is an ordered pair (x_0^1, y_0), $(x_0^2, y_0) \in M \cap p^{-1}(y_0)$ for some $y_0 \in Y_0$. Without loss of generality, we may assume that $(x_0^1, y_0) \gg (x_0^2, y_0)$. By the Zorn's lemma, one can find a maximal element (x_M, y_0) in $M \cap p^{-1}(y_0)$ with respect to '$\geq$'. Now let $t_n \to -\infty$ be a sequence such that $(x_M, y_0) \cdot t_n \to (x_0^2, y_0)$. By the lower continuity of q at y_0, there is a sequence $\{(x_n, y_0)\} \subset M \cap p^{-1}(y_0)$ such that $(x_n, y_0) \cdot t_n \to (x_0^1, y_0)$. Since $(x_0^1, y_0) \gg (x_0^2, y_0)$, there is an n_0 sufficiently large such that $(x_{n_0}, y_0) \cdot t_{n_0} > (x_M, y_0) \cdot t_{n_0}$. This implies that $(x_{n_0}, y_0) > (x_M, y_0)$, a contradiction with the maximality of (x_M, y_0).

Let $P(M)$ be the proximal relation on $(M, \mathbb{R})$. The above discussion implies that $O(M) \subset P(M)$. To see this, we let $(x_1, y), (x_2, y) \in O(M)$ with $(x_1, y) > (x_2, y)$. For a fixed $y_0 \in Y_0$, by the minimality of $(Y, \mathbb{R})$, there is a sequence $t_n \to +\infty$ such that $y \cdot t_n \to y_0$. Without loss of generality, we assume that $(x_i, y) \cdot t_n \to (x_i^0, y_0)$ $(i = 1, 2)$ as $n \to \infty$. Now, if $(x_1, y), (x_2, y)$ are distal, then $(x_1^0, y_0) \neq (x_2^0, y_0)$. Since Π is strongly order preserving and $O(M)$ is a closed relation, $(x_1^0, y_0) > (x_2^0, y_0)$, a contradiction to the non-ordering principle.

Next, using linear stability and the strong monotonicity of Π, one can show the following: There are ϵ_0, δ_0, $K > 0$ such that if (x_1, y), $(x_2, y) \in M$, $\|x_1 - x_2\| < \epsilon_0$, and $(x_1, y) \cdot t$, $(x_2, y) \cdot t$ are not ordered for t in a finite interval $[0, t_0]$, then

$$\|\phi(x_1, y, t) - \phi(x_2, y, t)\| \leq K e^{-\delta_0 t} \|x_1 - x_2\| \tag{3.9}$$

for all $t \in [0, t_0]$ (see [**SY4**]).

By (3.9), the non-ordering principle and the fact $O(M) \subset P(M)$, one can further show that $P(M)$ is an equivalence relation. Therefore the Ellis semigroup $E(M)$ contains a unique minimal (left) ideal I. Thus, $I \subset \alpha(e) \cap \omega(e)$, where e is the identity of $E(M)$. It follows that if $((x_1, y), (x_2, y)) \in P(M) \setminus O(M)$, then they are both positively and negatively proximal. Now let $t_n \to -\infty$ be such that $\|\phi(x_1, y, t_n) - \phi(x_2, y, t_n)\| \to 0$ as $n \to \infty$. Then for n sufficiently large, $\|\phi(x_1, y, t_n) - \phi(x_2, y, t_n)\| < \epsilon_0$ and $(x_1, y) \cdot (t + t_n)$, $(x_2, y) \cdot (t + t_n)$ are not ordered for $t \in [0, -t_n]$. By (3.9),

$$\|x_1 - x_2\| = \|\phi((x_1, y) \cdot t_n, -t_n) - \phi((x_2, y) \cdot t_n, -t_n)\| \leq \epsilon_0 K e^{\delta_0 t_n}.$$

We let $n \to \infty$ to conclude that $x_1 = x_2$. This shows that $O(M) = P(M)$. Therefore, both $O(M)$ and $P(M)$ are closed and invariant relations.

Consider $Y_* = M/P(M) = M/O(M)$ and $(Y_*, \mathbb{R})$ be the induced flow. A similar argument using (3.9) and the proof of Theorem 3.2 shows that $p_* : (Y_*, \mathbb{R}) \to (Y, \mathbb{R})$ is a distal and N to 1 extension for some positive integer N. Let $\tilde{p} : (M, \mathbb{R}) \to (Y_*, \mathbb{R})$ be the projection and let $Y_*^0 = \tilde{p}^{-1}(Y_0)$. Clearly, $\operatorname{card} M \cap \tilde{p}^{-1}(y) = 1$ $(y \in Y_*^0)$. □

By Remark 3.1 and Theorem 2.6, if $(Y, \mathbb{R})$ is almost periodic, then $(M, \mathbb{R})$ in the above theorem is almost automorphic (almost periodic) in the case of linear stability (uniform stability). We note that a linearly stable ω-limit set of Π need not be minimal.

All results above are sharp. In the case of strong monotonicity, a linearly stable minimal set need not be almost periodic and N can be bigger than 1 (see [**SY4**], [**SY5**]).

We refer the readers to [**SS**]-[**SS2**] for more discussions on almost periodic lifting dynamics and to [**SY**]-[**SY6**] for almost automorphic lifting dynamics in skew-product flows and semi-flows.

4. Almost Periodic Floquet Theory

In [**J1**], using topological dynamics techniques, Johnson derived an analogue of the classical Floquet theory for two dimensional, almost periodic linear system of ordinary differential equations which provided a clear qualitative picture of such a system. We now give a brief review on the Johnson's theory along with some discussions on a case which was left open in Johnson's original work [**J1**].

Consider

$$x' = a(y\cdot t)x, \quad x\in\mathbb{R}^2, \quad y\in Y_2, \tag{4.1$_y$}$$

where $tr(a)\equiv 0$, and $(Y,\mathbb{R})$ is an almost periodic minimal flow with compact metric phase space Y.

By using

$$a(y) = \begin{pmatrix} 0 & -\sigma(y) \\ \sigma(y) & 0 \end{pmatrix} + \begin{pmatrix} \delta(y) & \epsilon(y) \\ \epsilon(y) & -\delta(y) \end{pmatrix},$$

one can rewrite $(4.1)_y$ into the following polar coordinate form

$$r' = f(\theta, y\cdot t)r \tag{4.2$_{\theta,y}$}$$

$$\theta' = L(\theta, y\cdot t), \tag{4.3$_y$}$$

where $r = |x| \in \mathbb{R}^+$, $\theta = arg(x)\in S^1$ and

$$f(\theta,y) = \delta(y)\cos 2\theta + \epsilon(y)\sin 2\theta$$

$$L(\theta,y) = \sigma(y) + \epsilon(y)\cos 2\theta - \delta(y)\sin 2\theta.$$

Let $p_0 : S^1\times Y \to P^1\times Y$ be the natural projection, where P^1 denotes the projective 1-space (= the set of lines through the origin in $\mathbb{R}^2$). Then the relation $f_0(p_0(\theta,y))\equiv f(\theta,y)$, $L_0(p_0(\theta,y))\equiv L(\theta,y)$ induces functions f_0, L_0 on $P^1\times Y$. Therefore, $(4.3)_y$ generates skew-product flows $\tilde{\Pi}$ and Π on both $S^1\times Y$ and $P^1\times Y$ respectively. Let $p : P^1\times Y\to Y$, $\tilde{p} = p\circ p_0 : S^1\times Y \to Y$ be the natural projections.

DEFINITION 4.1. 1) The system

$$x' = (a(y\cdot t) - \lambda I)x \tag{4.1$_{y,\lambda}$}$$

is said to admits an *exponential dichotomy* (ED) over Y if there are constants $\alpha, K > 0$ and a continuous family of projections $P_\lambda(y) : \mathbb{R}^2\to\mathbb{R}^2$ such that for all $y\in Y$ one has

$$\|\Phi_\lambda(y,t)P_\lambda(y)\Phi_\lambda^{-1}(y,s)\| \le Ke^{-\alpha(t-s)}, \quad s\le t,$$

$$\|\Phi_\lambda(y,t)(I-P_\lambda(y))\Phi_\lambda^{-1}(y,s)\| \le Ke^{-\alpha(t-s)}, \quad t\le s,$$

where $\Phi_\lambda(y,t)$ denotes the principle matrix of $(4.1)_{y,\lambda}$.

2) The *dynamical spectrum* Σ of the linear skew-product flow generated by $(4.1)_y$ is $\Sigma = \{\lambda\in\mathbb{R} \mid (4.1)_{y,\lambda}$ does not admit an exponential dichotomy (ED) over $Y\} = \{\lambda\in\mathbb{R}\mid (4.1)_{y,\lambda}$ has a nontrivial bounded solution for some $y\in Y\}$.

According to a general spectral theory due to Sacker-Sell ([**SS3**]) and Selgrade ([**Se**]), either $\Sigma = \{-\beta,\beta\}$ for some $\beta>0$ or $[-\beta,\beta]$ for some $\beta\ge 0$.

DEFINITION 4.2. Let $(Y, \mathbb{R})$ be a minimal flow. A continuous map $U : Y \to GL(2, \mathbb{R})$ defines a *strong Perron transformation* if, for each $y \in Y$, the map $\mathbb{R} \to GL(2, \mathbb{R})$, $t \mapsto U(y \cdot t)$ is continuously differentiable, and the map $Y \to L(\mathbb{R}^2)$, $y \mapsto \frac{d}{dt}U(y \cdot t)|_{t=0}$ is continuous.

Similar to the classical Floquet theory, the almost periodic Floquet theory of Johnson concerns with finding a suitable strong Perron transformation which transforms $(4.1)_y$ into a canonical form (e.g., an upper triangular or a diagonal system). Such a theory depends on a detailed classification of minimal sets of $\Pi = (P^1 \times Y, \mathbb{R})$ in all situations of the dynamical spectrum Σ (see also [**NO**] for a measure theoretical classification of Π). To give an idea, let $M \subset P^1 \times Y$ be a minimal set of Π and $\tilde{M}$ be a minimal lift of M in $\tilde{\Pi} = (S^1 \times Y, \mathbb{R})$. Note that $\tilde{M}$ is either a 1-cover or a 2-cover of M. Define $\tilde{a} : \tilde{M} \to L(\mathbb{R}^2)$, $\tilde{a}(m) = a(\tilde{p}(m))$ $(m \in \tilde{M})$. Then

$$(4.4)_m \qquad x' = \tilde{a}(m \cdot t)x$$

coincides with $(4.1)_y$ if $y = \tilde{p}(m)$. Let $U : \tilde{M} \to SL(\mathbb{R}^2)$,

$$m = (\theta, y) \mapsto \begin{pmatrix} \cos\theta & -\sin\theta \\ \sin\theta & \cos\theta \end{pmatrix}.$$

Then U is a strong Perron transformation and $x = U(m \cdot t)z$ transforms $(4.4)_m$ into the form

$$(4.5)_m \qquad z' = \begin{pmatrix} u(m \cdot t) & v(m \cdot t) \\ 0 & -u(m \cdot t) \end{pmatrix} z.$$

If either for some $y \in Y$, $M \cap p^{-1}(y)$ admits a distal pair or Π admits precisely two minimal sets, then another strong Perron transformation would further transform $(4.5)_m$ into a diagonal system (see [**J1**]).

Differing from the classical Floquet theory of periodic systems, a fundamental fact of Johnson's theory is that a minimal set of Π is often almost automorphic (see [**J1**]-[**J2**]). Thus, an almost automorphic (not necessary almost periodic) strong Perron transformation has to be introduced to transform $(4.1)_y$ into an almost automorphic canonical form (which need not be a constant coefficient system either).

Case A: $\Sigma = \{-\beta\} \cup \{\beta\}$ for some $\beta > 0$. In this case, $(4.1)_y$ $(y \in Y)$ admits an ED with precisely two invariant line bundles $\mathcal{B}_1$, $\mathcal{B}_2$ of $P^1 \times Y$ associated to the ED (or to the Lyapunov exponents $-\beta$, β). That is, Π admits precisely two minimal sets $M_i = \{(\theta, y) \in P^1 \times Y |$ the line with direction θ is in $\mathcal{B}_i\}$ $(i = 1, 2)$ which are 1-covers of Y.

THEOREM 4.1. ([**J1**], 6.7) *Suppose that $\Sigma = \{-\beta, \beta\}$. There is an almost periodic minimal set $\tilde{M}$ of $\tilde{\Pi}$ which is either a 1-cover or a 2-cover of Y and a strong Perron transformation $x = U(m \cdot t)z$ $(m \in \tilde{M})$ which transforms* $(4.4)_m$ *into*

$$z' = \begin{pmatrix} b(m \cdot t) & 0 \\ 0 & -b(m \cdot t) \end{pmatrix} z.$$

Case B: $\Sigma = \{0\}$. By the definition of Σ, there is $y \in Y$ such that $(4.1)_y$ admits a nontrivial bounded solution. We therefore have the following two cases.

Case B$_1$: All solutions of $(4.1)_y$ for all y are bounded.

This case was originally studied by Cameron ([**C**]) for almost periodic systems and extensions were made in Johnson [**J1**] and Ellis and Johnson [**EJ**] for distal and recurrent systems respectively. We note that these results are valid for linear systems with arbitrary dimensions.

Restricting to two dimensions, then either $P^1 \times Y$ itself is minimal or it contains at least three minimal sets (therefore infinitely many minimal sets by [**J1**], [**SS2**]). Below we only summarize the result and outline a proof from [**J1**].

THEOREM 4.2. ([**C**], [**J1**], [**EJ**]) *Suppose we are in Case B_1. Then there is a strong Perron transformation $U : Y \to GL(2,\mathbb{R})$ such that the change of variable $x = U(y \cdot t)z$ takes $(4.1)_y$ into*

$$z' = \begin{pmatrix} 0 & -b(y\cdot t) \\ b(y\cdot t) & 0 \end{pmatrix} z. \tag{4.6$_y$}$$

PROOF. Define a flow $\hat{\Pi}_t$ on $SL(2,\mathbb{R}) \times Y = \{A \in GL(2,\mathbb{R}) | detA = 1\} \times Y$ as follows:

$$\hat{\Pi}_t(A,y) = (\Phi(y,t)A, y\cdot t),$$

where $\Phi(y,t)$ denotes the principle matrix of $(4.1)_y$. It can be shown that $\hat{\Pi}_t$ is distal. Therefore, for fixed $y_0 \in Y$, $X = cl\{\hat{\Pi}_t(I,y_0)\} \subset SL(2,\mathbb{R}) \times Y$ is minimal and distal. Let $\eta : X \to Y$, $(A,y) \mapsto y$ be the natural projection and $\eta_* : E(X) \to E(Y)$ be the projection induced by η according to Theorem 2.1, where $E(X)$ and $E(Y)$ are the Ellis semi-groups of X and Y respectively. Since both $E(X)$ and $E(Y)$ are groups (Theorem 2.3, 1)), $G = \{\gamma \in E(X) | \eta_*\gamma(y_0) = y_0\}$ is a subgroup, and therefore $G_0 = G|_{\eta^{-1}(y_0)}$ is a group of self maps of $X_0 = \eta^{-1}(y_0)$. Let $(A,y_0) \in X_0$ and let $\hat{\Pi}_{t_\alpha} \to \beta_0 \in G_0$ in $E(X)$. Assume that $\lim_\alpha \Phi(y_0,t_\alpha)A = A_{\beta_0}$. It can be shown that the map $G_0 \to \eta^{-1}(y_0) \subset SL(2,\mathbb{R})$, $\beta_0 \mapsto A_{\beta_0}$ is a group isomorphism. Thus $\eta^{-1}(y_0)$ is a compact subgroup of $SL(2,\mathbb{R})$. We identify G_0 with $\eta^{-1}(y_0)$.

Since X, Y are distal, the map $y \mapsto \eta^{-1}(y) : Y \to 2^X$ is continuous with respect to the Hausdorff metric (Theorem 2.4). Therefore, for $(A,y) \in \eta^{-1}(y)$, if $\{t_\alpha\}$ is such that $\hat{\Pi}_{t_\alpha}(I,y_0) \to (A,y)$, then $\eta^{-1}(y) = A \cdot G_0$.

Define inner product $\langle \cdot,\cdot \rangle_*$ on $\mathbb{R}^2$:

$$\langle x_1, x_2 \rangle_* = \int_{\eta^{-1}(y_0)} \langle Ax_1, Ax_2 \rangle dA,$$

where dA is the normalized Haar measure on $\eta^{-1}(y_0)$. Then $\eta^{-1}(y_0)$ is a subgroup of the unitary group with respect to $\langle \cdot,\cdot \rangle_*$ and there is a positive definite operator Q_0 on $\mathbb{R}^2$ such that $\langle Q_0x_1, Q_0x_2 \rangle_* = \langle x_1, x_2 \rangle$, $(x_1, x_2 \in \mathbb{R}^2)$.

Define $R : Y \to GL(2,\mathbb{R})$, $R(y) = AA_*$, where (A,y) is some element of $\eta^{-1}(y)$ and A_* is the adjoint of A with respect to $\langle \cdot,\cdot \rangle_*$. One can check that R is well defined, continuous, positive definite and self-adjoint with respect to $\langle \cdot,\cdot \rangle_*$. Let $U_1(y)$ be the unique positive definite, self-adjoint square root of $R(y)$ and let $U(y) = U_1(y)Q_0$ $(y \in Y)$. Then $U : Y \to GL(2,\mathbb{R})$ defines a strong Perron transformation. Consider the transformed equation

$$z' = \hat{a}(y\cdot t)z \tag{4.7$_y$}$$

under $x = U(y \cdot t)z$. If $z_1(t)$, $z_2(t)$ are two solutions of $(4.7)_y$, then it follows from the identity $R(y \cdot t) = \Phi(y,t)AA_*\Phi(y,t)_*$ (for some $A \in \eta^{-1}(y)$) that

$$\begin{aligned}\langle z_1(t), z_2(t)\rangle &= \langle Q_0^{-1}U_1^{-1}(y\cdot t)\Phi(y,t)x_1, Q_0^{-1}U_1^{-1}(y\cdot t)\Phi(y,t)x_2\rangle \\ &= \langle U_1^{-1}(y\cdot t)\Phi(y,t)x_1, U_1^{-1}(y\cdot t)\Phi(y,t)x_2\rangle_* \\ &= \langle R^{-1}(y\cdot t)\Phi(y,t)x_1, \Phi(y,t)x_2\rangle_* \\ &= \langle R^{-1}(y)x_1, x_2\rangle_* = \text{constant}.\end{aligned}$$

Thus, $\hat{a}(y)$ must be skew symmetric. □

In fact, if some equation $(4.1)_y$ admits an almost periodic solution, then all solutions to $(4.1)_y$ $(y \in Y)$ are almost periodic and $b(y)$ in $(4.6)_y$ may be assumed to be a constant b_0 ([**J1**]).

Case B$_2$: There is $y \in Y$ such that $(4.1)_y$ admits an unbounded solution.

In this case, Π admits no more than two minimal sets. Johnson ([**J1**]) had shown the following: If Π admits precisely two minimal sets M_i $(i = 1, 2)$, then $(M_i, \mathbb{R}) \to (Y, \mathbb{R})$ $(i = 1, 2)$ are almost automorphic extensions. If Π admits only one minimal set M which is not proximal extension of Y, then there is an almost periodic minimal flow $(Y_*, \mathbb{R})$, a 2-cover of $(Y, \mathbb{R})$, such that $(M, \mathbb{R}) \to (Y_*, \mathbb{R})$ is an almost automorphic extension.

THEOREM 4.3. ([**J1, 6.12**]) *Suppose we are in Case B_2 and let some fiber $p^{-1}(y)$ contain a distal pair. Then there is* (i) *an almost automorphic minimal set $\tilde{M}$ of $\tilde{\Pi}$ which is either an almost* 1*-cover or an almost* 2*-cover of Y;* (ii) *a strong Perron transformation $x = U(m \cdot t)z$ $(m \in \tilde{M})$ which transforms $(4.4)_m$ into*

$$z' = \begin{pmatrix} b(m\cdot t) & 0 \\ 0 & -b(m\cdot t)\end{pmatrix} z.$$

Case C: $\Sigma = [-\beta, \beta]$, $\beta > 0$. R. Johnson had shown the following.

THEOREM 4.4. (Johnson [**J1**], [**J2**]) *If $\Sigma = [-\beta, \beta]$, $\beta > 0$, then*
1) $\Pi = (P^1 \times Y, \mathbb{R})$ *admits a unique minimal set M;*
2) M *supports exactly two ergodic measures;*
3) $p : (M, \mathbb{R}) \to (Y, \mathbb{R})$ *is a proximal extension.*

By citing examples ([**J4**]) originally due to Millionšćikov and Vinograd, Johnson ([**J1**]) conjectured the following: If $\Sigma = [-\beta, \beta]$, with $\beta \geq 0$, and if Π admits a unique minimal set M with $p : (M, \mathbb{R}) \to (Y, \mathbb{R})$ being a proximal extension, then p is an almost automorphic extension. Later, Johnson ([**J3**]) proved that there is a large class of equations of form $(4.1)_y$ which admit an unbounded solution for all $y \in Y$, the corresponding flow $\Pi = (P^1 \times Y, \mathbb{R})$ is minimal and $p : (P^1 \times Y, \mathbb{R}) \to (Y, \mathbb{R})$ is a proximal extension. Therefore, the original conjecture in [**J1**] fails in general. Below, we will discuss some cases in which the conjecture of Johnson is still valid.

Let $(\tilde{M}, \mathbb{R})$ be a minimal lift of $(M, \mathbb{R})$ in $\tilde{\Pi} = (S^1 \times Y, \mathbb{R})$. If the conjecture were true, then $\tilde{p} : (\tilde{M}, \mathbb{R}) \to (Y, \mathbb{R})$ would be either an almost 1 to 1 or an almost 2 to 1 extension. Suppose that $(\tilde{M}, \mathbb{R})$ were also almost automorphic (thus $(\tilde{M}, \mathbb{R})$ would be an almost automorphic extension of either a 1-cover or a 2-cover of $(Y, \mathbb{R})$). Then there would exist an almost automorphic strong Perron transformation $x = U(m{\cdot}t)z$ $(m \in \tilde{M})$ which transforms

$$x' = a(\tilde{p}(m \cdot t))x$$

into an upper triangular form:

$$(4.8)_m \qquad z' = \begin{pmatrix} b(m \cdot t) & c(m \cdot t) \\ 0 & -b(m \cdot t) \end{pmatrix} z.$$

Even so, $(4.8)_m$ can not be of diagonal form for otherwise $p^{-1}(y)$ would admit a distal pair for some $y \in Y$.

As far as completion of the almost periodic Floquet theory is concerned, there are two fundamental questions which are related to the Johnson's conjecture: 1) *When is $(M, \mathbb{R})$ almost automorphic?* 2) *If $(M, \mathbb{R})$ were almost automorphic, would $(\tilde{M}, \mathbb{R})$ also be almost automorphic?*

The study of these issues is also interesting from other aspects (e.g., linear almost periodic oscillation problems). We note that since $\dfrac{\partial L(\theta, y)}{\partial \theta} \equiv -2f(\theta, y)$ in $(4.2)_{\theta,y}$ and $(4.3)_y$, both $(M, \mathbb{R})$ and $(\tilde{M}, \mathbb{R})$ admit a positive Lyapunov exponent if $\Sigma = [-\beta, \beta]$, with $\beta > 0$. Therefore, in the case of interval spectrum, even though both $(M, \mathbb{R})$ and $(\tilde{M}, \mathbb{R})$ are almost automorphic, they are not almost periodic by the non-unique ergodicity, and the flows on them can be complicated or even chaotic (see [**J4**] for an example of topological complication of such a minimal flow).

Below, we present some discussions related to the Johnson's conjecture. For this purpose, we parameterize both S^1 and P^1 by the angular variable θ. For simplicity, $\theta(y, t)$ will denote the solution of $(4.3)_y$ with $\theta(y, 0) = \theta$.

THEOREM 4.5. *The flow $(\tilde{M}, \mathbb{R})$ is almost automorphic if and only if $P(\tilde{M})$ is a closed relation.*

PROOF. We first note that since $tr(y) \equiv 0$, the following Liouville's formula holds

$$(4.9) \qquad e^{\int_0^t f(\theta_1(y,s), y\cdot s)ds} e^{\int_0^t f(\theta_2(y,s), y\cdot s)ds} \sin(\theta_1(y,t) - \theta_2(y,t)) \equiv \sin(\theta_1 - \theta_2),$$

$(t \in \mathbb{R}, (\theta_i, y) \in S^1 \times Y, i = 1, 2)$. Therefore, if $\theta_1, \theta_2 \in S^1$ is such that $0 < \theta_1 - \theta_2 < \pi$, then $0 < \theta_1(y, t) - \theta_2(y, t) < \pi$ for any $y \in Y$ and $t \in \mathbb{R}$.

By Theorem 2.6, if $(\tilde{M}, \mathbb{R})$ is almost automorphic, then $P(\tilde{M})$ is a closed relation. Conversely, suppose that $P(\tilde{M})$ is a closed relation. Let $Y_0 \subset Y$ be the residual set for which $h : Y \to 2^{\tilde{M}} : y \mapsto \tilde{M} \cap \tilde{p}^{-1}(y)$ is continuous in Hausdorff metric (see the proof of Theorem 3.3). Fix a $y_0 \in Y_0$. We first show that $\tilde{M} \cap \tilde{p}^{-1}(y_0)$ contains no more than two points. Suppose not. Then one can find two points $(\theta_1^0, y_0), (\theta_2^0, y_0)$ on $\tilde{M} \cap \tilde{p}^{-1}(y_0)$ with $0 < \theta_1^0 - \theta_2^0 < \pi$. Let $\tilde{I}$ be the unique minimal (left) ideal in $E(\tilde{M})$ and $u \in J(\tilde{I})$ be an idempotent point such that $u(\theta_1^0, y_0) = (\theta_1^0, y_0)$. Let σ denote the reflection: $S^1 \times Y \to S^1 \times Y$, $(\theta, y) \mapsto (\theta + \pi, y)$. Since $p : (M, \mathbb{R}) \to (Y, \mathbb{R})$ is a proximal extension, for any $(\theta, y_0) \in \tilde{M} \cap \tilde{p}^{-1}(y_0)$, either $u(\theta, y_0) = (\theta_1^0, y_0)$ or $u(\theta, y_0) = \sigma(\theta_1^0, y_0)$.

Let $\{t_\alpha | \alpha \in \Lambda\}$ be a net in $\mathbb{R}$ such that $\tilde{\Pi}_{t_\alpha} \to u$ in $E(\tilde{M})$. By the lower continuity of h at y_0, there is a net $\{(\theta_\alpha, y_0) | \alpha \in \Lambda\} \subset \tilde{M} \cap \tilde{p}^{-1}(y_0)$ with $(\theta_\alpha, y_0) \cdot t_\alpha \to (\theta_2^0, y_0)$. It follows that there is an α_0 such that $0 < \theta_1^0(y_0, t_\alpha) - \theta_\alpha(y_0, t_\alpha) < \pi$, $(\alpha > \alpha_0)$, that is, $0 < \theta_1^0 - \theta_\alpha < \pi$, $(\alpha > \alpha_0)$ by (4.9). By taking a subnet, we assume without loss of generality that $\{\theta_\alpha\}$ is (strictly) monotonically increasing with respect to the positive (counterclockwise) orientation of S^1, $0 < \theta_1^0 - \theta_\alpha < \pi$ for all α and $\theta_0 = \lim_\alpha \theta_\alpha$.

Fix an α. Since $0 < \theta_{\alpha'} - \theta_\alpha < \pi$, by (4.9), $0 < \theta_{\alpha'}(y_0, t_{\alpha'}) - \theta_\alpha(y_0, t_{\alpha'}) < \pi$, $(\alpha' > \alpha)$. It follows that $0 \leq \theta_2^0 - \lim_{\alpha'} \theta_\alpha(y_0, t_{\alpha'}) \leq \pi$. Therefore, $u(\theta_\alpha, y_0) \neq (\theta_1^0, y_0)$ for any α. This is already a contradiction if $\tilde{M}$ is a 1-cover of M.

In the case that $\tilde{M}$ is a 2-cover of M, we must have $u(\theta_\alpha, y_0) = \sigma(\theta_1^0, y_0)$ for all α. Without loss of generality, we assume that $0 < \theta_0 - \theta_\alpha < \pi$ for all α. By (4.9) again, $0 < \theta_0(y_0, t_\alpha) - \theta_\alpha(y_0, t_\alpha) < \pi$ for all α, that is, $0 \leq \lim_\alpha \theta_0(y_0, t_\alpha) - \theta_2^0 \leq \pi$. It follows that $u(\theta_0, y_0) = (\theta_1^0, y_0)$. Therefore, $((\theta_1^0, y_0), (\theta_0, y_0)) \in P(\tilde{M})$ and $(\sigma(\theta_1^0, y_0), (\theta_\alpha, y_0)) \in P(\tilde{M})$ for all α. By the closeness of $P(\tilde{M})$, $(\sigma(\theta_1^0, y_0), (\theta_0, y_0)) \in P(\tilde{M})$. On the other hand, since $P(\tilde{M})$ is also an equivalence relation (Theorem 2.3, 5)), $((\theta_1^0, y_0), (\sigma(\theta_1^0, y_0)) \in P(\tilde{M})$, a contradiction.

Above all, for any $y_0 \in Y_0$, since $p : (M, \mathbb{R}) \to (Y, \mathbb{R})$ is a proximal extension, $\tilde{M} \cap \tilde{p}^{-1}(y_0)$ either consists of one point $\{(\theta_0, y_0)\}$ or two points $\{(\theta_0, y_0), \sigma(\theta_0, y_0)\}$. By the definition of Y_0, it is easy to see that either card $\tilde{M} \cap \tilde{p}^{-1}(y_0) = 1$ for all $y_0 \in Y_0$ or card $\tilde{M} \cap \tilde{p}^{-1}(y_0) = 2$ for all $y_0 \in Y_0$.

Now, let $Y_* = \tilde{M}/P(\tilde{M})$. Then $(Y_*, \mathbb{R})$ is either a 1 to 1 or a 2 to 1 extension of $(Y, \mathbb{R})$ and $(\tilde{M}, \mathbb{R}) \to (Y^*, \mathbb{R})$ is an almost automorphic extension. □

By the above theorem, if $\tilde{M}$ is a 1-cover of M, then without any condition, $(\tilde{M}, \mathbb{R})$, $(M, \mathbb{R})$ are both almost automorphic extensions of $(Y, \mathbb{R})$. Also, the proof of the theorem implies that in the current case an almost automorphic minimal set $\tilde{M}$ of $\tilde{\Pi}$ is at most an almost 2-cover of Y. Thus, if the entire space $P^1 \times Y$ or $S^1 \times Y$ is minimal (see [**J2**]), neither $\tilde{\Pi}_t$ nor Π_t can be almost automorphic. We now restrict ourselves to the case that $\tilde{M}$ is a 2-cover of M.

THEOREM 4.6. *If $\tilde{M}$ is a 2-cover of M, then the following are equivalent:*

1) *$P(\tilde{M})$ is an equivalence relation;*
2) *For any $(\theta_1, y), (\theta_2, y) \in \tilde{M}$, $|\theta_1(y,t) - \theta_2(y,t)|$ either stay away from 0 for all t or from π for all t;*
3) *For any idempotent point $u \in E(M)$, $p_0^{-1}(u)$ contains a unique idempotent point in $E(\tilde{M})$.*

If, in addition, $\Sigma = [-\beta, \beta]$, $\beta > 0$, then any of 1)-3) *above implies that*

4) *$(\tilde{M}, \mathbb{R}) \to (Y, \mathbb{R})$ is an almost 2 to 1 extension and therefore $(M, \mathbb{R}) \to (Y, \mathbb{R})$ is an almost automorphic extension.*

PROOF. Since $(M, \mathbb{R}) \to (Y, \mathbb{R})$ is a proximal extension, $E(M)$ contains a unique minimal (left) ideal I (Theorem 2.3, 4)). It follows that if $\tilde{I} \subset E(\tilde{M})$ is a minimal (left) ideal, then for any $\gamma \in \tilde{I}$ and any (θ_1, y), $(\theta_2, y) \in \tilde{M}$, either $\gamma(\theta_1, y) = \gamma(\theta_2, y)$ or $\gamma(\theta_1, y) = \sigma\gamma(\theta_2, y)$, where σ is the reflection defined in the proof of Theorem 4.5. We note that an alternative of the condition in 2) is that for any (θ_1, y), $(\theta_2, y) \in \tilde{M}$, either (θ_1, y), (θ_2, y) are distal or (θ_1, y), $\sigma(\theta_2, y)$ are distal.

1) $\Longrightarrow$ 2): If there are (θ_1, y), $(\theta_2, y) \in \tilde{M}$ such that $(\theta_1, y), (\theta_2, y) \in P(\tilde{M})$ and $(\theta_1, y), \sigma(\theta_2, y) \in P(\tilde{M})$, then $(\theta_2, y), \sigma(\theta_2, y) \in P(\tilde{M})$, a contradiction.

2) $\Longrightarrow$ 1): Let u be any idempotent point in $E(\tilde{M})$. If $\{(\theta_1, y), (\theta_2, y)\} \in P(\tilde{M})$, then $(\theta_1, y), \sigma(\theta_2, y)$ are distal, that is, $u(\theta_1, y) \neq u\sigma(\theta_2, y)$. It follows that $\sigma u(\theta_1, y) = u\sigma(\theta_2, y) = \sigma u(\theta_2, y)$, that is, $u(\theta_1, y) = u(\theta_2, y)$. Clearly, $P(\tilde{M})$ is an equivalence relation.

1) $\iff$ 3): Define a relation $\sim$ on $E(\tilde{M})$ as follows: $\gamma_1 \sim \gamma_2$ if and only if for any $(\theta, y) \in \tilde{M}$ either $\gamma_1(\theta, y) = \gamma_2(\theta, y)$ or $\gamma_1(\theta, y) = \sigma\gamma_2(\theta, y)$. It turns out that $\sim$ is a closed, invariant (with respect to both $(E(\tilde{M}), \mathbb{R})$ and the semigroup structure on $E(\tilde{M})$) relation, and moreover $\gamma_1 \sim \gamma_2$ if and only if $p_0(\gamma_1) = p_0(\gamma_2)$. Now fix a $u \in J(I)$ and take a $\gamma \in p_0^{-1}(u)$ and an idempotent $\tilde{u} \in E(\tilde{M})$ which covers u. Then $\tilde{u} \sim \gamma \sim \gamma^2$. Note that for any $(\theta, y) \in \tilde{M}$, if $\tilde{u}(\theta, y) = \gamma(\theta, y)$, then $\tilde{u}\gamma(\theta, y) = \tilde{u}^2(\theta, y) = \tilde{u}(\theta, y) = \gamma(\theta, y)$, and if $\tilde{u}(\theta, y) = \sigma\gamma(\theta, y)$, then $\tilde{u}\gamma(\theta, y) = \sigma\tilde{u}^2(\theta, y) = \sigma\tilde{u}(\theta, y) = \gamma(\theta, y)$. That is, $\tilde{u}\gamma = \gamma$. A similar argument shows that $\gamma^3 = \gamma$. It follows that $p_0^{-1}(I)$ is fiber-wise distal and laminates into minimal (left) ideals. Now let $\tilde{I}$ be a minimal (left) ideal of $E(\tilde{M})$. Then $\tilde{I} \subset p_0^{-1}(I)$ and $G \equiv \tilde{I} \cap p_0^{-1}(u) = \{\gamma \in \tilde{u}\tilde{I} | \gamma^2 = \tilde{u}\}$, where $\tilde{u}$ is an idempotent point in G. Thus G is an abelian, (normal) subgroup of $\tilde{u}\tilde{I}$, and therefore G contains exactly one idempotent point $\tilde{u} \in J(\tilde{I})$ (by [**E**], G is in fact a topological group). We omit the rest of the proof by noting that $P(\tilde{M})$ is an equivalence relation if and only if $p_0^{-1}(I)$ is a minimal (left) ideal.

We now assume that $\Sigma = [-\beta, \beta]$, with $\beta > 0$, and show that if $P(\tilde{M})$ is an equivalence relation, then $(\tilde{M}, \mathbb{R}) \to (Y, \mathbb{R})$ is an almost 2 to 1 extension.

By Theorem 4.4 and [**NO**], $(M, \mathbb{R})$ admits two ergodic measures μ_+, μ_- which concentrate on two separate ergodic sheets. More precisely, there is a full measure set $Y_* \subset Y$ (with respect to the Haar measure μ_0) and measurable functions $\theta_{+,-} : Y \to M$ such that $\int_Y f(\theta_\pm(y), y) d\mu_0 = \pm\beta$ $(y \in Y_*)$. Note that $0 < |\theta_+(y) - \theta_-(y)| < \pi$ for all $y \in Y_*$.

We now fix a $y_* \in Y_*$ and without loss of generality assume that

$$0 < \theta_-(y_*) - \theta_+(y_*) < \pi.$$

Define I_i, $i = 1$ to 4 as follows:

$$\begin{aligned} I_1 =& ((\theta_+(y_*), y_*), (\theta_-(y_*), y_*)), \; I_2 = ((\theta_-(y_*), y_*), \sigma(\theta_+(y_*), y_*)), \\ I_3 =& (\sigma(\theta_+(y_*), y_*), \sigma(\theta_-(y_*), y_*)), \; I_4 = (\sigma(\theta_-(y_*), y_*), \theta_+(y_*), y_*)). \end{aligned}$$

Let $(\theta, y_*) \in I_1 \cup I_4$ and let $x(t)$, $x_\pm(t)$ be the solutions of $(4.1)_{y_*}$ with $\arg x(0) = \theta$, $\arg x_\pm(0) = \theta_\pm(y_*)$ respectively. Expressing $x_-(t)$ as a linear combination of $x_+(t)$ and $x(t)$, say, $x_-(t) = c_+ x_+(t) + c x(t)$, then $c_+ c < 0$. Since $\lim_{t \to \infty} \frac{\ln \|x_\pm(t)\|}{t} = \pm\beta$ (Birkhoff ergodic theorem), one has that $\lim_{t \to \pm\infty} \|x_\pm(t)\| = \infty$, $\lim_{t \to \mp\infty} \|x_\pm(t)\| = 0$. Applying (4.9) with $\theta_1 := \theta$, $\theta_2 := \theta_-(y_*)$ or $\theta_1 := \theta$, $\theta_2 := \theta_+(y_*)$, one also has $\lim_{t \to \pm\infty} \|x(t)\| = \infty$. Now

$$\begin{aligned} \|x_-(t)\|^2 =& c_+^2 \|x_+(t)\|^2 + c^2 \|x(t)\|^2 \\ & - 2|c| \cdot |c_+| \cdot \|x_+(t)\| \cdot \|x(t)\| \cos(\theta(y_*, t) - \theta_+(y_*, t)). \end{aligned}$$

One must have $\lim_{t \to +\infty} |\theta(y_*, t) - \theta_+(y_*, t)| = 0$. Similarly, $\lim_{t \to +\infty} |\theta(y_*, t) - (\theta_+ + \pi)(y_*, t)| = 0$, $((\theta, y_*) \in I_2 \cup I_3)$, $\lim_{t \to -\infty} |\theta(y_*, t) - \theta_-(y_*, t)| = 0$, $((\theta, y_*) \in I_1 \cup I_2)$, and $\lim_{t \to -\infty} |\theta(y_*, t) - (\theta_- + \pi)(y_*, t)| = 0$, $((\theta, y_*) \in I_3 \cup I_4)$. It follows that if $(\theta_1, y_*) \in I_1$, $(\theta_4, y_*) \in I_4$, then $\lim_{t \to +\infty} |\theta_1(y_*, t) - \theta_4(y_*, t)| = 0$ and $\lim_{t \to -\infty} |\theta_1(y_*, t) - \theta_4(y_*, t)| = \pi$. By 2), either $I_1 \cap \tilde{M} \cap \tilde{p}^{-1}(y_*) = \emptyset$ or $I_4 \cap \tilde{M} \cap \tilde{p}^{-1}(y_*) = \emptyset$. Without loss of generality, assume that $I_4 \cap \tilde{M} \cap \tilde{p}^{-1}(y_*) = \emptyset$. Let Y_0 be as in Theorem 4.5 and let $y_0 \in Y_0$. If $\operatorname{card} \tilde{M} \cap \tilde{p}^{-1}(y_0) > 2$, then a similar argument to the proof of Theorem 4.5 shows that card $\tilde{M} \cap \tilde{p}^{-1}(y_*) = \infty$, in particular, $I_1^* \equiv I_1 \cap \tilde{M} \cap \tilde{p}^{-1}(y_*) \neq \emptyset$ and $I_3^* = I_3 \cap \tilde{M} \cap \tilde{p}^{-1}(y_*) \neq \emptyset$. Now, a point in I_1^* is positively proximal to $(\theta_+(y_*), y_*)$ and negatively proximal to $(\theta_-(y_*), y_*)$. It follows that $(\theta_-(y_*), y_*)$, $(\theta_+(y_*), y_*)$ are proximal. Above all, two points (θ_i, y_*) $(i = 1, 2)$, are proximal if and only if they lie in the same compact set $I_1^* \cup \{(\theta_+(y_*), y_*)\} \cup \{(\theta_-(y_*), y_*)\}$ or $I_3^* \cup \{\sigma(\theta_+(y_*), y_*)\} \cup \{\sigma(\theta_-(y_*), y_*)\}$. Thus, $P(\tilde{M})$ when restricted to $\tilde{M} \cap \tilde{p}^{-1}(y_*)$ is closed. But an argument similar to the one in Theorem 4.5 would give a contradiction. Therefore, card $\tilde{M} \cap \tilde{p}^{-1}(y_0) = 2$, $(y_0 \in Y_0)$, that is, $(\tilde{M}, \mathbb{R}) \to (Y, \mathbb{R})$ is almost 2 to 1. □

We remark that, by the above proof, if one of the conditions 1)-3) holds, then the two ergodic sheets $\{(\theta_+(y), y) | y \in Y_*\}$, $\{(\theta_-(y), y) | y \in Y_*\}$ give an 'upper' and 'lower' bounds of $M \cap p^{-1}(y)$ $(y \in Y_*)$. This is a typical scenario of flow $\pi = (P^1 \times Y, \mathbb{R})$ in the interval spectrum case (see the example in [**J4**]). In general, by a perturbation theorem of dynamical spectrum ([**SS3**]), a perturbation of system $(4.1)_y$ often gives rise to a linear system with point spectrum $\Sigma_\epsilon = \{\beta + \epsilon, -(\beta + \epsilon)\}$, $(0 < |\epsilon| \ll 1)$ (see [**J5**], [**JY**]), in which the induced flow $\Pi_\epsilon = (P^1 \times Y, \mathbb{R})$ admits exactly two minimal sets M_i^ϵ $(i = 1, 2)$ which are all 1-covers of Y. When passing to a limit, M_i^ϵ clips together on Y_0 $(\mu_0(Y_0) = 0)$ and converge to the two ergodic sheets of M on Y_* $(\mu_0(Y_*) = 1)$.

Reviewing conditions in 2) of the above theorem, it seems that for $P(\tilde{M})$ to be an equivalence relation, solutions of $(4.1)_y$ should have certain 'regular' oscillatory behavior. Nevertheless, it is possible to show that $(M, \mathbb{R})$ is almost automorphic without knowing whether $P(\tilde{M})$ is an equivalence relation.

Since Σ contains 0, there is a $y \in Y$ such that $(4.1)_y$ admits a non-trivial bounded solution. This bounded solution must be unique up to linear dependence of solutions, for otherwise, all solutions of $(4.1)_y$ for all $y \in Y$ would be bounded.

Let $\hat{Y} = \{y \in Y | (4.1)_y$ admits a unique non-trivial bounded solution$\}$. Then $\hat{Y}$ is an invariant set of Y. In the case that $\Sigma = [-\beta, \beta]$ for $\beta > 0$, $\hat{Y}$ is both measure theoretically and topologically small (according to [**J**], there is an invariant residual subset $Y^* \subset Y$ such that for any $y \in Y^*$, solutions of $(4.1)_y$ all oscillate in magnitude between 0 and ∞. Let Y_* be the full measure set defined in the proof of Theorem 4.6. It is easy to see that Y_*, Y^* and $\hat{Y}$ are disjoint).

Now for $y \in \hat{Y}$, let $\hat{\theta}(y) \in P^1$ be such that solution $x(y, t)$ with $arg x(y, 0) = \hat{\theta}(y)$ is bounded. Using (4.9) and arguments of Theorem 4.6, 4), one can show that if θ_1, θ_2 lie in the same positively oriented interval $(\hat{\theta}(y), \hat{\theta}(y) + \pi)$ or $(\hat{\theta}(y) + \pi, \hat{\theta}(y))$, then $|\theta_1(y, t) - \theta_2(y, t)|$ stays away from π for all $t \in \mathbb{R}$, and if $\theta_1 \in (\hat{\theta}(y), \hat{\theta}(y) + \pi)$ and $\theta_2 \in (\hat{\theta}(y) + \pi, \hat{\theta}(y))$, then $|\theta_1(y, t) - \theta_2(y, t)|$ stays away from 0 for all $t \in \mathbb{R}$.

PROPOSITION 4.7. *If for some $\hat{y} \in \hat{Y}$, $(\hat{\theta}(\hat{y}), \hat{y}) \notin M$, then card $\tilde{M} \cap \tilde{p}^{-1}(y) = 2$ for all $y \in Y_0$, that is, $p : (M, \mathbb{R}) \to (Y, \mathbb{R})$ is an almost automorphic extension.*

PROOF. By the above discussion, it is easy to see that if $\hat{y} \in \hat{Y}$ is such that $(\hat{\theta}(\hat{y}), \hat{y}) \notin M$, then the restriction of $P(\tilde{M})$ to $\tilde{M} \cap \tilde{p}^{-1}(\hat{y})$ is closed. It follows from arguments in the proof of Theorem 4.5 that card $\tilde{M} \cap \tilde{p}^{-1}(y_0) = 2$ $(y_0 \in Y_0)$, that is, card $M \cap p^{-1}(\hat{y}) = 1$, $(y_0 \in Y_0)$. □

Of course, we do not know whether the condition in the above proposition is valid in general. Suppose it is valid for some systems, then it is still a question whether $P(\tilde{M})$ is an equivalence relation, or whether $(\tilde{M}, \mathbb{R})$ is almost automorphic (note this is needed as far as an almost automorphic strong Perron transformation is considered).

In summary, the study of the conjecture of Johnson leads to three dynamical questions which may also be of general interests.

Let $p : (X,T) \to (Y,T)$ be a homomorphism of minimal transformation groups.

QUESTION 1. If (Y,T) is almost periodic and p is an almost N to 1 ($N \geq 2$) extension, when is (X,T) almost automorphic?

QUESTION 2. If (Y,T) is almost automorphic and p is an N to 1 ($N \geq 2$) extension, when is (X,T) almost automorphic?

QUESTION 3. If $P(Y)$ is an equivalence (closed) relation and p is an N to 1 ($N \geq 2$) extension, when is $P(X)$ an equivalence (closed) relation?

References

[A] J. Auslander, *Minimal Flows and Their Extensions*, North-Holland, Amsterdam, New York, Oxford, 1985.

[B] S. Bochner, *Curvature and Betti Numbers in Real and Complex Vector Bundles,*, Universitá e Politecnico di Toorino, Rendiconti del seminario matematico **15** (1955-1956).

[C] R. H. Cameron, *Almost Periodic Properties of Bounded Solutions of Linear Differential Equations with Almost Periodic Coefficients*, J. Math. Phys. **15** (1936).

[E] R. Ellis, *Locally Compact Transformation Groups*, Duke Math. J. **24** (1957).

[E1] ———, *Lectures on Topological Dynamics*, W.A. Benjamin, Inc., NY, 1969.

[EJ] R. Ellis and R. A. Johnson, *Topological Dynamics and Linear Differential Systems*, J. Diff. Eq. **44** (1982), 21-39.

[Fl] P. Flor, *Rhythmische Abbildungen Abelscher Gruppen II*, Zeit. Wahrsch. **7** (1967).

[F] H. Furstenberg, *The Structure of Distal Flows*, Amer. J. Math. **85** (1963).

[FW] H. Furstenberg and B. Weiss, *On Almost* 1 - 1 *Extensions*, Israel J. Math. **65** (1989).

[Ha] J. K. Hale, *Theory of Functional Differential Equations*, Springer-Verlag, New York Heidelberg Berlin, 1977.

[He] D. Henry, *Geometric Theory of Semi-linear Parabolic Equations*, Lecture Notes in Math., vol. 840, Springer-Verlag, Berlin, 1981.

[J] R. A. Johnson, *Minimal Functions with Unbounded Integral*, Israel J. Math. **31** (1978).

[J1] ———, *On a Floquet Theory for Almost Periodic Two-dimensional Linear Systems*, J. Diff. Eq. **37** (1980).

[J2] ———, *Linear Differential Systems with Interval Spectrum*, Proc. Amer. Math. Soc. **80** (1980).

[J3] ———, *Two Dimensional, Almost Periodic Linear Systems with Proximal and Recurrent Behavior*, Proc. Amer. Math. Soc. **82** (1981).

[J4] ———, *On Almost-Periodic Linear Differential Systems of Millionščikov and Vinograd*, J. Math. Anal. Appl. **85** (1982).

[J5] ———, *Hopf Bifurcation from Non-periodic Solutions of Differential Equations, I. Linear Theory*, J. Dyn. Diff. Eq. **1** (1989).

[JY] R. A. Johnson and Y. Yi, *Hopf Bifurcation from Non-periodic Solutions of Differential Equations, II. Nonlinear Theory*, J. Diff. Eq. **107** (1994).

[MP] N. G. Markley and M. E. Paul, *Almost Automorphic Symbolic Minimal Sets without Unique Ergodicity*, Israel J. of Math. **34** (1979).

[M] R. K. Miller, *Almost Periodic Differential Equations as Dynamical Systems with Applications to the Existence of a. p. Solutions*, J. Diff. Eq. **1** (1965).

[NO] S. Novo and R. Obaya, *An Ergodic Classification of Bidimensional Linear Systems*, J. Dyn. Diff. Eq. (to appear).

[R] A. Reich, *Prakompakte Gruppen und Fastperiodizitat*, Math. Z. **116** (1970).

[SS] R. J. Sacker and G. R. Sell, *Skew-product Flows, Finite Extensions of Minimal Transformation Groups and Almost Periodic Differential Equations*, Bull. Amer. Math. Soc. **79** (1973).

[SS1] ———, *Finite Extensions of Minimal Transformation Groups*, Trans. Amer. Math. Soc. **190** (1974).

[SS2] ———, *Lifting Properties in Skew-Product Flows with Applications to Differential Equations*, Memoirs Amer. Math. Soc. **11** (1977).

[SS3] ———, *A Spectral Theory for Linear Differential Systems*, J. Diff. Eq. **27** (1978).

[Se] J. Selgrade, *Isolated Invariant Sets for Flows on Vector Bundles*, Trans. Amer. Math. Soc. **203** (1975).

[S1] G. R. Sell, *Non-autonomous Differential Equations and Topological Dynamics I, II*, Trans. Amer. Math. Soc. **127** (1967).

[S2] ———, *Topological Dynamics and Ordinary Differential Equations*, Van Nostrand Reinhold Company, 1971.

[SY] W. Shen and Y. Yi, *Dynamics of Almost Periodic Scalar Parabolic Equations*, J. Diff. Eq. **122** (1995).

[SY1] ———, *Asymptotic Almost Periodicity of Scalar Parabolic Equations with Almost Periodic Time Dependence*, J. Diff. Eq. **122** (1995).

[SY2] ———, *On Minimal Sets of Scalar Parabolic Equations with Skew-product Structures*, Trans. of Amer. Math. Soc. **347** (1995), 4413-4431.

[SY3] ———, *Ergodicity of Minimal Sets in Scalar Parabolic Equations*, J. Dyn. Diff. Eq. **8** (1996).

[SY4] ———, *Almost Automorphic and Almost Periodic Dynamics in Skew-product Semi-flows, part II. Skew-product Semi-flows*, Memoirs of Amer. Math. Soc. (to appear).

[SY5] ———, *Almost Automorphic and Almost Periodic Dynamics in Skew-product Semi-flows, part III. Applications to Differential Equations*, Memoirs of Amer. Math. Soc. (to appear).

[SY6] ———, *Convergence in Almost Periodic Fisher and Kolmogorov Models*, Submitted.

[T] R. Terras, *On Almost Periodic and Almost Automorphic Differences of Functions*, Duke Math. J. **40** (1973).

[T1] ———, *Almost Automorphic Functions on Topological Groups*, Indiana U. Math. J. **21** (1972).

[V] W. A. Veech, *Almost automorphic functions on groups*, Amer. J. Math. **87** (1965).

[V1] ———, *Point-distal flows*, Amer. J. Math. **92** (1970).

[V2] ———, *Topological Dynamics*, Bull. Amer. Math. Soc. **83** (1977).

[Y] Y. Yi, *Almost Periodic and Almost Automorphic Dynamics in Skew-product Semi-flows, Part I. Almost Automorphy and Almost Periodicity*, Memoirs of Amer. Math. Soc. (to appear).

School of Mathematics, University of Minnesota, MPLS, MN 55455
E-mail address: sell@math.umn.edu

Department of Mathematics, Auburn University, Auburn, AL 36849
E-mail address: ws@math.auburn.edu

School of Mathematics, Georgia Institute of Technology, Atlanta, GA 30332
E-mail address: yi@math.gatech.edu

Contemporary Mathematics
Volume **215**, 1998

An Ergodic and Topological Approach to Almost Periodic Bidimensional Linear Systems

S. Novo and R. Obaya

ABSTRACT. An ergodic classification of bidimensional linear systems has recently been developed by the authors. In this paper this classification and the topological one introduced by R. Johnson are reviewed and compared in detail. Examples to illustrate the differences with the periodic case are given.

1. Introduction

The principal purpose of this review is to study the projective flow induced by a family of almost periodic bidimensional linear systems. This flow contains valuable information on the structure and qualitative behavior of the solutions of the systems, with applications to spectral problems.

The qualitative study of a flow consists in the characterization of its invariant and minimal subsets, task in which topological dynamics and ergodic theory techniques are often used. We will consider in detail the topological classification, given by Johnson in [**J3**], and the ergodic one obtained by Novo and Obaya in [**NO1**]. We also show the relation between both classifications as well as known examples to illustrate all the possible situations.

When all the solutions are bounded (*bounded orbits case*), topological arguments give precise information on the qualitative behavior of the solutions. The flow is either minimal or laminates into minimal subsets, k-covers of the base from which the dynamical properties are inherited. If there are solutions with exponential growth (positive Lyapunov exponent) and the equations have exponential dichotomy (i.e. there is a splitting of the linear bundle), then the projective flow admits two minimal subsets, each isomorphic to the base.

The remaining cases are more complicated and ergodic techniques must be introduced to study the measurable structure of the flow. This study is specially important in the cases in which the Floquet theory does not apply and the systems are not reducible to constant coefficients. We are interested in the study of properties which are satisfied for almost every trajectory. The idea is that a collective property is easier to recognize than a single one, although several times the behavior may differ in a set of null measure. In the absolutely continuous case, we state

---

1991 *Mathematics Subject Classification.* Primary 58F11, 54H20; Secondary 34C35.

Partially supported by Junta de Castilla y León under project VA57/94.

precise results concerning the existence and means of construction of absolutely continuous invariant measures. The relation with the ergodic components of the flow is also obtained. In the singular case, the classification of the ergodic measures is simpler and bring to mind the description obtained when the Lyapunov exponent is positive. From a topological viewpoint, in the unbounded case now considered, only one or two minimal subsets can exist. The case of two minimal sets provide almost automorphic extensions of the base. The study of the unique minimal set case is not complete, although the characterization of the set is given in some cases.

This paper is arranged as follows: Section 2 and 3 review some basic notions and well-known results in ergodic theory and topological dynamics, as well as some facts concerning the projective flow induced by a family of bidimensional linear systems. The definition and properties of disintegration of a measure which will be needed in this paper are also detailed in these preliminary sections.

Section 4 deals with some differences between the periodic and almost periodic cases. Section 5 is devoted to the absolutely continuous case; we show that a linear invariant measure always exists and we relate this fact to the ergodic structure of the real and complex projective flows. The bounded orbits case is considered in Section 6. After stating precisely the topological structure of the flow, we show that a linear invariant measure with continuous density function always exists. We finish this section summarizing the ergodic classification and relating it to the topological one. In Section 7, the classification of the ergodic measures in the singular case is derived, and finally in Section 8, after reviewing the main results of the topological classification for almost periodic linear systems, the relation with the ergodic one is given.

2. Preliminaries

In order to make the paper reasonably self-contained we recall some definitions and results more or less standard in ergodic theory and topological dynamics.

Let X be a compact metric space. A real *continuous flow* on X is defined by a continuous mapping $\Phi : \mathbb{R} \times X \to X$, $(t, x) \mapsto \Phi(t, x)$ satisfying:

(i) $\Phi_0 = \mathrm{Id}$,

(ii) $\Phi_{t+s} = \Phi_t \circ \Phi_s$,

where $\Phi_t(x) = \Phi(t, x)$ for all $x \in X$ and $t \in \mathbb{R}$.

The set $\{\Phi_t(x) \,/\, t \in \mathbb{R}\}$ is called the *orbit* or the *trajectory* of the point x. The set X is a disjoint union of trajectories. If (Z, Ψ) is another continuous flow, a *flow homomorphism* from (X, Φ) to (Z, Ψ) is a continuous mapping $f : X \to Z$ such that $f(\Phi_t(x)) = \Psi_t(f(x))$ for all $x \in X$ and $t \in \mathbb{R}$.

We say that a subset $X_1 \subset X$ is Φ invariant if $\Phi_t(X_1) = X_1$ for every $t \in \mathbb{R}$. A mapping $f : X \to \mathbb{R}$ is Φ invariant if it is constant along the trajectories, i.e., $f(\Phi_t(x)) = f(x)$ for all $x \in X$ and $t \in \mathbb{R}$. A subset $X_1 \subset X$ is called *minimal* if is compact, Φ invariant and the only nonempty compact Φ invariant subset of it is itself. Every compact and Φ invariant set contains a minimal subset; in particular it is easy to prove that a compact Φ invariant subset is minimal if and only if every trajectory is dense. We say that the continuous flow (X, Φ) is *recurrent* or *minimal* if X is minimal.

Let d be a metric on X. We say that the flow (X, Φ) is *distal* when, for each pair x, y of different elements of X, there is a $\delta > 0$ such that $d(\Phi_t(x), \Phi_t(y)) > \delta$ for every $t \in \mathbb{R}$, i.e., $\inf\{d(\Phi_t(x), \Phi_t(y)) \,/\, t \in \mathbb{R}\} = 0$ if and only if $x = y$. We say that the flow (X, Φ) is *almost periodic* when for every $\varepsilon > 0$ there is a $\delta > 0$ such that, if x, y are elements of X with $d(x, y) < \delta$ then $d(\Phi_t(x), \Phi_t(y)) < \varepsilon$ for every $t \in \mathbb{R}$. If (X, Φ) is almost periodic, it is distal. The converse is not true; even if (X, Φ) is minimal and distal, it need not to be almost periodic. We say that x and y form a *proximal pair* if $\inf\{d(\Phi_t(x), \Phi_t(y)) \,/\, t \in \mathbb{R}\} = 0$, otherwise the pair is said to be *distal.* For the basic properties on almost periodic and distal flows we refer the reader to Ellis [**E**] and Sacker-Sell [**SS1**].

Let $\pi : (X, \Phi) \to (Z, \Psi)$ be a surjective flow homomorphism and suppose (X, Φ) is minimal. We say that (X, Φ) is an *almost automorphic extension* (a.a. extension) of (Z, Ψ) if there is $z \in Z$ such that $card(\pi^{-1}(z)) = 1$. We say that (X, Φ) is a *proximal extension* of (Z, Ψ) if any x, $y \in X$ with $\pi(x) = \pi(y)$ form a proximal pair. An a.a. extension is a proximal extension (see Veech [**Ve**]).

A Borel measure on X will be a finite regular measure defined on the Borel sets. Let μ be a normalized Borel measure on X, μ is Φ invariant(or *invariant under* Φ) if $\mu(\Phi_t(X_1)) = \mu(X_1)$ for every Borel subset $X_1 \subset X$ and every $t \in \mathbb{R}$. It is Φ invariant (or *ergodic under* Φ) if, in addition, $\mu(X_1) = 0$ or $\mu(X_1) = 1$ for every Φ invariant subset $X_1 \subset X$. We will denote by $\mathcal{M}_{\text{inv}}(X, \Phi)$ the set of positive and normalized Φ invariant measures on X. The Krylov-Bogoliubov theorem (see Nemytskii -Stepanoff [**NS**]) asserts that $\mathcal{M}_{\text{inv}}(X, \Phi)$ is nonempty when X is a compact metric space. The extremal points of the convex and weakly compact set $\mathcal{M}_{\text{inv}}(X, \Phi)$ are the Φ ergodic measures, from which it is deduced that also the set of Φ ergodic measures is nonempty. The decomposition of the flow (X, Φ) into ergodic components and the construction and representation theorems of Φ invariant measures from Φ ergodic measures are well known (see Phelps [**P**] and Mañé [**Ma**]). We say that (X, Φ) is *uniquely ergodic* (u.e.) if it has a unique normalized invariant measure which is then necessarily ergodic. If (X, Φ) is u.e. it is not necessarily minimal; however, if (X, Φ) is u.e. and $\mu(U) > 0$ for every non -empty open set U, then (X, Φ) is minimal. An almost periodic and minimal flow (X, Φ) is always u.e.

The theorem stated below is one of the most important in ergodic theory. Its proof can be found, among many others texts on ergodic theory, in Cornfeld -Fomin -Sinai [**CFS**].

Birkhoff Ergodic Theorem. Let (X, Φ) be a continuous flow, μ a normalized invariant measure and $f \in L^1(X, \mu)$. Then for almost every $x \in X$ (with respect to μ) the following limits

$$\lim_{T\to\infty} \frac{1}{T} \int_0^T f(\Phi_t(x))\, dt\,, \quad \lim_{T\to\infty} \frac{1}{2T} \int_{-T}^T f(\Phi_t(x))\, dt$$

exist, are equal and define a Φ invariant function f^* on the Φ invariant subset of convergence points. Moreover, $f^* \in L^1(X, \mu)$ and

$$\int_X f\, d\mu = \int_X f^*\, d\mu.$$

In particular, if μ is Φ ergodic then f^* is constant almost everywhere (with respect to μ) and its value coincides with the integral $\int_X f\, d\mu$.

The limits which appear in the Birkhoff ergodic theorem are called *time means* or *means along the trajectory*. A consequence of this theorem, concerning positive functions, is the following result.

PROPOSITION 2.1. *Let (X, Φ) be a continu ous flow, μ a normalized invariant measure and $f \geq 0$ a positive mapping on X. Then for almost every $x \in X$ (with respect to μ) the following limits*

$$\lim_{T\to\infty} \frac{1}{T} \int_0^T f(\Phi_t(x))\, dt\,, \quad \lim_{T\to\infty} \frac{1}{2T} \int_{-T}^T f(\Phi_t(x))\, dt$$

exist, are equal and define a Φ invariant function f^* on the Φ invariant subset of convergence points. Moreover, one of the following statements is satisfied:

1. *$f,\ f^* \in L^1(X,\mu)$ and $\int_X f\, d\mu = \int_X f^*\, d\mu$, or*
2. *$\int_X f\, d\mu = \int_X f^*\, d\mu = \infty$.*

Next we recall the notions of cocycle and cohomology of cocycles which are of great importance in the study of continuous and measurable flows.

Let (Ω, Ξ) be a continuous flow on a compact metric space Ω. We will denote the $\mathbb{R}$ action by $\Xi(t,\xi) = \xi{\cdot}t$ for every $\xi \in \Omega$ and $t \in \mathbb{R}$. For an integer $n \geq 1$ let $GL(\mathbb{R}^n)$ denote the group of nonsingular $n \times n$ real matrices. A *cocycle* on Ω is a continuous mapping $U : \mathbb{R} \times \Omega \to GL(\mathbb{R}^n)$ which satisfies

$$U(s+t,\xi) = U(s,\xi{\cdot}t)\, U(t,\xi)$$

for all $\xi \in \Omega$ and $s,\ t \in \mathbb{R}$. We note that U is a cocycle on Ω if and only if

$$\Phi(t,\xi,\mathbf{x}) = (\xi{\cdot}t, U(t,\xi)\, \mathbf{x})$$

is a *linear skew -product flow* on $V_{\mathbb{R}} = \Omega \times \mathbb{R}^n$, i.e., the first component of the flow is independent of $\mathbf{x}$ and the second one is linear in $\mathbf{x}$. We say that the flow $(V_{\mathbb{R}}, \Phi)$ is *smooth* if the mapping $S(\xi) = (d/dt)U(t,\xi)|_{t=0}$ exists and is continuous. In this case the cocycle U is simply the fundamental matrix solution of

$$\mathbf{x}' = S(\xi{\cdot}t)\, \mathbf{x}$$

satisfying $U(0,\xi) = \mathrm{Id}$. This is the prototypical example of a cocycle that will be also called *smooth.*

Let U_1 and U_2 be two cocycles on Ω with range in $GL(\mathbb{R}^n)$. We say that U_1 and U_2 are *cohomologous* if there is a continuous mapping $P : \Omega \to GL(\mathbb{R}^n)$ which satisfies

$$U_1(t,\xi) = P(\xi{\cdot}t) U_2(t,\xi) P^{-1}(\xi)$$

for all $\xi \in \Omega$ and $t \in \mathbb{R}$. It is shown in Ellis -Johnson **[EJ]** that every cocycle is cohomologous to a smooth one, which implies that many properties can be directly proved for this kind of cocycles.

Throughout the paper we will state and prove the results for linear skew - product flows but the above relations between them and smooth cocycles show that everything may be rephrased in terms of cocycle theory.

3. Bidimensional Linear Skew-Product Flows

In this section, we collect some basic notions and fundamental results concerning the projective flow induced by a family of bidimensional linear systems, which will be frequently used in the rest of the paper.

Let us consider the real linear system

$$\mathbf{z}' = \begin{pmatrix} a_0(t) & b_0(t) \\ c_0(t) & -a_0(t) \end{pmatrix} \begin{pmatrix} z_1 \\ z_2 \end{pmatrix} = S_0(t)\mathbf{z} \tag{3.1}$$

whose matrix of coefficients is almost periodic and with trace zero.

Let Ω be the hull of S_0, namely, the closure of the set of mappings $t \to S_{0,t}$, with $S_{0,t}(s) = S_0(t+s)$, in the topology of uniform convergence on compact sets. Clearly $\Omega \subset C(\mathbb{R}, L(\mathbb{R}^2))$; we represent by $\xi \in \Omega$ any of its elements. The translation $\mathbb{R} \times \Omega \to \Omega$, $(t, \xi) \to \xi \cdot t$, with $\xi \cdot t(s) = \xi(t+s)$ defines a continuous flow Ξ on Ω. From the almost periodicity of S_0 it can be deduced that the flow (Ω, Ξ) is almost periodic and minimal, and consequently also u.e. and distal. Actually, Ω is a compact group and its Haar measure m_0 is the unique Ξ invariant (hence Ξ ergodic) measure.

Let $S : \Omega \to L(\mathbb{R}^2)$ be the operator which takes ξ to $\xi(0)$. Consider the family of linear systems

$$\mathbf{z}' = \begin{pmatrix} a(\xi\cdot t) & b(\xi\cdot t) \\ c(\xi\cdot t) & -a(\xi\cdot t) \end{pmatrix} \mathbf{z} = S(\xi\cdot t)\,\mathbf{z}\,, \quad \xi \in \Omega \tag{3.2}$$

and notice that, if $\xi = S_0$, system (3.2) coincides with system (3.1) because $S(\xi{\cdot}t) = \xi(t) = S_0(t)$.

Equations 3.2 induce a linear skew -product flow in the linear bundle $V_{\mathbb{C}} = \Omega \times \mathbb{C}^2$ ($V_{\mathbb{R}} = \Omega \times \mathbb{R}^2$). This flow takes $(t, \xi, \mathbf{z}_0)$ to $(\xi\cdot t, \mathbf{z}(t, \xi, \mathbf{z}_0))$ where $\mathbf{z}(t, \xi, \mathbf{z}_0)$ satisfies the equation defined by (3.2) along the trajectory passing through ξ with the initial data $\mathbf{z}(0, \xi, \mathbf{z}_0) = \mathbf{z}_0$. By linearity on the fiber, the application $\Pi : V_{\mathbb{C}} \to K_{\mathbb{C}} = \Omega \times P^1(\mathbb{C})$, $(\xi, \mathbf{z}) \mapsto (\xi, z_2/z_1)$, transports this flow to the projective bundle. The symbol Φ will represent the flow application in any invariant subset of $K_{\mathbb{C}}$ that is considered.

As usual, $P^1(\mathbb{C})$ represents the space of complex lines through the origin in $\mathbb{C}^2$; analogously $P^1(\mathbb{R})$ represents the space of real lines through the origin in $\mathbb{R}^2$. We can identify $P^1(\mathbb{C})$ with the Riemann sphere and consider $P^1(\mathbb{R})$ as a great circle of $\mathbb{S}^2$; also, we may visualize $P^1(\mathbb{C})$ as the extended complex plane $\hat{\mathbb{C}} = \mathbb{C} \cup \{\infty\}$ by means of the stereographic projection, which makes $P^1(\mathbb{R}) = \mathbb{R} \cup \{\infty\}$. Taking the complex coordinate $Z = z_2/z_1$ in (3.2), we obtain the Riccati equations

$$Z' = c(\xi\cdot t) - 2a(\xi\cdot t)Z - b(\xi\cdot t)Z^2. \tag{3.3}$$

Let $Z(t, \xi, Z_0)$ be the solution of the equation (3.3) with the initial data $Z(0, \xi, Z_0) = Z_0$: then $\Phi(t, \xi, Z_0) = (\xi\cdot t, Z(t, \xi, Z_0))$ defines the solution curve of the equation (3.3) on $K_{\mathbb{C}}$.

We will denote by $\mathcal{M}_{m_0}(K_{\mathbb{C}})$ the set of Borel measures on $K_{\mathbb{C}}$ projecting into m_0, and by $\mathcal{M}_{\mathrm{inv}}(K_{\mathbb{C}}, \Phi)$ the subset of measures from $\mathcal{M}_{m_0}(K_{\mathbb{C}})$ which are Φ invariant. We shall consider, unless otherwise indicated, all the Φ invariant measures to be positive and normalized.

It is obvious that $K_{\mathbb{R}} = \Omega \times P^1(\mathbb{R})$ is a closed and Φ invariant subset of $K_{\mathbb{C}}$. Writing the real solutions of (3.2) in polar symplectic coordinates $\varphi = \cot^{-1}(x_2/x_1)$, $\rho = (x_1^2 + x_2^2)/2$ we obtain

$$\begin{aligned} \varphi' &= f(\xi\cdot t, \varphi) = b(\xi\cdot t)\cos^2\varphi - c(\xi\cdot t)\sin^2\varphi + 2a(\xi\cdot t)\sin\varphi\cos\varphi \\ &= \frac{1}{2}(b(\xi\cdot t) - c(\xi\cdot t)) + a(\xi\cdot t)\sin 2\varphi + \frac{1}{2}(b(\xi\cdot t) + c(\xi\cdot t))\cos 2\varphi, \\ \rho' &= -\frac{\partial f}{\partial\varphi}(\xi\cdot t, \varphi)\rho = (-2a(\xi\cdot t)\cos 2\varphi + (b(\xi\cdot t) + c(\xi\cdot t))\sin 2\varphi)\rho. \end{aligned} \tag{3.4}$$

We denote by $\varphi(t,\xi,\varphi_0)$ both the solution of (3.4) as a function on $\mathbb{R}$ or on $P^1(\mathbb{R})$ with initial condition $\varphi(0,\xi,\varphi_0)=\varphi_0$: the context will give, in each case, the exact meaning of the symbol. The mapping $\Phi(t,\xi,\varphi_0)=(\xi{\cdot}t,\varphi(t,\xi,\varphi_0))$ is precisely the restriction of the flow to $K_{\mathbb{R}}$. The relation $X=\cot\varphi$ gives us the change between the systems of coordinates that we have introduced. The real solution of (3.2) along the trajectory that passes through ξ with initial data $\mathbf{x}(0,\xi,\varphi)=(\sin\varphi,\cos\varphi)^t$ will be denoted by $\mathbf{x}(t,\xi,\varphi)$.

We will use the symbol r_1 to denote the normalized Lebesgue measure on $P^1(\mathbb{R})$ and $m_1=m_0\otimes r_1$ for the complete product measure on $K_{\mathbb{R}}$. Analogously to the complex case mentioned before, we will denote by $\mathcal{M}_{m_0}(K_{\mathbb{R}})$ the set of Borel measures on $K_{\mathbb{R}}$ which project into m_0 and by $\mathcal{M}_{\rm inv}(K_{\mathbb{R}},\Phi)$ the subset of measures from $\mathcal{M}_{m_0}(K_{\mathbb{R}})$ which are Φ invariant.

We will consider separately two cases in our study: the *absolutely continuous case* when there exists a Φ invariant measure on $K_{\mathbb{R}}$ absolutely continuous with respect to m_1, and the *singular case* when every Φ invariant measure is singular with respect to m_1.

Let $U(t,\xi)$ be the fundamental matrix of the equations (3.2) with $U(0,\xi)=\mathrm{Id}$. The *Lyapunov exponent* is defined as

$$\gamma=\lim_{t\to\infty}\frac{1}{t}\log\|U(t,\xi)\|$$

where the limit exists and takes the same real value for almost every ξ with respect to m_0 (see Craig -Simon [**CS**]).

We denote by $\mathbb{Y}$ either $\mathbb{R}$ or $\mathbb{C}$ to consider simultaneously both projective spaces, and $K_{\mathbb{Y}}=\Omega\times P^1(\mathbb{Y})$. We always consider the set of Borel measures $\mathcal{M}(P^1(\mathbb{Y}))$ with the weak topology.

We recall the notion and properties of disintegration of a measure that will be of special importance in our paper (see Bourbaki [**B**] and Ionescu -Tulcea [**IT**]). Let μ be a Borel measure on $K_{\mathbb{Y}}$ and $(\mu_\xi)_{\xi\in\Omega}$ a collection of Borel measures on $P^1(\mathbb{Y})$. We say that μ *disintegrates* into $(\mu_\xi)_{\xi\in\Omega}$ with respect to m_0 if

(i) the family $(\mu_\xi)_{\xi\in\Omega}$ is m_0 -*adequate*, that is, for every $\varepsilon>0$ there is a compact set $\Omega_\varepsilon\subset\Omega$ with $m_0(\Omega_\varepsilon)>1-\varepsilon$, such that the map $\Omega_\varepsilon\to\mathcal{M}(P^1(\mathbb{Y}))$, $\xi\to\mu_\xi$ is continuous, and

(ii) for every $g\in C(K_{\mathbb{Y}})$ one has

$$\int_{K_{\mathbb{Y}}} g\,d\mu=\int_\Omega\left[\int_{P^1(\mathbb{Y})} g(\xi,\varphi)\,d\mu_\xi(\varphi)\right]dm_0. \tag{3.5}$$

Notice that when the collection of measures $(\mu_\xi)_{\xi\in\Omega}$ is m_0 adequate, then relation (3.5) defines a Borel measure μ on $K_{\mathbb{Y}}$ which projects into m_0 and disintegrates into $(\mu_\xi)_{\xi\in\Omega}$. The following converse result is also known. For any Borel measure μ on $K_{\mathbb{Y}}$ projecting into m_0, there exists a disintegration $(\mu_\xi)_{\xi\in\Omega}$ of μ with respect to m_0 such that for every $g\in L^1(K_{\mathbb{Y}},\mu)$ the map $\xi\to\int_{P^1(\mathbb{Y})}g(\xi,\varphi)\,d\mu_\xi(\varphi)$ is defined almost everywhere, is m_0 -integrable and (3.5) is satisfied.

Let μ be a Borel measure on $K_{\mathbb{Y}}$ which disintegrates into $(\mu_\xi)_{\xi\in\Omega}$ with respect to m_0. We say that μ is *absolutely continuous* (resp. *singular*) if μ_ξ is absolutely continuous (resp. singular) with respect to the Lebesgue measure r_1 for almost every $\xi\in\Omega$. Let μ be a singular Borel measure on $K_{\mathbb{R}}$ which disintegrates into $(\mu_\xi)_{\xi\in\Omega}$

with respect to m_0. We say that μ is *singular continuous* (resp. *purely discontinuous*) if μ_ξ is singular continuous (resp. purely discontinuous) with respect to r_1 for almost every $\xi \in \Omega$. We will denote by $\mathcal{M}^{\text{a.c.}}_{m_0}(K_\mathbb{R})$, $\mathcal{M}^{\text{s.c.}}_{m_0}(K_\mathbb{R})$ and $\mathcal{M}^{\text{p.d.}}_{m_0}(K_\mathbb{R})$ the above sets of measures on $K_\mathbb{R}$. It is easy to see that each Φ ergodic measure belongs to one of these sets. Moreover, $\mathcal{M}_{m_0}(K_\mathbb{R}) = \mathcal{M}^{\text{a.c.}}_{m_0}(K_\mathbb{R}) \oplus \mathcal{M}^{\text{s.c.}}_{m_0}(K_\mathbb{R}) \oplus \mathcal{M}^{\text{p.d.}}_{m_0}(K_\mathbb{R})$.

If $\mu \in \mathcal{M}_{m_0}(K_\mathbb{R})$ and disintegrates into $(\mu_\xi)_{\xi\in\Omega}$, then a straightforward calculation shows that μ is Φ invariant if and only if, for each $t \in \mathbb{R}$, the relation

$$\mu_\xi([\varphi_1, \varphi_2]) = \mu_{\xi\cdot t}([\varphi_t(\xi, \varphi_1), \varphi_t(\xi, \varphi_2)]) \tag{3.6}$$

is satisfied for every φ_1, $\varphi_2 \in P^1(\mathbb{R})$ and for almost every $\xi \in \Omega$ (see Furstenberg [**F**]). By modifying the definition of μ_ξ on a subset of null measure, relation (3.6) can be satisfied for all $\xi \in \Omega$ and φ_1, $\varphi_2 \in P^1(\mathbb{R})$.

If $\mu \in \mathcal{M}_{\text{inv}}(K_\mathbb{R}, \Phi)$ and disintegrates into $(\mu_\xi)_{\xi\in\Omega}$, we deduce from (3.6) that $\mu_\xi(P^1(\mathbb{R})) = 1$ for almost every $\xi \in \Omega$. This means that normalized Φ invariant measures disintegrate into normalized ones.

Let $\mathcal{A}_0$ be the completion of the σ -algebra of the Borel sets with respect to m_0. We always consider $\mathcal{P}(K_\mathbb{C})$ with the Hausdorff metric. Let M be a Φ invariant subset of $K_\mathbb{C}$, $\pi_M : M \to \Omega$ the projection on the base. We say that M is a *measurable k-sheet* if there is a Ξ -invariant subset $\Omega_0 \subset \Omega$ with $m_0(\Omega_0) = 1$ such that

(i) card $(\pi_M^{-1}(\xi)) = k$ for every $\xi \in \Omega_0$, and
(ii) the function $\Omega_0 \to \mathcal{P}(K_\mathbb{C})$, $\xi \mapsto \pi_M^{-1}(\xi) = \{(\xi, Z_1(\xi)), \dots, (\xi, Z_k(\xi))\}$ is $\mathcal{A}_0$ -measurable.

For $\xi \in \Omega_0$ we consider a purely discontinuous measure μ_ξ with $\mu_\xi(Z_j(\xi)) = 1/k$ for $j = 1, \dots, k$. The previous condition (ii) means exactly that the family $(\mu_\xi)_{\xi\in\Omega}$ is m_0 -adequate. Thus the flow $(K_\mathbb{C}, \Phi)$ possesses an invariant measure $\mu \in \mathcal{M}^{\text{p.d.}}_{m_0}(K_\mathbb{C})$, satisfying $\mu(M) = 1$, defined by

$$\int_{K_\mathbb{C}} g\, d\mu = \frac{1}{k}\sum_{j=1}^{k} \int_\Omega g(\xi, Z_j(\xi))\, dm_0$$

for every $g \in C(K_\mathbb{C})$. Obviously μ disintegrates into $(\mu_\xi)_{\xi\in\Omega}$.

We say that M is an *ergodic k-sheet* if μ is the only element of $\mathcal{M}_{\text{inv}}(K_\mathbb{C}, \Phi)$ which is concentrated in M. If M is a compact subset of $K_\mathbb{C}$ and card$(\pi_M^{-1}(\xi)) = k$ for every $\xi \in \Omega$, we say that M is a *closed k-sheet.* For $k = 1$ we simply refer to M as a *measurable*, *ergodic*, or *closed* sheet.

Most of the times the set Ω_0 appears in a natural way along the process of construction of M. It is even possible to modify M in such a way that it remains Φ invariant and card$(\pi_M^{-1}(\xi)) = k$ for every $\xi \in \Omega$. For this reason, we sometimes will consider $\Omega_0 = \Omega$ if we do not have previous information about the exact definition of M. However, M is not necessarily a Borel subset of $K_\mathbb{C}$. Proposition 2.3.9 of Veech [**Ve**] asserts that if a minimal subset K contains a Borel ergodic sheet defined everywhere, then it is an almost automorphic extension of Ω. From this we conclude that none of the ergodic sheets given in the example of Section 4, in which $K_\mathbb{R}$ is minimal, can be extended to the whole base Ω as a Borel ergodic sheet.

The following theorem of Furstenberg [**F**] points out the importance of the measurable k-sheets.

THEOREM 3.1. *Let us suppose that* $(K_{\mathbb{R}}, \Phi)$ *admits an invariant measure* $\mu \in \mathcal{M}^{\text{a.c.}}_{m_0}(K_{\mathbb{R}}) \cup \mathcal{M}^{\text{s.c.}}_{m_0}(K_{\mathbb{R}})$.

(i) *If* μ *is* Φ *ergodic, then it is the only* Φ *ergodic measure on* $K_{\mathbb{R}}$.
(ii) *If* μ *is not* Φ *ergodic, then there is* $k \in \mathbb{N}$ *such that every* Φ *ergodic measure on* $K_{\mathbb{R}}$ *is concentrated in an ergodic* k*-sheet.*

We say that the systems (3.2) admit *exponential dichotomy* (see [**SS2**]) if there is a splitting $V_{\mathbb{C}} = V^+ \oplus V^-$, where V^+ and V^- are Φ invariant one-dimensional closed subbundles of $V_{\mathbb{C}}$, and constants C and β such that

$$\|\Phi(t,\xi,\mathbf{z})\| \le C e^{-\beta t}\|\mathbf{z}\| \;\text{ for } t \ge 0, \quad (\xi,\mathbf{z}) \in V^+ ,$$
$$\|\Phi(t,\xi,\mathbf{z})\| \le C e^{\beta t}\|\mathbf{z}\| \;\text{ for } t \le 0, \quad (\xi,\mathbf{z}) \in V^- .$$

In this case the Lyapunov exponent is positive and the projections of the above subbundles $\Pi(V^+)$ and $\Pi(V^-)$ define two closed ergodic sheets in which the singular Φ ergodic measures are concentrated.

4. Some Differences Between the Periodic and Almost Periodic Cases

In the periodic case, Floquet theory allows us to describe easily the measurable and topological structure of the projective flow and both of these structures coincide. However this is not so in the general almost periodic case.

Let $S_0(t)$, the matrix of coefficients of system (3.1), be periodic of period 1. Thus Ω, the hull of S_0, is a circle and the projective bundle $K_{\mathbb{R}}$ a 2 -torus. The Floquet theorem allows us to take (3.1) into a constant coefficients system and the same may be said, by translation, of the whole family of systems (3.2) defined in the hull Ω. We can identify Ω with the quotient space $\mathbb{R}/\mathbb{Z}$ and the continuous flow on Ω with the translation $\Xi(\xi,t) = \xi{\cdot}t = \xi + t$ for every $\xi \in \Omega$ and $t \in \mathbb{R}$. Then we can write (3.2) in the form

$$\mathbf{z}' = S_0(\xi{\cdot}t)\,\mathbf{z}\,, \quad \xi \in \Omega \tag{4.1}$$

where (3.1) is obtained by taking $\xi = 0$. Three cases may be distinguished:

(i) *All the solutions are bounded.* Then there is a continuous 1-periodic matrix $P(t)$ such that the periodic family of transformations $\mathbf{z} = P(\xi{\cdot}t)\,\mathbf{y}$ takes the systems (4.1) into the constant coefficients one $\mathbf{y} = \begin{pmatrix} 0 & b \\ -b & 0 \end{pmatrix} \mathbf{y}$ for every $\xi \in \Omega$. Writing down the real solutions in polar symplectic coordinates we get $\varphi' = b$, $\rho' = 0$. We will denote the transformed flow by Ψ. We are in the absolutely continuous case because m_1 is a Ψ invariant measure. The study of the transformed systems depends now on the value of the real constant b. We obtain three subcases:

(i.1) When $b \in \pi\mathbb{Z}$ the projective bundle $K_{\mathbb{R}}$ laminates into minimal subsets which are closed ergodic sheets. In fact, in the new flow $(K_{\mathbb{R}}, \Psi)$ the closed ergodic sheets are $N_\alpha = \{(\xi, b\,\xi + \alpha) \,/\, \xi \in \Omega\}$ for $\alpha \in P^1(\mathbb{R})$.
(i.2) If $b \in \pi(\mathbb{Q} - \mathbb{Z})$ then $K_{\mathbb{R}}$ laminates into minimal subsets which are closed ergodic k-sheets where $k > 1$ is the minimum integer number such that $bk \in \pi\mathbb{Z}$. The ergodic k-sheets in $(K_{\mathbb{R}}, \Psi)$ are $N_\alpha = \{(\xi, \varphi_j^\alpha(\xi)) \,/\, j = 1, 2, \cdots, k,\; \xi \in \Omega\}$ with $\varphi_j^\alpha(\xi) = b\,\xi + \alpha + (j-1)\pi/k$.

(i.3) If $b \in \pi(\mathbb{R}-\mathbb{Q})$ then the trajectories are dense and the flow $(K_{\mathbb{R}}, \Phi)$ is u.e. and minimal.

(ii) *There are unbounded solutions which do not grow exponentially.* Then there is a family of 1 or 2-periodic changes of variables $\mathbf{z} = P(\widetilde{\xi}\cdot t)\,\mathbf{y}$ defined in the new base $\widetilde{\Omega}$ (a 1 or 2 -cover of Ω) which takes $\mathbf{z}' = S_0(\widetilde{\xi}\cdot t)\,\mathbf{z}$ with $\widetilde{\xi} \in \widetilde{\Omega}$ into $\mathbf{y}' = \begin{pmatrix} 0 & 1 \\ 0 & 0 \end{pmatrix} \mathbf{y}$, which in polar symplectic coordinates takes the form $\varphi' = \cos^2\varphi$, $\rho' = \sin 2\varphi\,\rho$. The transformed flow $(\widetilde{K}_{\mathbb{R}} = \widetilde{\Omega} \times P^1(\mathbb{R}), \Psi)$ is u.e. and the unique minimal subset is the closed ergodic sheet $N = \{(\widetilde{\xi}, \pi/2) \,/\, \widetilde{\xi} \in \widetilde{\Omega}\}$, from which we can deduce that $(K_{\mathbb{R}}, \Phi)$ is also u.e. with a unique singular measure concentrated in a closed ergodic sheet that is the unique minimal subset.

(iii) *Some solutions grow exponentially.* After a 1 or 2-periodic change of variables (4.1) becomes $\mathbf{y}' = \begin{pmatrix} \beta & 0 \\ 0 & -\beta \end{pmatrix} \mathbf{y}$ for some $\beta \neq 0$. In this case we obtain two closed ergodic sheets which are the minimal subsets in which the singular Φ invariant measures are concentrated. This is the only case in which the Lyapunov exponent is positive.

Notice that all the ergodic sheets appearing in the periodic case are closed, which also means that all the Φ invariant one-dimensional subbundles of $V_{\mathbb{R}}$ are continuous. In the quasiperiodic case the possibility of measurable subbundles must be taken into account.

In [**J1**] Johnson gives an example of a quasiperiodic linear system with bounded solutions whose projective flow foliates into measurable, non-closed ergodic sheets. He derives, by means of a simple modification of a previous example due to Furstenberg [**F**], a non-continuous function $R \in L^2(\Omega, m_0)$ (in fact, $e^{\pm R} \in L^r(\Omega, m_0)$ for all $r < \infty$ as shown in [**NO1**]) and an analytic function b on Ω such that

$$R(\xi\cdot t) - R(\xi) = \int_0^t b(\xi\cdot s)\,ds\,,$$

that is, R is a solution along the flow of the differential equation $R' = b(\xi\cdot t)$.

First of all we will recall the definition and some of the properties of the functions R and b introduced in [**J1**]. Let Ω be the 2-torus $\mathbb{T}^2$ and m_0 the Lebesgue measure on Ω. The irrational (Kronecker) flow (Ω, Ξ) is defined by $\Xi_t(\theta_1, \theta_2) = (\theta_1 + \alpha t, \theta_2 + t)$ for $(\theta_1, \theta_2) \in \Omega$, where α is the irrational number

$$\alpha = \sum_{k=1}^{\infty} 2^{-\nu_k}\,, \qquad \text{with } \nu_1 = 1\,, \quad \text{and } \nu_{k+1} = 2^{\nu_k} + \nu_k + 1 \quad \text{for } k > 1.$$

Let $n_k = 2^{\nu_k}$, $m_k = -[n_k\alpha]$ (where $[x]$ denotes the largest integer not exceeding x) and $n_{-k} = n_k$, $m_{-k} = m_k$. The functions R and b are defined by

$$R(\theta_1, \theta_2) = \sum_{k\neq 0} \frac{1}{|k|} e^{2\pi i(n_k\theta_1 + m_k\theta_2)}\,,$$

$$b(\theta_1, \theta_2) = \sum_{k\neq 0} \frac{1}{|k|} 2\pi i(n_k\alpha + m_k) e^{2\pi i(n_k\theta_1 + m_k\theta_2)}\,,$$

for each $(\theta_1, \theta_2) \in \Omega$. Then R is a measurable, non-continuous function defined almost everywhere on Ω and b is analytic on Ω. Although b has mean value zero, the function $\int_0^t b(\xi\cdot s)\,ds$ is unbounded (i.e. not almost periodic) for every $\xi \in \Omega$. We consider the family of systems

$$\mathbf{z}' = \begin{bmatrix} 0 & b(\xi\cdot t) \\ -b(\xi\cdot t) & 0 \end{bmatrix} \mathbf{z}, \qquad \xi \in \Omega\,. \tag{4.2}$$

Let us denote by $(K_{\mathbb{R}}, \Phi)$ the skew-product flow induced by them. It is known that m_1 is a Φ invariant measure. If we write down the real solutions of (4.2) in polar symplectic coordinates we get $\varphi' = b(\xi\cdot t)$, $\rho' = 0$, which implies that the solutions are bounded, and the sets $M_\alpha = \{(\xi, R(\xi) + \alpha) \,/\, \xi \in \Omega\}$, for $\alpha \in P^1(\mathbb{R})$, provide a partition of the real projective bundle $K_{\mathbb{R}}$ into non-closed ergodic sheets. On the other hand, $(K_{\mathbb{R}}, \Phi)$ is minimal and we are thus led to an example in which the measurable and topological structures do not agree.

5. Invariant Measures in the Absolutely Continuous Case

We start this section by recalling most of the results of Obaya-Paramio [**OP**] and Novo-Obaya [**NO1**] concerning the existence and means of construction of absolutely continuous invariant measures. In these papers the so called *linear invariant measures* are introduced; these form a class of absolutely continuous invariant measures which are directly associated to the ergodic sheets of the complex bundle.

As in Section 3, we consider the projective flows $(K_{\mathbb{R}}, \Phi)$ and $(K_{\mathbb{C}}, \Phi)$ induced by the family of systems (3.2). The next result, proved in [**OP**], characterizes the density function of the Φ invariant measures which are absolutely continuous with respect to m_1.

PROPOSITION 5.1. *Let $p \in L^1(K_{\mathbb{R}}, m_1)$ be a positive function. The following statements are equivalent:*

(i) *The measure $d\mu = p\,dm_1$ is Φ invariant.*

(ii) *p satisfies the functional equation*

$$p(\Phi_t(\xi,\varphi)) = p(\xi,\varphi)\exp\Big\{-\int_0^t \frac{\partial f}{\partial \varphi}(\Phi_s(\xi,\varphi))ds\Big\}. \tag{5.1}$$

Let μ be an absolutely continuous Φ invariant measure with $d\mu = p\,dm_1$. We consider $D = \{(\xi,\varphi) \in K_{\mathbb{R}} \,/\, p(\xi,\varphi) > 0\}$ and $q = (1/p)\,\chi_D$. Then q is a non-negative solution of the functional equation

$$q(\Phi_t(\xi,\varphi)) = q(\xi,\varphi)\exp\Big\{\int_0^t \frac{\partial f}{\partial \varphi}(\Phi_s(\xi,\varphi))\,ds\Big\}. \tag{5.2}$$

DEFINITION 5.2. Let q be a measurable solution of the equation (5.2). We say that q is *quadratic on the fiber* if there exist measurable functions A, B, $C : \Omega \to \mathbb{R}$ and a Ξ -invariant subset Ω_0 with $m_0(\Omega_0) = 1$ such that

$$q(\xi,\varphi) = A(\xi)\cos^2\varphi + B(\xi)\sin^2\varphi + 2C(\xi)\sin\varphi\cos\varphi$$

for every $(\xi,\varphi) \in \Omega_0 \times P^1(\mathbb{R})$.

DEFINITION 5.3. Let μ be a Φ invariant measure with $d\mu = (1/q)\,dm_1$. We say that μ is a *linear Φ invariant measure* if q is a quadratic on the fiber solution of the equation (5.2).

The existence of Φ invariant measures absolutely continuous with respect to m_1 depends on the behavior of the families of functions $\{p_T\}_{T>0}$ and $\{q_T\}_{T>0}$ defined, for each $T>0$, by the expressions

$$\begin{aligned} p_T(\xi,\varphi) &= \frac{1}{2T}\int_{-T}^{T} \exp\left\{\int_0^t \frac{\partial f}{\partial \varphi}(\Phi_s(\xi,\varphi))\,ds\right\}, \\ q_T(\xi,\varphi) &= \frac{1}{2T}\int_{-T}^{T} \exp\left\{-r\int_0^t \frac{\partial f}{\partial \varphi}(\Phi_s(\xi,\varphi))\,ds\right\}dt \\ &= \frac{1}{2T}\int_{-T}^{T} |\mathbf{x}(t,\xi,\varphi)|^{2r}\,dt\,, \end{aligned}$$

where r is a fixed real positive constant. It is easy to show that for every measurable subset $A \subset K_{\mathbb{R}}$ the following relation is satisfied

$$\int p_T\,\chi_A\,dm_1 = \frac{1}{2T}\int_{-T}^{T} m_1(\Phi_t(A))\,dt\,. \tag{5.3}$$

THEOREM 5.4. *The following facts are equivalent:*

(i) *$(K_{\mathbb{R}},\Phi)$ admits an invariant measure absolutely continuous with respect to m_1 (resp. equivalent to m_1).*
(ii) *There exists a Φ invariant subset $D \subset K_{\mathbb{R}}$ with $m_1(D)>0$ (resp. $m_1(D)=1$), such that the limit $p(\xi,\varphi)=\lim_{T\to+\infty} p_T(\xi,\varphi)$ exists and is a positive real number for every $(\xi,\varphi)\in D$.*

Moreover, under these conditions, the measure $d\mu = (1/m_1(D))\,p\,\chi_D\,dm_1$ is normalized and Φ invariant.

PROOF. (ii) $\Rightarrow$ (i). First of all, it is easy to show that for every $0<l<T$ we have

$$\frac{T-l}{T}p_{T-l}(\xi,\varphi) \le p_T(\Phi_l(\xi,\varphi))\exp\left\{\int_0^l \frac{\partial f}{\partial \varphi}(\Phi_s(\xi,\varphi))\,ds\right\} \le \frac{T+l}{T}p_{T+l}(\xi,\varphi)\,.$$

Taking limits as $T\to\infty$ we obtain

$$p(\Phi_l(\xi,\varphi)) = p(\xi,\varphi)\exp\left\{-\int_0^l \frac{\partial f}{\partial \varphi}(\Phi_s(\xi,\varphi))ds\right\}$$

for every $(\xi,\varphi)\in D$. Defining p as zero in $K_{\mathbb{R}}-D$ we get a solution of (5.1). Next we have to check that $p\,\chi_D \in L^1(K_{\mathbb{R}},m_1)$ which is a consequence of Fatou's lemma because

$$\int p\,\chi_D\,dm_1 = \int \liminf_{T\to\infty} p_T\,\chi_D\,dm_1 \le \liminf_{T\to\infty}\int p_T\,\chi_D\,dm_1\,,$$

and since from (5.3) it is obtained that $\int p_T\,\chi_A\,dm_1 = m_1(A)$ for every $T>0$ and every Φ invariant subset A, we conclude that $\lambda = \int p\,\chi_D\,dm_1 \le m_1(D)$. Therefore, from Proposition 5.1 it is deduced that $d\mu = (1/\lambda)\,p\,\chi_D\,dm_1$ is a normalized Φ invariant measure equivalent to m_1 in D.

We will show that the family $\{p_T\}_{T>0}$ is uniformly integrable. As a matter of fact, for every $\varepsilon > 0$ there exists $\delta_1 > 0$ such that, if $A \subset D$ is a measurable set with $\mu(A) < \delta_1$, then $m_1(A) < \varepsilon$. In an analogous way there exists $\delta_2 > 0$ such that, if $A \subset D$ is a measurable set with $m_1(A) < \delta_2$, then $\mu(A) < \delta_1$. Hence, if $A \subset D$ and $m_1(A) < \delta_2$, then $\mu(A) = \mu(\Phi_t(A)) < \delta_1$ and finally $m_1(\Phi_t(A)) < \varepsilon$ for every $t \in \mathbb{R}$. From this and (5.3) we obtain that $\int p_T\, \chi_A\, dm_1 < \varepsilon$ for every $T \in \mathbb{R}$. We may now apply Vitali's theorem to get

$$0 < \lambda = \int p\, \chi_D\, dm_1 = \lim_{T\to\infty} \int p_T\, \chi_D\, dm_1 = m_1(D)$$

which completes this part of the proof.

(i) $\Rightarrow$ (ii). We denote by $d\mu = \widetilde{p}\, dm_1$ the normalized Φ invariant measure and let $C = \{(\xi,\varphi) \in K_{\mathbb{R}} \,/\, \widetilde{p}(\xi,\varphi) > 0\}$. It is obvious that $(1/\widetilde{p})\, \chi_C \in L^1(K_{\mathbb{R}},\mu)$. Birkhoff's ergodic theorem guarantees the existence of a Φ invariant function $p^* \in L^1(K_{\mathbb{R}},\mu)$ such that

$$p^*(\xi,\varphi) = \lim_{T\to\infty} \frac{1}{2T}\int_{-T}^{T} \frac{1}{\widetilde{p}}\, \chi_C(\Phi_t(\xi,\varphi))\, dt = \lim_{T\to\infty} \frac{p_T(\xi,\varphi)}{\widetilde{p}(\xi,\varphi)}\, \chi_C(\xi,\varphi)$$

almost everywhere and $\int p^*\, d\mu = \int (1/\widetilde{p})\, \chi_C\, d\mu$, i.e., $\int p^* \widetilde{p}\, \chi_C\, dm_1 = m_1(C)$. In fact, this last equality holds in every Φ invariant subset of C. This implies that for almost every $(\xi,\varphi) \in K_{\mathbb{R}}$ with respect to μ

$$\lim_{T\to\infty} p_T(\xi,\varphi)\, \chi_C(\xi,\varphi) = p^*(\xi,\varphi)\widetilde{p}(\xi,\varphi) = p(\xi,\varphi)\,.$$

Finally, in the set $D \subset C$ of all the points of convergence where p does not vanish

$$m_1(D) = \int p^*\widetilde{p}\, \chi_D\, dm_1 = \int p^*\widetilde{p}\, \chi_C\, dm_1 = m_1(C) > 0$$

which proves the claim. □

THEOREM 5.5. *The following facts are equivalent:*

(i) $(K_{\mathbb{R}},\Phi)$ *admits an absolutely continuous invariant measure* $d\mu = p\, dm_1$ *with* $p \in L^{1+r}(K_{\mathbb{R}}, m_1)$.
(ii) *There exists a* Φ *invariant subset* $D \subset K_{\mathbb{R}}$ *with* $m_1(D) > 0$, *such that the limit* $q(\xi,\varphi) = \lim_{T\to+\infty} q_T(\xi,\varphi)$ *exists and is a positive real number for every* $(\xi,\varphi) \in D$.

Moreover, under these conditions, $(1/q)^{1/r}\chi_D \in L^{1+r}(K_{\mathbb{R}}, m_1)$ *and if we take* $\lambda = \left(\int_{K_{\mathbb{R}}} (1/q)^{1/r}\chi_D\, dm_1\right)^{-1}$, *the measure* $d\mu{=}\lambda\, (1/q)^{1/r}\chi_D\, dm_1$ *is normalized and* Φ *invariant.*

PROOF. (ii) $\Rightarrow$ (i). It is easy to check, as in the previous theorem, that $\widetilde{q} = (1/q)^{1/r}\chi_D$ satisfies the functional equation (5.1). Moreover, for $r \geq 1$ the relation $\|\cdot\|_1 \leq \|\cdot\|_r$ and Schwarz's inequality provide

$$\Big(\frac{1}{q_T(\xi,\varphi)}\Big)^{1/r} = \frac{1}{\left(\frac{1}{2T}\int_{-T}^{T} |\mathbf{x}(t,\xi,\varphi)|^{2r}\, dt\right)^{1/r}} \leq \frac{1}{\frac{1}{2T}\int_{-T}^{T} |\mathbf{x}(t,\xi,\varphi)|^{2}\, dt}$$

$$\leq \frac{1}{2T}\int_{-T}^{T} \frac{1}{|\mathbf{x}(t,\xi,\varphi)|^2}\, dt = p_T(\xi,\varphi)\,;$$

while for $0 < r < 1$, from Schwarz's inequality and $\|\cdot\|_r \leq \|\cdot\|_1$ we also obtain

$$\begin{aligned}\left(\frac{1}{q_T(\xi,\varphi)}\right)^{1/r} &\leq \left(\frac{1}{2T}\int_{-T}^{T}\frac{1}{|\mathbf{x}(t,\xi,\varphi)|^{2r}}\,dt\right)^{1/r} \\ &\leq \frac{1}{2T}\int_{-T}^{T}\frac{1}{|\mathbf{x}(t,\xi,\varphi)|^{2}}\,dt = p_T(\xi,\varphi)\,.\end{aligned}$$

Thus, as in the previous proof, Fatou's lemma shows that $\widetilde{q} \in L^1(K_{\mathbb{R}}, m_1)$. Hence, if $1/\lambda = \int \widetilde{q}\, dm_1$, then $d\mu = \lambda\,\widetilde{q}\, dm_1$ is a normalized Φ invariant measure. It remains to prove that $\widetilde{q} \in L^{1+r}(K_{\mathbb{R}}, m_1)$. According to Proposition 2.1 (a consequence of Birkhoff's ergodic theorem) we find that

$$\begin{aligned}&\lim_{T\to\infty}\frac{1}{2T}\int_{-T}^{T}\frac{1}{q}\,\chi_D(\Phi_t(\xi,\varphi))\,dt \\ &= \lim_{T\to\infty}\frac{1}{2T}\int_{-T}^{T}\frac{1}{q(\xi,\varphi)}\,\chi_D(\xi,\varphi)\exp\left\{-r\int_0^t\frac{\partial f}{\partial\varphi}(\Phi_s(\xi,\varphi))\,ds\right\}dt = \chi_D(\xi,\varphi)\end{aligned}$$

for almost every $(\xi,\varphi) \in K_{\mathbb{R}}$. Since $\chi_D(\xi,\varphi) \in L^1(K_{\mathbb{R}},\mu)$ it is deduced that also $(1/q)\,\chi_D \in L^1(K_{\mathbb{R}},\mu)$, and moreover $\int(1/q)\,\chi_D\, d\mu = \int \chi_D\, d\mu = \mu(D) = 1$, that is $\lambda\int(1/q)^{1+1/r}\,\chi_D\, dm_1 = 1$, which implies that $\widetilde{q} = (1/q)^{1/r}\chi_D \in L^{r+1}(K_{\mathbb{R}}, m_1)$. This completes (ii) $\Rightarrow$ (i).

We omit (i) $\Rightarrow$ (ii) which is similar to the corresponding one in the above theorem and can be found in [**OP**]. □

The case $r = 1$, where the limit of the family $\{q_T\}_{T>0}$ gives rise to a Φ invariant measure with square integrable density function, is of great interest. Notice that this measure is a linear Φ invariant one.

Let $\widetilde{S} : \Omega \to L(\mathbb{R}^3)$ be a measurable map such that $t \to \widetilde{S}(\xi\cdot t)$ is continuous for almost every $\xi \in \Omega$. We say that a measurable map $\mathbf{X} : \Omega \to \mathbb{R}^3$ is a *solution along the flow* of the linear systems $\mathbf{X}' = \widetilde{S}(\xi\cdot t)\,\mathbf{X}$, if there is a Ξ -invariant subset $\Omega_0 \subset \Omega$ with $m_0(\Omega_0) = 1$ such that, for every $\xi \in \Omega_0$, $t \to \mathbf{X}(\xi\cdot t)$ is a C^1 -map and $\mathbf{X}'(\xi\cdot t) = \widetilde{S}(\xi\cdot t)\,\mathbf{X}(\xi\cdot t)$ for every $t \in \mathbb{R}$. From relation (5.2) the following characterization of the quadratic on the fiber solutions is easily obtained:

PROPOSITION 5.6. *Let us consider the measurable map* $\mathbf{X} = (A,B,C)^t : \Omega \to \mathbb{R}^3$. *The following statements are equivalent:*

(i) *The function*

$$q(\xi,\varphi) = A(\xi)\cos^2\varphi + B(\xi)\sin^2\varphi + 2C(\xi)\sin\varphi\cos\varphi$$

is a quadratic on the fiber solution of (5.2).

(ii) $\mathbf{X}$ *is a solution along the flow of the systems*

$$\mathbf{X}' = \begin{pmatrix} 2a(\xi\cdot t) & 0 & -2b(\xi\cdot t) \\ 0 & -2a(\xi\cdot t) & -2c(\xi\cdot t) \\ -c(\xi\cdot t) & -b(\xi\cdot t) & 0 \end{pmatrix}\mathbf{X}, \quad \xi \in \Omega\,. \tag{5.4}$$

The ergodic sheets are directly related to the quadratic on the fiber solutions of (5.2) and consequently to the linear Φ invariant measures, as expressed in the following theorem.

THEOREM 5.7. *Let $\Omega_0 \subset \Omega$ be a Ξ -invariant subset with $m_0(\Omega_0) = 1$.*

(i) *If $M = \{(\xi, X(\xi) + i\,Y(\xi)) \;/\; \xi \in \Omega_0\}$ is a complex ergodic sheet of $K_{\mathbb{C}}$, with $Y(\xi) > 0$ for every $\xi \in \Omega_0$, and we define*

$$p(\xi,\varphi) = \left(\frac{1}{Y(\xi)}\cos^2\varphi + \frac{X^2(\xi)+Y^2(\xi)}{Y(\xi)}\sin^2\varphi - \frac{2X(\xi)}{Y(\xi)}\sin\varphi\cos\varphi\right)^{-1}$$

for every $(\xi,\varphi) \in \Omega_0 \times P^1(\mathbb{R})$, then $d\mu = p\,dm_1$ is a linear Φ invariant measure.

(ii) *If $M_i = \{(\xi, X_i(\xi)) \;/\; \xi \in \Omega_0\}$ $i = 1,2$ are two real ergodic sheets of $K_{\mathbb{C}}$, and we define*

$$A(\xi) = \frac{2}{X_1(\xi) - X_2(\xi)}, \quad B(\xi) = \frac{2X_1(\xi)X_2(\xi)}{X_1(\xi) - X_2(\xi)}, \quad C(\xi) = \frac{X_2(\xi) + X_1(\xi)}{X_2(\xi) - X_1(\xi)},$$

then

$$q(\xi,\varphi) = A(\xi)\cos^2\varphi + B(\xi)\sin^2\varphi + 2C(\xi)\sin\varphi\cos\varphi$$

is a quadratic on the fiber solution of (5.2). Moreover $A(\xi)B(\xi) - C^2(\xi) = -1$ for every $\xi \in \Omega_0$.

Let us assume that $(K_{\mathbb{R}}, \Phi)$ admits a linear invariant measure μ_0. Then the description of the ergodic structures of $(K_{\mathbb{R}}, \Phi)$ and $(K_{\mathbb{C}}, \Phi)$ can be carried out and we now summarize it. If the measure μ_0 is not ergodic there is $k \in \mathbb{N}$ such that $K_{\mathbb{R}}$ decomposes into ergodic k-sheets. If the measure μ_0 is ergodic we set $k = \infty$ to unify the notation. $K_{\mathbb{C}}$ decomposes into ergodic components: it contains two ergodic sheets and its other components are exact copies of those found on $K_{\mathbb{R}}$. The ergodic structure of $K_{\mathbb{C}}$ depends on k and we frequently use this dependence throughout this paper.

Next we will show that in the absolutely continuous case a linear Φ invariant measure always exists, from which it is deduced that the above description of the ergodic structures may be applied to this case. The proof is based in the introduction of a new projective bundle whose ergodic measures are concentrated into ergodic sheets. We take the new base $\widetilde{\Omega} = K_{\mathbb{R}}$ and $\widetilde{\Xi} = \Phi|_{\widetilde{\Omega}}$; the flow $(\widetilde{\Omega}, \widetilde{\Xi})$ extends (Ω, Ξ) and allows to lift the equations (3.2) to $\widetilde{\Omega} \times \mathbb{C}^2$. The new coefficients are defined by the relations $\widetilde{a}(\xi,\varphi) = a(\xi)$, $\widetilde{b}(\xi,\varphi) = b(\xi)$, $\widetilde{c}(\xi,\varphi) = c(\xi)$, for every $(\xi,\varphi) \in K_{\mathbb{R}}$. The solutions of the corresponding Riccati equations (3.3) define the new flow $\widetilde{\Phi}$ on the projective bundles $\widetilde{K}_{\mathbb{R}} = \widetilde{\Omega} \times P^1(\mathbb{R})$ and $\widetilde{K}_{\mathbb{C}} = \widetilde{\Omega} \times P^1(\mathbb{C})$. It is obvious that

$$\widetilde{M} = \{(\xi,\varphi,\varphi) \,/\, (\xi,\varphi) \in K_{\mathbb{R}}\}$$

defines a closed ergodic sheet for every ergodic measure on $\widetilde{\Omega}$. By taking any Φ invariant compact subset $\widehat{\Omega} \subset K_{\mathbb{C}}$ as a new base, similarly extended projective flows can be introduced. Let $\widetilde{m}_0$ be a Φ ergodic measure on $K_{\mathbb{R}}$ and $\widetilde{m}_1 = \widetilde{m}_0 \otimes r_1$ the product measure on $\widetilde{K}_{\mathbb{R}}$. From the characterization of the Φ invariant measures given in Theorem 5.4, in terms of the family $\{p_T\}_{T>0}$, it is easy to show (see [**NO1**] for details) that $(K_{\mathbb{R}}, \Phi)$ admits an invariant measure absolutely continuous with respect to m_1 if and only if $(\widetilde{K}_{\mathbb{R}}, \widetilde{\Phi})$ admits an invariant measure absolutely continuous with respect to $\widetilde{m}_1$. However, the corresponding ergodic structures are different. The existence of an absolutely continuous or singular continuous measure in $\widetilde{K}_{\mathbb{R}}$ assures the decomposition of $\widetilde{K}_{\mathbb{R}}$ into ergodic sheets and then the existence of linear $\widetilde{\Phi}$ invariant measures.

THEOREM 5.8. *Let us assume that* $(K_{\mathbb{R}}, \Phi)$ *admits an invariant measure* $\mu \in \mathcal{M}^{\text{a.c.}}_{m_0}(K_{\mathbb{R}}) \cup \mathcal{M}^{\text{s.c.}}_{m_0}(K_{\mathbb{R}})$. *Then it admits a linear invariant measure.*

PROOF. We only give the main ideas of the proof and we refer the reader to [**NO1**] for more details. It is not hard to show that μ can be considered equivalent to m_1; thus let $d\mu = p(\xi, \varphi)\, dm_1$. If $K_{\mathbb{R}}$ contains an ergodic sheet then $K_{\mathbb{R}}$ decomposes into ergodic sheets and admits infinite linear Φ invariant measures.

If $K_{\mathbb{R}}$ does not contain an ergodic sheet we introduce the extended flow $(\widetilde{K}_{\mathbb{R}}, \widetilde{\Phi})$. We represent by $\langle \widetilde{m}_0, \mu\rangle$ the Borel measure on $\widetilde{K}_{\mathbb{R}}$ defined by

$$(5.5)\qquad \int_{\widetilde{K}_{\mathbb{R}}} g\, d\langle \widetilde{m}_0, \mu\rangle = \int_{K_{\mathbb{R}}} \left[\int_{P^1(\mathbb{R})} g(\xi, \widetilde{\varphi}, \varphi)\, p(\xi, \varphi)\, d\varphi \right] d\widetilde{m}_0, \qquad \forall\, g \in C(\widetilde{K}_{\mathbb{R}}).$$

From (3.6) we claim that $\langle \widetilde{m}_0, \mu\rangle \in \mathcal{M}_{\text{inv}}(\widetilde{K}_{\mathbb{R}}, \widetilde{\Phi})$ and projects into $\widetilde{m}_0$. Moreover, it is an absolutely continuous measure with density function $p(\xi, \widetilde{\varphi})$, i.e., $d\langle \widetilde{m}_0, \mu\rangle = p(\xi, \widetilde{\varphi})\, d\widetilde{m}_1$. Hence, as we have remarked before, $\widetilde{K}_{\mathbb{R}}$ decomposes into ergodic sheets and admits linear $\widetilde{\Phi}$ invariant measures. Let $\widetilde{\mu}$ be one of these measures with $d\widetilde{\mu} = (1/\widetilde{q})\, d\widetilde{m}_1$ where $\widetilde{q}$ is given by

$$\widetilde{q}(\xi, \widetilde{\varphi}, \varphi) = A(\xi, \widetilde{\varphi}) \cos^2\varphi + B(\xi, \widetilde{\varphi}) \sin^2\varphi + 2\, C(\xi, \widetilde{\varphi}) \sin\varphi \cos\varphi\,,$$

and the coefficients are continuous functions in $\widetilde{\varphi}$ for almost every $\xi \in \Omega$. The map $\widetilde{\mathbf{X}} : K_{\mathbb{R}} \to \mathbb{R}^3$, $(\xi, \widetilde{\varphi}) \mapsto (A(\xi, \widetilde{\varphi}), B(\xi, \widetilde{\varphi}), C(\xi, \widetilde{\varphi}))^t$ is, from Proposition 5.6, a solution along the flow of the linear systems

$$(5.6)\qquad \widetilde{\mathbf{X}}' = \begin{pmatrix} 2\,\widetilde{a}(\Phi_t(\xi, \widetilde{\varphi})) & 0 & -2\,\widetilde{b}(\Phi_t(\xi, \widetilde{\varphi})) \\ 0 & -2\,\widetilde{a}(\Phi_t(\xi, \widetilde{\varphi})) & -2\,\widetilde{c}(\Phi_t(\xi, \widetilde{\varphi})) \\ -\widetilde{c}(\Phi_t(\xi, \widetilde{\varphi})) & -\widetilde{b}(\Phi_t(\xi, \widetilde{\varphi})) & 0 \end{pmatrix} \widetilde{\mathbf{X}}\,, \qquad (\xi, \widetilde{\varphi}) \in \widetilde{\Omega}\,.$$

Let us assume that $\widetilde{m}_0$ disintegrates into the family $(\widetilde{m}_{0,\xi})_{\xi\in\Omega}$ with respect to m_0. We define

$$A(\xi) = \int_{P^1(\mathbb{R})} A(\xi, \widetilde{\varphi})\, d\widetilde{m}_{0,\xi}(\widetilde{\varphi})\,, \quad B(\xi) = \int_{P^1(\mathbb{R})} B(\xi, \widetilde{\varphi})\, d\widetilde{m}_{0,\xi}(\widetilde{\varphi}) \quad \text{and}$$
$$C(\xi) = \int_{P^1(\mathbb{R})} C(\xi, \widetilde{\varphi})\, d\widetilde{m}_{0,\xi}(\widetilde{\varphi})\,.$$

The map $\mathbf{X} : \Omega \to \mathbb{R}^3$, $\xi \mapsto (A(\xi), B(\xi), C(\xi))^t$ is now a solution along the flow of the linear systems (5.4). Consequently,

$$q(\xi, \varphi) = A(\xi) \cos^2\varphi + B(\xi) \sin^2\varphi + 2\, C(\xi) \sin\varphi \cos\varphi$$

is a non-trivial quadratic on the fiber solution of (5.2). Since $K_{\mathbb{R}}$ does not contain real ergodic sheets there is a real constant $\lambda > 0$ with $A(\xi)B(\xi) - C^2(\xi) = \lambda^2$ almost everywhere in Ω. Finally, $d\mu_1 = (\lambda/q)\, dm_1$ is a normalized linear Φ invariant measure. □

Another proof of the above result can be given in terms of the *conformal baricenter* introduced by Douady and Earle in [**DE**]. This concept is used by Thieullen [**T**] and Oseledets [**O**] in the classification of discrete $GL(2, \mathbb{R})$ -valued cocycles.

We start by recalling the definition of conformal baricenter. We denote by D the unit disk $D = \{w \in \mathbb{C} \,/\, |w| < 1\}$. For each Borel measure μ on $\mathbb{S}^1$ satisfying $\mu(\{w\}) < 1/2$ for every $w \in \mathbb{S}^1$, there exists a unique point $w_0 \in D$ such that $(1-|w_0|^2)\int_{\mathbb{S}^1}(\zeta - w_0)/(1-\bar{w}_0\zeta)\,d\mu(\zeta) = 0$. The value w_0 is called the *conformal baricenter of* μ and denoted by $b(\mu)$. An important property of the conformal baricenter is that, for every conformal automorphism g of the unit disk D, we have $b(g\mu) = gb(\mu)$ where $g\mu$ denotes the measure on $\mathbb{S}^1$ induced by g in the usual way, i.e., $g\mu(A) = \mu(g^{-1}(A))$ for every measurable subset A.

Alternative Proof of Theorem 5.8. We will give a sketch of the proof. The conformal mapping of $\mathbb{C}$, $K(Z) = (Z-i)/(1-iZ)$ which transforms the upper half-plane into the unit disk, transports the skew-product flow on $K_{\mathbb{C}}$ to the compact bundle $K_{\mathbb{S}} = \Omega \times \mathbb{S}^1$. The symbol $\widehat{\Phi}$ represents the flow application in any invariant subset of $K_{\mathbb{S}}$. Notice that $K(\cot\varphi) = \exp(2i\varphi)$, so we get another coordinatization of the flow on $K_{\mathbb{S}}$ induced by the one on $K_{\mathbb{R}}$ in polar symplectic coordinates. The absolutely continuous Φ invariant measure μ on $K_{\mathbb{R}}$ induce a $\widehat{\Phi}$ -invariant measure ν on $K_{\mathbb{S}}$, absolutely continuous with respect to the complete product measure on $K_{\mathbb{S}}$. Denoting by $(\nu_\xi)_{\xi\in\Omega}$ the disintegration of ν, each ν_ξ is a measure on $\mathbb{S}^1$ with no atoms and we can define $Z(\xi) = K^{-1}(b(\nu_\xi))$. Finally, from the properties of the conformal baricenter it is easy to show that $Z(\xi\cdot t) = \Phi(t,\xi,Z(\xi))$, i.e., $\{(\xi, Z(\xi)) \,/\, \xi \in \Omega\}$ is a complex ergodic sheet of $K_{\mathbb{C}}$, and from Theorem 5.7 we conclude that a linear Φ invariant measure exists. □

Summing up, if μ is a Φ invariant measure absolutely continuous with respect to m_1 then two different options are possible:

(a.1) μ is Φ ergodic, which implies that it is necessarily a linear Φ invariant measure, and the only Φ invariant measure (case $k = \infty$).

(a.2) μ is not Φ ergodic and every Φ ergodic measure on $K_{\mathbb{R}}$ is concentrated in an ergodic k-sheet. Moreover, $K_{\mathbb{R}}$ decomposes into ergodic k-sheets.

Theorem 5.7 asserts that linear Φ invariant measures are directly provided by complex ergodic sheets. Conversely, if $d\mu = (1/q)dm_1$ is a linear Φ invariant measure then the equation $q = 0$ defines two complex ergodic sheets. Consequently, $(K_{\mathbb{R}}, \Phi)$ admits an absolutely continuous invariant measure if and only if $K_{\mathbb{C}}$ contains a complex ergodic sheet. The construction of a linear Φ invariant measure from three different real ergodic sheets is also possible. In the case $k > 1$, in which real ergodic sheets do not exist, the unique normalized linear Φ invariant measure can be constructed from two different real ergodic k-sheets (see [**NO1**] for the proof):

THEOREM 5.9. *Let M_1 and M_2 be two real ergodic k-sheets $(k > 1)$*

$$M_j = \{(\xi, \varphi_i^j(\xi)) \,/\, \xi \in \Omega_0,\ i = 1,2,\dots,k\}, \quad j = 1,2,$$

where $(\xi, \varphi_i^2(\xi))$ (resp. $(\xi, \varphi_i^1(\xi))$) is the first element (resp. the $(i-1)$ -element) that we find when we move clockwise from $(\xi, \varphi_i^1(\xi))$ (resp. from $(\xi, \varphi_1^1(\xi))$) and $m_0(\Omega_0) = 1$. Let $X_i(\xi) = \cot\varphi_i^1(\xi)$ and $Y_i(\xi) = \cot\varphi_i^2(\xi)$ for $i = 1,2,\dots,k$. We *define*

$$A(\xi) = \frac{1}{l}\sum_{i=1}^{l} \frac{2}{X_i(\xi) - Y_i(\xi)}, \quad B(\xi) = \frac{1}{l}\sum_{i=1}^{l} \frac{2X_i(\xi)Y_i(\xi)}{X_i(\xi) - Y_i(\xi)} \quad \text{and}$$

$$C(\xi) = \frac{1}{l}\sum_{i=1}^{l} \frac{X_i(\xi) + Y_i(\xi)}{Y_i(\xi) - X_i(\xi)}.$$

There is a constant $c > 0$ such that $A(\xi)B(\xi) - C^2(\xi) = c^2$ almost everywhere. Taking $q(\xi, \varphi) = A(\xi)\cos^2\varphi + B(\xi)\sin^2\varphi + 2C(\xi)\sin\varphi\cos\varphi$ we obtain that $d\mu_0 = (c/q)\,dm_1$ is a normalized linear invariant measure of $(K_{\mathbb{R}}, \Phi)$.

6. The Bunded Orbits Case

This section deals with the case in which all the solutions of the family (3.2) are bounded. To ensure this for every $\xi \in \Omega$, it is enough to have it for some $\xi_0 \in \Omega$. We recall that we are considering the almost periodic case for which the base (Ω, Ξ) is almost periodic, minimal, u.e. and distal.

Notice that if $\xi \in \Omega$ and $t \in \mathbb{R}$, then for each pair $\varphi_1, \varphi_2 \in P^1(\mathbb{R})$ there exists $\varphi^* \in P^1(\mathbb{R})$ with

$$\varphi(t, \xi, \varphi_2) - \varphi(t, \xi, \varphi_1) = \exp\left\{\int_0^t \frac{\partial f}{\partial \varphi}(\Phi_s(\xi, \varphi^*))\, ds\right\}.$$

If the solutions of (3.2) are bounded, there exist positive constants m, M such that

$$m\,|\varphi_2 - \varphi_1| \le |\varphi(t, \xi, \varphi_2) - \varphi(t, \xi, \varphi_1)| \le M\,|\varphi_2 - \varphi_1| \tag{6.1}$$

for any pair of points (ξ, φ_1), $(\xi, \varphi_2) \in K_{\mathbb{R}}$. Since (Ω, Ξ) is a distal flow the relation (6.1) assures that the flow $(K_{\mathbb{R}}, \Phi)$ is also distal. From that it is known that if $(K_{\mathbb{R}}, \Phi)$ is not minimal it laminates into minimal subsets [**E**]. Moreover, we can state precisely the type of minimal subsets [**SS1**]:

THEOREM 6.1. *Let us assume that all the solutions of (3.2) are bounded for some $\xi_0 \in \Omega$. Then $(K_{\mathbb{R}}, \Phi)$ is minimal or there is a $k \in \mathbb{N}$ such that it laminates into closed ergodic k-sheets M_j, i.e., $K_{\mathbb{R}} = \cup_{j\in J} M_j$. Moreover, for each $j \in J$ the flow (M_j, Φ) is almost periodic.*

PROOF. We will only give a sketch of the proof. It is known [**E**] that for each $(\xi, \varphi) \in K_{\mathbb{R}}$ the set $\mathrm{cl}\,\{\Phi_t(\xi, \varphi) \,/\, t \in \mathbb{R}\}$ is minimal, from which it is deduced that $K_{\mathbb{R}}$ is minimal or laminates into minimal subsets. Let us now prove that if $K_{\mathbb{R}}$ is not minimal, the minimal subsets are closed ergodic k-sheets.

We begin by showing that if M is a minimal subset and $\xi \in \Omega$, the section $M_\xi = \{\varphi \in P^1(\mathbb{R}) \,/\, (\xi, \varphi) \in M\}$ is $K_{\mathbb{R}}$ or a finite set. If M_ξ is not a finite set, let $(\varphi_j)_{j\in\mathbb{N}}$ be a sequence of points in M_ξ with $\varphi_j \to \varphi_0 \in M_\xi$ as $j \to \infty$. We show now that M_ξ coincides with its derived set M'_ξ. Since $M = \mathrm{cl}\,\{\Phi_t(\xi, \varphi_0) \,/\, t \in \mathbb{R}\}$, every point (ξ, φ_0^*) can be obtained as $\lim_{n\to\infty} \Phi_{t_n}(\xi, \varphi_0)$ for some sequence $(t_n)_{n\in\mathbb{N}}$, i.e., $\xi{\cdot}t_n \to \xi$ and $\varphi(t_n, \xi, \varphi_0) \to \varphi_0^*$ as $n \to \infty$. Moreover, from relation (6.1) we deduce that the family of maps $\Psi_n : P^1(\mathbb{R}) \to P^1(\mathbb{R})$, $\varphi \mapsto \varphi(t_n, \xi, \varphi)$ is equicontinuous and the Arzela -Ascoli Theorem asserts that, in this case, there is a convergent subsequence. We assume that the complete sequence converges and consider

$$\begin{aligned} \Psi : P^1(\mathbb{R}) &\longrightarrow P^1(\mathbb{R}) \\ \varphi &\longmapsto \lim_{n\to\infty} \Psi_n(\varphi). \end{aligned}$$

It is not hard to show that Ψ is a homeomorphism of $P^1(\mathbb{R})$ preserving the orientation. Consequently, $\varphi_j^* = \Psi(\varphi_j) \to \Psi(\varphi_0) = \varphi_0^*$ which means that $\varphi_0^* \in M'_\xi$. It is also easy to see that we can approach each point on M'_ξ always from the same side in $P^1(\mathbb{R})$, from which we conclude that $M_\xi = P^1(\mathbb{R})$.

To show that if for some $\xi_0 \in \Omega$ the set M_{ξ_0} is finite, the same happens for every $\xi \in \Omega$ and with the same number of points, we construct a homeomorphism of $P^1(\mathbb{R})$ taking M_{ξ_0} into M_ξ. To do that it is enough to consider a sequence of times $(t_n)_{n\in\mathbb{N}}$ satisfying $\xi_0 \cdot t_n \to \xi$ as $n \to \infty$ (the base (Ω, Φ) is minimal) and define as before the map Ψ from the sequence $(t_n)_{n\in\mathbb{N}}$.

Similar arguments prove that if $M \neq K_\mathbb{R}$ and N is another minimal subset, the points of M_ξ and N_ξ separate each other, from which it is deduced that $card M_\xi = card N_\xi$ for every $\xi \in \Omega$. Therefore, there exists $k \in \mathbb{N}$ such that all the minimal subsets are closed k-sheets. Finally, one of the lifting theorems given by Sacker -Sell in [**SS1**] says that if the flow (Ω, Ξ) is almost periodic, and M is a closed k-sheet, then the flow (M, Φ) is also almost periodic. Thus, (M, Φ) is u.e. and therefore a closed ergodic k -sheet. □

The following proposition asserts that the bounded orbits case is always in the absolutely continuous one.

PROPOSITION 6.2. *Let us assume that the solutions of (3.2) are bounded for some $\xi_0 \in \Omega$. Then $(K_\mathbb{R}, \Phi)$ admits a linear invariant measure with continuous density function.*

PROOF. If $(K_\mathbb{R}, \Phi)$ is not minimal it follows from Theorem 6.1 that it decomposes into closed k-sheets. In the case that $k = 1$ we apply Theorem 5.7 to a closed complex sheet of $K_\mathbb{C}$ and obtain a linear Φ invariant measure with continuous density function. If $k > 1$ we obtain the result by applying Theorem 5.9 to two different closed k-sheets.

Let us now assume that $(K_\mathbb{R}, \Phi)$ is minimal. First we show that there exists a Φ invariant measure with continuous density function. We are then interested in continuous positive solutions of the equation (5.1); so if we take logarithms and write $k = \log p$, we obtain the new relation

$$k(\Phi_t(\xi, \varphi)) - k(\xi, \varphi) = -\int_0^t \frac{\partial f}{\partial \varphi}(\Phi_s(\xi, \varphi))\, ds\,.$$

Next, since $(K_\mathbb{R}, \Phi)$ is a minimal flow, $\partial f/\partial\varphi(\xi, \varphi) \in C(K_\mathbb{R})$ and for $(\xi_0, \varphi_0) \in K_\mathbb{R}$ the function $t \mapsto \int_0^t \partial f/\partial\varphi(\Phi_s(\xi_0, \varphi_0))\, ds$ is bounded, the Gottschalk-Hedlund Theorem [**GH**] (also proved in [**FKS**]) guarantees the existence of a continuous solution $k \in C(K_\mathbb{R})$ of (6.2), from which we obtain a Φ invariant measure μ with continuous density function $p(\xi, \varphi) = \exp k(\xi, \varphi)$, i.e., $d\mu = p\, dm_1$.

Now we take the proof to the extended projective flow $(\widetilde{K}_\mathbb{R}, \widetilde{\Phi})$ which decomposes into closed ergodic sheets. Let $d\widetilde{\mu}_0 = (1/q)\, d\widetilde{m}_1$ be a linear $\widetilde{\Phi}$ invariant measure with continuous density function where

$$q(\xi,\widetilde{\varphi},\varphi) = A(\xi,\widetilde{\varphi})\cos^2\varphi + B(\xi,\widetilde{\varphi})\sin^2\varphi + 2C(\xi,\widetilde{\varphi})\sin\varphi\cos\varphi\,.$$

The new coefficients

$$A(\xi) = \int_{P^1(\mathbb{R})} A(\xi,\widetilde{\varphi})\,p(\xi,\widetilde{\varphi})\,d\widetilde{\varphi}\,,\quad B(\xi) = \int_{P^1(\mathbb{R})} B(\xi,\widetilde{\varphi})\,p(\xi,\widetilde{\varphi})\,d\widetilde{\varphi}\,,$$

$$C(\xi) = \int_{P^1(\mathbb{R})} C(\xi,\widetilde{\varphi})\,p(\xi,\widetilde{\varphi})\,d\widetilde{\varphi}$$

are continuous functions on Ω. Let

$$q(\xi,\varphi) = A(\xi)\cos^2\varphi + B(\xi)\sin^2\varphi + 2C(\xi)\sin\varphi\cos\varphi\,.$$

Finally, arguing as in Theorem 5.8, we obtain a positive constant c such that $d\mu_0 = (c/q)\,dm_1$ is the required normalized linear Φ invariant measure. □

Since the flow (Ω,Ξ) is distal and the solutions of (3.2) are bounded, there exists a strong Perron transformation which takes the systems (3.2) into skew-symmetric form. In fact, this is a general result for n-dimensional linear systems (see for instance Cameron [**C**] and Johnson [**J3**]). Moreover, in the bidimensional case we can construct explicitly the transformation in terms of the coefficients of the linear Φ invariant measure with continuous density function as shown in Alonso-Obaya [**AO**].

Summarizing, we obtain the following options for the bounded orbits case:

(b.1) $(K_{\mathbb{R}},\Phi)$ is minimal and u.e. with a unique normalized ergodic measure which is linear and has continuous density function.

(b.2) $(K_{\mathbb{R}},\Phi)$ is minimal and decomposes into measurable ergodic k-sheets in which the ergodic measures are concentrated. Thus, the topological and ergodic structures do not agree. In this case, apart from the normalized linear Φ invariant measure with continuous density function, Φ invariant measures with discontinuous density function can exist. An example is the one given by Johnson in [**J1**] that we have mentioned in Section 4.

(b.3) $(K_{\mathbb{R}},\Phi)$ is not minimal and laminates into closed ergodic k-sheets which are almost periodic and concentrate the Φ ergodic measures. Consequently, the topological and ergodic structures coincide. In this case all the solutions are almost periodic.

We want also to remark that the existence of an almost periodic solution implies the almost periodicity of all the solutions and of the flow (see Johnson [**J2**]). From that we conclude that, in situation (b.2), no solution to any equation is almost periodic.

7. Invariant Measures in the Singular Case

In this section, we intend to determine precisely the number and type of Φ ergodic measures on the complex projective bundle in the singular case, that is, when every Φ invariant measure on $K_{\mathbb{R}}$ is singular with respect to m_1. This means that every Φ invariant measure is purely discontinuous. The dynamics on the projective flow is completely different from the one appearing in the absolutely continuous case.

The multiplicative ergodic theorem proved by Oseledets provides the ergodic measures on the projective flow when the Lyapunov exponent of(3.2) is positive: they turn out to be two singular ones concentrated in ergodic sheets. This case may also appear when the Lyapunov exponent vanishes.

PROPOSITION 7.1. *Let us assume that there is a* Φ *invariant measure* $\mu \in \mathcal{M}_{m_0}(K_{\mathbb{C}})$ *with* $\mu(K_{\mathbb{R}}) = 0$*; then* $(K_{\mathbb{R}}, \Phi)$ *admits a linear invariant measure.*

PROOF. We take the new base $\widetilde{\Omega} = K_{\mathbb{C}}$ and the extended flow $(\widetilde{K}_{\mathbb{C}}, \widetilde{\Phi})$. We assume that μ is a Φ ergodic measure on $K_{\mathbb{C}}$. The set

$$M = \{(\xi, \widetilde{Z}, \widetilde{Z}) \,/\, (\xi, \widetilde{Z}) \in K_{\mathbb{C}}\}$$

defines a complex ergodic sheet with respect to μ. It follows from Theorem 5.7 that $(\widetilde{K}_{\mathbb{R}}, \widetilde{\Phi})$ admits a linear $\widetilde{\Phi}$ invariant measure, from which we conclude, as remarked in Section 5, that $(K_{\mathbb{R}}, \Phi)$ admits an absolutely continuous invariant measure. Finally, we deduce from Theorem 5.8 that $(K_{\mathbb{R}}, \Phi)$ admits a linear invariant measure. □

In the singular case every Φ ergodic measure is concentrated on the real projective bundle. We now inquire for the number of such measures.

PROPOSITION 7.2. *Let us assume that* $(K_{\mathbb{R}}, \Phi)$ *admits three different ergodic measures* $\mu_i \in \mathcal{M}_{m_0}(K_{\mathbb{R}})$, $i = 1, 2, 3$. *Then it admits a linear invariant measure.*

PROOF. We assume that $\mu_i \in \mathcal{M}^{\mathrm{p.d.}}_{m_0}(K_{\mathbb{R}})$, because the result has already been proved in the remaining cases. Then μ_i is concentrated in an ergodic k_i-sheet M_i for $i = 1, 2, 3$, and $M = \cup_{i=1}^{3} M_i$ represents a measurable k-sheet with $k \geq 3$. We take the new base $\widetilde{\Omega} = K_{\mathbb{R}}$ and the extended flow $(\widetilde{K}_{\mathbb{R}}, \widetilde{\Phi})$. The set $N = \{(\xi, \widetilde{\varphi}, \widetilde{\varphi}) \,/\, (\xi, \widetilde{\varphi}) \in K_{\mathbb{R}}\}$ defines a closed ergodic sheet. Given $(\xi, \widetilde{\varphi}) \in K_{\mathbb{R}}$, we denote by $(\xi, \varphi_j(\xi, \widetilde{\varphi}))$ the j-th element of M which we find when we move from $(\xi, \widetilde{\varphi})$ clockwise on $P^1(\mathbb{R})$ for each $j = 1, 2 \ldots, k$. The sets $M_j = \{(\xi, \widetilde{\varphi}, \varphi_j(\xi, \widetilde{\varphi})) \,/\, (\xi, \widetilde{\varphi}) \in K_{\mathbb{R}}\}$ represent k different ergodic sheets with respect to μ_1. Under these conditions we know that $\widetilde{K}_{\mathbb{R}}$ decomposes into ergodic sheets and $(\widetilde{K}_{\mathbb{R}}, \widetilde{\Phi})$ admits linear $\widetilde{\Phi}$-invariant measures with respect to μ_1. Consequently, $(K_{\mathbb{R}}, \Phi)$ also admits linear invariant measures and the result is proved. Another proof of this result can also be given in terms of the conformal baricenter as in Theorem 5.8 □

Referring to the continuous structure we know that if $K_{\mathbb{R}}$ contains three minimal subsets or a closed k-sheet with $k \geq 3$ then all the solutions of (3.2) are bounded. A similar result follows for the measurable structure. Proposition 7.2 says that when there exist three different Φ ergodic measures or a measurable k-sheet with $k \geq 3$, we are in the absolutely continuous case.

Summing up, the description of $\mathcal{M}_{\mathrm{inv}}(K_{\mathbb{R}}, \Phi)$ is easier in the singular case where only three different options are possible:

(s.1) $(K_{\mathbb{R}}, \Phi)$ admits a unique ergodic measure concentrated in an ergodic sheet.
(s.2) $(K_{\mathbb{R}}, \Phi)$ admits two different ergodic measures, each of them concentrated in an ergodic sheet.
(s.3) $(K_{\mathbb{R}}, \Phi)$ admits a unique ergodic measure concentrated in an ergodic 2 - sheet.

Examples of linear systems whose projective bundles contain a unique ergodic sheet are well known; in fact, this first option (s.1) with a minimal projective flow is generic in the closure of a large class of systems with null Lyapunov exponent (see Novo -Obaya [**NO2**]). A bidimensional linear system with positive Lyapunov exponent presents singular dynamics with two ergodic sheets, that is, the situation (s.2). Examples in these conditions with non-closed sheets are given by Millionshchikov in [**Mi**] and Vinograd in [**Vi**]. An example of a flow with a unique minimal subset and an ergodic 2-sheet can be found in Johnson [**J5**]. From this example Novo -Obaya in [**NO3**] construct an example of a uniquely ergodic flow with a closed 2-sheet, i.e. situation (s.3), and show the completeness of the classification.

8. Topological Structure of the Projective Flow

The aim of this section is to review the main results of the topological classification for almost periodic bidimensional linear systems given by Johnson in [**J3**], and to relate to the ergodic one, already recalled in the previous sections.

The relation between both classifications has already been analyzed in the periodic case in Section 4, in which they coincide, and in the bounded one in Section 6. Thus, in this section, we will always consider the unbounded orbits case. As we have previously remarked, the existence of three or more minimal subsets implies the boundedness of the solutions. Therefore, it remains to analyze the cases in which one or two minimal subsets of $K_{\mathbb{R}}$ exist.

Let us start with the case of two minimal subsets N_1 and N_2 with projections on the base $\pi_i : N_i \to \Omega$, $i = 1, 2$. If the Lyapunov exponent is positive, the flows (N_1, Φ) and (N_2, Φ) are u.e. and isomorphic to (Ω, Ξ). We are in the singular case with two closed ergodic sheets N_1, N_2, i.e., option (s.2) of the ergodic classification. Moreover, the Sacker-Sell spectral theorem [**SS2**] provides exponential dichotomy, that is the splitting of $V_{\mathbb{R}}$ in two one-dimensional invariant subbundles . In this case there is an almost periodic change of variables, defined on a 1-cover or a 2-cover of (Ω, Ξ), which takes (3.2) into diagonal form.

Now we consider the case of null Lyapunov exponent. The next proposition asserts that one of the minimal subsets is an ergodic sheet.

PROPOSITION 8.1. *Let N_1 and N_2 be the two minimal subsets of $K_{\mathbb{R}}$. Then there is an $i \in \{1, 2\}$ such that $card\pi_i^{-1}(\xi) = 1$ for almost every $\xi \in \Omega$.*

PROOF. Let $\xi \in \Omega$ and assume that $card\pi_i^{-1}(\xi) \geq 2$ for $i = 1, 2$. We consider four points (ξ, φ_1), $(\xi, \psi_1) \in N_1$ and (ξ, φ_2), $(\xi, \psi_2) \in N_2$ and the corresponding solutions of (3.2), $\mathbf{x}(t, \xi, \varphi_i)$ and $\mathbf{x}(t, \xi, \psi_i)$ for $i = 1, 2$.

If $\mathbf{x}(t, \xi, \varphi_1)$ and $\mathbf{x}(t, \xi, \psi_1)$ are bounded, then all the solutions of (3.2) are bounded which contradicts our assumptions, so we conclude that at least one of the solutions, say $\mathbf{x}(t, \xi, \varphi_1)$, is unbounded. Thus, let us assume that $\mathbf{x}(t_n, \xi, \varphi_1) \to \infty$ as $n \to \infty$. Since N_1 and N_2 are distinct, an easy argument using the constancy of the determinants of fundamental matrices (recall that the trace of the matrix of coefficients is 0) shows that $\mathbf{x}(t_n, \xi, \varphi_2) \to 0$ and $\mathbf{x}(t_n, \xi, \psi_2) \to 0$ as $n \to \infty$. From this it is deduced that the determinant of the fundamental matrix $U(t, \xi)$ with columns $\mathbf{x}(t, \xi, \varphi_2)$, $\mathbf{x}(t, \xi, \psi_2)$ cannot be constant and a contradiction is obtained. Consequently, we have proved that, for any fixed $\xi \in \Omega$, there is at least one $i \in \{1, 2\}$ such that $card\pi_i^{-1}(\xi) = 1$.

Now we consider the Ξ -invariant subsets $\Omega_i = \{\xi \in \Omega \,/\, card\pi_i^{-1}(\xi) = 1\}$ for $i = 1, 2$. Since $\Omega = \Omega_1 \cup \Omega_2$ and m_0 is Ξ- ergodic we conclude that the measure of one of them is 1 and the proposition is proved. $\square$

As a corollary of the previous proposition we deduce that one of the minimal subset is an a.a. extension of (Ω, Ξ). In fact, the same may be said of the other minimal subset.

PROPOSITION 8.2. *Both minimal subsets N_1 and N_2 define almost automorphic extensions of* (Ω, Ξ).

PROOF. As we have just remarked, one of them, say (N_1, Φ), is an a.a. extension of (Ω, Ξ). Suppose that (N_2, Φ) is not an a.a. extension of (Ω, Φ). This means by the proof of Proposition 8.1 that for every $\xi \in \Omega$ we have $card\pi_1^{-1}(\xi) = 1$. Let $(\xi, \varphi) \in N_1$ and consider the solution $\mathbf{x}(t, \xi, \varphi)$ of (3.2). Since there are two points (ξ, φ_1), $(\xi, \varphi_2) \in N_2$ distally related to (ξ, φ), arguing as in Proposition 8.1 we find that $\mathbf{x}(t, \xi, \varphi)$ is bounded above. Moreover, $|\mathbf{x}(t, \xi, \varphi)|^2 = 2\rho(t, \xi, \varphi) = \exp\left\{-\int_0^t \partial f/\partial\varphi\,(\Phi_s(\xi, \varphi))\,ds\right\}$, which implies that $-\int_0^t \partial f/\partial\varphi\,(\Phi_s(\xi, \varphi))\,ds$ is bounded above for every $(\xi, \varphi) \in N_1$. Now Proposition 1.13 of [**J3**] asserts that this integral is uniformly bounded, above and below for all $(\xi, \varphi) \in N_1$ and $t \in \mathbb{R}$.

Again using a fundamental matrix argument we see that the solutions $\mathbf{x}(t, \xi, \varphi)$ of (3.2) for $(\xi, \varphi) \in N_2$ must be bounded. Then, since $card\pi_2^{-1}(\xi) \geq 2$, we conclude that all the solutions are bounded which contradicts the hypothesis. Thus, (N_2, Φ) is an a.a. extension of (Ω, Ξ). $\square$

An example of a quasiperiodic linear system with unbounded solutions such that the projective flow $(K_{\mathbb{R}}, \Phi)$ contains two minimal subsets which are closed ergodic sheets, and decomposes into non-closed measurable ergodic sheets, (i.e. absolutely continuous dynamics (a.2) with $k = 1$) is given in [**NO1**].

The construction of a bidimensional linear system whose projective flow contains two minimal subsets, one of which is not a copy of the base is given in [**J4**]. In the singular case both minimal subsets are always ergodic sheets, which is option (s.2) of the ergodic classification of Section 7.

PROPOSITION 8.3. *Let N_1 and N_2 be the two minimal subsets of $K_{\mathbb{R}}$ and assume that all the Φ invariant measures are singular with respect to m_1. Then N_1 and N_2 are ergodic sheets.*

PROOF. It follows from Proposition 8.1 that one of these minimal subsets, for instance N_1, satisfies $card\pi_1^{-1}(\xi) = 1$ for almost every $\xi \in \Omega$, i.e., N_1 is an ergodic sheet. We denote its elements by $(\xi, \varphi_1(\xi))$ for every $\xi \in \Omega_0$ with $m_0(\Omega_0) = 1$.

Now we consider the Ξ invariant subset $\Omega_1 = \{\xi \in \Omega_0 \,/\, card\pi_2^{-1}(\xi) \geq 2\}$. We claim that $m_0(\Omega_1) = 0$. Let us suppose, on the contrary, that $m_0(\Omega_1) = 1$. Then, for each $\xi \in \Omega_1$ we denote by $(\xi, \varphi_2(\xi))$ (resp. $(\xi, \varphi_3(\xi))$) the first element of N_2 that we find when we move from $(\xi, \varphi_1(\xi))$ clockwise (resp. counterclockwise) on $P^1(\mathbb{R})$. The sets $M_j = \{(\xi, \varphi_j(\xi)) \,/\, \xi \in \Omega_1\}$, $j = 2, 3$ represent two more different ergodic sheets and, as proved in Proposition 7.2, from three ergodic sheets we deduce that we are in the absolutely continuous case which contradicts our assumptions. Therefore, $m_0(\Omega_1) = 0$ and N_2 is an ergodic sheet. $\square$

Finally, we consider the case in which $K_{\mathbb{R}}$ contains a unique minimal subset N with projection on the base $\pi_N : N \to \Omega$. In this context, all the ergodic situations described in Sections 5 and 7 can occur. Johnson [**J3**] proves that if a distal pair exists, then the subflow (N, Φ) is an a.a. extension of a 2-cover of (Ω, Ξ). The next proposition characterizes the minimal subset in the singular case if a distal pair exists. It turns out to be a *distal measurable* 2-*sheet*, i.e. a real measurable 2-sheet $M = \{(\xi, \varphi_i(\xi)) \,/\, i = 1, 2,\ \xi \in \Omega\}$ such that the points $(\xi, \varphi_1(\xi))$, $(\xi, \varphi_2(\xi))$ form a distal pair for almost every $\xi \in \Omega$. An example of this situation with a closed 2-sheet can be found, as we have already remarked, in [**NO3**].

PROPOSITION 8.4. *Let N be the unique minimal subset of $K_{\mathbb{R}}$. Suppose that some fiber of $K_{\mathbb{R}}$ contains a distal pair and all the Φ invariant measures are singular with respect to m_1. Then N is a distal measurable 2-sheet.*

PROOF. Since distal pairs move along the trajectories and are preserved on the limits, it is easy to obtain for every $\xi \in \Omega$ a new distal pair $(\xi, \varphi_1(\xi))$, $(\xi, \varphi_2(\xi))$ of elements of N; thus $card\pi_N^{-1}(\xi) \geq 2$ for every $\xi \in \Omega$. In addition, it follows from Lemma 7.3 of [**J3**] that, if $card\pi_N^{-1}(\xi) > 2$, there exists an element $\varphi_0(\xi) \in P^1(\mathbb{R})$ with $(\xi, \varphi_0(\xi)) \in N$ such that

(i) if $(\xi, \varphi_1) \in N$ and $\varphi_1 \neq \varphi_0(\xi)$ then $(\xi, \varphi_0(\xi))$, (ξ, φ_1) form a distal pair,
(ii) if $(\xi, \varphi_i) \in N$ and $\varphi_i \neq \varphi_0(\xi)$, $i = 1, 2$ then the pair (ξ, φ_1), (ξ, φ_2) is proximal.

Thus, if $card\pi_N^{-1}(\xi) > 2$ for almost every $\xi \in \Omega$ then it is easy to prove that the map $\xi \mapsto \varphi_0(\xi)$ is measurable and $\{(\xi, \varphi_0(\xi)) \,/\, \xi \in \Omega\}$ is a Φ invariant set, i.e., it defines an ergodic sheet. Finally, similar arguments to the ones used in Proposition 8.3 allow us to conclude that we are in the absolutely continuous case and the projective flow decomposes into ergodic sheets. Therefore, as $(K_{\mathbb{R}}, \Phi)$ has singular dynamics, $card\pi_N^{-1}(\xi) = 2$ almost everywhere and N is a distal measurable 2-sheet. □

As an easy consequence of the previous result we deduce that, in the singular case, if $(K_{\mathbb{R}}, \Phi)$ is u.e. with a unique ergodic measure concentrated in an ergodic sheet, that is option (s.1) of the ergodic classification, the flow $(K_{\mathbb{R}}, \Phi)$ is a proximal extension of (Ω, Ξ). Finally, it can be also proved that the same happens when the Lyapunov exponent is positive (case (s.2) of the classification) and the projective flow $(K_{\mathbb{R}}, \Phi)$ contains a unique minimal subset (see [**J3**]).

References

[AO] A. I. Alonso and R. Obaya, *Ergodic structure of bidimensional linear systems*, Israel J. Math. **92** (1995), 185–205.

[B] N. Bourbaki, *Integration*, Chpt. VI, 1st ed., Herman, Paris, 1959.

[C] R. H. Cameron, *Almost periodic properties of bounded solutions of linear differential equations with almost periodic coefficients*, J. Math. Phys. **15** (1936), 73–81.

[CFS] I. P. Cornfeld, S. V. Fomin and Ya. G. Sinai, *Ergodic Theory*, Springer-Verlag, Berlin, Heidelberg, New York,, 1982.

[CS] W. Craig and B. Simon, *Subharmonicity of the Lyapunov index*, Duke Math. J. **50 No. 2 yr 1983**, 551–560.

[DE] A. Douady and C. J. Earle, *Conformally natural extension of homeomorphism of the circle*, Acta Math. **157** (1986), 23–48.

[E] R. Ellis, *Lectures on Topological Dynamics*, Benjamin, New York, 1969.

[EJ] R. Ellis and R. Johnson, *Topological dynamics and linear differential systems*, J. Differential Equations **44** (1982), 21–39.

[F] H. Furstenberg, *Strict ergodicity and transformations of the torus*, Amer. J. Math. **85** (1961), 573–601.
[FKS] H. Furstenberg, H. Keynes and L. Shapiro, *Prime flows in topological dynamics*, Israel J. Math. **14** (1973), 26–38.
[GH] W. Gottschalk and G. Hedlund, *Topological Dynamics*, Amer. Math. Soc. Collq. Publ. **36** (1955).
[IT] A. and C. Ionescu-Tulcea, *Topics in the Theory of Lifting*, Springer-Verlag, New York, 1969.
[J1] R. Johnson, *Measurable subbundles in linear skew-product flows*, Illinois J. Math. **23** (1979), 183–198.
[J2] same, *Analyticity of spectral subbundles*, J. Differential Equations **35** (1980)), 366–387.
[J3] ———,, *On a Floquet theory for almost-periodic two-dimensional linear systems*, J. Differential Equations **37** (1980), 184–205.
[J4] ———, *A linear, almost periodic equation with an almost automorphic solution*, Proc. Amer. Math. Soc. **82 No. 2** (1981), 199–205.
[J5] ———, *Un'equazione lineare quasi-periodica con una propietà inconsueta*, Bolletino U.M.I. (6) **2-A** (1983), 115–121.
[Ma] R. Mañé, *Ergodic Theory and Differentiable Dynamics* (1987), Springer Verlag, Berlin, Heidelberg, New York.
[Mi] V. M. Millionshchikov, *Proof of the existence of irregular systems of linear differential equations with almost periodic coefficients*, Diff. Urav. **4 No. 3** (1968), 391–396.
[NS] V. Nemytskii and V. Stepanoff, *Qualitative Theory of Differential Equations*, Princeton University Press, Princeton, NJ, 1960.
[NO1] S. Novo and R. Obaya, *An ergodic classification of bidimensional linear systems*, J. Dynamics Differential Equations **8 No. 3** (1996), 373–406.
[NO2] ———,, *Bidimensional linear systems with singular dynamics*, Proc. Amer. Math. Soc. **124 No. 10** (1996), 3163–3172.
[NO3] ———, *On the dynamical behaviour of an almost periodic linear system with an ergodic 2-sheet studied by R. Johnson*, Univ. of Valladolid preprint (1996).
[OP] R. Obaya and M. Paramio, *Directional differentiability of the rotation number for the almost periodic Schrödinger equation jour Duke Math. J.* **66** (1992), 521–552.
[O] V. I. Oseledets, *Classification of $GL(2,\mathbb{R})$-valued cocycles of dynamical systems*, Moscow State University preprint, 1995.
[P] R. Phelps, *Lectures on Choquet's Theory*, Van Nostrand Mathematical Studies, American Book Co., New York, 1966.
[SS1] R. J. Sacker and G. R. Sell, *Lifting properties in skew-products flows with applications to differential equations*, Mem. Amer. Math. Soc. Providence **190** (1977).
[SS2] ———, *A spectral theory for linear differential systems*, J. Differential Equations **27** (1978), 320–358.
[T] P Thieullen, *Ergodic reduction of random products of two -by -two matrices*, Université Paris -Sud preprint (1995).
[Ve] W. A. Veech, *Topological dynamics*, Bull. A.M.S. **83** (1977), 775–830.
[Vi] R. E. Vinograd, *A problem suggested by N. P. Erugin*, Diff. Urav. **11 No. 4** (1975), 632–638.

Departamento de Matemática Aplicada a la Ingeniería, E.T.S. de Ingenieros Industriales, Universidad de Valladolid, 47011 Valladolid, Spain.
E-mail address: `sylnov@wmatem.eis.uva.es`

Departamento de Matemática Aplicada a la Ingeniería, E.T.S. de Ingenieros Industriales, Universidad de Valladolid, 47011 Valladolid, Spain.
E-mail address: `rafoba@wmatem.eis.uva.es`

Contemporary Mathematics
Volume **215**, 1998

An Application of Topological Dynamics to Bifurcation Theory

R. Johnson

ABSTRACT. We use techniques of topological dynamics and ergodic theory to study a bifurcation problem which caricatures the stochastic Duffing-van der Pol oscillator.

1. Introduction

This paper represents another contribution to the rapidly developing area of random bifurcation theory. Papers treating this subject have been written by, among others, L. Arnold and his collaborators at Bremen (including P. Boxler, K. Xu, R. Schenk-Hoppé, and B. Schmalfuß; (see [A], [AB], [AX], [AS], [H1], [H2]) and by Johnson and Yi ([J], [JY]). In these papers, one studies problems of the form

$$x' = f_\mu(t, x) \qquad (x \in \mathbf{R}^2), \tag{1}$$

where μ is a real parameter, $f_\mu(t, 0) = 0$ for all t and μ, and the time-dependence of f_μ is non-periodic. (Thus most classical methods of bifurcation theory do not apply.) Under appropriate assumptions, the solution $x = 0$ is asymptotically stable for some parameter values, say for $\mu < 0$, but it is no longer stable for $\mu > 0$. One seeks information regarding the existence and properties of new asymptotically stable solutions, or "attractors".

In the papers of the Bremen school, the time dependence of f_μ is frequently due to the presence of a stochastic process which modifies the coefficients of a deterministic differential equation $x' = f_*(x)$. On the other hand, in [J] and [JY], the family of functions $\{f(t, \cdot) \mid t \in \mathbf{R}\}$ is determined by a minimal and uniquely ergodic (but not periodic) real flow, for example a rotation of a torus. Thus in the first case the differential equations exhibit "very high" randomness, while in the second case one is so to speak at the other extreme in the category of nonperiodic processes.

Before discussing in detail the contents of the present paper, we wish to mention two phenomena discovered and underlined in the works mentioned above. We

---

1991 *Mathematics Subject Classification.* 28D10, 34A47.

Research partially supported by grants from the M.U.R.S.T.

believe that these phenomena occur rather frequently in random bifurcation problems.

The first phenomenon is a "two-step" Hopf-type bifurcation. This particular bifurcation scenario was observed numerically in a stochastically driven Duffing-van der Pol oscillator by Schenk-Hoppé ([H1], [H2]):

$$x'' = (\alpha + \sigma_1 \xi_t^{(1)})x + \beta x' - x^3 - x^2 x' + \sigma_2 \xi_t^{(2)}.$$

Here $\xi_t^{(1)}$ and $\xi_t^{(2)}$ are real or white noise processes, and σ_1, σ_2 are parameters. Roughly speaking, if α is a negative constant and β varies, then for appropriate values of σ_1 and σ_2, the "random point" $x = 0$ loses stability at a certain critical value $\beta = \beta_c$ to a random combination of two Dirac measures supported in the $x - x'$ space $\mathbf{R}^2$. Upon further increase of β, there is a second critical value of β at which the random combination of two Dirac measures yields stability to a random invariant circle. One speaks of a "Hopf bifurcation" because, at the end, one has a (random) invariant circle which is asymptotically stable. See [A], [H1].

The second phenomenon is that of parameter intermittency. It has been studied rigorously when the time-dependence in (1) is minimal and uniquely ergodic ([J], [JY]). There actually are at least two types of parameter intermittency. First of all, there may be a "discontinuous family" A_μ of attractors for (1) as $\mu \to 0^+$ ([JY]). Second, as pointed out in [J], an attractor may appear only for $\mu \geq \mu_1$, where $\mu_1 > 0$. The interval $(0, \mu_1)$ then represents a transition regime.

We can now discuss in more detail the goal and content of this paper. Motivated by Schenk-Hoppés work, we will study the problem

$$x'' = (\alpha + \xi_t)x + \beta x' - x^2 x' - x^3 \tag{2}$$

when the process $\{\xi_t\}$ is determined by a real, minimal uniquely ergodic flow $(\Omega, \{\tau_t\}, \nu)$. That is, Ω is a compact metric space, $\{\tau_t \mid t \in \mathbf{R}\}$ is a 1-parameter group of homeomorphisms of Ω, ν is an ergodic probability measure on Ω, and

$$\xi_t(\omega) = g(\tau_t(\omega)),$$

where $g : \Omega \to \mathbf{R}$ is a continuous function. We will give conditions under which the first step of the two-step random Hopf bifurcation can be rigorously proved to occur.

Such conditions have already been given in [J]. In general those conditions are consistent with the second type of parameter intermittency mentioned above. Here we will modify (2)—thereby rendering it simpler in certain respects—by supposing that fast oscillations appear at the parameter value $\beta = \beta_c$ at which $x = 0$ loses stability. (The equations which we will actually study are written out in $(7)_\mu$ below). We will find that, if appropriate conditions are satisfied, an attractor A_μ appears for small $\mu > 0$. This attractor is an analog of the random combination of Dirac measures alluded to above. There is no parameter intermittency.

We remark that, as yet, we have no rigorous result concerning the second step of the two-step Hopf pattern. Using a Lyapounov function technique, Schenk-Hoppé [H2] has given conditions under which, in the case when $\{\xi_t\}$ is real or white noise, a random attractor exists in an annulus in $x - x'$ space. The annulus does not contain the origin; thus the second step of the two-step bifurcation process has (presumably) occurred. However, the structure of the attractor still remains to be studied.

The author wishes to thank L. Arnold, B. Schmalfuß, R. Schenk- Hoppé, Y. Yi, and other colleagues for stimulating discussions on the subject of random bifurcation theory. This paper was written on the occasion of the retirement of Prof. Robert Ellis. The author dedicates it to him with respect and gratitude.

2. Analysis and Results

We begin our analysis of equation (2) by putting it in system form. Writing $y = x'$ we obtain

$$(3)\qquad \begin{pmatrix} x \\ y \end{pmatrix}' = \begin{pmatrix} 0 & 1 \\ \alpha + \xi_t & \beta \end{pmatrix} \begin{pmatrix} x \\ y \end{pmatrix} - \begin{pmatrix} 0 \\ x^2 y + x^3 \end{pmatrix}.$$

As discussed in the Introduction, we impose an hypothesis of minimality and unique ergodicity on the process $\{\xi_t\}$.

ASSUMPTION 2.1. There is a uniquely ergodic, minimal real flow $(\Omega, \{\tau_t\}, \nu)$ and a continuous real function $g : \Omega \to \mathbf{R}$ such that

$$\xi_t(\omega) = g(\tau_t(\omega)) \qquad (t \in \mathbf{R},\ \omega \in \Omega).$$

(The space Ω is taken to be a compact metric space, and ν is a Radon probability measure on Ω which is ergodic with respect to the flow $\{\tau_t\}$.)

Of course the process $\{\xi_t\}$ now has a structure much less random than that of a stochastic process. On the other hand, there is ample evidence that, even in the quasi-periodic case, the phenomenon of stability break down is very complex.

Now fix the parameters α and β. Note that (3) depends on ω via the process $\{\xi_t\}$. The equations $(3){=}(3)_\omega$ induce a nonlinear semiflow $\{\tau_t^n\}$ on $\Omega \times \mathbf{R}^2$ in the following way: if $(x_0, y_0)^t \in \mathbf{R}^2$ and $\omega \in \Omega$, define

$$\tau_t^n\left(\omega, \begin{pmatrix} x_0 \\ y_0 \end{pmatrix}\right) = \left(\tau_t(\omega), \begin{pmatrix} x(t) \\ y(t) \end{pmatrix}\right),$$

where $(x(t), y(t))^t$ is the solution to $(3)_\omega$ with initial condition $(x_0, y_0)^t$. One has only a semiflow because, in general, solutions to $(3)_\omega$ do not exist for all time.

We agree to incorporate α in ξ_t from now on, or equivalently to hold α at some fixed value, say $\alpha = -1$. We will need to consider the linearized equations

$$(4)_\omega\qquad \begin{pmatrix} x \\ y \end{pmatrix}' = \begin{pmatrix} 0 & 1 \\ \xi_t & \beta \end{pmatrix} \begin{pmatrix} x \\ y \end{pmatrix}.$$

Observe that equations $(4)_\omega$ induce a flow $\{\tau_t^\ell\}$ on $\Omega \times \mathbf{R}^2$, as follows:

$$\tau_t^\ell\left(\omega, \begin{pmatrix} x_0 \\ y_0 \end{pmatrix}\right) = \left(\tau_t(\omega), \Phi_\omega(t) \begin{pmatrix} x_0 \\ y_0 \end{pmatrix}\right)$$

for $\omega \in \Omega$ and $(x_0, y_0)^t \in \mathbf{R}^2$. Here $\Phi_\omega(t) \equiv \Phi_{\omega,\beta}(t)$ is the 2×2 matrix of $(4)_\omega$ which satisfies $\Phi_\omega(0) = \begin{pmatrix} 1 & 0 \\ 0 & 1 \end{pmatrix} = I$. The flow $\{\tau_t^\ell\}$ is frequently called a linear skew-product flow ([SS]).

Standard theory (e.g., [W]) shows that, for fixed β, the linear equations $(4)_\omega$ admit two Lyapounov exponents $\lambda_1(\beta) \geq \lambda_2(\beta)$. One has for ν-a.a. $\omega \in \Omega$:

$$\lambda_1(\beta) = \lim_{t \to \infty} \frac{1}{t} \ln \|\Phi_{\omega,\beta}(t)\|.$$

It will be convenient for us to consider the "traceless" linear system

$$(5)_\omega \qquad \begin{pmatrix} x \\ y \end{pmatrix}' = \begin{pmatrix} -\beta/2 & 0 \\ \xi_t & \beta/2 \end{pmatrix} \begin{pmatrix} x \\ y \end{pmatrix},$$

obtained by subtracting the constant matrix $I \cdot \beta/2$ from the coefficient matrix in $(4)_\omega$. Let $\lambda_0(\beta)$ be the largest Lyapounov exponent of the traceless equations $(5)_\omega$. We clearly have

$$(6) \qquad \lambda(\beta) = \lambda_0(\beta) + \beta/2.$$

Here and below, we agree to write $\lambda(\beta) = \lambda_1(\beta)$. Both the quantities λ and λ_0 are of importance in our theory.

At this point we recall a basic

DEFINITION 2.2. Equations $(5)_\omega$ are said to have an exponential dichotomy (ED) if there are constants $K > 0$, $\gamma > 0$ and continuous invariant vector subbundles $B^-, B^+ \subset \Omega \times \mathbf{R}^2$ such that:

1. $B^- \oplus B^+ = \Omega \times \mathbf{R}^2$ (Whitney sum);
2. if $(\omega, u) \in B^+$, then $\|\Phi_\omega(t)u\| \leq Ke^{-\gamma t}\|u\|$ for all $\omega \in \Omega$ and $t \geq 0$;
3. if $(\omega, u) \in B^-$, then $\|\Phi_\omega(t)u\| \leq Ke^{\gamma t}\|u\|$ for all $\omega \in \Omega$ and $t \leq 0$.

Here β is fixed, and $\|\cdot\|$ refers to some norm on $\mathbf{R}^2$. The invariance of B^+ and B^- is with respect to the linear skew-product flow on $\Omega \times \mathbf{R}^2$. The letter u refers to a generic vector $(x, y)^t$ in $\mathbf{R}^2$.

We see that equations $(5)_\omega$ have an ED if and only if the "solution space" $\Omega \times \mathbf{R}^2$ admits a hyperbolic splitting. Note that, if equations $(5)_\omega$ admit an ED, then the bundles $B^\pm$ must have fiber dimension one, because $\operatorname{tr} \begin{pmatrix} -\beta/2 & 1 \\ \xi_t & \beta/2 \end{pmatrix} = 0$.

Now we introduce another key hypothesis.

ASSUMPTION 2.3. There is a (necessarily negative) critical value $\beta = \beta_c \leq 0$ with the following properties. (i) If $\beta < \beta_c$, then $\lambda_1(\beta) \equiv \lambda(\beta) < 0$. (ii) There is a number $\beta_* > \beta_c$ such that $\lambda_2(\beta) < 0 < \lambda(\beta)$ if $\beta_c < \beta < \beta_*$. (iii) If $\beta = \beta_c$, the traceless equations $(5)_\omega$ admit an exponential dichotomy.

REMARKS 2.4. (i) The condition (iii) actually implies, via standard perturbation results ([C], [P]), that equations $(5)_\omega$ have an ED for all β near β_c. Thus we can assume that equations $(5)_\omega$ have an ED for $\beta_c \leq \beta < \beta_*$.

(ii) It follows from 2.3 (iii) that $\lambda(\beta)$ and $\lambda_0(\beta)$ are continuous (exercise). Thus $\lambda(\beta_c) = 0$ and, by (6), $\lambda_0(\beta_c) = -\beta_c/2$. This explains why β_c must be negative: condition 2.3 (iii) implies that $\lambda_0(\beta_c) > 0$.

(iii) The significance of condition 2.3.(ii) is the following. If $\beta < \beta_c$, then $\lambda(\beta) < 0$, and hence (using ODE techniques) one can show that the set $A_0 = \Omega \times \{0\} \subset \Omega \times \mathbf{R}^2$ is an invariant and asymptotically stable subset of $\Omega \times \mathbf{R}^2$ with respect to the nonlinear semiflow $\{\tau_t^n\}$. That is, A_0 is an attractor if $\beta < \beta_c$. However, A_0 is no longer stable if $\beta > \beta_c$, because $\lambda(\beta) > 0$. Thus A_0 loses stability at $\beta = \beta_c$, and the natural question arises as to the existence of a new attractor for $\beta > \beta_c$. We will investigate this question below.

(iv) We will not discuss general conditions on $\{\xi_t\}$ which ensure that Assumption 2.3 is valid. (It is not hard to find such processes.) We justify the omission by referring to the goal of this paper: find conditions under which the first step of the numerically observed "two-step" bifurcation pattern can be rigorously proved to occur.

Next we formulate precisely the hypothesis—to which we referred in the Introduction—to the effect that equation (2) displays fast oscillations at the critical value $\beta = \beta_c$. Write $\mu = \beta - \beta_c$, and consider the following variant of equation (3):

$$(7)_\mu \qquad \begin{pmatrix} x \\ y \end{pmatrix}' = \mu\left[\begin{pmatrix} 0 & 1 \\ \xi_t & \mu + \beta_c \end{pmatrix}\begin{pmatrix} x \\ y \end{pmatrix} - \begin{pmatrix} 0 \\ x^2 y + x^3 \end{pmatrix}\right].$$

These are the equations which we will study in detail. Note the factor of μ in front of the right-hand side. It is not present in the original system (3). Its presence indicates that there are fast oscillations at the critical value $\mu = 0$ where $x = y = 0$ loses stability, as one sees by substituting $\mu t = s$ in equations $(7)_\mu$.

Our analysis of equations $(7)_\mu$ will consist of the following steps. We will write equations $(7)_\mu$ in polar coordinates, then "localize" the equations around the unstable bundle B^-. We will apply the method of averaging to prove our main result (Theorem 2.9). As we will see, it will be necessary to impose certain conditions other than those already discussed. Among these are a "vague attractor" condition and a " large deviations" condition which must hold at $\mu = 0$. These conditions will be discussed in detail as we go along.

In polar coordinates (r, θ), equations $(7)_\mu$ take the form

$$(8)_\mu \qquad \begin{aligned} r' &= \mu[\ell(\xi_t, \theta(t))r - b(\xi_t, \theta(t))r^3] \\ \theta' &= \mu[c(\xi_t, \theta(t)) - d(\xi_t, \theta(t))r^2], \end{aligned}$$

where

$$\begin{aligned} \ell(\xi_t, \theta) &= \frac{\xi_t + 1}{2}\sin 2\theta + (\mu + \beta_c)\sin^2\theta \\ b(\xi_t, \theta) &= \cos^2\theta\sin^2\theta + \cos^3\theta\sin\theta \\ c(\xi_t, \theta) &= \xi_t\cos^2\theta - \sin^2\theta + (\mu + \beta_c)\sin\theta\cos\theta \\ d(\xi_t, \theta) &= \cos^4\theta + \cos^3\theta\sin\theta. \end{aligned}$$

We have abused notation slightly in that ℓ, b, c, d are actually functions of (ω, θ); e.g. $\ell(\omega, \theta) = \dfrac{g(\omega) + 1}{2}\sin 2\theta + (\mu + \beta_c)\sin^2\theta$.

We localize equations $(8)_\mu$ around the unstable bundle B^-. To explain what this means, let $K \subset \mathbf{R}^2$ be the unit circle, and let $T = (\Omega \times K) \cap B^-$ be the "trace" of B^- on $\Omega \times K$. Then T is a 2-cover of Ω. Let $\widehat{\Omega}$ be a connected component of T. Since Ω is connected, $\widehat{\Omega}$ is either a 2-cover of Ω (in which case $\widehat{\Omega} = T$), or $\widehat{\Omega}$ is homeomorphic to Ω. There is a natural flow $\{\widehat{\tau}_t \ (t \in \mathbf{R})\}$ on $\widehat{\Omega}$ defined by the linear equations (4), and the natural projection $\pi : \widehat{\Omega} \to \Omega$ defines a flow homomorphism. If $\widehat{\Omega}$ is a 1-cover of Ω, then π is an isomorphism of flows. There is a unique probability measure $\widehat{\nu}$ on $\widehat{\Omega}$ which is ergodic with respect to the flow $\{\widehat{\tau}_t\}$, and $\pi(\widehat{\nu}) = \nu$.

Now let $\widehat{\omega} \in \widehat{\Omega}$, and let $\theta^-(\widehat{\omega})$ be a determination of the polar angle of the second component of $\widehat{\omega}$. Then $\theta^-(\widehat{\omega})$ is only defined $\bmod 2\pi$. However it will only appear below as an argument of 2π-periodic functions. This remark will justify all our uses of $\theta^-(\widehat{\omega})$ in what follows.

Fix $\widehat{\omega} \in \widehat{\Omega}$, write

$$\theta = \psi + \theta^-(t)$$

where $\theta^-(t) \equiv \theta^-(\widehat{\tau}_t(\widehat{\omega}))$, and express equations $(8)_\mu$ in terms of the variables (r, ψ). We get

$$(9)_\mu \qquad \begin{aligned} r' =& \mu[\ell(\xi_t, \theta^-(t))r + \ell_\theta(\xi_t, \theta^-(t))r\psi - b(\xi_t, \theta^-(t))r^3 \\ & + O(r\psi^2) + O(r^3\psi)] \\ \theta' =& \mu[c_\theta(\xi_t, \theta^-(t))\psi - d(\xi_t, \theta^-(t))r^2 + O(\psi^2) + O(r^2\psi)], \end{aligned}$$

where the subscripts in ℓ_θ and c_θ indicate the partial derivatives with respect to θ, and the terms $O(r\psi^2)$, $O(r^3\psi)$, $O(\psi^2)$ and $O(r^2\psi)$ are uniform in $\widehat{\omega}$ as $\psi \to 0$ and $r \to 0$. We have used the standard "big-O" notation; thus e.g., the quantity $O(\psi^2)$ divided by ψ^2 remains bounded as $r \to 0$ and as $\psi \to 0$.

LEMMA 2.5. *For each $\widehat{\omega} \in \widehat{\Omega}$:*

$$\text{(a)} \lim_{t \to \to \infty} \frac{1}{t} \int_0^t \ell(\xi_s, \theta^-(s))\, ds = \int_{\widehat{\Omega}} \ell(\widehat{\omega})\, d\widehat{\nu}(\widehat{\omega}) = \lambda(\mu)$$

$$\text{(b)} \lim_{t \to \to \infty} \frac{1}{t} \int_0^t c_\theta(\xi_s, \theta^-(s))\, ds = \int_{\widehat{\Omega}} c_\theta(\widehat{\omega})\, d\widehat{\nu}(\widehat{\omega}) = -2\lambda_0(\mu).$$

Recall that $\mu = \beta - \beta_c$.

PROOF. Part (a) follows from the r-relation in $(9)_\mu$ and the Birkhoff ergodic theorem for uniquely ergodic flows [W]. To prove (b), we observe that, by direct calculation,

$$c_\theta(\widehat{\omega}) = \beta - 2\ell(\widehat{\omega})$$

for each $\widehat{\omega} \in \widehat{\Omega}$. Then (b) follows from (6), part (a), and the Birkhoff ergodic theorem for uniquely ergodic flows. $\square$

We have now localized equations $(8)_\mu$ around B^-. The result is equations $(9)_\mu$, which we now proceed to study using the method of averaging [F]. Write

$$\overline{\ell} = \lim_{t \to \infty} \frac{1}{t} \int_0^t \ell(\xi_s, \theta^-(s))\, ds;$$

$$\overline{\ell}_\theta = \lim_{t \to \infty} \frac{1}{t} \int_0^t \ell_\theta(\xi_s, \theta^-(s))\, ds;$$

$$\overline{b} = \lim_{t \to \infty} \frac{1}{t} \int_0^t b(\xi_s, \theta^-(s))\, ds;$$

$$\overline{c}_\theta = \lim_{t \to \infty} \frac{1}{t} \int_0^t c_\theta(\xi_s, \theta^-(s))\, ds;$$

$$\overline{d} = \lim_{t \to \infty} \frac{1}{t} \int_0^t d(\xi_s, \theta^-(s))\, ds.$$

The limits exist, uniformly in t and $\widehat{\omega}$, and are equal to the appropriate space averages with respect to $\widehat{\nu}$. All of these quantities are, in general, functions of μ because θ^- is a function of μ as well as of $\widehat{\omega}$. Note that, by Lemma 2.5, $\overline{\ell} = \lambda(\mu)$ and $\overline{c}_\theta = -2\lambda_0(\mu)$.

Now define

$$\overline{R}_1(r,\psi) = \overline{\ell}r + \overline{\ell}_\theta r\psi - \overline{b}r^3$$
$$\overline{\Psi}_1(r,\psi) = \overline{c}_\theta\psi - \overline{d}r^3.$$

Then set

$$R_1(\widehat{\omega},r,\psi) = \ell(\widehat{\omega})r + \ell_\theta(\widehat{\omega})r\psi - b(\widehat{\omega})r^3$$
$$\Psi_1(\widehat{\omega},r,\psi) = c_\theta(\widehat{\omega})\psi - d(\widehat{\omega})r^2.$$

Here and below, we suppress the μ-dependence in these quantities. Finally, write

$$R_\mu(\widehat{\omega},r,\psi) = \int_0^\infty e^{-\mu s}[R_1(\widehat{\tau}(\widehat{\omega}),r,\psi) - \overline{R}_1(r,\psi)]\,ds$$
$$\Psi_\mu(\widehat{\omega},r,\psi) = \int_0^\infty e^{-\mu s}[\Psi_1(\widehat{\tau}_s(\widehat{\omega}),r,\psi) - \overline{\Psi}_1(r,\psi)]\,ds.$$

We recall a variant of a well-known result which is central to the averaging method. See [F], [JY] for the proof.

LEMMA 2.6. *Let $h = h(\mu,\widehat{\omega})$ denote one of the functions $\ell, \ell_\theta, b, c_\theta, d$ and let $\overline{h}(\mu)$ denote its mean value with respect to $\widehat{\nu}$. Then*

$$\mu\int_0^\infty e^{-\mu s}[h(\mu,\widehat{\tau}_s(\widehat{\omega})) - \overline{h}(\mu)]\,ds$$

tends to zero as $\mu \to 0^+$, uniformly with respect to $\widehat{\omega} \in \widehat{\Omega}$.

For later use we write

$$\eta(\mu) = \sup_{\widehat{\omega}\in\widehat{\Omega}} \mu\left|\int_0^\infty e^{-\mu s}[\ell(\mu,\widehat{\tau}_s(\widehat{\omega})) - \overline{\ell}(\mu)]\,ds\right|, \tag{10}$$

and observe that, by Lemma 2.6, $\eta(\mu) \to 0$ as $\mu \to 0^+$. That is, using the standard little-o notation:

$$\eta(\mu) = o(1) \qquad \text{as } \mu \to 0^+.$$

We are now ready to carry out the averaging procedure in equations $(9)_\mu$. Fix $\overline{\omega} \in \overline{\Omega}$, and write

$$R_\mu(t,r,\psi) \equiv R_\mu(\widehat{\tau}_t(\widehat{\omega}),r,\psi) = e^{\mu t}\int_t^\infty e^{-\mu s}[R_1(\widehat{\tau}_s(\widehat{\omega}),r,\psi) - \overline{R}_1(r,\psi)]\,ds$$
$$\Psi_\mu(t,r,\psi) \equiv \Psi_\mu(\widehat{\tau}_t(\widehat{\omega}),r,\psi) = e^{\mu t}\int_t^\infty e^{-\mu s}[\Psi_1(\widehat{\tau}_s(\widehat{\omega}),r,\psi) - \overline{\Psi}_1(r,\psi)]\,ds.$$

Introduce the following change of variables in equations $(9)_\mu$:

$$r = \rho + \mu R_\mu(t,\rho,\varphi)$$
$$\psi = \varphi + \mu\Psi_\mu(t,\rho,\varphi).$$

Because of Lemma 2.6, this transformation is invertible on some neighborhood of $r = \psi = 0$ which does not depend on $t \in \mathbf{R}$. To express equations $(9)_\mu$ with respect to the variables ρ, φ, write

$$z = \begin{pmatrix} \rho \\ \varphi \end{pmatrix}$$
$$I = \begin{pmatrix} 1 & 0 \\ 0 & 1 \end{pmatrix}$$
$$F_\mu = \begin{pmatrix} R_\mu \\ \Psi_\mu \end{pmatrix}$$
$$f = \begin{pmatrix} \ell_\rho + \ell_\theta \rho\varphi - b\rho^3 \\ c_\theta \varphi - d\rho^2 \end{pmatrix}$$
$$\begin{pmatrix} J_1 \\ J_2 \end{pmatrix} = -\frac{\partial f}{\partial z}(t, z) \cdot \mu F_\mu(t, z).$$

Then

$$(11)_\mu \qquad z'\left[I - \mu \frac{\partial F_\mu}{\partial z}(t, z)\right] = \begin{pmatrix} \overline{\ell}_\rho + \overline{\ell}_\theta \rho\varphi - b\rho^3 + \mu R_\mu(t, \rho, \varphi) \\ +J_1(t, \rho, \varphi) + \widetilde{J}_1(t, \rho, \varphi) \\ \overline{c}_\theta \varphi - \overline{d}\rho^2 + \mu \Psi_\mu(t, \rho, \varphi) \\ +J_2(t, \rho, \varphi) + \widetilde{J}_2(t, \rho, \varphi) \end{pmatrix}$$

where the terms in $\widetilde{J}_1$ and $\widetilde{J}_2$ will play no essential role in what follows. We obtain from $(11)_\mu$:

$$(12)_\mu \qquad \begin{aligned} \rho' &= \overline{\ell}_\rho + \overline{\ell}_\theta \rho\varphi - \overline{b}\rho^3 + \mu R_\mu(t, \rho, \varphi) + J_1(t, \rho, \varphi) + \widetilde{J}_1 + o_1(1) \\ \varphi' &= \overline{c}_\theta \varphi - \overline{d}\rho^2 + \mu \Psi_\mu(t, \rho, \varphi) + J_2(t, \rho, \varphi) + \widetilde{J}_2 + o_2(1), \end{aligned}$$

where the terms $o_1(1)$ and $o_2(1)$ are of the form

$$o_1(1) = o(1)[\overline{\ell}_\rho + \overline{\ell}_\theta \rho\varphi - \overline{b}\rho^3 + \mu R_\mu + J_1 + \widetilde{J}_1]$$
$$o_2(1) = o(1)[\overline{c}_\theta \varphi - \overline{d}\rho^2 + \mu \Psi_\mu + J_2 + \widetilde{J}_2]$$

and $o(1) \to 0$ as $\mu \to 0^+$.

The goal now is to analyze equation $(12)_\mu$. First of all, we impose a "vague attractor" condition which in the present circumstances is simply

ASSUMPTION 2.7. The coefficient $\overline{b}$ is positive: $\overline{b} > 0$.

It is sufficient for our purposes to assume that $\overline{b}(0) > 0$ because $\mu \to \overline{b}(\mu)$ is continuous. For each sufficiently small $\mu > 0$, it can be arranged that $\overline{b} = 1$ by a simple scaling $\rho \to s\widetilde{\rho}$ where $s = \overline{b}^{-1/2}$. We will assume that $\overline{b} = 1$ in what follows.

Recall now that $\overline{\ell} = \lambda(\mu)$, $\overline{c}_\theta = -2\lambda_0(\mu)$. We make a final change of variables:

$$\rho = v + \sqrt{\lambda(\mu)}, \qquad \varphi = \varphi.$$

The equations $(12)_\mu$ become

$$(13)_\mu \qquad \begin{aligned} v' &= -\lambda(\mu)v + \overline{\ell}_\theta \sqrt{\lambda(\mu)}\varphi + \overline{\ell}_\theta v\varphi - 3\sqrt{\lambda(\mu)}v^2 - v^3 \\ &\qquad + \mu R_\mu + J_1 + \widetilde{J}_1 + o_1(1) \\ \varphi' &= -2\lambda_0(\mu)\varphi - \overline{d}\lambda(\mu) - 2\overline{d}\sqrt{\lambda(\mu)}\varphi - \overline{d}v^2 \\ &\qquad + \mu\Psi_\mu + J_2 + \widetilde{J}_2 + o_2(1), \end{aligned}$$

where the arguments of $\mu R_\mu, \mu\Psi_\mu, J_1, \widetilde{J}_1, J_2, \widetilde{J}_2, o_1(1)$ and $o_2(1)$ are $(t, \rho = v + \sqrt{\lambda(\mu)}, \varphi)$. To study these equations, we will use an hypothesis of large deviations type.

ASSUMPTION 2.8. Let $\eta(\mu)$ be the function defined in (10). Assume that there is a constant c (to be determined more precisely later) such that $\eta(\mu) \le c\beta(\mu)$ for small $\mu > 0$.

As will be seen later, our choice of c will depend on the quantities $\overline{\ell}_\theta, \overline{d}$ and λ_0 at $\mu = 0$. However there is a necessary upper bound on c (if our methods are to be applied). Namely set

$$|\ell|_\infty = \sup_{\widehat{\omega}\in\widehat{\Omega}} |\ell(\widehat{\omega})|,$$

where the supremum is calculated at $\mu = 0$. Then it is necessary that

$$c \le (1 + |\ell|_\infty)^{-1}.$$

We are now ready to formulate our main result.

THEOREM 2.9. *Suppose that:*

(i) $(\Omega, \{\tau_t \mid t \in \mathbf{R}\}, \nu)$ *is minimal and uniquely ergodic;*
(ii) *the traceless linear equations* $(5)_\omega$ *admit an exponential dichotomy at* $\mu = 0$;
(iii) *the vague attractor condition (2.7) holds;*
(iv) *the large deviations condition (2.8) holds for small* $\mu > 0$ *with* $c = [22(1 + |\ell|_\infty)]^{-1}$;
(v) $\lambda_0 > 22\, \overline{d}\, \overline{\ell}_\theta$ *at* $\mu = 0$ *(and hence for small positive* μ*).*

Then for all sufficiently small $\mu > 0$, *equations (3) admit a compact attractor* $A_\mu \subset \Omega \times \mathbf{R}^2$. *The attractor* A_μ *tends to* $\Omega \times \{0\} \subset \Omega \times \mathbf{R}^2$ *in the Hausdorff sense as* $\mu \to 0^+$.

Recall that a set A_μ is an attractor for the nonlinear semiflow $\{\tau_t^n\}$ on $\Omega \times \mathbf{R}^2$ if (i) for all $t \in \mathbf{R}$ and all $(\omega, u) \in A_\mu$, $\tau_t^n(\omega, u)$ is defined and lies in A_μ; (ii) there is a neighborhood W of A_μ in $\Omega \times \mathbf{R}^2$ such that $\bigcap_{T\ge 0}\bigcup_{t\ge T} \tau_t^n(W) = A_\mu$.

Before beginning the proof, we note that the values of c and of $\overline{d}\,\overline{\ell}_\theta/\lambda_0$ may be varied to a certain extent without altering the validity of the proof.

PROOF. Fix $\mu > 0$ so small that conditions (ii)-(v) hold. Let $\widehat{B}$ be the space of continuous maps $(v, \varphi) : \Omega \to \mathbf{R}^2$ with the metric $m((v_1, \varphi_1), (v_2, \varphi_2)) = |v_1 - v_2|_\infty + |\varphi_1 - \varphi_2|_\infty$; here $|\cdot|_\infty$ denotes the usual sup-norm. Let $B \subset \widehat{B}$ be the closed subset consisting of points (v, φ) such that

$$|v|_\infty \le p\sqrt{\lambda(\mu)}, \qquad |\varphi|_\infty \le \frac{q\lambda(\mu)}{\lambda_0(\mu)}.$$

Before long we will choose $p = 1/10$ and $q = \overline{d}$ (we assume throughout that $\overline{d} \ne 0$, and leave it to the reader to choose an appropriate q if $\overline{d} = 0$). These values are not uniquely determined; however we will see that–if our methods are to work—one must choose $p < 1/6$ and $q > \overline{d}/2$.

Define the following nonlinear mapping T of $\widehat{B}$ to itself:

$$T(v, \varphi) = \int_{-\infty}^{0} \begin{pmatrix} e^{\lambda(\mu)s} F_1\big(\widehat{\tau}_s(\widehat{\omega}) v(\widehat{\tau}_s(\widehat{\omega})), \varphi(\widehat{\tau}_s(\widehat{\omega}))\big) \\ e^{2\lambda_0(\mu)s} F_2\big(\widehat{\tau}_s(\widehat{\omega}), v(\widehat{\tau}_s(\widehat{\omega})), \varphi(\widehat{\tau}_s(\widehat{\omega}))\big) \end{pmatrix} ds,$$

where

$$\begin{aligned} F_1(\widehat{\omega}, v, \varphi) &= \overline{\ell}\sqrt{\lambda(\mu)}\varphi + \overline{\ell}_\theta v\varphi - 3\sqrt{\lambda(\mu)}v^2 - v^3 \\ &\quad + \mu R_\mu + J_1 + \widetilde{J}_1 + o_1(1) \\ F_2(\widehat{\omega}, v, \varphi) &= -\overline{d}\lambda(\mu) - 2\overline{d}\sqrt{\lambda(\mu)}v - \overline{d}v^2 \\ &\quad + \mu\Psi_\mu + J_2 + \widetilde{J}_2 + o_2(1), \end{aligned}$$

where the arguments of $\mu R_\mu, J_1, \widetilde{J}_1, o_1(1), \mu\Psi_\mu, J_2, \widetilde{J}_2$ and $o_2(1)$ are $(\widehat{\omega}, v+\sqrt{\lambda(\mu)}, \varphi)$. Observe that (v, φ) is a fixed point of T if and only if the set

$$A_\mu = \{(\pi(\widehat{\omega}), v(\widehat{\omega}), \varphi(\widehat{\omega})) \mid \widehat{\omega} \in \widehat{\Omega}\}$$

is a compact invariant subset of $\Omega \times \mathbf{R}^2$ such that the coordinates (ω, v, φ) of each point in A_μ satisfy $|v| \leq p\sqrt{\lambda(\mu)}$ and $|\varphi| \leq \dfrac{q\lambda(\mu)}{\lambda_0(\mu)}$. Here (v, φ) are related to the polar coordinates (r, θ) in $\mathbf{R}^2$ via the coordinate transformations introduced above.

Now set $p = 1/10$, $q = \overline{d}$. We will prove that T admits a fixed point in B. First of all, simple estimates together with (iv) and (v) show that T does map B into B. Condition (iv) is needed because of the v-terms in μR_μ and in J_1.

If $\overline{\ell}_\theta \neq 0$, the map T is not a contraction on B, at least not with respect to the metric m. This is easily verified and is due to the presence of the terms $\overline{\ell}_\theta v\varphi$ and $\overline{\ell}_\theta \varphi\sqrt{\lambda(\mu)}$ in F_1. However, we will now show that T admits enough "contraction-like" properties to assure that it has a unique fixed point in B.

Let us write

$$T(v, \varphi) = (T_1(v, \varphi), T_2(v, \varphi)),$$

and set $B_1 = \{v : \widehat{\Omega} \to \mathbf{R} \mid |v|_\infty \leq \frac{1}{10}\sqrt{\lambda(\mu)}\}$, $B_2 = \{\varphi : \widehat{\Omega} \to \mathbf{R} \mid |\varphi|_\infty \leq \frac{\overline{d}\lambda(\mu)}{\lambda_0(\mu)}\}$. Then $B = B_1 \times B_2$. The following assertion is easily verified: for sufficiently small positive μ, the map $T_1 : B_1 \to B_1$ is a contraction for each fixed $\varphi \in B_2$. In fact, the contraction constant is less than one uniformly in $\varphi \in B_2$.

Using this observation, we define a mapping $\widehat{T} : B \to B$ in the following way. Let $(v, \varphi) \in B$, and let $v_* = v_*(\varphi)$ be the unique fixed point of T_1 when its second argument is held fixed at φ. Then set $\varphi_* = T_2(v_*, \varphi)$ and define

$$\widehat{T}(v, \varphi) = (v_*, \varphi_*).$$

We will show that $\widehat{T}$ is uniformly contracting with respect to the φ-variable. It will follow from this that $\widehat{T}$ admits a unique fixed point $(v_\infty, \varphi_\infty) \in B$, and this fixed point will define the attractor A_μ which we seek.

First of all, let $\varphi_1, \varphi_2 \in B_2$, and let v_1, v_2 be the corresponding fixed points of T_1 in B_1. We have by direct analysis of T_1:

$$\begin{aligned} &|v_1 - v_2|_\infty = |T_1(v_1, \varphi_1) - T_1(v_2, \varphi_2)|_\infty \\ &\leq |T_1(v_1, \varphi_1) - |T_1(v_2, \varphi_1)|_\infty + |T_1(v_2, \varphi_1) - T_1(v_2, \varphi_2)|_\infty \\ &\leq (6p + 3p^2 + o(1))|v_1 - v_2|_\infty + \frac{\overline{\ell}_\theta q}{\lambda_0(\mu)}|v_1 - v_2|_\infty + \frac{(1+p)\overline{\ell}_\theta + o(1)}{\sqrt{\lambda(\mu)}}|\varphi_1 - \varphi_2|_\infty. \end{aligned}$$

Since $6p + 3p^2 + \dfrac{\bar{\ell}_\theta q}{\lambda_0(\mu)} + o(1) < 0.7$ for sufficiently small μ, we have

$$|v_1 - v_2|_\infty \le \frac{10}{3}\left[\frac{(1+p)\bar{\ell}_\theta + o(1)}{\sqrt{\lambda(\mu)}}\right] |\varphi_1 - \varphi_2|_\infty. \tag{14}$$

Next consider the points $\widetilde{\varphi}_1 = T_2(v_1, \varphi_1)$, $\widetilde{\varphi}_2 = T_2(v_2, \varphi_2)$. We have

$$\begin{aligned} |\widetilde{\varphi}_1 - \widetilde{\varphi}_2|_\infty &\le |T_2(v_1,\varphi_1) - T_2(v_1,\varphi_2)|_\infty + |T_2(v_1,\varphi_2) - T_2(v_2,\varphi_2)|_\infty \\ &\le \frac{\bar{d}\sqrt{\lambda(\mu)}}{\lambda_0(\mu)}|v_1 - v_2|_\infty + \frac{\bar{d}p}{\lambda_0(\mu)}|v_1 - v_2|_\infty \\ &\quad + o(1)|v_1 - v_2|_\infty + o(1)|\varphi_1 - \varphi_2|_\infty. \end{aligned} \tag{15}$$

Combining (14) and (15), we have

$$\begin{aligned} |\widetilde{\varphi}_1 - \widetilde{\varphi}_2|_\infty &\le \frac{10}{3}\frac{\bar{d}(1+p)}{\lambda_0(\mu)}[(1+p)\bar{\ell}_\theta + o(1)]|\varphi_1 - \varphi_2|_\infty + \\ &\quad + o(1)|\varphi_1 - \varphi_2| \end{aligned}$$

as $\mu \to 0^+$. The conditions $p = \frac{1}{10}$, $q = \bar{d}$ and (v) now imply that

$$|\widetilde{\varphi}_1 - \widetilde{\varphi}_2|_\infty \le K|\varphi_1 - \varphi_2|$$

for small $\mu > 0$, where $K < 1$ is a constant independent of φ_1 and φ_2.

Now let (v_0, φ_0) be a fixed initial point in B, and define $(v_n, \varphi_n) = \widehat{T}^n(v_0, \varphi_0)$ $(n = 1, 2, \dots)$. For $n > m \ge 1$ we have

$$|\varphi_n - \varphi_m|_\infty \le \left(\sum_{\ell=m+1}^{n} K^\ell\right)|\varphi_1 - \varphi_0|_\infty,$$

hence $\{\varphi_n\}$ is a Cauchy sequence in B_2. Let φ_∞ be the limit of this Cauchy sequence. Using the uniform contractivity of T_1, we see that, if $v_\infty \in B_1$ is the unique fixed point of $v \to T_1(v, \varphi_\infty)$, then $v_n \to v_\infty$.

We now have

$$\widehat{T}(v_\infty, \varphi_\infty) = \lim_{n\to\infty} \widehat{T}(v_n, \varphi_n) = \lim_{n\to\infty}(v_{n+1}, \varphi_{n+1}) = (v_\infty, \varphi_\infty).$$

Thus $(v_\infty, \varphi_\infty)$ is a fixed point of $\widehat{T}$. Using the contractivity properties of T_1 and T_2, one can show that A_μ is an attractor; we leave the details of this to the reader. This completes the proof of Theorem 2.9. □

REMARK 2.10. (i) If $\widehat{\Omega}$ is a 2-cover of Ω, then A_μ is also a 2-cover of Ω. This situation is analogous to the numerical picture derived by Schenk-Hoppé. On the other hand, if $\widehat{\Omega}$ is a 1-cover of Ω, then the proof of Theorem 2.9 actually gives two attractors for each small $\mu > 0$. This is because $\widehat{\Omega}$ is a 1-cover of Ω if and only if the line bundle B^- is orientable. If B^- is orientable, we can carry out the construction in the proof of Theorem 2.9 in a neighborhood of each of the two half-line (ray) bundles defined by B^-.

(ii) In some situations assumptions (iv) and (v) of Theorem 2.9 can be eliminated. This is true if, for example, $\{\xi_t\}$ is a periodic process. We plan to investigate this question (for nonperiodic processes, of course) in another publication.

References

[A] L. Arnold, *Six lectures on random dynamical systems*, Springer-Verlag, Lecture Notes in Math. 1609, R. Johnson (ed.), Berlin, 1995, pp. 1–43.

[AB] L. Arnold and P. Boxler, Progress in Probability **27** (1992), M. Pinsky and V. Wihstutz (eds.), Birkhauser, Boston, 241–256.

[AS] L. Arnold and B. Schmalfuß, *Fixed points and attractors for random dynamical systems, Proc. of IUTAM Symposium on Advances in Nonlinear Stochastic Mechanics*, A. Naess (ed.), Kluwer Academic Publishers, Dordrecht, 1995.

[AX] L. Arnold and K. Xu, *Invariant measures for random dynamical systems, and a necessary condition for stochastic bifurcation from a fixed point*, Random and Computational Dynamics **2** (1994), 165–182.

[C] A. Coppel, *Dichotomies in Stability Theory*, Springer-Verlag, Lecture Notes in Math. 629, Berlin, 1978.

[CL] E. Coddington and N. Levinson, *Theory of Ordinary Differential Equations*, McGraw-Hill, New York, 1955.

[E] R. Ellis, *Lectures on Topological Dynamics*, Benjamin, New York, 1967.

[F] A. Fink, *Almost Periodic Differential Equations*, Springer-Verlag, Lecture Notes in Math. 377, Berlin, 1974.

[H1] K. Schenk-Hoppé,, *Bifurcation scenarios of the noisy Duffing-van der Pol oscillator*, Nonlinear Dynamics (1996. To appear).

[H2] K. Schenk-Hoppé, *The stochastic Duffing-van der Pol equation, doctoral dissertation* (1996), Universität, Bremen.

[J] R. Johnson, *On a random bifurcation problem, manuscript..*

[JY] R. Johnson and Y. Yi, *Hopf bifurcation from nonperiodic solutions of differential equations II*, Diff. Eqns. **107** (1994), 310–340.

[P] K. Palmer, *On Cherry's example of a nonintegrable Hamiltonian system*, Diff. Eqns. **65** (1986), 321–360.

[SS] R. Sacker and G. Sell, *A spectral theory for linear differential systems*, Diff. Eqns. **27** (1978), 429–458.

[W] P. Walters, *Lectures on Ergodic Theory*, Springer-Verlag, Lecture Notes in Math. 458, Berlin, 1975.

Dipartimento di Sistemi e Informatica, Università di Firenze, Via di Santa Marta 3, 50139 Firenze, Italy.

E-mail address: `johnson @ ingfil.ing.unifi.it`

Selected Titles in This Series

(*Continued from the front of this publication*)

187 **G. W. Brumfiel and H. M. Hilden,** $SL(2)$ representations of finitely presented groups, 1995

186 **Shreeram S. Abhyankar, Walter Feit, Michael D. Fried, Yasutaka Ihara, and Helmut Voelklein, Editors,** Recent developments in the inverse Galois problem, 1995

185 **Raúl E. Curto, Ronald G. Douglas, Joel D. Pincus, and Norberto Salinas, Editors,** Multivariable operator theory, 1995

184 **L. A. Bokut′, A. I. Kostrikin, and S. S. Kutateladze, Editors,** Second International Conference on Algebra, 1995

183 **William C. Connett, Marc-Olivier Gebuhrer, and Alan L. Schwartz, Editors,** Applications of hypergroups and related measure algebras, 1995

182 **Selman Akbulut, Editor,** Real algebraic geometry and topology, 1995

181 **Mila Cenkl and Haynes Miller, Editors,** The Čech Centennial, 1995

180 **David E. Keyes and Jinchao Xu, Editors,** Domain decomposition methods in scientific and engineering computing, 1994

179 **Yoshiaki Maeda, Hideki Omoro, and Alan Weinstein, Editors,** Symplectic geometry and quantization, 1994

178 **Hélène Barcelo and Gil Kalai, Editors,** Jerusalem Combinatorics '93, 1994

177 **Simon Gindikin, Roe Goodman, Frederick P. Greenleaf, and Paul J. Sally, Jr., Editors,** Representation theory and analysis on homogeneous spaces, 1994

176 **David Ballard,** Foundational aspects of "non"standard mathematics, 1994

175 **Paul J. Sally, Jr., Moshe Flato, James Lepowsky, Nicolai Reshetikhin, and Gregg J. Zuckerman, Editors,** Mathematical aspects of conformal and topological field theories and quantum groups, 1994

174 **Nancy Childress and John W. Jones, Editors,** Arithmetic geometry, 1994

173 **Robert Brooks, Carolyn Gordon, and Peter Perry, Editors,** Geometry of the spectrum, 1994

172 **Peter E. Kloeden and Kenneth J. Palmer, Editors,** Chaotic numerics, 1994

171 **Rüdiger Göbel, Paul Hill, and Wolfgang Liebert, Editors,** Abelian group theory and related topics, 1994

170 **John K. Beem and Krishan L. Duggal, Editors,** Differential geometry and mathematical physics, 1994

169 **William Abikoff, Joan S. Birman, and Kathryn Kuiken, Editors,** The mathematical legacy of Wilhelm Magnus, 1994

168 **Gary L. Mullen and Peter Jau-Shyong Shiue, Editors,** Finite fields: Theory, applications, and algorithms, 1994

167 **Robert S. Doran, Editor,** C^*-algebras: 1943–1993, 1994

166 **George E. Andrews, David M. Bressoud, and L. Alayne Parson, Editors,** The Rademacher legacy to mathematics, 1994

165 **Barry Mazur and Glenn Stevens, Editors,** p-adic monodromy and the Birch and Swinnerton-Dyer conjecture, 1994

164 **Cameron Gordon, Yoav Moriah, and Bronislaw Wajnryb, Editors,** Geometric topology, 1994

163 **Zhong-Ci Shi and Chung-Chun Yang, Editors,** Computational mathematics in China, 1994

162 **Ciro Ciliberto, E. Laura Livorni, and Andrew J. Sommese, Editors,** Classification of algebraic varieties, 1994

161 **Paul A. Schweitzer, S. J., Steven Hurder, Nathan Moreira dos Santos, and José Luis Arraut, Editors,** Differential topology, foliations, and group actions, 1994

160 **Niky Kamran and Peter J. Olver, Editors,** Lie algebras, cohomology, and new applications to quantum mechanics, 1994

159 **William J. Heinzer, Craig L. Huneke, and Judith D. Sally, Editors,** Commutative algebra: Syzygies, multiplicities, and birational algebra, 1994

(See the AMS catalog for earlier titles)